Iron-Group Metal Compound Nanomaterials

Iron-Group Metal Compound Nanomaterials

Preparation, Characterization, Structure, and Applications

Jinlong Zheng

WILEY-VCH

Author

Jinlong Zheng
School of Advanced Materials
Innovation
University of Science and
Technology Beijing
Beijing, 100083
China

Cover Design: Wiley
Cover Images: Courtesy of Jinlong Zheng

Typesetting: Straive, Chennai, India
Printing and Binding: CPI Group (UK) Ltd, Croydon CR0 4YY

Library of Congress Card No.: applied for

British Library Cataloguing-in-Publication Data A catalogue record for this book is available from the British Library.

Bibliographic information published by the Deutsche Nationalbibliothek The Deutsche Nationalbibliothek lists this publication in the Deutsche Nationalbibliografie; detailed bibliographic data are available on the Internet at <http://dnb.d-nb.de>.

Print ISBN: 978-3-527-357109
ePDF ISBN: 978-3-527-858583
ePub ISBN: 978-3-527-858576
oBook ISBN: 978-3-527-858590

C9783527357109_020326

Contents

About this Book

This is a comprehensive monograph on the preparation, characterization, and multi-field applications of iron-group (Fe, Co, and Ni) metal compound nanomaterials, advancing sustainable energy and environmental technology.

This monograph systematically presents the cutting-edge research progress of iron-group (Fe, Co, and Ni) metal compound nanomaterials, covering their fundamental preparation methods, advanced characterization techniques, structural regulation mechanisms, and diverse applications in energy storage (batteries and supercapacitors), electrocatalysis, photocatalysis, biomass upgrading, pollutant degradation, functional devices (sensors and microwave absorbers), and biomedical engineering (drug delivery). By integrating theoretical principles, experimental methods, and practical cases, the book establishes a complete knowledge system from material design to performance optimization, providing in-depth insights for the development of low-cost, high-efficiency functional nanomaterials.

The book is structured logically and comprehensively: the opening chapters focus on the synthesis strategies of iron-group metal compound nanomaterials, detailing the key parameters and mechanism of each preparation technology. Subsequent chapters elaborate on eight core characterization techniques, explaining their principles, application scenarios, and structure–performance correlation analysis methods. The core application chapters cover key fields, such as energy storage systems, electrocatalytic energy conversion, photocatalytic utilization, biomass electrocatalytic upgrading, pollutant degradation, functional devices, and biomedical engineering.

This monograph includes key information on:

- Precise synthesis and structural regulation of iron-group metal oxides, sulfides, phosphides, nitrides, selenides, and their composite materials.
- Multi-scale characterization techniques for analyzing crystal structure, morphology, surface properties, and electronic structure.
- Application of iron-group metal compounds in batteries and strategies to improve battery capacity, cycle life, and safety.
- Design of electrode materials with high specific capacitance, optimization of charge storage mechanisms, and strategies to enhance rate performance and long-cycle stability.

- Design principles and performance optimization strategies of electrocatalysts for energy conversion reactions, as well as the integration of electrocatalytic systems with renewable energy sources.
- Mechanisms of photocatalytic solar energy conversion and the synergy between iron-group metals and photocatalysts to enhance visible light response.
- Efficient degradation pathways of pollutants and environmental remediation applications, including magnetic recyclable catalyst design.
- Applications in sensors, microwave absorption, and drug delivery.

As an essential reference, this book is suitable for researchers and engineers engaged in materials science, energy chemistry, environmental engineering, catalytic technology, and battery research, as well as graduate students majoring in related disciplines. It provides a comprehensive technical guide for both beginners to quickly grasp the field and senior researchers to inspire innovative ideas, promoting the practical application of iron-group metal compound nanomaterials in sustainable energy, environmental governance, and energy storage fields.

Preface

Over the past few decades, the global challenges of energy shortage, environmental pollution, and the urgent demand for advanced functional devices have driven unprecedented interest in iron-group (Fe, Co, and Ni) metal compound nanomaterials. These materials are endowed with earth-abundant reserves, tunable electronic structures, versatile chemical compositions (oxides, sulfides, phosphides, nitrides, etc.), and exceptional physicochemical properties, such as high catalytic activity, controllable magnetic response, and excellent electrochemical performance. These nanomaterials have emerged as a cornerstone for breakthroughs in sustainable energy conversion, environmental remediation, energy storage, functional devices, and even biomedical engineering. Their unique ability to bridge the gap between low-cost synthesis and high-performance applications makes them indispensable in fields ranging from electrocatalytic water splitting to targeted drug delivery, inspiring continuous exploration of their structural design, performance optimization, and practical deployment.

Against this backdrop, the present monograph endeavors to provide a comprehensive and systematic overview of iron-group metal compound nanomaterials for researchers, engineers, and students in materials science, energy chemistry, environmental engineering, and related disciplines. Our goal is to establish a cohesive knowledge framework that connects fundamental principles (synthesis and characterization) to cutting-edge applications, while highlighting the critical structure–activity relationships that govern these materials' performance. Whether you are new to the field, seeking foundational insights, or a seasoned researcher pursuing innovative directions, this book intends to serve as a valuable roadmap for navigating the rapidly evolving landscape of iron-group metal compound nanomaterials.

This book mainly includes three parts. Part I covers the fundamentals of iron-group metal compound nanomaterials (Chapters 1–3). It begins with an introduction to their background, significance, classification, structural features, and research challenges (Chapter 1). Subsequently, various preparation methods are systematically elaborated, including hydrothermal/solvothermal synthesis, electrodeposition, and template methods, with explanations of their principles, pros and cons, and application scenarios (Chapter 2). The third chapter then focuses on key characterization techniques, including X-Ray Diffraction (XRD), Scanning

Electron Microscopy (SEM), Transmission Electron Microscopy (TEM), and X-Ray Photoelectron Spectroscopy (XPS), clarifying their working mechanisms and roles in material analysis (Chapter 3).

Part II is dedicated to the structural characteristics of iron-group metal compound nanomaterials (Chapter 4). It details the structural morphologies across different dimensions, from zero-dimensional (0D) quantum dots and one-dimensional (1D) nanowires/rods/tubes to two-dimensional (2D) nanosheets and three-dimensional (3D) hierarchical structures. The chapter also analyzes how dimensionality influences material properties and lays the foundation for their subsequent applications.

Part III comprehensively summarizes the diverse applications of these nanomaterials (Chapters 5–10). It explores their roles in energy storage (rechargeable batteries and supercapacitors), catalytic reactions (electrocatalysis, photocatalysis, and organic oxidation), and emerging fields such as sensors, microwave absorption, and drug delivery. Each application chapter highlights performance, mechanisms, and current research progress. Finally, the prospects and challenges in this field are discussed to provide insights for future research (Chapter 11).

The book will appeal to both students and scientists in materials science, energy chemistry, environmental engineering, and related disciplines, as well as to those seeking the basic knowledge of synthesis, characterization, structures, and potential applications of iron-group metal compound nanomaterials.

Jinlong Zheng
Beijing, China
January 2026

Acknowledgments

This work was financially supported by the National Natural Science Foundation of China (52271200), the Guangdong Basic and Applied Basic Research Foundation (2024A1515010393), the Fundamental Research Funds for the Central Universities, and the USTB MatCom of Beijing Advanced Innovation Center for Materials Genome Engineering.

Abbreviations and Symbols

- 0D (Zero-Dimensional): A classification of low-dimensional nanomaterials, referring to zero-dimensional nanomaterials (e.g., quantum dots).
- 1D (One-Dimensional): A classification of low-dimensional nanomaterials, referring to one-dimensional nanomaterials (e.g., nanowires and nanorods).
- 2D (Two-Dimensional): A classification of low-dimensional nanomaterials, referring to two-dimensional nanomaterials (e.g., nanosheets and MOFs).
- 3D (Three-Dimensional): A classification of nanomaterials, referring to three-dimensional nanomaterials (e.g., heterostructures and porous frameworks).
- AEM (Adsorbate Evolution Mechanism): A mechanism for OER involving the evolution of adsorbed species.
- AM (Alternating Magnetic): Related to magnetic hyperthermia and magnetic targeting, using alternating magnetic fields.
- AMF (Alternating Magnetic Field): Used in magnetic response drug release and magnetic hyperthermia applications.
- API (Active Pharmaceutical Ingredients): The core component of drug delivery systems, with therapeutic effects.
- APS (Advanced Oxidation Processes): A process for pollutant degradation using advanced oxidation technologies.
- ASC (Asymmetric Supercapacitor)/Asymmetric/Hybrid Supercapacitors: A type of supercapacitor device with asymmetric electrode structures.
- ATAP (Amphiphilic Tail-Anchoring Peptide): A modifier for the surface of drug carriers, with amphiphilic tail-anchoring properties.
- BBB (Blood-Brain Barrier): A barrier related to the targeted delivery of brain tumors, which affects the entry of drugs into the brain.
- BDC (1,4-Dicarboxybenzene): A ligand for MOF synthesis used to form the framework structure of MOFs.
- BET (Brunauer-Emmett-Teller): A method for analyzing specific surface areas to determine the surface area of materials.
- BMO (Binary Transition Metal Oxide): A category of electrode materials composed of two transition metal oxides.
- CE (Counter Electrode): A component of solar cells (DSSC) used to complete the circuit.

- CF (Carbon Fiber): A carbon-based carrier material with a fibrous structure.
- CFP (Carbon Fiber Paper): An electrode substrate material made of carbon fibers in paper form.
- CMC (Carboxymethyl Cellulose): A reagent for stabilizing nanomaterials, with good water solubility and dispersibility.
- CNTF (Carbon Nanotube Film): A formed material made of carbon nanotubes, with film-like morphology.
- CNTs (Carbon Nanotubes): Carbon-based one-dimensional materials with a tubular structure.
- CO_2RR (Carbon Dioxide Reduction Reaction): A reaction for carbon resource conversion involving the reduction of carbon dioxide.
- COF (Covalent Organic Framework): A material used in photocatalysis and drug delivery, with a covalent framework structure.
- CpG-ODN (Cytosine Phosphate-Guanosine Oligonucleotide): An adjuvant and reagent for tumor immunotherapy, with immunomodulatory effects.
- CV (Cyclic Voltammetry): A method for testing electrochemical performance to study redox reactions and capacitance characteristics.
- CVD (Chemical Vapor Deposition): A material preparation technology for depositing thin films or nanomaterials using chemical reactions in the gas phase.
- DCE (Dichloroethane): A halogenated hydrocarbon pollutant target for degradation.
- DFT (Density Functional Theory): A theoretical calculation method used to study electronic structures and reaction mechanisms.
- DHMF (2,5-Dihydroxymethylfuran): A reduction product of HMF used in the synthesis of fuels and chemicals.
- DHMTHF (2,5-Dimethyltetrahydrofuran): A biomass-derived fuel precursor used in the preparation of biofuels.
- DMF (Dimethylformamide)/DMF (N,N-Dimethylformamide): A solvent used in material synthesis, with good solubility for organic and inorganic compounds.
- DOS (Density of States): A result of electronic structure calculation, representing the number of electronic states at a certain energy level.
- DSSC (Dye-Sensitized Solar Cell): A solar energy conversion device based on dye sensitization.
- $E_{1/2}$ (Half-Wave Potential): An indicator for evaluating electrocatalytic performance (e.g., ORR), representing the potential at which the current reaches half of the limiting current.
- εd (d-Band Center): A parameter related to the regulation of the adsorption strength of catalytic intermediates, representing the center of the d-band of metals.
- EDLC (Electric Double-Layer Capacitor)/Electric Double-Layer Capacitors: A type of supercapacitor based on the electric double-layer principle.
- EDS (Energy Dispersive Spectroscopy)/EDX (Energy Dispersive X-ray Spectroscopy): A method for analyzing element distributions and compositions to determine the elemental composition of materials.

- EELS (Electron Energy Loss Spectroscopy): A method for analyzing material microstructures to study the electronic state and chemical composition of materials
- EG (Ethylene Glycol): A raw material for biomass polyol oxidation, used in the synthesis of organic acids.
- EIS (Electrochemical Impedance Spectroscopy): A method for testing kinetics and charge transfer resistance to analyze electrochemical reaction processes.
- EOR (Ethanol Oxidation Reaction): A reaction for electrocatalytic biomass upgrading involving the oxidation of ethanol.
- EPR (Enhanced Permeability and Retention Effect): A mechanism for passive targeting of tumors, enabling drugs to accumulate in tumor tissues.
- EXAFS (Extended X-ray Absorption Fine Structure): A characterization technique for analyzing the local structure and atomic distances beyond the absorption edge.
- FDCA (2,5-Furandicarboxylic Acid): A biomass-derived polymer monomer used in the synthesis of biodegradable plastics.
- FE (Faradaic Efficiency): An indicator for evaluating the selectivity of electrocatalytic reactions, representing the ratio of electrons used for the target reaction to the total electrons.
- FFT (Fast Fourier Transform): A method for analyzing material structures to process and analyze structural data.
- FTIR (Fourier Transform Infrared Spectroscopy): A method for detecting reaction intermediates and characterizing structures to analyze functional groups and chemical bonds.
- FTO (Fluorine-Doped Tin Oxide): A conductive substrate (for electrodes and solar cells) with good electrical conductivity and light transmittance.
- GA (Glycolic Acid): An oxidation product of ethylene glycol used in the cosmetics and pharmaceutical industries.
- GAD (Glyceraldehyde): An oxidation product of glycerol used in the synthesis of fine chemicals.
- g-C_3N_4 (Graphitic Carbon Nitride): A photocatalytic material with a graphite-like structure.
- GCD (Galvanostatic Charge–Discharge)/Galvanostatic Charging/Discharging: A method for testing capacitance/capacity by charging and discharging at a constant current.
- GNA (Gluconic Acid): An oxidation product of glucose used in the food and pharmaceutical industries.
- GO (Graphene Oxide): A carrier material and modifier for composite materials, with oxide groups on the surface.
- GRA (Glucaric Acid): An oxidation product of glucose used in the synthesis of polymers
- HAADF-STEM (High-Angle Annular Dark-Field Scanning Transmission Electron Microscopy): A tool for characterizing microstructures and element distributions to observe heavy atom distributions.
- HEA (High-Entropy Alloy): A type of alloy material with high-entropy characteristics.

- HER (Hydrogen Evolution Reaction): A core electrocatalytic reaction (used in water electrolysis for hydrogen production) involving the evolution of hydrogen.
- HMF (5-Hydroxymethylfurfural): A biomass platform molecule used in the synthesis of high-value chemicals.
- HMFCA (5-Hydroxymethyl-2-Furancarboxylic Acid): An oxidation intermediate of HMF used in the synthesis of fine chemicals.
- HMFOR (5-Hydroxymethylfurfural Oxidation Reaction): A reaction for the conversion of biomass platform molecules involving the oxidation of 5-hydroxymethylfurfural.
- HPLC (High-Performance Liquid Chromatography): A method for quantitative analysis of products to determine the content of target products.
- HRTEM (High-Resolution Transmission Electron Microscopy)/HR-TEM (High-Resolution Transmission Electron Microscopy): A tool for fine characterization of microstructures to observe lattice fringes and crystal defects.
- HSAB (Hybrid Sodium-Air Battery): A composite air battery system combining sodium-air battery technology.
- IARC (International Agency for Research on Cancer): An organization for classifying the carcinogenicity of formaldehyde responsible for evaluating the carcinogenic risk of chemicals.
- ICG (Indocyanine Green): A reagent used in fluorescence imaging and photothermal therapy.
- ICP (Inductively Coupled Plasma): A method for analyzing element contents to determine the concentration of trace elements.
- IDC (Interdigital Capacitor): A capacitor device with an interdigital electrode structure used in ethanol gas sensing.
- IDLH (Immediately Dangerous to Life or Health): A safety threshold for toxic VOCs, representing the concentration that is immediately dangerous to life and health.
- In situ XRD (in situ X-Ray Diffraction): A method for dynamic structure characterization to study structural changes of materials in situ.
- IONP (Iron Oxide Nanoparticle): A material used in magnetic resonance imaging and magnetic hyperthermia.
- j_K (Kinetic Current Density): An indicator for evaluating the kinetics of electrocatalytic reactions, representing the current density related to the reaction rate.
- j_L (Limiting Current Density): An indicator for evaluating electrocatalytic activity, representing the maximum current density during the reaction.
- KIB (Potassium-Ion Battery)/KIBs (Potassium-Ion Batteries): A new-type battery system based on potassium ion intercalation/deintercalation.
- K-L (Koutecky-Levich): An equation for calculating the number of electrons transferred used to study the reaction mechanism.
- LD (Low-Dimensional): A general term for the dimension of nanomaterials, encompassing low-dimensional nanomaterials.
- LDH (Layered Double Hydroxide)/LDHs (Layered Double Hydroxides): Core electrode materials with a layered structure (e.g., NiFe-LDH).

- LFP (Lithium Iron Phosphate): A cathode material for lithium-ion batteries with good safety performance.
- LIB (Lithium-Ion Battery)/LIBs (Lithium-Ion Batteries): A mainstream battery system based on lithium-ion intercalation/deintercalation.
- Li–S (Lithium–Sulfur Battery)/Li–S Batteries: A high-energy-density battery system based on the reaction between lithium and sulfur.
- LOM (Lattice Oxygen-Mediated Mechanism): A mechanism for OER involving lattice oxygen.
- LSV (Linear Sweep Voltammetry): A method for testing electrocatalytic activity (e.g., polarization curves) by sweeping the potential linearly.
- MA (Maleimide): A modifier for the surface of nanoenzymes, used for functional modification.
- M-AOP (Metal-Activated Advanced Oxidation Process): A process for pollutant degradation activated by metals.
- MCMB (Mesocarbon Microbeads): A carbon-based anode material with microbead-like morphology.
- MOCVD (Metal–Organic Chemical Vapor Deposition): A technology for preparing carbon/magnetic metal composite materials using metal–organic compounds as precursors.
- MOF (Metal–Organic Framework)/MOFs (Metal–Organic Frameworks): Precursor materials and electrode materials with a framework structure composed of metals and organic ligands.
- MOR (Methanol Oxidation Reaction): A reaction for electrocatalytic biomass upgrading involving the oxidation of methanol.
- MRI (Magnetic Resonance Imaging): A technology for diagnostic imaging and integrated diagnosis and treatment used to observe tissue structures.
- MS (Mass Spectrometry): A method for qualitative and quantitative analysis of products to identify the composition and structure of products.
- MS (Micro-Supercapacitor)/MSCs (Micro-Supercapacitors): A micro-energy storage device with supercapacitor characteristics.
- MWNT (Multi-Walled Carbon Nanotube): A carbon nanotube material with multiple walls, used in microwave absorption.
- NASICON (Sodium Superionic Conductor): An ion-conducting material with high sodium ion conductivity.
- N–C (Nitrogen-Doped Carbon): A carrier material and electrocatalytic material doped with nitrogen.
- NCM ($LiNi_{1-x-y}Mn_xCo_yO_2$): A ternary cathode material for lithium-ion batteries, namely lithium nickel manganese cobalt oxide.
- N-CNTs (Nitrogen-Doped Carbon Nanotubes): Modified carbon nanotube materials doped with nitrogen.
- NRR (Nitrogen Reduction Reaction): A reaction for electrocatalytic nitrogen fixation to produce ammonia involving the reduction of nitrogen.
- NTs (Nanotubes): One-dimensional nanomaterials with a tubular morphology.
- NVFP ($Na_{3.5}V_{1.5}Fe_{0.5}(PO_4)_3$): A polyanion-type cathode material for sodium-ion batteries, namely sodium vanadium iron phosphate.

- NWs (Nanowires): One-dimensional nanomaterials with a wire-like morphology.
- OER (Oxygen Evolution Reaction): A core electrocatalytic reaction (used in water electrolysis for hydrogen production and overall water splitting) involving the evolution of oxygen.
- OPM (Oxidation Pathway Mechanism): A mechanism for OER involving oxidation pathways.
- ORR (Oxygen Reduction Reaction)/O_2RR (Oxygen Reduction Reaction): A core electrocatalytic reaction (used in fuel cells and metal-air batteries) involving the reduction of oxygen.
- OWS (Overall Water Splitting): A coupled reaction of HER and OER for water electrolysis to produce hydrogen.
- PANI (Polyaniline): A conductive polymer used as a composite electrode material and carrier modifier.
- PBA (Prussian Blue Analogue)/PBAs (Prussian Blue Analogues): A category of electrode materials with a structure similar to Prussian Blue.
- PCE (Power Conversion Efficiency): An indicator for evaluating the performance of solar cells, representing the efficiency of converting light energy into electrical energy.
- PDA (Polydopamine): A material surface modifier with good adhesion and biocompatibility.
- PDOS (Partial Density of States)/pDOS (Projected Density of States): A detailed result of electronic structure analysis, representing the density of states of specific orbitals or atoms.
- PDT (Photodynamic Therapy): A method of tumor phototherapy using photosensitizers and light.
- PEDOT (Polythiophene): A conductive polymer used as a composite electrode material.
- PEF (Polyethylene Furanoate): A bio-based plastic used in the production of environmentally friendly packaging materials.
- PEMFC (Proton Exchange Membrane Fuel Cell): An energy conversion device that converts chemical energy into electrical energy through proton exchange.
- PET (Polyethylene Terephthalate): A plastic material (with FDCA as an alternative raw material) used in the production of bottles and fibers.
- p-n (p-Type/n-Type Semiconductor): Related to heterojunction materials and sensors, referring to p-type and n-type semiconductors.
- PPy (Polypyrrole): A conductive polymer used as a composite electrode material and carrier modifier.
- PSA (Prostate-Specific Antigen): A cancer biomarker used for the diagnosis of prostate cancer.
- PTB (Poly(Thiophene-Benzothiadiazole)): A conductive polymer used in solar cells, with good photoelectric conversion performance.
- PTT (Photothermal Therapy): A method of tumor phototherapy using photothermal conversion materials.
- PV (Photovoltaic): Related to solar energy conversion, referring to the conversion of light energy into electrical energy.

- QD (Quantum Dot): Zero-dimensional nanomaterials (e.g., $Ni(OH)_2$ QDs) with quantum confinement effects.
- Q-ME-PBA (Quaternary Medium-Entropy Prussian Blue Analogue)/Q-ME-PBAs (Quaternary Medium-Entropy Prussian Blue Analogues): Cathode materials for KIBs, which are Prussian Blue analogues with quaternary medium-entropy characteristics.
- RDS (Rate-Determining Step): A key step in reaction mechanism analysis, which determines the overall reaction rate.
- rGO (Reduced Graphene Oxide): A conductive composite matrix obtained by reducing graphene oxide.
- RHE (Reversible Hydrogen Electrode): A reference for electrochemical potentials used to calibrate electrode potentials.
- RL (Reflection Loss): An indicator for evaluating microwave absorption performance, representing the loss of microwave energy.
- ROS (Reactive Oxygen Species): Related to photodynamic therapy and nanoenzyme catalysis, which are reactive oxygen molecules with biological activity.
- RRDE (Rotating Ring-Disk Electrode): A tool for analyzing ORR pathways to study the reaction mechanism of oxygen reduction.
- SA (Sodium Alginate): A reagent for stabilizing nanomaterials, with good biocompatibility.
- SAC (Single-Atom Catalyst): High-efficiency catalytic materials with single-atom active sites.
- SAED (Selected Area Electron Diffraction): A method for characterizing crystal structures to determine the crystalline phase of materials.
- SB (Sick Building Syndrome): A disease related to formaldehyde pollution caused by indoor air pollution.
- SDS (Sodium Dodecyl Sulfate): A surfactant used in material synthesis to regulate particle size and morphology.
- SEI (Solid Electrolyte Interphase): A film/phase at the electrode interface of batteries, which affects battery performance and cycle life.
- SEM (Scanning Electron Microscopy): A tool for characterizing morphologies to observe the surface morphology of materials.
- SI (Selectivity Index): An indicator for evaluating drug toxicity, representing the selectivity of drugs to target cells.
- SIB (Sodium-Ion Battery)/SIBs (Sodium-Ion Batteries): A new-type battery system based on sodium ion intercalation/deintercalation.
- SS (Specific Surface Area): An indicator for evaluating material performance (number of active sites), representing the surface area per unit mass of materials.
- STEM (Scanning Transmission Electron Microscopy): A tool for analyzing material microstructures and element mapping to study element distributions.
- Tafel (Tafel Plot): A method for evaluating electrocatalytic reaction kinetics to determine the reaction rate and overpotential.
- TEM (Transmission Electron Microscopy)/TEMS (Transmission Electron Microscopy): A tool for characterizing microstructures to observe the internal structure of materials.

- TEP (Threshold Thermodynamic Equilibrium Potential): A potential reference for overall water splitting reactions, representing the thermodynamic equilibrium potential.
- TLV (Threshold Limit Value): An exposure safety threshold for toxic VOCs, representing the maximum allowable exposure concentration.
- TM (Transition Metal): A category of elements composing materials (e.g., Fe, Co, and Ni).
- TM-N-C (Transition Metal-Nitrogen-Carbon): Catalysts for oxygen reduction reactions, composed of transition metals, nitrogen, and carbon.
- TOF (Turnover Frequency): An intrinsic indicator for evaluating catalytic activity, representing the number of reactant molecules converted per active site per unit time.
- UOR (Urea Oxidation Reaction): A reaction used in pollutant degradation and hydrogen production coupling involving the oxidation of urea.
- UPS (Underpotential Deposition): A characterization technique for studying the deposition behavior of metals on electrode surfaces.
- VOC (Volatile Organic Compound): The target of gas sensing and pollutant degradation, which is a volatile organic pollutant.
- WO_3 (Tungsten Trioxide): A material used in photocatalysis and gas sensing.
- XAFS (X-ray Absorption Fine Structure): A method for analyzing the coordination environment of metals to study the local structure of materials.
- XANES (X-ray Absorption Near-Edge Structure): A characterization technique for analyzing the electronic state and coordination environment of atoms near the absorption edge.
- XPS (X-ray Photoelectron Spectroscopy): A method for characterizing element valences and surface compositions to analyze the chemical state of materials.
- XRD (X-ray Diffraction): A method for analyzing crystal structures and phases to determine the crystalline properties of materials.
- ZnO (Zinc Oxide): A material used in photocatalysis and gas sensing.

Part I

Fundamentals of Iron-Group Metal Compound Nanomaterials

1

Introduction

1.1 Background and Significance of Iron-Group Metal Compound Nanomaterials

1.1.1 Fundamental Properties of Iron-Group Metals

As core members of the Group VIII transition metals in the periodic table, iron-group metals (Fe, Co, and Ni) possess unique electronic configurations and physicochemical properties. Their valence electron configurations are Fe ($3d^6 4s^2$), Co ($3d^7 4s^2$), and Ni ($3d^8 4s^2$), respectively. The rich valence state changes (Fe^{2+}/Fe^{3+}, Co^{2+}/Co^{3+}, and Ni^{2+}/Ni^{3+}) endow their compounds with excellent redox activity, while the presence of unpaired d electrons confers unique magnetic properties. In terms of resource distribution, iron is one of the relatively abundant elements in the Earth's crust. Although cobalt and nickel are strategically scarce metals, global reserves can still meet the needs of large-scale applications, and resource recycling can be achieved through secondary recovery technology. Compared with noble metals (Pt, Ir, and Ru), iron-group metal compounds have significantly lower raw material costs, providing a resource foundation for the development of low-cost functional materials [1].

At the macroscale, iron-group metal compounds have been widely used in traditional fields such as metallurgy and chemical engineering (Figure 1.1). However, the size effect, surface effect, and quantum size effect at the nanoscale lead to a qualitative leap in their properties. When the particle size is reduced to the nanoscale range, the specific surface area of the material is significantly higher than that of bulk materials, and the proportion of surface atoms increases remarkably, resulting in an exponential growth in the number of surface-active sites. This structural evolution enables iron-group metal compound nanomaterials to exhibit excellent performance in areas such as energy, environment, and catalysis. Examples include the wide application of ternary lithium batteries and the high catalytic hydrogen evolution activity of Co–P nanosheets, opening up a broad space for their applications in emerging fields.

Iron-Group Metal Compound Nanomaterials: Preparation, Characterization, Structure, and Applications, First Edition. Jinlong Zheng.

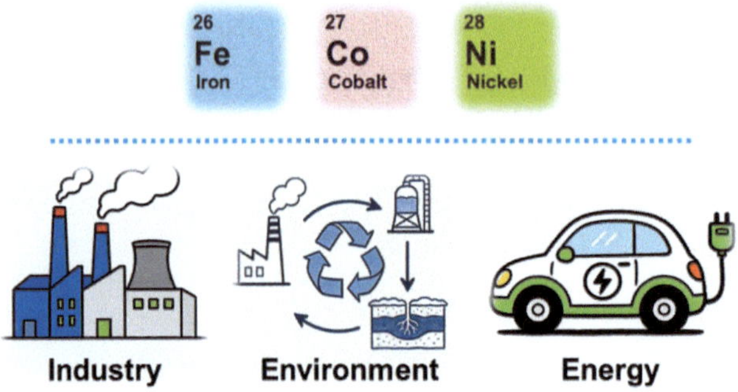

Figure 1.1 Wide applications of iron-group metal (Fe, Co, and Ni) materials in industrial, environmental, and energy fields.

1.1.2 Energy Innovation and Green Environmental Protection

Under the guidance of the "dual carbon" goals, the global energy structure transformation and environmental governance have become important development issues. With their multifunctional characteristics, iron-group metal compound nanomaterials have become core materials supporting the development of new energy, energy conservation, environmental protection, and other technologies. In the field of energy storage, ternary lithium batteries are widely used; in the field of environmental governance, iron-group metal compound nanomaterials have become core materials for heavy metal pollution remediation and organic wastewater purification due to their efficient catalytic degradation and adsorption properties. These strategic needs provide a strong driving force for the research and application of iron-group metal compound nanomaterials, making them a research hotspot in the field of materials science and engineering.

1.1.3 Interdisciplinary Research Value and Industrial Application Potential

The research on iron-group metal compound nanomaterials integrates multiple disciplinary fields such as materials science, chemistry, physics, environmental science, and energy engineering. Its development promotes the innovation of interdisciplinary theories and technologies. At the basic research level, the research on the size effect, interface effect, and structure–activity relationship of such materials provides important theoretical support for nanomaterial science. At the application technology level, their applications in energy conversion, environmental governance, electronic information, and other fields promote the technological upgrading and structural optimization of related industries [2].

In terms of industrial applications, iron-group metal compound nanomaterials have formed a diversified application pattern. In the energy field, Fe_3O_4/carbon composites are widely used as anode materials for lithium-ion batteries. In the catalytic field, Fe/Co–N–C single-atom catalysts exhibit catalytic activity comparable to Pt/C in the oxygen reduction reaction (ORR) of fuel cells. In the environmental field, sulfidated nanoscale zero-valent iron (S-nZVI) exhibits extremely high removal efficiency for Cr(VI) and can convert highly toxic Cr(VI) into low-toxic Cr(III), which has been applied in engineering at multiple chromium-contaminated sites. These application achievements fully demonstrate the important industrial value of iron-group metal compound nanomaterials, and their further development will provide key technical support for solving global energy and environmental problems.

1.2 Classification and Structural Features of Iron-Group Metal Compound Nanomaterials

The versatile physical and chemical properties of iron-group metal (Fe, Co, and Ni) compound nanomaterials originate from their diverse chemical compositions and tunable structural characteristics. As the core foundation for exploring their functional mechanisms and expanding application scenarios, classifying these materials based on chemical bonding types and structural configurations not only helps clarify the intrinsic correlation between "composition-structure-property" but also provides a systematic framework for the design and optimization of advanced functional materials.

In this chapter, we focus on five representative categories of iron-group metal compound nanomaterials with extensive research and practical value: (hydr)oxides, sulfides and phosphides, carbides and nitrides, high-entropy compounds, and coordination compounds (Figure 1.2). This classification follows a logical progression from traditional binary compounds to emerging multicomponent/functionalized systems: (hydr)oxides, as the most fundamental and widely studied type, rely on ionic–covalent hybrid bonding between metal ions and O/OH to form diverse crystal structures (e.g., layered, spinel, and orthorhombic), laying the foundation for basic property research; sulfides and phosphides feature unique electronic structures derived from metal–chalcogen/metal–pnictogen interactions, which endow them with excellent electrochemical activity; carbides and nitrides combine high conductivity, mechanical stability, and noble metal-like catalytic properties due to their covalent-metallic hybrid bonding; high-entropy compounds break through the limitations of traditional binary/ternary systems, leveraging the "high-entropy effect" and lattice distortion to achieve synergistic enhancement of multiple properties; coordination compounds, constructed via coordination bonds between metal centers and ligands (N, P, and S-containing), exhibit unparalleled structural designability and metal–ligand synergy.

By systematically elaborating on the crystal structures, morphological characteristics, electronic properties, and key structural regulation strategies of each category, this chapter aims to establish a comprehensive understanding of iron-group metal compound nanomaterials, providing theoretical support for subsequent discussions on their applications in energy conversion, environmental remediation, catalysis, and other fields.

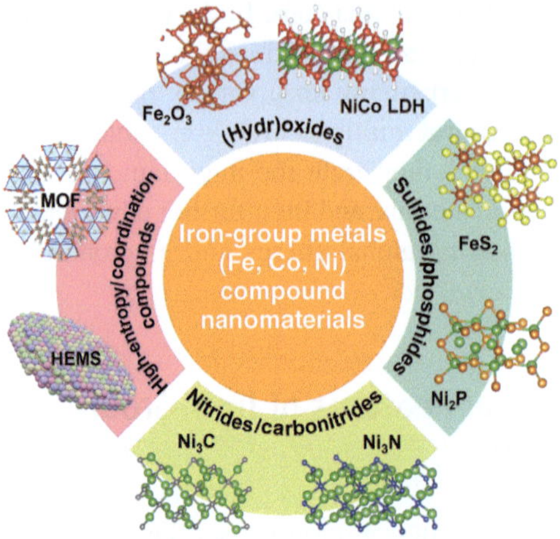

Figure 1.2 Classification of iron-group metal (Fe, Co, and Ni) compound nanomaterials.

1.2.1 (Hydro)oxide Nanomaterials

Iron-group metal (hydro)oxides are the most widely studied iron-group metal compound nanomaterials. The extensively researched structures include spinel structure, perovskite structure, hydrotalcite structure, etc. Their crystal structures and morphologies have a decisive impact on performance.

1.2.1.1 Metal Oxides

Iron oxides mainly include FeO (rock salt structure), α-Fe_2O_3 (corundum structure), γ-Fe_2O_3 (spinel structure), and Fe_3O_4 (inverse spinel structure). The structure of Fe_3O_4 can be expressed as $Fe^{3+}[Fe^{2+}Fe^{3+}]O_4$, which is formed by the ionic bonding of Fe^{2+}, Fe^{3+}, and O^{2-}, with a unique electron conduction path [3]. Cobalt oxides are mainly CoO (rock salt structure) and Co_3O_4 (spinel structure). In Co_3O_4, Co^{2+} occupies tetrahedral interstitial sites and Co^{3+} occupies octahedral interstitial sites, which endows it with both redox activity and electrical conductivity. Nickel oxides mainly include NiO (rock salt structure), Ni_2O_3 (corundum structure), and NiO_2 (rutile structure). NiO is a p-type semiconductor with a certain band gap, and its photocatalytic activity can be significantly improved through nanonization. The electronic structure of iron-group metal oxides is closely related to their valence state changes. The electron transfer between Fe^{2+} and Fe^{3+} in Fe_3O_4 gives it a certain electrical conductivity; the coexistence of Fe^{2+} and Fe^{3+} in Co_3O_4 endows it with excellent redox performance, which can transfer electrons through valence state transformation in catalytic reactions [4]; the electronic structure of NiO has obvious p-d hybridization characteristics, and the surface state density increases after nanonization, leading to a significant improvement in the separation efficiency of photogenerated carriers [5].

In addition, various (hydro)oxides composed of multiple iron-group metal elements have been widely studied, with typical representatives including spinel-type (such as $CoFe_2O_4$ and $NiFe_2O_4$) and perovskite-type (such as $LaFeO_3$ and $LaCoO_3$). The general formula of spinel-type composite oxides is AB_2O_4, where A is divalent metal ions such as Fe^{2+}, Co^{2+}, and Ni^{2+}, and B is trivalent metal ions such as

Fe^{3+} and Co^{3+}. This structure endows them with magnetism, catalytic activity, and stability [6]. $CoFe_2O_4$ nanoparticles have high saturation magnetization and coercivity within a specific nanoscale range, making them suitable for high-density magnetic storage and microwave absorption materials. Perovskite-type composite oxides have an ABO_3 structure, where A is a rare earth metal ion, and B is an iron-group metal ion such as Fe, Co, or Ni. The flexibility of their crystal structure allows precise regulation of their electronic structure through element doping. $LaCoO_3$ nanoparticles exhibit excellent catalytic activity in the ORR.

1.2.1.2 Metal Hydroxides

Iron-group metal hydroxides have rich structures, including layered structures (such as $Ni(OH)_2$ and layered double hydroxides [LDHs]), orthorhombic structures (such as α-FeOOH), and dense structures (such as β-$Co(OH)_2$), with high specific surface area, excellent redox activity, and structural tunability. They have low raw material costs and simple preparation processes, and their performance can be precisely optimized through crystal phase regulation, doping modification, defect engineering, and other means, showing irreplaceable application value in catalysis, energy storage, environmental remediation, and other fields.

LDHs have a typical layered structure, consisting of positively charged metal hydroxide layers and interlayer anions, with the general formula $[M^{2+}{}_{1-x}M^{3+}{}_{x}(OH)_2]^{x+}(A^{n-})_{x/n}\cdot mH_2O$ (M^{2+} is Fe^{2+}, Co^{2+}, or Ni^{2+}; M^{3+} is Fe^{3+} or Co^{3+}). NiFe LDH nanosheets have a suitable interlayer spacing and large specific surface area, and the exchangeability of interlayer anions gives them unique advantages in catalysis and adsorption fields. Through the synergistic effect of carbonate and sulfate anions, NiFe LDH can also exhibit high stability in seawater electrolysis.

1.2.2 Sulfide and Phosphide Nanomaterials

The chemical compositions of iron-group metal sulfide nanomaterials mainly include MS, M_3S_4, and M_7S_8 (M is Fe, Co, or Ni), and their crystal structures are mostly cubic or hexagonal; iron-group metal phosphides (MP_x, M = Fe, Co, or Ni; $x = 1$ or 2) are a class of functional materials with metalloid characteristics whose crystal structures are mainly orthorhombic, cubic, or hexagonal. Their unique electronic structures give them broad application prospects in electrocatalysis, energy storage, and other fields.

1.2.2.1 Sulfide Nanomaterials

Iron sulfides mainly include FeS (hexagonal), FeS_2 (pyrite structure, cubic), and Fe_3S_4 (spinel structure). In the crystal structure of FeS_2, Fe^{2+} is located at the vertices of the face-centered cubic lattice, and $S_2{}^{2-}$ ions occupy tetrahedral interstitial sites. This structure endows FeS_2 with good electrical conductivity and catalytic activity. Cobalt sulfides include CoS (hexagonal), Co_3S_4 (spinel structure), and Co_9S_8 (cubic). In Co_3S_4, Co^{2+} and Co^{3+} occupy tetrahedral and octahedral interstitial sites, respectively, and the electron transfer between Co ions of different valence states endows it with excellent redox activity. Nickel sulfides mainly include

NiS (hexagonal), Ni_3S_2 (hexagonal), and Ni_3S_4 (cubic). The crystal structure of N_3S_2 has a unique tunnel structure, which is conducive to ion transport and exhibits high-capacity characteristics in electrochemical energy storage.

1.2.2.2 Phosphide Nanomaterials

Iron phosphides mainly include FeP (orthorhombic), Fe_2P (hexagonal), and Fe_3P (cubic). In the crystal structure of FeP, Fe atoms form a distorted tetrahedral structure, and P atoms are located in tetrahedral interstitial sites. This structure gives FeP good electrical conductivity and chemical stability. Cobalt phosphides include CoP (orthorhombic), Co_2P (hexagonal), and Co_3P (cubic). The Co—P bond in CoP has partial covalent bond characteristics, and the electron cloud distribution is uniform, which is conducive to the adsorption and activation of reactants in catalytic reactions. Nickel phosphides mainly include NiP (orthorhombic), Ni_2P (hexagonal), and Ni_3P (cubic). In the crystal structure of Ni_2P, Ni atoms form a hexagonal close packing, and P atoms occupy octahedral interstitial sites. Its surface has highly active Ni–P sites, showing excellent catalytic performance in the hydrogen evolution reaction.

The electronic structure of iron-group metal phosphides has unique advantages. The electronegativity of the P element is lower than that of Fe, Co, and Ni, leading to the transfer of electrons from metal atoms to P atoms, making the surface of metal atoms positively charged, which is conducive to the adsorption and reduction of H^+ [7]. In addition, the presence of P atoms can adjust the d-band center of metal atoms, optimize the adsorption energy, and improve the catalytic activity. For example, the d-band center of Ni_2P is close to the ideal d-band center value of hydrogen evolution reaction (HER) catalytic materials, giving it intrinsic HER catalytic activity comparable to that of Pt.

1.2.3 Nitride and Carbide Nanomaterials

1.2.3.1 Nitride Nanomaterials

The crystal structures of iron-group metal nitrides (MN_x, M = Fe, Co, and Ni) are mainly cubic (such as Fe_4N, Co_4N, and Ni_3N) and hexagonal (such as Fe_2N and Co_2N), which have high electrical conductivity, high hardness, and good chemical stability. The electronic structure of nitrides has unique advantages. The electronegativity of N atoms is higher than that of iron-group metals, leading to the transfer of electrons from metal atoms to N atoms, forming electron-deficient centers on the surface of metal atoms, which is conducive to the adsorption and activation of reactants. In addition, the presence of N atoms can adjust the d-band center of metal atoms and optimize the energy barrier of catalytic reactions.

1.2.3.2 Carbide Nanomaterials

Iron-group metal carbides (MC, M = Fe, Co, and Ni) mainly include Fe_3C (cementite, orthorhombic), Co_3C (hexagonal), and Ni_3C (hexagonal), which have electronic structures and catalytic properties similar to noble metals. In the crystal structure of Fe_3C, Fe atoms form a chain arrangement, and C atoms are located in the interchain interstitial sites. This structure gives Fe_3C good electrical conductivity and catalytic activity. The electronic structure of carbides is similar to that of noble metals, and their d-band centers are close to those of noble metals such as Pt and Pd, giving them noble metal-like catalytic activity.

1.2.4 High-Entropy and Coordination Compound Nanomaterials

1.2.4.1 High-Entropy Compound Materials

High-entropy compounds construct multicomponent systems by introducing nonmetallic elements such as O, S, P, and OH^- with iron-group metals, inheriting core characteristics such as high-entropy effect and lattice distortion effect. At the same time, the synergistic effect between nonmetallic elements and metals expands the structural regulation dimension and functional scenarios, becoming a research hotspot in the field of functional materials. The crystal structure of high-entropy compounds can be regulated through component design and synthesis processes, such as rock salt type, spinel type, and layered structure. The strong interaction between multicomponent metals and nonmetals not only optimizes the electronic structure but also further improves the material performance through lattice distortion, defect engineering, etc., showing excellent potential in energy conversion, energy storage, catalysis, and other fields [8].

Through the metal synergy to optimize electron transport and structural stability, the iron-group metal-based high-entropy material system further strengthens the low-cost and high-performance advantages of iron-group metal-based materials in the synergistic development of energy, catalysis, energy storage, and other fields, providing rich ideas for the design and application of new functional materials.

1.2.4.2 Coordination Compound Nanomaterials

Iron-group metal coordination compounds form stable structures through the coordination interaction between metal central ions and ligands (containing coordinating atoms such as N, P, S, and O), with both structural designability and functional diversity. The central ions of such compounds can flexibly present valence states such as Fe^{2+}/Fe^{3+}, Co^{2+}/Co^{3+}, and Ni^{2+}/Ni^{3+}, and the coordination numbers cover various types such as 2, 4, 5, and 6, corresponding to rich spatial configurations such as linear, tetrahedral, square planar, and octahedral. For example, $Ni(CN)_4^{2-}$ is a square planar dsp^2 hybrid inner-orbital structure, and FeF_6^{3-} is an octahedral sp^3d^2 hybrid outer-orbital structure. The structural modification of ligands (such as the regulation of aryl substituents of phenanthroline ligands) and the optimization of the coordination environment can precisely adjust the electronic structure and spin state of the metal center, providing possibilities for the development of materials in the field of energy conversion.

1.2.5 Structure–Activity Relationship Between Structural Features and Performance

The performance of iron-group metal compound nanomaterials is closely related to their structural features, such as chemical composition, crystal structure, morphology, size, and defects. The revelation of the structure–activity relationship is the key to material design and performance optimization.

1.2.5.1 Crystal Structure and Performance

The crystal structure determines the electronic structure and physicochemical properties of the material. Spinel-structured Fe_3O_4 and Co_3O_4 have good electrical conductivity and magnetism, making them suitable for magnetic storage and catalytic fields; layered LDHs and sulfides have large specific surface areas and ion exchange capacities, making them suitable for adsorption and electrochemical energy storage. Orthorhombic FeP and CoP have noble metal-like electronic structures and excellent catalytic activity. Lattice distortion in the crystal structure can adjust the electronic structure. For example, the lattice distortion effect in high-entropy alloys improves electron transport efficiency and significantly enhances catalytic activity.

1.2.5.2 Morphology and Size Effect

Morphology regulates performance by affecting specific surface area, exposure degree of active sites, and mass transfer efficiency. Nanosheets, nanorods, porous structures, etc., have large specific surface areas, which can expose more active sites and improve catalytic and adsorption performance; hollow structures and core–shell structures can alleviate volume expansion and improve the cycle stability of electrochemical energy storage. The size effect is a core characteristic of nanomaterials. Reducing the particle size can significantly increase the specific surface area and the proportion of surface atoms, but excessively small particle size is prone to agglomeration and performance degradation. For example, Fe_3O_4 nanoparticles have the highest saturation magnetization within a specific particle size range, and further reduction in particle size will lead to a decrease in magnetization due to the enhancement of superparamagnetic effects [9].

1.2.5.3 Defect and Interface Effect

Defects (such as vacancies, dislocations, and grain boundaries) and interfaces are key structural features to improve material performance. Defects such as sulfur vacancies and oxygen vacancies can increase the number of active sites, adjust the electronic structure, and improve catalytic activity; grain boundaries and dislocations can accelerate electron transfer and improve electrical conductivity. The interface effect is particularly important in composite materials. Heterogeneous interfaces significantly improve performance through synergistic effects such as electron transfer and charge separation.

1.3 Significance of Iron-Group Metal Compound Nanomaterials in Energy, Environment, and Catalysis Fields

Iron-group metal (Fe, Co, and Ni) compound nanomaterials, featuring tunable electronic structures, abundant active sites, and excellent stability, have emerged as

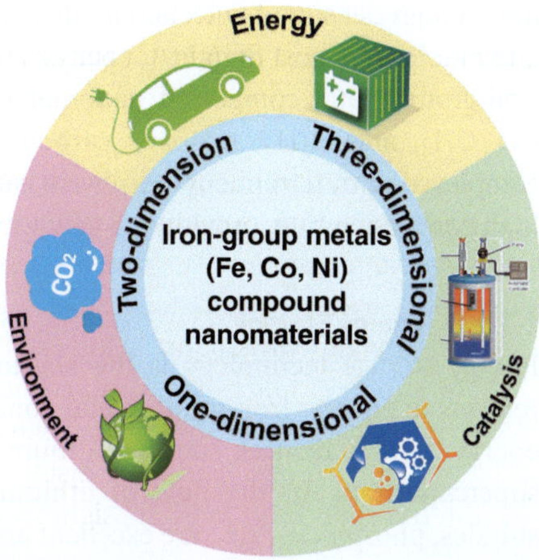

Figure 1.3 Application significance of iron-group metal (Fe, Co, Ni) compound nanomaterials in energy, environmental, and catalytic fields.

pivotal functional materials in addressing global challenges such as energy scarcity, environmental pollution, and inefficient catalysis. Their unique physicochemical properties—derived from diverse compositions (e.g., (hydr)oxides, sulfides, and phosphides) and structural designs—enable targeted adaptations to the core demands of energy conversion/storage, environmental remediation, and catalytic reactions.

This chapter focuses on the practical significance and application value of these nanomaterials in three key fields: energy, environment, and catalysis (Figure 1.3). It synthesizes the detailed explorations of their roles in optimizing energy conversion efficiency (e.g., electrolysis and energy storage), enhancing environmental remediation efficacy (e.g., pollutant degradation and heavy metal removal), and boosting catalytic activity/selectivity (e.g., electrocatalysis and heterogeneous catalysis). By highlighting their irreplaceable contributions to advancing sustainable development and industrial upgrading, this chapter underscores the profound implications of iron-group metal compound nanomaterials for technological innovation and global sustainability.

1.3.1 Application Significance in Energy Field

1.3.1.1 Energy Conversion

Energy conversion is a key approach to solving the energy crisis. Iron-group metal compound nanomaterials play a core role in energy conversion technologies such as water electrolysis for hydrogen production, fuel cells, and CO_2 reduction. In the field of water electrolysis for hydrogen production, HER and oxygen evolution reaction (OER) are core half-reactions. Iron-group metal compound nanomaterials have become ideal alternative materials for noble metal catalysts due to their excellent catalytic activity. CoMnP nanomaterials exhibit better HER performance than commercial Pt/C catalysts in alkaline medium [10]. In the field of fuel cells, the ORR is the core reaction of the cathode. Iron-group metal nitrides, carbides, and single-atom catalysts exhibit excellent ORR activity. Fe–N–C single-atom catalysts have a high density of ORR active sites, high maximum power density in fuel cells, and much lower cost than Pt/C catalysts [11]. In the field of CO_2 reduction reaction, iron-group metal compound nanomaterials can convert CO_2 into fuels such as CO, CH_4, and C_2H_4, realizing carbon cycle utilization. Iron catalyst materials complexed with tetraphenylporphyrin can efficiently convert CO_2 into CH_4 with a high conversion rate, providing a new way to mitigate the greenhouse effect [12].

1.3.1.2 Energy Storage

Energy storage technology is the key to ensuring the large-scale application of renewable energy. Iron-group metal compound nanomaterials are widely used in energy storage devices such as lithium-ion batteries, sodium-ion batteries, and supercapacitors. In the field of lithium-ion batteries, iron-group metal oxides, sulfides, phosphides, etc., are excellent anode materials. Fe_3O_4 hollow nanospheres have a high theoretical specific capacity, and their actual specific capacity is significantly improved by compounding with graphene, which can still maintain a high capacity after multiple cycles [13]. In the field of sodium-ion batteries, iron-group metal compound nanomaterials have become a research hotspot for sodium-ion battery anode materials due to their abundant resources and low cost. Fe–Mn-based composite materials exhibit excellent rate performance in sodium-ion batteries and can still maintain a certain specific capacity at a high current density [14]. In the field of supercapacitors, iron-group metal oxides and hydroxides are commonly used electrode materials. $Ni(OH)_2$ nanosheets have an extremely high specific capacity, and their cycle stability and rate performance are significantly improved by compounding with $Co(OH)_2$ [15].

1.3.2 Application Significance in Environmental Field

1.3.2.1 Water Pollution Control

Iron-group metal compound nanomaterials have important application value in water pollution control fields such as organic pollutant degradation, heavy metal ion removal, and water purification. In terms of organic pollutant degradation, iron-group metal oxides, sulfides, etc., have excellent photocatalytic and Fenton-like catalytic activities and can efficiently degrade organic pollutants such as dyes, antibiotics, and phenols. Co_3O_4/graphene composite materials can efficiently degrade tetracycline in Fenton-like reactions with a high degradation rate, and the activity does not decrease significantly after multiple cycles of use [16]. In terms of heavy metal ion removal, iron-group metal compound nanomaterials can effectively remove heavy metal ions such as Cr(VI), Pb^{2+}, and Cd^{2+} from water through adsorption, reduction, and other effects. Sulfidated nanoscale zero-valent iron (S-nZVI) inhibits side reactions through the surface FeS_x layer, significantly improving the Cr(VI) removal efficiency and maintaining stable activity within a wide pH range [17]. In terms of water purification, iron-group metal compound nanomaterials can be used for sterilization, algae removal, etc. Ag/Ni/Fe_3O_4 composite nanomaterials have both magnetic responsiveness and bactericidal activity, which allows them to be quickly separated and recovered by an external magnetic field while still maintaining high bactericidal efficiency after multiple cycles of use [18].

1.3.2.2 Air Pollution Control

Iron-group metal compound nanomaterials have broad application prospects in air pollution control fields such as waste gas treatment and air purification. In terms of automobile exhaust purification, nanomaterials such as Co_3O_4 and Fe_2O_3 can be used as catalysts to efficiently catalyze the oxidation of pollutants such as CO and HC and reduce NO at the same time, with high purification efficiency [19]. In terms of industrial waste gas treatment, iron-group metal compound nanomaterials can effectively remove pollutants such as SO_2 and volatile organic compounds (VOCs). Co_3O_4 nanomaterials exhibit high activity in catalyzing VOCs (such as toluene and formaldehyde), with a high toluene conversion rate and low light-off temperature [20]. In terms of air purification, iron-group metal compound nanomaterials can be used for indoor air purification to remove harmful gases such as formaldehyde and benzene. MIL-88B(Fe) nanomaterials can catalyze the oxidation of formaldehyde at room temperature with a high degradation rate [21].

1.3.3 Application Significance in Catalysis Field

Iron-group metal compound nanomaterials have rich active sites, tunable electronic structures, and excellent redox performance, giving them significant applications in heterogeneous catalysis, electrocatalysis, photocatalysis, and other fields, and making them a research hotspot in the catalytic field.

1.3.3.1 Heterogeneous Catalysis

In the field of heterogeneous catalysis, iron-group metal compound nanomaterials are widely used in reactions such as hydrogenation, dehydrogenation, and oxidation. In hydrogenation reactions, Ni_2P nanoparticles have high catalytic activity and selectivity for benzene hydrogenation to cyclohexane; In_2O_3/Co_3O_4 has excellent catalytic activity for CO_2 hydrogenation to methanol, with a high methanol yield. In dehydrogenation reactions, Fe_3O_4/carbon composite materials have high catalytic activity and selectivity for ethanol dehydrogenation to acetaldehyde. In oxidation reactions, $Cu_{0.7}Co_{0.3}Fe_2O_4$ nanomaterials have high catalytic activity and selectivity for toluene oxidation to benzaldehyde; Co_3C nanoparticles can efficiently convert CO and H_2 into long-chain hydrocarbons in Fischer–Tropsch synthesis, and Fe_3O_4 nanoparticles exhibit excellent catalytic activity in ammonia synthesis, which can replace traditional iron-based catalysts.

1.3.3.2 Electrocatalysis

Electrocatalysis is the core technology of energy conversion and storage. Iron-group metal compound nanomaterials exhibit excellent performance in electrocatalytic reactions such as HER, OER, ORR, and CO_2 reduction. In addition to the HER, OER, and ORR catalytic performance mentioned in the energy conversion section, iron-group metal compound nanomaterials also have important applications in other electrocatalytic reactions. Pt/Co_3O_4 has significant electrocatalytic activity for the dehydrogenation of isopropanol to acetone, providing a new way for the recycling of carbon resources [22].

1.3.3.3 Photocatalysis

Photocatalytic technology can convert solar energy into chemical energy, which has important applications in energy and environmental fields. Iron-group metal compound nanomaterials are an important class of photocatalytic materials. α-Fe_2O_3 is a typical visible-light-responsive photocatalytic material with a suitable band gap, which can absorb visible light and exhibit excellent performance in photocatalytic water splitting for hydrogen production and organic pollutant degradation. By compounding with g-C_3N_4, the separation efficiency of photogenerated carriers in the α-Fe_2O_3/g-C_3N_4 heterostructure is significantly improved, and the photocatalytic hydrogen evolution rate is high [23].

1.4 Challenges and Research Motivations

1.4.1 Main Challenges

Although iron-group metal compound nanomaterials exhibit excellent performance and broad application prospects in many fields, they still face many challenges in practical applications, mainly including structure and performance stability, large-scale preparation, and adaptability to actual working conditions.

1.4.1.1 Challenges in Structure and Performance Stability

The structural and performance instability of iron-group metal compound nanomaterials remains the primary bottleneck to their practical application. In electrochemical energy storage and catalytic applications, the volume expansion, agglomeration, and oxidative passivation of materials seriously affect their cycle stability. For example, Fe_3O_4 nanoparticles have obvious volume expansion during lithium-ion battery cycling, leading to the collapse of the electrode structure and significant capacity attenuation after multiple cycles. Nanocatalysts are easily oxidized and corroded by reverse current in the hydrogen evolution reaction, resulting in gradual attenuation of catalytic activity and significant activity loss after multiple cycles.

In environmental applications, the long-term stability of iron-group metal compound nanomaterials is insufficient. Nanoscale zero-valent iron (nZVI) is easily oxidized in air to form a Fe_xO_y passivation layer, leading to a rapid decrease in its reduction activity. Fe-based nanoparticles are prone to agglomeration, and the particle size increases significantly, resulting in a substantial decrease in their adsorption and catalytic performance.

1.4.1.2 Challenges in Large-Scale Preparation Technology

At present, the preparation of iron-group metal compound nanomaterials is mainly concentrated at the laboratory scale, and the large-scale preparation technology remains immature, with problems such as low yield, high cost, and poor product consistency. Common laboratory preparation methods, such as the sol–gel method, hydrothermal method, and thermal decomposition method, are difficult to realize in large-scale industrial production. For example, the thermal decomposition method requires high-temperature conditions, high equipment cost, low yield, and high product preparation cost. The nanomaterials prepared by the coprecipitation method have low purity and wide particle size distribution, which can lead to obvious fluctuations in material performance.

In the large-scale preparation process, it is difficult to precisely control the morphology, size, and structure of the material. For example, in the large-scale production of iron-based nanocrystalline ribbons, it is difficult to precisely control parameters such as the cooling rate and annealing temperature of the melt spinning furnace, leading to large fluctuations in the magnetic properties of the product. The agglomeration problem of nanoparticles is more prominent in large-scale production, requiring a large number of surfactants for modification, which increases production costs and environmental burdens. In addition, the treatment of wastewater and waste gas in the large-scale preparation processes is also an important issue to be solved. Traditional preparation methods often produce a large number of pollutants, which do not meet the requirements of green and low-carbon development.

1.4.1.3 Challenges in Adaptability to Actual Working Conditions

Iron-group metal compound nanomaterials exhibit excellent performance under ideal laboratory conditions, but their performance often degrades significantly under actual complex working conditions, making it difficult to meet practical application requirements. In the energy field, electrolytes used for water electrolysis to produce hydrogen often contain impurity ions (such as Cl^- and SO_4^{2-}), which can cause corrosion and activity attenuation of iron-group metal compound catalysts; under high-temperature and high frequency conditions, the magnetic properties of iron-based nanocrystalline materials in the on-board chargers of new energy vehicles will degrade significantly, and the loss will increase.

In the environmental field, the pH value of actual wastewater fluctuates greatly, and the composition of pollutants is complex, which seriously affects the adsorption and catalytic performance of iron-group metal compound nanomaterials. NO_3^- in groundwater will occupy the active sites of nanoscale zero-valent iron, significantly reducing the Cr(VI) removal efficiency; when organic pollutants and heavy metal ions coexist in industrial wastewater, they compete for adsorption sites, reducing the removal efficiency of the material. In addition, in the seawater environment, high salinity will accelerate the corrosion of iron-group metal compound nanomaterials, and the corrosion rate surges under high Cl^- concentration conditions, leading to rapid failure of the material.

1.4.2 Research Motivations and Development Directions

In response to the above challenges faced by iron-group metal compound nanomaterials, conducting relevant research has important scientific significance and practical application value. The main research motivations and development directions are as follows.

1.4.2.1 Structural Design and Performance Optimization

Through precise structural design and performance optimization, improving the structure and performance stability of iron-group metal compound nanomaterials is the key to solving their application bottlenecks. In terms of structural design, core–shell structures, porous structures, heterostructures, etc., can be constructed to alleviate volume expansion and inhibit agglomeration and oxidative passivation. The shell layer of the core–shell structure can protect the core material from oxidation and enhance electrical conductivity; the porous structure can provide sufficient buffer space to alleviate volume expansion and improve cycle stability.

In terms of performance optimization, the electronic structure of the material can be adjusted via element doping, defect engineering, interface engineering, or surface modification to improve catalytic activity and stability. Element doping can optimize the d-band center of the material and enhance the adsorption and activation ability of reactants; defect engineering (such as vacancies and dislocations) can increase the number of active sites and improve the intrinsic catalytic activity; interface engineering can facilitate electron transfer at metal sites, adjust the electronic structure of the active phase, optimize the adsorption energy of reaction intermediates, and improve material performance; surface modification (such as sulfidation and phosphidation) can form a protective layer, inhibit oxidative passivation, and improve stability.

1.4.2.2 Development of Green Large-Scale Preparation Technology

Developing green, efficient, and low-cost large-scale preparation technology is a prerequisite for the industrial application of iron-group metal compound nanomaterials. Green preparation technology should meet the requirements of low energy consumption, low pollution, and high yield, which can be achieved by optimizing existing preparation methods and developing new preparation technologies. For example, nanoscale iron can be prepared from steel plant pickling waste liquor, and waste resource utilization can be realized at the same time.

New preparation technologies, such as low-temperature solution method, template method, and self-assembly method, have the advantages of mild preparation conditions, high product purity, and easy scale-up, and are expected to replace traditional preparation methods. The template method can prepare nanomaterials with specific morphologies and structures, such as porous and hollow structures, and is easy to scale-up production. In addition, developing continuous preparation processes (such as continuous hydrothermal methods and continuous spray pyrolysis methods) can greatly improve production efficiency, reduce production costs, and promote the industrialization process of iron-group metal compound nanomaterials.

1.4.2.3 Improvement of Adaptability to Actual Working Conditions and Application Expansion

Improving the adaptability of iron-group metal compound nanomaterials under actual complex working conditions and expanding their application fields are the keys to realizing their practical application value. According to the characteristics of working conditions in different application scenarios, special materials and technical solutions should be developed. In the energy field, develop water electrolysis catalysts resistant to high salt and high temperature to meet the needs of seawater electrolysis and high-temperature energy storage devices.

In the environmental field, develop pollution remediation materials with broad-spectrum adaptability to meet the needs of wastewater treatment with different pH values and different pollutant compositions; improve the selectivity of materials for target pollutants through surface modification and composite modification; and inhibit competitive reactions.

In addition, the application fields of iron-group metal compound nanomaterials should be expanded by developing multifunctional composite materials and realizing "one thing for multiple uses." For example, develop composite materials with catalytic, adsorption, and magnetic separation functions to achieve efficient pollutant removal and rapid material recovery in wastewater treatment; develop materials with both energy storage and sensing functions for smart wearable devices. At the same time, strengthen the life cycle assessment of materials to ensure their environmental friendliness during preparation, use, and disposal, and promote sustainable development.

References

1 Bhavyesh, D., Soliya, S., Konakanchi, R. et al. (2024). The recent advances in iron-catalyzed C(sp^3)-H functionalization. *Chemistry, an Asian Journal* 19: e202301056.

2 Olusegun, S.J., Souza, T.G.F., Souza, G.d.O. et al. (2023). Iron-based materials for the adsorption and photocatalytic degradation of pharmaceutical drugs: a comprehensive review of the mechanism pathway. *Journal of Water Process Engineering* 51: 103457.

3 Gareev, K.G. (2023). Diversity of iron oxides: mechanisms of formation, physical properties and applications. *Magnetochemistry* 9: 119.

4 Hu, Z., Hao, L., Quan, F., and Guo, R. (2022). Recent developments of Co_3O_4-based materials as catalysts for the oxygen evolution reaction. *Catalysis Science & Technology* 12: 436–461.

5 Bhattacharya, C., Saji, S.E., Mohan, A. et al. (2020). Sustainable nanoplasmon-enhanced photoredox reactions: synthesis, characterization, and applications. *Advanced Energy Materials* 10: 2002402.

6 Jiang, P., Xu, Q., Tran, N. et al. (2022). Boosted microwave absorption properties of $CoFe_2O_4$ with extraordinary 3D morphologies. *Ceramics International* 48: 13541–13550.

7 Liu, Y., McCue, A., and Li, D. (2021). Metal phosphides and sulfides in heterogeneous catalysis: electronic and geometric effects. *ACS Catalysis* 11: 9102–9127.

8 Perivoliotis, D.K., Ekspong, J., Zhao, X. et al. (2023). Recent progress on defect-rich electrocatalysts for hydrogen and oxygen evolution reactions. *Nano Today* 50: 101883.

9 Chen, R., Zhang, Y., Zhao, J., and Wei, Y. (2010). Synthesis of variable-sized Fe_3O_4 nanocrystals and their magnetic properties. *Chinese Journal of Inorganic Chemistry* 26: 1207–1212.

10 Xu, Y., Wei, S., Gan, L. et al. (2022). Amorphous carbon interconnected ultrafine CoMnP with enhanced Co electron delocalization yields Pt-like activity for alkaline water electrolysis. *Advanced Functional Materials* 32: 2112623.

11 Xie, Z., Yan, J., Zheng, X. et al. (2025). Kinetic analysis of Fe leaching and stability in Fe–N–C catalysts during oxygen reduction. *Chem Catalysis* 6: 101525.

12 Rao, H., Schmidt, L.C., Bonin, J., and Robert, M. (2017). Visible-light-driven methane formation from CO_2 with a molecular iron catalyst. *Nature* 548: 74–77.

13 Jiang, F., Yan, X., Du, R. et al. (2019). Fe_3O_4 hollow nanosphere-coated spherical-graphite composites: a high-rate capacity and ultra-long cycle life anode material for lithium ion batteries. *Nanomaterials* 9: 996.

14 Peng, B., Zhou, Z., Shi, J. et al. (2024). Earth-abundant Fe-Mn-based compound cathodes for sodium-ion batteries: challenges and progress. *Advanced Functional Materials* 34: 2311816.

15 Li, Y., Zhang, J., Yang, W. et al. (2025). Sandwich-like activated carbon/$Co(OH)_2$/$Ni(OH)_2$ composites prepared by electrodeposition for high-performance supercapacitors. *Journal of Energy Storage* 112: 115529.

16 Ma, R., Zhou, G., Gu, M. et al. (2024). Cobalt leaching inhibition: transforming coordination polymers into spherical Co_3O_4@NC catalysts for accelerated tetracycline degradation via enhanced PMS activation. *Applied Surface Science* 648: 158980.

17 Alazaiza, M.Y.D., Albahnasawi, A., Copty, N.K. et al. (2022). Nanoscale zero-valent iron application for the treatment of soil, wastewater and groundwater contaminated with heavy metals: a review. *Desalination and Water Treatment* 253: 194–210.

18 Hoang, V.H., Phan, T.N.B., Nguyen, V.T. et al. (2025). One-pot green synthesis of Ag/Ni/Fe_3O_4-activated carbon beads for recyclable photo-Fenton antibiotic removal and antibacterial action: mechanistic study and optimization. *RSC Advances* 15: 13478–13496.

19 Zhang, H., Tang, J., Huang, B. et al. (2025). Highly dispersed nano-Fe_2O_3 anchored on natural octahedral molecular sieve manganese oxide (OMS-2) enhance simultaneous removal of NO and Hg^0 at low temperature. *Journal of Environmental Chemical Engineering* 13: 118942.

20 Xiao, M., Zhao, T., Li, Y. et al. (2024). Co_3O_4-based catalysts for the low-temperature catalytic oxidation of VOCs. *ChemCatChem* 16: e202301524.

21 Zhang, S., Zhuo, Y., Ezugwu, C.I. et al. (2021). Synergetic molecular oxygen activation and catalytic oxidation of formaldehyde over defective MIL-88B(Fe) nanorods at room temperature. *Environmental Science & Technology* 55: 8341–8350.

22 Yang, T., Kastenmeier, M., Ronovsky, M. et al. (2021). Selective electrooxidation of 2-propanol on Pt nanoparticles supported on Co_3O_4: an in-situ study on atomically defined model systems. *Journal of Physics D: Applied Physics* 54: 164002.

23 Li, Y.-P., Li, F.-T., Wang, X.-J. et al. (2017). Z-scheme electronic transfer of quantum-sized α-Fe_2O_3 modified g-C_3N_4 hybrids for enhanced photocatalytic hydrogen production. *International Journal of Hydrogen Energy* 42: 28327–28336.

2

Preparations of Iron (Fe, Co, and Ni)-Group Metal Compound Nanomaterials

Scientists around the world have unanimously turned their attention to a completely new type of material, that is, nanomaterials, and predict that the application of nanotechnology marks the entry of human science and technology into a new era, namely the era of nanoscience and technology. The 1st International Conference on Nanoscience and Technology, held in 1990, officially announced "nanomaterial science" as a new branch of materials science to the world, signifying the birth of a relatively independent new discipline. Nanoscience researches the so-called "mesoscopic" field between "macroscopic" and "microscopic," where the size of the basic building blocks of the constructed materials ranges from 0.1 to 100 nm. The reason why nanocatalytic materials exhibit unique catalytic performance primarily arises from the aforementioned "small-size effect" and "surface effect." In addition, the anisotropy of the crystals used to prepare nanocatalysts is the essential reason for their oriented growth. Crystals have different crystal planes, and the growth rates of each crystal plane of anisotropic crystals vary under certain environments. Different growth rates or stacking rates of grains on different crystal planes will lead to the formation of crystals with different microscopic morphologies. Only by providing a suitable growth environment for crystals can their oriented growth become possible. The growth environment of crystals can be regulated by changing some external factors, such as reaction temperature, reactant concentration, solvent, pH, and the addition of stabilizers (or surfactants). Several commonly used methods for the preparation of nanocatalysts are introduced below [1–3]. Among them, iron-based metal compound nanomaterials, as a type of material with important application potential, possess excellent magnetic, optical, and electrical properties, and their applications in fields such as catalysis have become increasingly widespread. For various manufacturing methods of heterogeneous catalysts, it is customary to name each nanocatalyst preparation method after the key and distinctive operating unit involved [4]. This chapter will summarize and elaborate on the preparation methods of iron-based metal compound nanomaterials, including the precipitation/coprecipitation method, the hydrothermal/solvothermal method, the impregnation method, the sol–gel method, the electrodeposition method, the vapor deposition method, the template method, the high-temperature pyrolysis method, the Joule heating method, and so on.

Iron-Group Metal Compound Nanomaterials: Preparation, Characterization, Structure, and Applications, First Edition. Jinlong Zheng.

2.1 Precipitation/Coprecipitation Method

The precipitation method is a manufacturing process that uses precipitation as its key and special step, and it is one of the most commonly used methods for preparing solid catalysts, widely used for producing high-content non-precious metals, metal oxides, metal salts, catalysts, or catalyst supports.

The general procedure of the precipitation method involves adding an alkaline substance (precipitating agent) to an aqueous solution of metal salts under stirring, followed by washing, filtering, drying, and calcining the resulting precipitate to produce the desired powdered precursor of the catalyst [5]. In large-scale production, metal salts are made into aqueous solutions for economic reasons; in some special cases, nonaqueous solutions such as acids, bases, or organic solvents can also be used. The precipitation method includes single-component precipitation, coprecipitation, uniform precipitation, impregnation precipitation, and crystal-guided precipitation, all of which share the common characteristic of being simple and convenient to operate. The key equipment in the precipitation method is generally the precipitation tank, which has a structure similar to that of a typical stirred tank reactor.

Single-component precipitation method is a process where a precipitating agent interacts with a solution to be precipitated to prepare a single-component precipitate. This is one of the most commonly used methods in catalyst preparation. Since the precipitate contains only one component, the operation is not very difficult. It can be used to prepare single-component catalysts or catalyst supports for non-precious metals. When combined with mechanical mixing and other unit operations, it can also be used to prepare multicomponent catalysts.

The coprecipitation method, also known as the chemical precipitation method, is a process that simultaneously precipitates two or more components required for the catalyst. In this method, a precipitating agent is added to a solution of metal salts, and the resulting precipitate is processed through washing, drying, and calcination to obtain semiconductor materials. The specific principle involves adding a precipitating agent to a soluble salt solution containing one or more ions, which hydrolyzes at a certain temperature to form insoluble hydroxides, hydrated oxides, or salts that precipitate out of the solution. The solvent and original anions in the solution are then washed away, and the desired oxide nanopowder is obtained through thermal decomposition or dehydration. This method is commonly used to prepare high-content multicomponent catalysts or catalyst supports. In the field of catalyst preparation, the coprecipitation method involves mixing solutions of one or more active components with a carrier component solution, followed by the addition of a precipitating agent to form a co-precipitate. Sometimes, only two or more active component solutions are mixed without the carrier component solution, and a precipitating agent is added to achieve coprecipitation. Subsequent steps include aging, washing, drying, shaping, and calcination to obtain the co-precipitated catalyst. The process of mixing carrier component solutions, precipitation, aging, washing, drying, crushing, shaping, and calcination is followed. Its characteristic is that multiple catalyst components can be obtained simultaneously in a mixture, with relatively constant proportions and uniform distribution among the components.

If solid solutions form between the components, the resulting dispersion and uniformity are even more ideal. This preparation method is simple and easy to implement, representing one of the earlier chemical synthesis methods for metal oxide nanomaterials in the liquid phase. Its key features include easy control of nucleation, simple addition of trace components with uniform composition, and the ability to obtain high-purity nano-composite oxides. Its advantage lies in the potential to achieve high-performance nanopowders without requiring harsh physical conditions. The method offers wide raw material availability, low cost, small equipment investment, and high powder yield, making it a primary approach for reducing the cost of nanopowders. The chemical precipitation method is divided into two types: conventional chemical precipitation (adding the precipitating agent to the metal salt solution) and reverse chemical precipitation (adding the metal salt solution to the precipitating agent). Comparing the nanopowders obtained by these two methods, the reverse chemical precipitation method yields nanomaterials with finer particle size and more uniform particles. The superior dispersion and uniformity of the coprecipitation method represent its greatest advantage over methods such as mixing. However, the structure and physicochemical properties of the precipitate are influenced by multiple factors, including the selection of precursor metal salts, precipitating agents, and solvents, solution concentration, as well as operational conditions such as temperature, solution pH, supersaturation level, mixing method of reagents, feeding sequence, dropping rate, and stirring rate. By adjusting these conditions, the structure and properties of the precipitate—such as crystal phase, chemical composition, purity, particle size, surface area, and pore structure—can be controlled.

The two precipitation methods mentioned above often encounter issues during operation, such as uneven mixing of the precipitating agent with the components to be precipitated, inconsistent precipitate particle sizes, and significant impurity incorporation. The homogeneous precipitation method can overcome these shortcomings. In this method, the precipitating agent is not directly added to the solution containing the components to be precipitated, nor does precipitation occur immediately after adding the precipitating agent. Instead, the metal salt solution to be precipitated is thoroughly mixed with a precursor of the precipitating agent first, creating a highly uniform system in advance. Then, conditions conducive to precipitation are gradually established by adjusting temperature and time, slowly increasing the pH, or gradually generating the precipitating agent within the system. This allows the precipitation to proceed slowly, resulting in a precipitate with highly uniform particles and high purity. Homogeneous precipitation is not limited to neutralization reactions; it can also utilize the hydrolysis of esters or other organic compounds, the decomposition of complexes, or redox reactions.

The impregnation precipitation method is a novel technique developed by combining conventional impregnation with precipitation. Specifically, after the impregnation operation with the salt solution is completed, a precipitating agent is added to deposit the components to be precipitated onto the support. Crystal-guided precipitation is an efficient and rapid method that utilizes a crystal-guiding agent (seed crystals) to induce the transformation of amorphous precipitates into crystalline precipitates.

The most commonly used precipitating agents are ammonium salts such as NH_3, NH_4OH, and $(NH_4)_2CO_3$, as they are easily removed during subsequent washing and heat treatment without leaving residues. When using KOH or NaOH, consideration must be given to the potential adverse effects of K^+ or Na^+ residues on catalyst performance. Generally, using salt-based precipitating agents often results in crystalline precipitates, and KOH is relatively expensive. However, if permissible, using NaOH or Na_2CO_3 to provide OH^- or CO_3^{2-} is generally a good choice. Particularly, the latter is not only inexpensive and readily available but also often forms crystalline precipitates that are easy to wash. Lower precipitation temperatures favor nucleus formation while slowing crystal growth, facilitating the production of fine particles. Solution concentration significantly influences the precipitation process. Generally, precipitation from dilute solutions requires a longer time for precipitate formation but yields larger crystals. There are two feeding sequences: one where the precipitating agent is added to the active component solution (conventional addition), and the other where the active component solution is added to the precipitating agent (reverse addition). The feeding sequence can influence the structure of the precipitate, thereby affecting the catalyst's properties. Stirring enhances solution turbulence, favoring crystal growth over nucleation, which tends to produce larger crystals. The effect of pH on precipitation is complex; different pH levels influence the precipitation rates of different ions, leading to precipitates with varying structures and significantly impacted properties. To control the precipitate's structure and particle size uniformity, maintaining a relatively constant pH during precipitation is essential. Therefore, the method of simultaneously feeding the active component solution and the precipitating agent is often employed.

The precipitation method allows for the controllable synthesis of metallic nanomaterials with different parameters, such as composition, morphology, thickness, size, and crystal phase. Batch operation is generally employed for catalyst preparation, requiring particular attention to the solution mixing method/sequence. There are generally three mixing methods: (i) adding the alkaline solution to the salt solution; (ii) adding the salt solution to the alkaline solution; (iii) simultaneous addition of the salt and alkaline solutions. When the alkaline solution is added to the salt solution, the pH gradually increases with the amount added. In the early stages of precipitation, the low pH may result in the product containing a large number of anions from the salt precursor. In this case, sufficient aging of the precipitate is necessary to replace the precursor salt anions. During the co-precipitation of multiple metal ions, since each metal ion precipitates at a different pH, the gradual addition of alkali to the salt solution causes the metal ions to precipitate separately at different pH values, leading to an inhomogeneous distribution of the precipitate. This issue can be addressed by adding the salt solution to the alkaline solution. In both of the aforementioned mixing methods, the pH of the system changes, potentially causing variations in the structure and properties of the precipitate at different pH levels, making it difficult to ensure the uniformity of the precipitate structure. The method of simultaneously adding the alkaline and salt solutions maintains a constant pH of the system, mitigating the effects of pH variation to some extent. In practice, different mixing methods for the alkaline and salt solutions can lead to significant differences in the catalyst's structure and performance. In addition, temperature and pH also greatly influence the physical properties of the resulting product. Our research focus is on applying the precipitation method to fabricate iron-based metallic heterostructures with high-quality interfaces to enhance the catalytic performance of the catalyst. Currently, the most common preparation process in experiments involves the initial reaction of a solid solution at high temperature/room temperature and atmospheric pressure. Specifically, several required soluble metal salts or oxides are selected and metered according to the desired composition to prepare a solution. This solution is then added to a conical flask or round-bottom flask. After heat treatment under atmospheric pressure, the resulting product undergoes subsequent secondary processing to ultimately obtain high-quality heterogeneous catalysts. For example, Lassoued et al. [6] used $FeCl_3$ as the iron source and NH_4OH as the precipitating agent, employing the precipitation method to synthesize Fe_2O_3 nanoparticles (NPs) with sizes ranging from 21 to 82 nm. They found that as the concentration of Fe^{3+} ions increased, the size of the NPs became larger, and the optical band gap of the Fe_2O_3 NPs subsequently decreased.

The precipitation method is by far the most common approach for synthesizing metallic iron NPs and iron oxides (Fe_2O_3, Fe_3O_4, and FeOOH) [7–10]. Typically, catalyst precipitates are generated from aqueous iron salt solutions by adding an alkaline precipitating agent at room temperature or elevated temperature under an inert atmosphere. Similarly, the coprecipitation method is widely used for synthesizing composite materials (such as $Fe_{3-x}Co_xO_4$, $Fe_{3-x}Cu_xO_4$, and Fe_3O_4-CeO_2) through the precipitation of multiple metal precursors [11–13]. According to this method, Fe^0 NPs can be prepared directly from an aqueous $FeSO_4$ solution by adding potassium borohydride without further processing [14, 15]. Fe_3O_4 magnetic NPs (MNPs) can also be obtained via the coprecipitation of ferric and ferrous ions in an alkaline solution under an inert atmosphere [16]. Similarly, Gogoi et al. [13] used $FeCl_2{\cdot}4H_2O$, $FeCl_3{\cdot}6H_2O$, and $Ce(NO_3)_3{\cdot}6H_2O$ as metal precursors and a 25% ammonia solution as the precipitating agent.

Hou et al. [17] innovatively obtained a single-phase bimetallic precursor via a boiling-recrystallization method, successfully growing different oxygen-containing sulfide heterostructures on flexible carbon fiber cloth (as shown in Figure 2.1a). The synergistic effect of the oxygen-containing CoS_2–MoS_2 (CoMoS) nanosheet heterostructure and its unique sheet-like structure generated abundant active catalytic sites. This efficient bifunctional electrocatalyst, constructed by interface engineering, exhibited excellent catalytic performance and stability for both the hydrogen evolution reaction (HER) and oxygen evolution reaction (OER) in alkaline solution, showing low overpotentials of 97 and 272 mV, respectively, at 10 mA cm^{-2}. Furthermore, Liu and colleagues synthesized CoFe-LDHs supports at room temperature via coprecipitation (as shown in Figure 2.1b), which were subsequently subjected to stirring treatment to obtain the single-atom catalyst Ru/CoFe-LDHs. The interfacial interaction between the highly dispersed Ru atoms and the CoFe-LDHs nanosheet support provided pathways for electron transfer, optimized the electronic states of individual atoms, and endowed the catalyst with

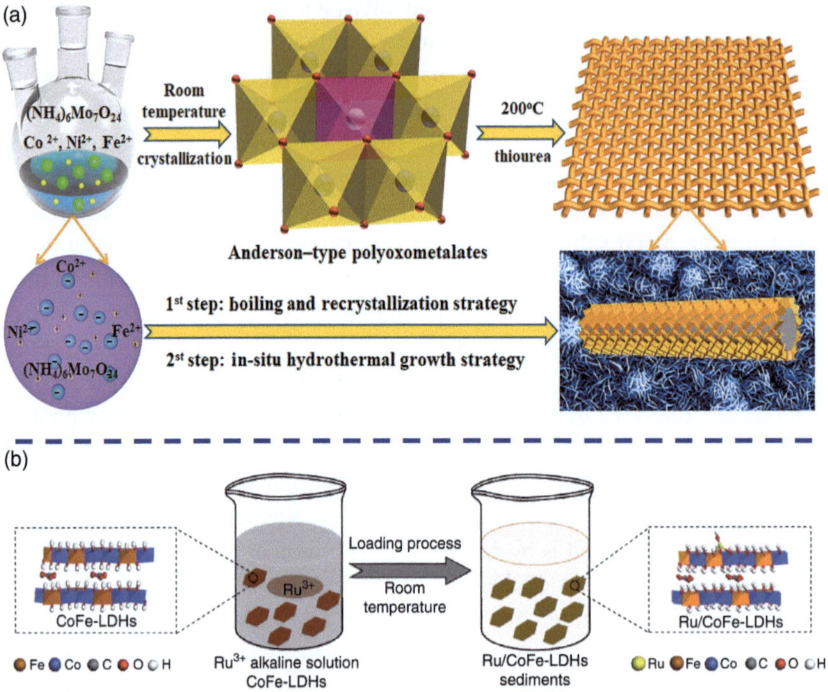

Figure 2.1 (a) Schematic illustration of the preparation of O-CoMoS. *Source:* Reproduced from Ref. [17] / with permission of American Chemical Society. (b) Synthesis and structural characterization of Ru/CoFe-LDHs. *Source:* Ref. [18] / Springer Nature / CC BY 4.0.

excellent OER catalytic activity, demonstrating a low overpotential of 198 mV to achieve a current density of $10\,mA\,cm^{-2}$ [18].

2.2 Hydrothermal/Solvothermal Synthesis

Hydrothermal synthesis is an important method for preparing oxide nanocrystals. It involves the reaction of starting mixtures (including alkoxides and inorganic salts) in a closed system using water as the solvent, under specific temperatures and the self-generated pressure of water. Typically, this process is carried out in stainless steel autoclaves, which often feature liners or cups made of heat-resistant and chemically inert polytetrafluoroethylene (PTFE). The reaction conditions generally require heating temperatures exceeding 100°C and pressures exceeding 101.3 kPa. Hydrothermal methods broadly refer to chemical reactions conducted in water or steam and other fluids under high temperature and pressure. Compared to precipitation methods, the hydrothermal method directly yields the desired products, such as oxides or sulfides, avoiding the calcination step required in precipitation methods to convert precursors into oxides. This results in synthesized oxide powders with good dispersibility, controllable size, minimal agglomeration, well-crystallized grains, and fully exposed crystal facets.

In this closed system, the pressure primarily depends on the system's composition and temperature. Under hydrothermal conditions, particles undergo nucleation and growth, forming ultrafine powders with controllable morphology and size. The characteristics of the resulting powders include fully developed crystals, small and uniformly distributed crystal sizes, absence of agglomeration, and no need for a calcination step. Important parameters affecting the hydrothermal process include the pH of the solution, solution concentration, hydrothermal temperature, and reaction time. Here, pressure is not an independent variable, but depends on the system composition and temperature. These factors can influence whether the product is amorphous or crystalline, as well as the type of crystal form, morphology, and size.

Solution concentration primarily determines the equilibrium process of hydrolysis and the nucleation rate, and may also influence the crystal form of the final product. The hydrothermal temperature is generally controlled between 110°C and 300°C. Reactions below 100°C are conducted under atmospheric pressure and are typically referred to as hot-fluid methods. Temperatures exceeding 300°C are often unnecessary, as they impose higher demands on the autoclave and approach the critical point of water. The reaction time depends on the reactant concentration and the hydrothermal temperature; higher temperatures correspond to shorter reaction times. Longer hydrothermal durations favor the formation of well-defined nanocrystals with relatively uniform sizes, analogous to the "aging" step in traditional catalyst precipitation methods.

On the other hand, nonaqueous solvothermal synthesis, which replaces water with organic solvents (such as formic acid, ethanol, benzene, ethylenediamine, and carbon tetrachloride) as the medium, represents a significant improvement of the hydrothermal method. In the reaction process, nonaqueous solvents not only act as pressure-transmitting media but also function as mineralizers. By selecting appropriate solvents and reaction conditions, the solvothermal method can effectively control the morphology and size of NPs. Hydrothermal/solvothermal synthesis is a relatively common type of wet-chemical synthesis, typically employing high-pressure autoclaves as sealed reaction vessels [4, 19–21]. The medium for the chemical reaction can be water or other organic solvents in which the precursors are dissolved. The temperature and pressure inside the autoclave are usually higher than the external environment, often resulting in experimental temperatures exceeding the solvent's boiling point. When the temperature in the closed system rises above the boiling point of the solvent system, the reaction proceeds under high pressure. The high-pressure environment inside the autoclave promotes material crystallization and enhances its crystallinity. The hydrothermal/solvothermal method is a simple, high-yield, and low-cost technique widely used for preparing inorganic nanomaterials. When employing this method, it is often combined with other preparation techniques, such as oriented adsorption growth. Chemical reagents, reducing agents, surfactants, etc., are added to the aqueous or organic solvent system, which is then homogenized via sonication. Subsequently, the reaction system is heated and pressurized to create a relatively high-temperature, high-pressure environment, enabling substances to react and crystallize anew, thereby yielding unique material structures. Researchers typically optimize the final product's structure, performance, and other parameters by adjusting factors such as the reaction solvent, reaction temperature, and pH. Electrocatalytic nanomaterials obtained through hydrothermal/solvothermal synthesis exhibit advantages such as controllable morphology, excellent growth orientation, good crystallinity, small size, and strong stability.

For example, Li et al. [22] mixed $Ni(NO_3)_2{\cdot}6H_2O$, $Fe(NO_3)_2{\cdot}9H_2O$, $RuCl_3{\cdot}xH_2O$, NF, and other substances and placed them in a Teflon-steel autoclave inside an oven, which was then heated at 120°C for 12 hours. The heterostructure NiFeRu-LDH/NF was prepared via this simple hydrothermal method, and the target product Ni_2P–Fe_2P–Ru_2P/NF was subsequently obtained through a typical phosphidation treatment using NaH_2PO_2 in a tube furnace, as shown in Figure 2.2a. The combination of multiple components and the synergistic involvement of the noble metal Ru endowed the material with excellent performance as a bifunctional catalyst under alkaline conditions. At a current density of 10 mA cm^{-2}, the Ni_2P–Fe_2P–Ru_2P/NF heterostructure exhibited low overpotentials of 78.6 mV for HER and 195 mV for OER, respectively. Similarly, Jiao et al. [23] also synthesized a $NiP_2/NiSe_2$ heterostructure electrocatalyst using a simple two-step method. First, $Ni(OH)_2$

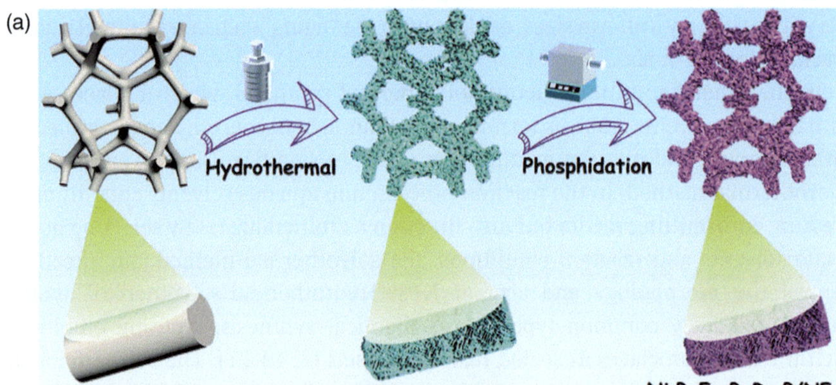

Figure 2.2 (a) Schematic illustration of the preparation of the Ni_2P–Fe_2P–Ru_2P/NF electrocatalyst. *Source:* Reproduced from Ref. [22] / with permission of The Royal Society of Chemistry. (b) Schematic illustration of the preparation of the $NiP_2/NiSe_2$ heterojunction electrocatalyst. *Source:* Reproduced from Ref. [23] / with permission of Elsevier.

nanosheets were grown on carbon cloth via a hydrothermal method ($Ni(OH)_2/CFC$). Then, a novel vapor-phase technique was employed to combine phosphidation and selenization treatments, resulting in the final product—a dual-phase heterostructure comprising phosphide and selenide (as shown in Figure 2.2b). The synergistic coupling interaction at the heterointerface between NiP_2 and $NiSe_2$ enhanced the electrical conductivity of the material. In 1 M KOH, the $NiP_2/NiSe_2$ heterostructure exhibited a low overpotential of 160 mV at 100 mA cm^{-2}.

2.3 Impregnation Method

The impregnation method, which uses impregnation as its key and distinctive step, is a widely applied technique in the industrial production of catalysts. This method is based on the principle of impregnating a porous support with a salt solution containing the active component (including promoters), allowing it to penetrate the internal surfaces to form an efficient catalyst. Typically, a liquid containing the active substance is used to impregnate various supports. After impregnation, equilibrium is reached, and the excess liquid is removed, followed by post-treatment steps such as drying, calcination, and activation, similar to those in the precipitation method. During drying, water evaporates, leaving the salts of the active components deposited on the internal surface of the support. These salts of metals and metal oxides are uniformly distributed within the fine pores of the support and, after thermal decomposition and activation, yield a highly dispersed supported catalyst. The active solution must thoroughly wet the support. Commonly used porous supports include alumina, silica, activated carbon, aluminum silicate, diatomite, pumice, asbestos, clay, magnesium oxide, and activated clay, which can be in powdered form or preformed granules. Oxide supports like alumina and silica, similar to most activated carbons with adsorptive surfaces, are readily wetted by aqueous solutions. Furthermore, capillary forces ensure the liquid is drawn into the entire porous structure; even one-end sealed capillaries can be filled. The dissolution of gases in the liquid also facilitates the process. However, some supports are difficult to wet, such as highly graphitized carbons or those without chemisorbed oxygen. In such cases, organic solvents or impregnation under vacuum can be employed.

The impregnation method offers the following advantages: first, the loaded component in most cases is distributed only on the surface of the support, resulting in high utilization efficiency, low required amounts, and reduced cost. This is particularly significant for supported catalysts containing precious metals like platinum, rhodium, palladium, and iridium, resulting in considerable cost savings. Second, commercially available, preformed, standardized support materials can be used, eliminating the need for a separate catalyst shaping step. Third, by selecting an appropriate support, the desired physical structural properties for the catalyst, such as specific surface area, pore radius, mechanical strength, and thermal conductivity, can be provided. Thus, the impregnation method is a simple, practical, and economical technique. It is widely used for preparing supported catalysts, especially those with low precious metal loadings. Its main disadvantage is that the calcination (thermal decomposition) step often generates waste gases, causing pollution.

The impregnation method can be divided into two processes: powder support impregnation and granular support impregnation. For granular supports, they are typically preformed into a specific shape, evacuated, then contacted with the solution, and an appropriate amount of a competitive adsorbent is added. Alternatively, the active component solution can be sprayed onto the support as it tumbles in a rotating container. Excess solution can then be removed by methods such as filtration, decantation, or centrifugation. Powder support impregnation is similar to granular support impregnation but requires additional shaping steps such as tableting, extrusion, or pelletization. The impregnation method significantly influences the catalyst's performance. When impregnating granular supports, the catalyst's surface structure depends on the surface structure of the support particles, such as specific surface area, porosity, and pore size distribution. Different catalytic reaction rates impose different requirements on the surface structure of the catalyst. The final dispersion of the metal deposited on the catalyst support depends on the interplay of many factors, including the impregnation method, the strength of adsorption, the extent to which the metal compound exists as occluded solute compared to species adsorbed on the pore walls, and the chemical changes that occur during heating and drying. Although most metal reagents can be adsorbed onto the support to varying degrees during impregnation, the adsorption process is quite complex, and different types of adsorption can occur. This can involve the adsorption of metal ions onto surfaces containing hydroxyl groups, or cation exchange on surfaces containing alkali metal or alkaline earth metal ions. The surface structure of the support may also be altered by different impregnation steps, thereby modifying its adsorption characteristics. These aspects must be considered during process implementation. If the support undergoes etching, the situation becomes more complex. Silica gel can be etched at high pH, while high-surface-area alumina can be etched at both excessively high and low pH values. During the impregnation of an alumina support with an acidic liquid, partial dissolution of the alumina may occur initially, followed by precipitation as the pH increases. It is preferable to use a buffer to control this effect.

For example, the platinum reagent chloroplatinic acid (H_2PtCl_6) is often adsorbed onto alumina or activated carbon, but not onto silica gel. Using the incipient wetness method (or dry impregnation) can result in the deposition of a thin shell of platinum on the exterior of alumina particles, which is beneficial for preventing diffusion limitations in fast reactions. To achieve a more uniform distribution, a competitive adsorption method can be employed. This involves adding nitric acid or hydrochloric acid to the solution to reduce the adsorption strength of chloroplatinate ions, thereby promoting more uniform deposition. On the other hand, platinum in anionic form interacts with alumina differently; it is less readily adsorbed onto alumina but can be strongly adsorbed onto silica gel. If the goal is to prepare a halogen-free catalyst, compounds such as diaminodinitroplatinum can be selected. Alternatively, by adding organic acids, such as citric acid, to the impregnating solution, the catalytic material can be embedded slightly deeper within the catalyst particles. This helps inhibit the deposition of poisons on the outer surface of the catalyst support, thereby increasing the catalyst's lifespan. Supported platinum catalysts used for oxidizing pollutants in automotive engine exhaust serve as an example. Generally, if the reagent has sufficient time to diffuse and side reactions are not dominant, the method of impregnation using an excess solution can lead to a largely uniform deposition of the adsorbate. If the initial adsorption is nonuniform and not too strong, diffusion may continue even after the support pellets are removed from the solution, leading to a more homogeneous distribution.

The fundamental principles of the impregnation method lie, on the one hand, in the capillary pressure generated by surface tension when solid pores come into contact with a liquid, causing the liquid to penetrate into the capillaries. On the other hand, it involves the adsorption of active components onto the surface of the support. To increase the impregnation amount or depth, the air inside the support can sometimes be evacuated beforehand using vacuum impregnation. Increasing the temperature of the impregnating solution (to reduce its viscosity) and enhancing agitation have similar effects. Although the operation of the impregnation method is straightforward, many complex issues are often encountered during the preparation process. For instance, during catalyst drying, the migration of catalytically active species toward the outer surface can sometimes lead to a decrease in the concentration of active substances on part of the internal surface, or even leave the support uncovered. The uniform or nonuniform distribution of active substances across the support's cross section is also a topic worthy of in-depth investigation. For certain reactions, it is not always necessary for the catalyst's active material to be uniformly dispersed over the entire internal surface; instead, having a higher concentration of active material in the surface and near-surface layers may be sufficient. The distribution of the active component across the support cross section can be classified into different types. The method for preparing catalysts with various types of cross-sectional distributions is the competitive adsorption method. According to this method, an appropriate amount of a second component, known as a competitive adsorbent, is added to the impregnating solution alongside the active component. During impregnation, the support adsorbs both the active component and the second component simultaneously. Due to differences in the probability and depth of adsorption of the two components on the support surface, competitive adsorption occurs. By selecting different competitive adsorbents and appropriately adjusting the impregnation process and conditions, the distribution type and depth of the active component on the support can be controlled. For example, using lactic acid, hydrochloric acid, or monochloroacetic acid as competitive adsorbents can result in a thickened egg-shell distribution. Furthermore, the depth of impregnation of the active component can be controlled by using different amounts and concentrations of the competitive adsorbent. The impregnation method is categorized into excess solution impregnation, incipient wetness impregnation, multiple impregnation, etc.

In the excess solution impregnation method, the support is immersed in an excess volume of the impregnating solution (where the volume of the impregnating solution exceeds the absorbable volume of the support). The excess solution is then drained off, and the catalyst's final product is obtained after drying and activation. The practical operational steps of the excess solution impregnation method are relatively straightforward. For example, the dried support is first placed in a stainless steel or enameled container and immersed in an aqueous solution of the active material, whose pH has been adjusted. The air within the fine pores of the support is displaced by the capillary pressure of the liquid, and pre-evacuation is generally not necessary. The excess aqueous solution is removed using methods such as filtration, decantation, or centrifugation. Following impregnation, subsequent steps generally include drying and calcination, similar to those used in precipitation methods. The surplus impregnation solution can typically be recovered and reused, either without treatment or after minor adjustments.

The incipient wetness impregnation method involves mixing the support with a volume of impregnating solution exactly equal to its pore volume. Since the volume of the impregnating solution is equivalent to the pore volume of the support, thorough mixing ensures that the solution just saturates the support particles without any excess, thereby eliminating the need for filtration and recovery of waste impregnation solution. However, it is important to note that the volume of solution absorbed is a function of the nature of the impregnating compound and the viscosity of the impregnating solution. Determining the volume of the impregnating solution requires preliminary experimental measurement. Incipient wetness impregnation can be carried out continuously or batch-wise. It features low equipment investment and high production capacity and allows precise adjustment of the loading amount, making it widely adopted in industry. In practical operation, this method involves preparing an aqueous solution containing the required amount of the active substance and then impregnating a specific quantity of the support with it. This process is typically performed using a spraying method, where the solution containing the active substance is sprayed onto the support contained in a rotating vessel. This method is suitable for situations where the support has a strong adsorption capacity for the active substance. However, regarding the uniform distribution of the active substance on the support, this method is inferior to the excess solution impregnation method. For the impregnation of multiple active substances, it must be considered that the coexistence of two or more solutes may alter the distribution of any single active substance on the support. In such cases, a specific substance is often added to seek the maximum catalytic activity. Alternatively, stepwise impregnation can be employed: one active substance is impregnated first, followed by drying and calcination, and then another active substance is impregnated. Sometimes, multiple active substances can be prepared as a mixed solution and then used for impregnation. When a uniform distribution of the active substance over the entire internal surface of the support is required, the support should undergo vacuum treatment prior to impregnation to remove internal gases. Simultaneously, increasing the temperature of the impregnating solution can enhance the depth of impregnation. The impregnation time for the support depends on factors such as the support's structure, solution concentration, and temperature, and typically ranges from 30 to 90 minutes.

To prepare catalysts with a high content of active substances, the process of impregnation, drying, and calcination can be repeated multiple times, which is referred to as the multiple impregnation method. The reasons for adopting the multiple impregnation method are twofold: first, the solubility of the impregnating compound is low, resulting in a small loading amount per impregnation cycle, thus necessitating repeated impregnations. Second, to avoid competitive adsorption among the components of a multicomponent impregnating compound, each component should be impregnated sequentially. After each impregnation, it is essential to carry out drying and calcination to convert the compounds into insoluble substances. This prevents the compounds deposited on the support during the previous impregnation from redissolving into the solution in the subsequent impregnation step and also increases the absorption capacity of the support for the next impregnation. The multiple impregnation process is complex, has low labor efficiency, and has high production costs. Its use should be avoided unless specifically necessary for the reasons mentioned above.

Bao et al. [24] employed a rapid and efficient room-temperature impregnation method to grow sulfur-doped oxyhydroxide S-MNiOOH (M=Fe, Co, Ni, and Cu) nanosheet arrays on nickel foam, which exhibited excellent HER and OER catalytic activities, as shown in Figure 2.3a. The study found that S-FeNiOOH exhibits the lowest density of states peak near the Fermi level, the strongest electron delocalization, and the best electrical conductivity, facilitating electron transport. At 100 mA cm^{-2} in 1 M KOH, the HER overpotential was 248 mV with a ΔGH* of 0.077 eV; the OER overpotential was 293 mV, reducing the energy barrier from O* to OOH* (η = 0.593), providing an ideal reaction interface for OER. The overall water splitting voltage for S-FeNiOOH reached 1.81 V, which is 130 mV, 185 mV, and 152 mV lower than that of S-CoNiOOH, S-NiNiOOH, and S-CuNiOOH,

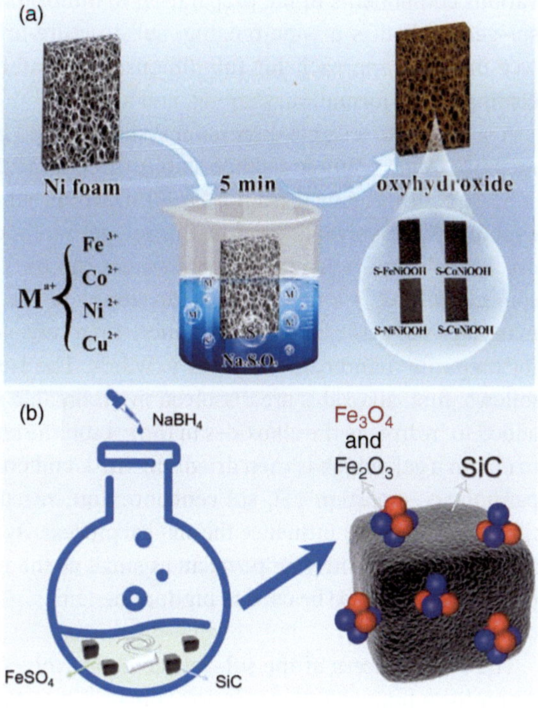

Figure 2.3 (a) Schematic illustration of the synthesis of S-MNiOOH (M=Fe, Co, Ni, and Cu) nanosheet arrays. *Source:* Reproduced from Ref. [24] / with permission of Elsevier. (b) Schematic diagram of the preparation of Fe/SiC. *Source:* Ref. [25] / John Wiley & Sons / CC BY 4.0.

respectively. It maintained a current density of $80\,mA\,cm^{-2}$ for 100 hours continuously at 1.9 V. Wang et al. [25] demonstrated that Fe/SiC synthesized by the impregnation-chemical reduction method is an excellent visible-light-driven CO_2 reduction photocatalyst (as shown in Figure 2.3b), particularly exhibiting high CH_4 selectivity. The uniformly distributed iron oxide catalyst effectively enhanced the photocatalytic activity and CH_4 selectivity.

2.4 Sol–Gel Method

The sol–gel method is a versatile wet-chemical technique capable of preparing materials across all dimensions, from zero to three dimensions. First, because the sol particles themselves are extremely fine, powdered materials obtained through this technique can easily reach the nanoscale, exhibiting high reactivity and lower sintering temperatures, while the strength and toughness of the resulting materials are significantly enhanced. Second, under certain conditions, the sol solution exhibits excellent fiber-forming properties, making it suitable for producing oxide fibers, particularly refractory oxide fibers. Furthermore, preparing thin films using sols via methods such as dip-coating, spray-coating, and tape casting is very convenient, with catalyst thickness adjustable from several nanometers to the micrometer range. Finally, during hydrolysis and polycondensation in the sol–gel process, reactants can be fully dispersed and mixed in the liquid phase, thereby potentially greatly improving the dispersion and compatibility of various components in the preparation of multiphase ceramic materials. Thus, the sol–gel method is a wide-ranging, substructure-oriented, and broadly applicable wet-chemical approach for full-dimensional material preparation, encompassing coatings, fiber formation, shaping, and aging.

A sol refers to a stable dispersion system formed by colloidal particles in a liquid medium. A gel refers to a shape-stable, three-dimensionally interconnected porous solid network obtained from a liquid medium, whose size is limited only by the container. The solvents used in the sol–gel method are water, alcohols, or other organic solvents, while the gelling sources can be alkoxides or inorganic salts. The sol–gel method is a new process developed in the 1960s for preparing inorganic materials such as glass and ceramics. In recent years, it has been widely used for preparing nano-oxide films or powders. The basic steps of this method are as follows: first, alkoxides are dissolved in an organic solvent; then, distilled water is added to hydrolyze the alkoxides to form a sol; the sol undergoes gelation treatment to obtain a gel, which is then dried, calcined, and crushed to yield the powder. Four parameters—solution pH, sol concentration, reaction temperature, and reaction time—significantly influence the sol–gel process. By appropriately controlling these four parameters, ultrafine powders as small as the nanoscale can be prepared, and this method can also be used to produce ceramics, glass, fibers, films, and bulk solid materials.

The basic process of the sol–gel method involves several steps. The first step is preparing a homogeneous solution containing metal alkoxides and water to ensure the hydrolysis of the alkoxides occurs at a molecularly uniform level. Since metal alkoxides have limited solubility in water, alcohols are generally chosen as solvents. The amounts of alcohol and water added should be appropriate, typically measured by the molar ratio of water to alkoxide. Catalysts significantly impact the hydrolysis rate, condensation rate, and the structural evolution of the sol and gel during aging. Commonly used acidic and alkaline catalysts are HCl and NH_4OH, respectively, and the amount of catalyst added is often measured by the molar ratio of catalyst to alkoxide. Because alcohol is a product of alkoxide hydrolysis and inhibits the hydrolysis reaction, vigorous stirring is required during preparation to maintain the homogeneity of the initial solution. To prevent the loss of volatile components, which could alter the composition, a reflux condenser is generally used. The second step is the preparation of the sol. There are two methods for preparing sols: the polymeric route and the particulate route, which differ primarily in the amount of water added. The polymeric sol is formed under controlled hydrolysis conditions where hydrolysis products and partially unhydrolyzed alkoxide molecules continue to polymerize, thus requiring minimal water addition. In contrast, the particulate sol is formed by adding a large amount of water to ensure complete hydrolysis of the alkoxides. The hydrolysis and condensation reactions of metal alkoxides are the fundamental reasons for the transition from a homogeneous solution to a sol. Controlling the conditions for alkoxide hydrolysis and condensation, such as the amount of water, catalyst, solution pH, and hydrolysis temperature, is essential for producing high-quality sols. The third step involves the process of aging the sol to form a wet gel. When the sol is left in an open or sealed container, the gradual transformation into a gel occurs due to solvent evaporation or continued condensation reactions. This process is often accompanied by Ostwald ripening, where the difference in solubility between small and large particles leads to an increase in average particle size. During aging, colloidal particles gradually aggregate to form a network structure, causing the entire system to lose its fluidity. The sol transitions from a Newtonian fluid to a Bingham plastic, exhibiting significant thixotropy. Shaping processes such as fiber spinning, coating, and casting can be performed during this stage. The fourth step is the drying of the gel. The wet gel contains a large amount of solvent and water, and the drying process often involves significant volume shrinkage, making it prone to cracking. Preventing gel cracking during drying is a critical yet challenging aspect of the sol–gel process, especially for larger bulk materials. This requires strict control of drying conditions, the use of chemical drying control additives, or the application of supercritical drying techniques. Finally, the dried gel undergoes heat treatment. The purpose is to remove pores in the xerogel and ensure the phase composition and microstructure meet the required product performance. During heating, the xerogel first removes water and alcohol adsorbed on the surface at low temperatures. Oxidation of –OR groups occurs between 260°C and 300°C, and structural –OH groups are removed above 300°C. Since heat treatment involves considerable volume shrinkage and the release of various gases (CO_2, H_2O, and ROH), and because –OR groups may carbonize into carbonaceous particles if not fully oxidized, the heating rate should not be too rapid. During heat treatment, sintering occurs, leading to densification of the gel. Due to the high specific surface area and high reactivity of the gel, its sintering temperature is several hundred degrees lower than that of conventional powder compacts. Using hot-pressing sintering can shorten the sintering time and improve product quality. With multiple controllable parameters in the above steps—such as the choice of precursor, solvent, temperature, hydrolysis and condensation catalysts, aging, and drying conditions—this method offers unique advantages in controlling the catalyst's structure, texture, and compositional homogeneity.

The sol–gel method for preparing nano-oxides typically involves the transformation of metal organic or inorganic compounds through solution, sol, and gel stages, followed by solidification and heat treatment to form oxides. Compared to traditional solid-phase sintering methods for oxides, the sol–gel method offers several advantages in the preparation of metal oxides:

1) **High Purity:** For instance, raw materials for the sol–gel method can be purified by distillation or crystallization to ensure their purity. The absence of mechanical grinding processes prevents the introduction of impurities, resulting in high-purity materials.
2) **Uniform Chemical Composition:** Components are mixed at the molecular level, enabling the production of multicomponent solid solutions with accurate chemical composition and homogeneous phase structure.
3) **Lower Processing Temperatures:** The lower synthesis temperature results in a uniform and fine particle size distribution of the product. The lower sintering temperature effectively controls the volatilization of certain volatile components.
4) Controllable particle size.
5) Simple process operation, no need for expensive equipment, and ease of industrialization.

Although the sol–gel method for preparing nano-oxides has many advantages and has been adopted in some industrial production, it still suffers from drawbacks such as long process cycles, insufficient utilization of raw materials, and a tendency for product agglomeration. Therefore, shortening the process cycle, improving raw material utilization, and addressing product agglomeration are important issues to be researched and resolved in the future.

Wang et al. [26] prepared Fe_7Co_3/carbon nanocomposites by embedding Fe_7Co_3 NPs in a porous carbon matrix using a co-gelation method. Kim et al. [27] prepared Fe_2O_3/multi-walled carbon nanotube (Fe_2O_3-MWNTs) nanocomposites using a modified sol–gel method. Self-assembly is a term used to describe the process by which a disordered system of pre-existing components forms an organized structure or pattern due to specific, local interactions among the components themselves, without external direction. Important applications of self-assembly in materials science include the formation of molecular crystals, colloids, lipid bilayers, phase-separated polymers, and self-assembled monolayers. Wen et al. [28] proposed a novel sol–gel method for the large-scale preparation of Ir single atoms supported on spinel-type $Ni_xMn_{3-x}O_4$ solid solutions. Due to the surface-exposed Ir single atoms, $Ir_1/Ni_{1.6}Mn_{1.4}O_4$ exhibited enhanced OER performance, achieving overpotentials of 330 and 350 mV at current densities of 100 and 200 mA cm^{-2}, respectively, in alkaline seawater, providing an effective method for the large-scale production of efficient alkaline seawater electrocatalysts (Figure 2.4a). Talifhani

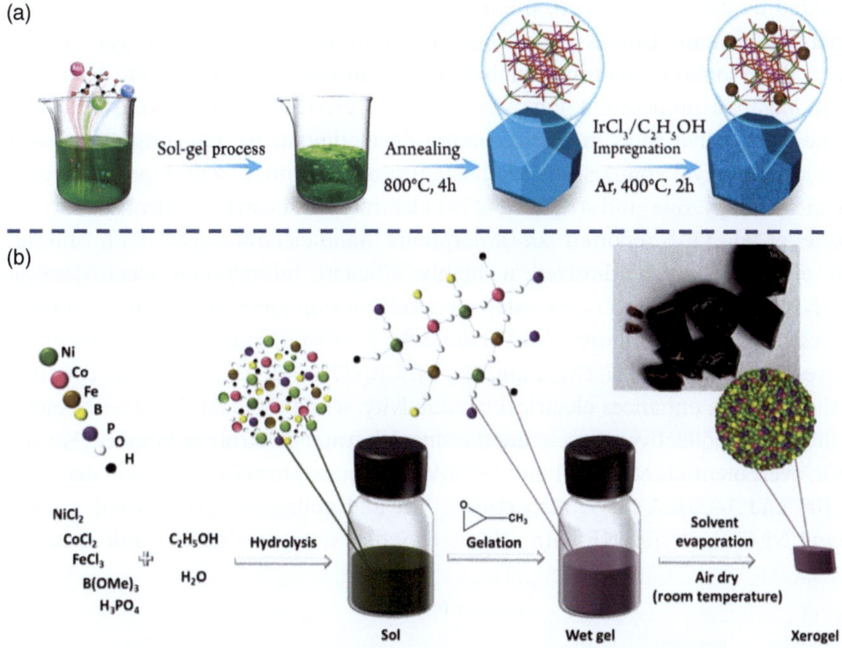

Figure 2.4 (a) Schematic diagram of the $Ir_1/Ni_{1.6}Mn_{1.4}O_4$ synthesis process. *Source:* Ref. [28] / John Wiley & Sons / CC BY 4.0". (b) Schematic illustration of the multielement electrocatalyst synthesis process. *Source:* Reproduced from Ref. [29] / with permission of John Wiley & Sons.

Mushiana et al. [29] developed a simple and effective sol–gel method, involving acid-catalyzed gelation at room temperature, to synthesize homogeneous high-entropy metal (HEM) electrocatalysts for methanol oxidation reaction (MOR) and urea oxidation reaction (UOR). Using Ni as the primary component for MOR and UOR, Co can reduce the overpotential, while Fe can enhance catalytic activity and durability, as shown in Figure 2.4b.

In addition to utilizing surface chemical reactions to connect magnetic substances with carbon materials, many research groups have also obtained magnetic carbon hybrid materials through static electronic interactions (one of the most important interactions between two molecules). The unique characteristics of the self-assembly process allow for the control of material properties. However, a major drawback of this method is the difficulty in adjusting the uniformity of the product.

2.5 Electrodeposition Method

The electrodeposition method involves using electrochemical techniques to deposit metals from an electrolyte, followed by calcination to obtain metal oxides. This process requires specific electrolytes and operating conditions and is influenced by numerous factors such as pH, temperature, and current density. In catalysts prepared by electrodeposition, there are no binders between the deposit and the electrode substrate. Due to the excellent adhesion, which prevents catalyst detachment, these catalysts maintain stable performance even under severe operating conditions. The preparation of nanomaterials via electrodeposition is currently one of the most active areas in electrocatalyst development. By adjusting parameters such as electrolyte composition, voltage/current magnitude, and deposition time, the morphology, size, and structure of the electrocatalyst can be controlled.

Che et al. [30] reported an amorphous nano-electrocatalyst with interfacial effects. They synthesized a highly efficient bifunctional electrocatalyst (Fe–Ni)Co_x–OH/Ni_3S_2 via a two-step electrodeposition method, with the synthesis process illustrated in Figure 2.5a. The material forms a unique dual-interface effect between (Fe–Ni)Co_x–OH/Ni_3S_2 and (Fe–Ni)Co_x–OH/NF. The electronic coupling at the interface enhances electrical conductivity, while the disordered S-vacancies in the sulfide collectively accelerate the sluggish water electrolysis kinetics. For the HER, overpotentials of only 91 and 145 mV are required to achieve current densities of 100 and 1400 mA cm^{-2}, respectively. Wu and colleagues synthesized a compound $Ni(OH)_2$@NiFe/NF with a Mott–Schottky structure (metal-semiconductor

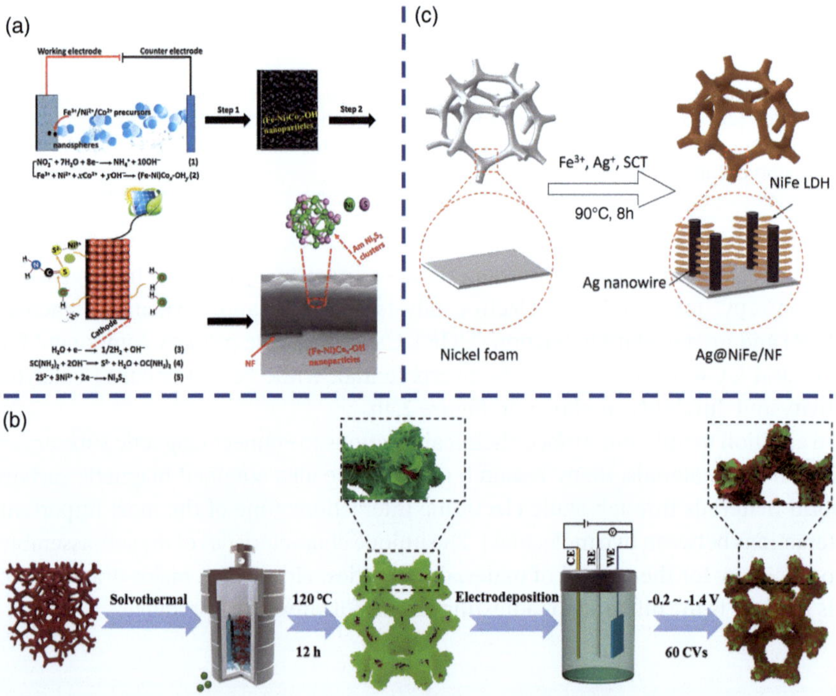

Figure 2.5 (a) Catalytic cycle mechanism for water-hydrogen/oxygen conversion in alkaline media and the preparation process of amorphous (Fe–Ni)Co_x–OH/Ni_3S_2 nanohybrids. *Source:* Reproduced from Ref. [30] / with permission of Elsevier. (b) Schematic diagram of the preparation of (Fe–Ni)Co_x–OH/NF-60. *Source:* Reproduced from Ref. [31] / with permission of American Chemical Society. (c) Schematic illustration of the fabrication route for the 3D Ag@NiFe/NF electrode. *Source:* Reproduced from Ref. [32] / with permission of Elsevier.

interface) by combining solvothermal and electrodeposition methods (Figure 2.5b). During the electrodeposition experiments, by adjusting the CV deposition cycles and the Ni/Fe ratio, they ultimately obtained a product exhibiting excellent activity in the HER, OER, and UOR [32]. Additionally, a heterostructured nanoarray comprising amorphous $NiFe(OH)_x$ and crystalline pyrite (Ni, Fe)Se_2 has been obtained via electrochemical deposition [33]. The electrodeposition method is feasible for increasing the electrochemically active surface area (ECSA), which can significantly enhance the number of active sites per unit surface area of the catalyst. Furthermore, electrodeposition is more suitable for scalable production, holding significant importance for advancing the commercialization of the electrochemical industry [34–36].

Besides the electrodeposition method, other electrochemical techniques, such as electrochemical etching, offer advantages over chemical etching, including more uniform etching, applicability to large areas, and reduced environmental pollution. Xiao and co-authors proposed a novel hierarchical structure, Ag@NiFe/NF, as an efficient bifunctional electrocatalyst for HER/OER (Figure 2.5c). Regarding the preparation of Ag@NiFe LDH, the nickel foam surface first undergoes an oxidation step, followed by an electrochemical etching step in a solution containing Fe^{3+} and Ag. Ni^{2+} and Fe^{3+} ions are deposited as NiFe LDHs on the surface of the NF skeleton, forming a dense interfacial structure. X-ray photoelectron spectroscopy (XPS) and density functional theory (DFT) indicate that the NiFe LDHs shell provides electron transport pathways to the Ag nanowire core. Electron transfer at the NiFe LDH and Ag interface enhances the adsorption energy of reaction intermediates, thereby boosting the catalytic activity of the catalyst [31].

2.6 Gas-Phase Deposition Method

The gas-phase deposition (GPD) method is a process that utilizes gaseous substances to undergo chemical reactions on a solid surface, resulting in the formation of a solid deposit. As a commonly used deposition technique, it is employed to prepare thin films, NPs, and nanostructured materials. This method utilizes atoms, molecules, or compounds from the gas phase to deposit material onto a substrate surface, thereby forming the desired thin film or nanostructure. GPD methods include chemical vapor deposition (CVD), physical vapor deposition (PVD), thermochemical evaporation, etc. [37, 38].

GPD methods offer a high degree of controllability and scalability, leading to their widespread application in fields such as materials science, nanotechnology, thin-film technology, and electronic device manufacturing [39–41]. By adjusting reaction conditions and selecting appropriate precursor substances, precise control over film properties such as composition, thickness, and structure can be achieved, thereby meeting the requirements of various applications.

2.6.1 Physical Vapor Deposition

Physical vapor deposition (PVD) is an advanced technology widely used in the field of thin film preparation. It involves transforming a material from a solid source into a gaseous phase and depositing it onto a substrate surface in a vacuum environment, enabling precise control over the film. The fundamental principle of PVD lies in converting solid materials into the gas phase through heating or other means. The evaporated or sputtered atoms or molecules condense onto the substrate surface under vacuum or an inert atmosphere to form a thin film. This process can be achieved through various methods, the most common of which include evaporation, sputtering, and laser pyrolysis. It is often used to prepare metal films, semiconductor films, ceramic films, etc.

Electron beam evaporation involves using an electron beam to heat the source material, causing it to evaporate and deposit onto the substrate surface. The concentrated energy and high controllability of the electron beam make it suitable for producing high-quality, uniform films, commonly used in optical coatings and other fields. Laser pyrolysis deposition utilizes a laser to locally heat the source material to high temperatures, achieving precise control over the film. Laser pyrolysis is widely used for preparing complex structures and functional films, such as defect repair in optoelectronic devices.

Magnetron sputtering is a method where, under the influence of a magnetic field, ions bombard the target material and sputter it into particles, forming a thin film (see Figure 2.6a). Magnetron sputtering is widely used to prepare metal, alloy, nitride, and other films, offering high deposition rates and good adhesion. Magnetron sputtering is a novel low-temperature sputtering coating method. Films prepared by this method exhibit advantages such as high quality, high density, good adhesion, and strength. Due to its stable equipment performance, ease of operation, controllable process parameters, good production reproducibility, suitability for large-area film deposition, and facilitation of continuous and semicontinuous production, this method has been used in the manufacturing of self-cleaning architectural glass.

Magnetron sputtering is a commonly used PVD method with relatively simple operation [44]. The traditional working principle involves argon ions bombarding the target material under the interaction of magnetic and electric fields in a low-pressure argon atmosphere, causing sputtered particles to uniformly disperse on the substrate. This improves the uniformity of the catalyst and indirectly reduces the thickness of the catalytic layer. In recent years, it has been widely applied in the preparation of heterogeneous electrocatalysts. Lang et al. [42] successfully synthesized a high oxidation state FeCo compound. The synchronous construction of FeCo oxide was achieved using RF-magnetron sputtering technology combined with thermal annealing treatment (Figure 2.6b). Experiments confirmed that the synergistic effect of the ternary oxide endows the material with a unique electronic structure, thereby improving the OER reaction kinetics and enhancing the catalytic performance of the catalyst. Inamdar et al. [43] prepared a ternary NiFeCo oxide catalyst thin film via RF magnetron sputtering, which exhibits excellent performance as an OER catalyst. This sputtering deposition method provides a viable pathway for fabricating highly efficient and durable multimetal oxide electrodes, applicable in areas such as water splitting and electrochemical energy storage (Figure 2.6c).

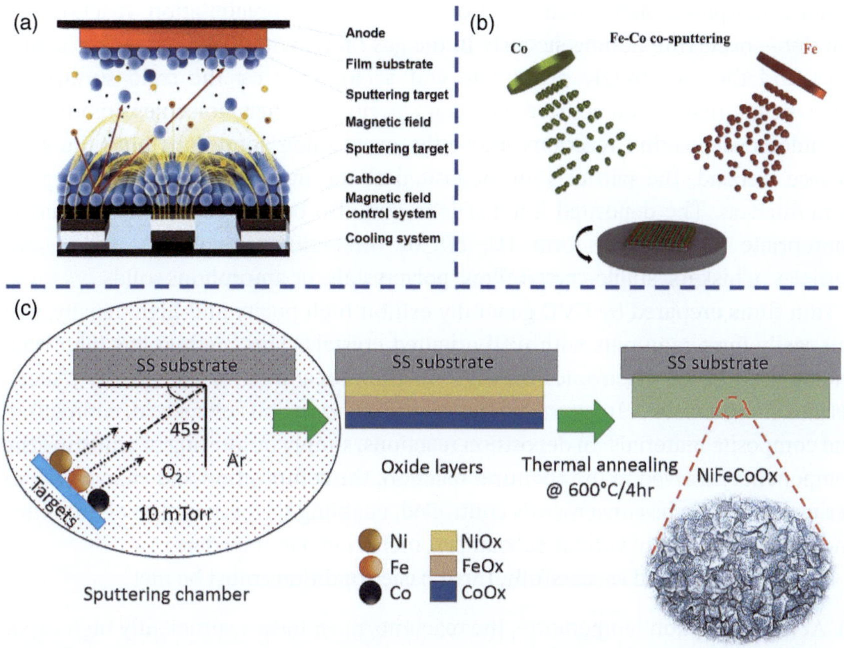

Figure 2.6 (a) Schematic diagram of the magnetron sputtering method. (b) Schematic diagram of the preparation of the FeCo catalyst via magnetron sputtering conversion film. *Source:* Reproduced from Ref. [42] / with permission of Springer Nature. (c) Schematic diagram of the synthesis of trimetallic NiFeCo oxide thin films via radio frequency magnetron sputtering technology. *Source:* Reproduced from Ref. [43] / with permission of John Wiley & Sons.

2.6.2 Chemical Vapor Deposition

Chemical vapor deposition (CVD) is a chemical process used to synthesize high-purity, high-performance solid materials within a reaction chamber [45, 46]. It is a preparation method where, under high-temperature physical evaporation or gaseous reaction conditions, reactive substances undergo chemical reactions and are deposited onto a substrate surface with the assistance of a flowing carrier gas. This process allows gases containing the deposition material to be introduced into the reaction chamber and subsequently deposited onto the substrate to form a coating. In CVD, gaseous precursor substances (typically gases or volatile compounds) are exposed to the substrate surface inside the reaction chamber, initiating chemical reactions to form the desired thin film material. CVD technology has various variants, such as thermal CVD, plasma-enhanced CVD (PECVD), and metal-organic CVD (MOCVD), and can be used to prepare two-dimensional (2D) materials, metal films, nitrides, carbides, and other materials. CVD is one of the most commonly used techniques for manufacturing one-dimensional (1D) nanomaterials. Due to its simple operation and potential for mass production, it is also a general method for synthesizing branched carbon nanotubes (CNTs) or encapsulating other "foreign" components.

Since gas-phase deposition reactions differ from precipitation reactions in solution—occurring homogeneously in the gas phase where one or two molecules react and then separately precipitate and accumulate on the solid surface—it follows that, first, it can achieve ultra-high purity, as other molecules are unlikely to undergo deposition reactions under the exact same conditions onto the solid surface. Second, the particles are deposited at the molecular level, allowing for ultra-fineness. The deposited fine particles can also be guided on the solid using appropriate processes to form 1D, 2D, or three-dimensional (3D) small-sized particles, whiskers, single-crystal films, polycrystals, or amorphous solids.

Thin films prepared by CVD generally exhibit high purity and high density, and can easily form materials with well-oriented crystals. This method has long been widely used in the electronics industry for the preparation of high-purity materials and single crystals. It is convenient for preparing various elemental, compound, and composite materials. In deposition reactions, simply by altering or adjusting the components involved in the chemical reaction, the composition and characteristics of the deposit can be conveniently controlled, enabling the preparation of thin films and other materials of various substances, including nanopowders.

For CVD to proceed successfully, three basic conditions must be met:

1) At the deposition temperature, the reactants must have a sufficiently high vapor pressure. If all reactants are gaseous at room temperature, the deposition setup is relatively simple. If a reactant has very low volatility at room temperature, it needs to be heated to volatilize it.
2) The reaction products, except for the desired solid deposit, must all be gaseous.
3) The vapor pressure of the deposit itself must be sufficiently low to ensure it remains fixed on the heated substrate throughout the deposition process.

The structure of a CVD system primarily consists of a reactor, a gas delivery system, and a heating system. The reactor is the most fundamental part of the CVD apparatus; its type and construction materials are determined by the physical and chemical characteristics of the system and the process parameters. Basic reactor types are vertical, horizontal, and bell-jar. The walls of horizontal and bell-jar reactors can be cooled using circulating water. Horizontal reactors have a relatively simple structure, but the uniformity of the deposited film is often poorer due to the influence of the gas flow pattern. Other reactor types, although structurally more complex, yield better film uniformity. Commonly used reactors are made of quartz tubes, and their walls can be maintained hot or cold to reduce or prevent deposition on the walls. The reactor is typically heated by resistance heating, but thermal radiation or high-frequency induction heating can also be used. Factors affecting the quality of chemically vapor-deposited films mainly include the deposition temperature, the ratio of reactant gases, and the substrate.

Currently, many research groups are developing CVD for preparing magnetic carbon nano- or micro-composites. Qi et al. successfully constructed a high-loading (up to 4.8 wt%) atomically dispersed Ni/Fe dual-metal site catalyst on a Fe_2O_3/ZIF-8 precursor via a gas-phase chemical vapor deposition (CVD) method (Figure 2.7a). This catalyst demonstrated excellent performance in the electrochemical CO_2 reduction

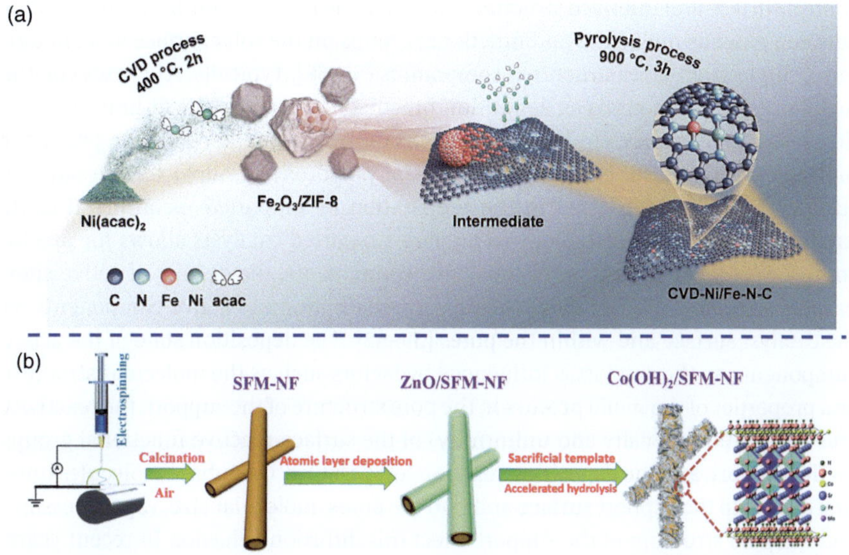

Figure 2.7 (a) Schematic of the CVD process to prepare high-density atomic dispersed Ni/Fe dual metal site catalysts. *Source:* Reproduced from Ref. [47] / with permission of The Royal Society of Chemistry. (b) Schematic diagram of the synthesis of $Co(OH)_2$/SFM-NF. *Source:* Reproduced from Ref. [48] / with permission of The Royal Society of Chemistry.

to CO, achieving a CO Faradaic efficiency of 96% at a current density of 700 mA cm^{-2} in a near-neutral electrolyte [47]. Flahaut et al. [49] reported the preparation of very small cobalt NPs within graphitic carbon shells using catalytic CVD. Wang et al. [50] systematically studied carbon-encapsulated cobalt NPs prepared via CVD using cobalt materials as precursors and carbon monoxide (CO) as the carrier gas. Wei et al. [51] developed a magnetic-field-assisted CVD method to synthesize branched or iron-encapsulated CNTs, creating branched or encapsulated nanostructures. El-Gendy et al. [52] have produced iron carbide, cobalt, and nickel NPs using CVD under high-pressure conditions. In these composites, the size of the Fe, Co, and Ni NPs ranges from 2 to 100 nm, and the composites exhibit a core/shell structure where one or more particles are formed within specific shells. Building on this, they extended the chemical vapor method to prepare carbon-encapsulated NiRu NPs with controllable morphology, structure, and magnetic properties. Sengupta and colleagues [53] utilized a simple mixture of Ni(salen) and a photoresist, combined with atmospheric pressure CVD using propane on Si at 850°C, to synthesize Ni-filled multi-walled CNTs. The CVD method enables precise control over carbon nanostructures, enabling the production of desired products. However, the synthesis of nanomaterials using such methods involves high costs, complex equipment, and high energy consumption. Furthermore, the thickness of the material can be controlled by adjusting the deposition time and the flow rate of the carrier gas in CVD. Atomic layer deposition (ALD) technology, also known as atomic layer epitaxy (ALE), is a special type of CVD technique [54]. Due to its unique technical characteristics and advantages, ALD enables researchers to flexibly and precisely control surface and interface structures. Its principle relies on self-limiting reactions between gaseous molecules and functional groups on the solid surface to form uniform, single-atom-layer structured compounds [55–58]. Typically, ALD uses volatile metal-organic compounds as deposition precursor molecules. A self-limiting reaction refers to a process, similar to a condensation reaction, between the precursor molecules and the functional groups (usually hydroxyl groups) on the support surface. ALD is widely used in the preparation of supported metal/metal oxide catalysts. Using ALD technology to prepare supported catalysts allows for precise control over the composition of the active components, the density of active sites, particle size, etc., and facilitates the uniform distribution of active components on the catalyst surface and within the pores [59–62]. The deposition state of the active components on the support is influenced by factors such as the molecular structure and properties of the metal precursor, the pore structure of the support, the reactivity and distribution (density and uniformity) of the surface-reactive functional groups on the support, and the deposition reaction temperature. Gas-phase molecules must first diffuse to the support surface and into the pores; molecular size, vapor pressure, and the pore structure of the support affect this diffusion behavior. In recent years, the application of interface engineering using ALD to prepare heterogeneous nano-electrocatalysts has been a key focus for researchers. Similar to CVD, ALD allows for more meticulous control of the deposited film and is characterized by deposition uniformity and consistency. For instance, Lin et al. [63] used the ALD method to uniformly deposit a 25-nm thick α-Fe_2O_3 layer on the surface of a $TiSi_2$ nanonet, and the resulting heterostructure material exhibited excellent photoelectrochemical water splitting performance. As another example, Yang et al. [64] prepared heterogeneous nanomaterials Ni_xFe_{1-x}–AHNAs using a modified one-step chemical deposition method. He et al. [48] constructed an amorphous $Co(OH)_2$ shell on the surface of an SFM perovskite using ALD technology, creating a core-shell heterostructure. The interfacial effects of the core–shell structure endowed it with excellent bifunctional OER/HER performance under alkaline conditions (Figure 2.7b).

In addition to CVD and PVD, there are other variants of vapor-phase deposition methods, such as thermochemical evaporation and low-pressure CVD (LPCVD). These methods may be more suitable for preparing specific materials or structures in particular application scenarios.

2.7 Template Method

The template method is an effective strategy for preparing nanomaterials with diverse sizes, shapes, and spatial arrangements. It involves using pre-existing nanostructures as templates to confine or guide the growth of specific nanostructured materials, thereby achieving control over the material's structure and morphology [65, 66]. In nanomaterial synthesis, a template can, theoretically, be any substance that possesses nanostructural features. Therefore, various types of materials can be selected as templates, some occurring naturally, such as nanostructured minerals, biomolecular assemblies, and organelles. Others are synthetic nanostructures, such as soft templates from molecular assemblies (colloidal particles, layered micelles, etc.) and hard templates (porous membranes, patterned solid surfaces, etc.). The advantage of the template method lies in the ability to pre-design templates with the target size and shape, thereby guiding the preparation of zero-dimensional (0D) NPs and 1D nanowires, as well as 2D and 3D hierarchical nanostructured materials [67–71].

Different morphologies impart different modification characteristics to materials. To obtain high-quality magnetic carbon composites, researchers have developed the template method, which involves carbonizing carbon sources within confined spaces to form magnetic substances with unique morphologies. Dong et al. [72] proposed a method using mesoporous silica (SBA-15) as a template to prepare γ-Fe_2O_3/carbon composites with high γ-Fe_2O_3 content, large specific surface area, large pore volume, and uniform pore size. Park et al. [73] incorporated metal nitrates into the SBA-15 mesoporous silica template body and carbonized polyfurfuryl alcohol to synthesize ordered mesoporous carbon containing Co, Ni, and Fe NPs. Lee et al. [74] used SBA-15 silica as a template, pyrrole monomer, and $FeCl_3$ as the carbon source and magnetic precursor, respectively, to prepare magnetic ordered mesoporous carbon rods. Liu et al. [75] prepared different magnetic carbon nanostructures, including CNTs grown from iron NP single-walled spheres, carbon-encapsulated iron NPs, iron NP-decorated multi-walled CNTs, and iron NP-filled multi-walled CNTs through pyrolysis at different sublimation temperatures. Besides mesoporous silica, anodic aluminum oxide (AAO) is also widely used as a hard template for synthesizing nanomaterials with desired structures. For magnetic carbon hybrid composites, Jang and colleagues [76] proposed a precursor impregnation method using AAO templates to obtain graphitic CNTs and CNTs containing magnetic iron oxide. An increasing number of research groups are utilizing AAO templates to prepare magnetic carbon nanocomposites.

Lou et al. [77] synthesized NiFe layered double hydroxide (LDH) nanocages with tunable shell layers using a self-templating method. By rationally adjusting the ethanol-to-water volume ratio, the number of shell layers can be precisely controlled. Due to their large ECSA and fine shell structure, the NiFe–LDH nanocages exhibited excellent electrocatalytic activity for the OER in alkaline electrolytes. Particularly, the double-shelled NiFe–LDH nanocages achieved a current density of 20 $mA\,cm^{-2}$ at a low overpotential of 246 mV, with a Tafel slope of only 71 $mV\,dec^{-1}$, along with good stability.

Chen et al. [78] successfully designed and synthesized a novel dual-site electrocatalyst using a template method. This catalyst features atomically dispersed Fe/Mn–N_x–C bimetallic sites and demonstrates state-of-the-art oxygen reduction reaction (ORR) activity, with a half-wave potential ($E_{1/2}$) of 0.88 V (vs. RHE), alongside excellent stability (Figure 2.8a). Ma et al. [79] developed a self-sacrificing template strategy to prepare Fe_1–N–C hollow microspheres (Fe_1/N-HCMS) assembled from ultrathin nanosheets by rationally carbonizing Fe^{3+}-chelated polydopamine-coated melamine-cyanuric acid complexes. The shell of Fe_1/N-HCMS is composed of ultrathin nanosheets with a thickness of only 2 nm,

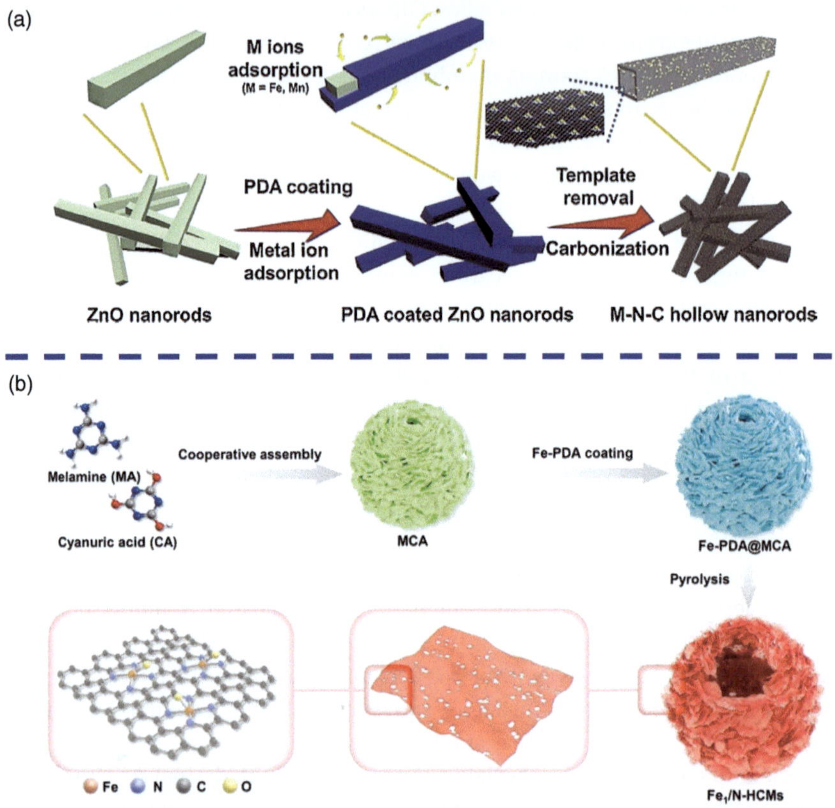

Figure 2.8 (a) Schematic diagram of the formation process of M−N_x−C hollow nanorods. *Source:* Reproduced from Ref. [78] / with permission of Elsevier. (b) Schematic diagram of the Fe_1/N-HCMS preparation process. *Source:* Reproduced from Ref. [78] / with permission of Elsevier.

which is considered an ideal platform for isolating and fully exposing single metal atoms (Figure 2.8b).

2.8 High-Temperature Pyrolysis Method

The high-temperature pyrolysis method involves using substances containing metal salts as precursors. Leveraging their inherent thermal instability, these precursors are subjected to high-temperature annealing under protective atmospheres such as N_2, Ar, or H_2. Decomposition occurs at the weak junctions of the precursor, followed by recombination into thermally stable solids. This is a widely used method for preparing electrocatalysts. The initial temperature of the pyrolysis process varies depending on the composition of the precursor. Different temperature ranges involve different reaction processes, resulting in products with varying compositions [80–84]. In essence, pyrolysis involves heating metal salt molecules to cleave them into smaller molecules that are released, encompassing many

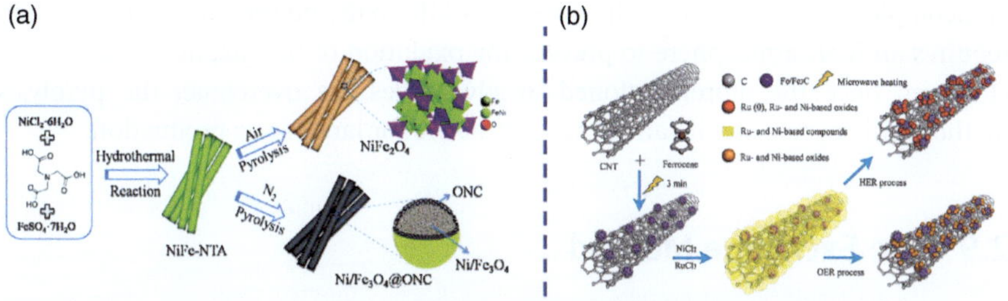

Figure 2.9 (a) Schematic diagram of the preparation process for $NiFe_2O_4$ and Ni/Fe_3O_4@ONC nanorods. *Source:* Reproduced from Ref. [85] / with permission of The Royal Society of Chemistry. (b) Schematic diagram of the rapid preparation process for Ni–Ru@Fe/C@CNT. *Source:* Reproduced from Ref. [86] / with permission of Elsevier.

complex physical and chemical processes. Precursors can be organic compounds or carbon-based materials like ZIFs and polymers. As shown in Figure 2.9a, Li et al. used a NiFe-based coordination polymer as a precursor and successfully synthesized mixed Ni/Fe_3O_4 nanocomposites encapsulated within N-doped carbon (ONC) nanorods via pyrolysis. The heterostructure material, regulated by interfacial effects, possesses high electrical conductivity, making it an excellent OER catalyst [85]. Furthermore, Xiao et al. synthesized carbon-encapsulated Fe/Fe_3C NPs grown on the surface of CNTs, creating an Fe/C@CNT heterostructure, through rapid pyrolysis of ferrocene. Subsequently, they introduced Ni and Ru compounds to increase the catalytic active sites of the catalyst, with the synthesis schematic shown in Figure 2.9b. For HER and OER, Ni–Ru@Fe/C@CNT required low overpotentials of only 32 mV and 246 mV, respectively, to achieve a current density of 10 mA cm^{-2} [86]. In the field of composite materials, CNT are among the most promising novel nano-reinforcements due to their excellent mechanical and conductive properties, as well as unique structural and geometric characteristics. Similarly, Song and colleagues proposed a new experimental scheme using guar gum and melamine as precursors. During pyrolysis, the C, N, and Ni elements decomposed from them and regrew on nickel foam, resulting in a hierarchical structure with Ni NPs encapsulated in vertically aligned N-doped CNTs. The synthesized compound, NCNT-NP@NF, exhibits multi-interface effects that accelerate electrocatalytic kinetics and demonstrate excellent HER/OER performance [87]. Beyond single pyrolysis reactions, researchers often combine "in-situ" polymerization processes with pyrolysis to prepare the target products. Liu et al. adopted a polymerization-pyrolysis process to synthesize a hybrid heterostructure, Co_4S_3/Mo_2C–N S C, successfully combining active HER and OER components through interface engineering. The optimized electronic structure and multielement synergistic effects make it an excellent non-precious metal bifunctional electrocatalyst [88].

The pyrolysis method is characterized by its suitability for large-scale production, but it offers very limited control over the structure of magnetic carbon nanomaterials. A major drawback of pyrolysis is the frequent formation of substantial amounts of carbon black, an unwanted byproduct resulting from the thermal decomposition or incomplete combustion of hydrocarbons. Additionally, conducting pyrolysis often requires an inert atmosphere to prevent the oxidation of the magnetic substances. However, once the aforementioned disadvantages are overcome, the pyrolysis method will become significantly more attractive for large-scale production.

2.9 Ion Exchange Method

The ion exchange method, as an advanced "soft chemistry" synthesis strategy, holds a significant position in the field of nanomaterials science, particularly suitable for the precise preparation of iron-, cobalt-, and nickel-based functional nanomaterials. Its core principle involves utilizing ions in a solution to undergo a replacement reaction with the original ions in pre-synthesized template nanocrystals, thereby achieving a fundamental transformation of the material's chemical composition while maximally retaining the original morphology and structural features of the template. This process can be understood as a metathesis reaction at the solid–liquid interface. For instance, a general reaction equation can be represented as a precursor A_mX_n reacting with ions B^{m+} in solution to generate the target product B_nX_m, while releasing A^{n+} ions. Whether this process occurs, and to what extent, is primarily governed by two major factors: thermodynamics and kinetics. Thermodynamically, the driving force for the spontaneous reaction typically comes from the product having a lower solubility product than the reactant, meaning the product is more stable. Kinetically, it is significantly influenced by the diffusion rate of ions within the solid, the condition of the reaction interface, and external conditions such as temperature, solvent, and concentration.

In the specific preparation of iron, cobalt, and nickel-based compounds, the application of the ion-exchange method demonstrates great flexibility and powerful structure–performance control capability. For instance, in anion exchange, pre-synthesized cobalt oxide nanosheets can be placed in a solution containing a sulfur or selenium source. Through the exchange of oxygen ions with sulfide or selenide ions, the oxide is successfully converted into a sulfide or selenide while perfectly inheriting its 2D sheet-like structure. Such materials show significant application potential in the electrocatalytic OER. Cation exchange offers another pathway. For example, easily synthesized cadmium sulfide nanorods can be used as templates. Leveraging the significant difference in solubility products between nickel ions and cadmium ions, the cadmium ions are replaced, resulting in well-defined nickel sulfide nanorods that are difficult to prepare by direct synthesis methods. The process of simultaneous cation and anion exchange is more complex, which enables the one-step replacement of multiple ions simultaneously. For instance, starting from a cobalt carbonate precursor and simultaneously introducing nickel ions and sulfide ions, bimetallic nickel cobalt sulfide compounds, which are of great interest in the supercapacitor field, can be synthesized.

Cheng et al. reported a low-temperature pyrolysis metal transfer synthesis strategy for preparing high-loading single-atom catalysts (SACs) via cation exchange reactions in a molten salt medium (Figure 2.10a). This method uses

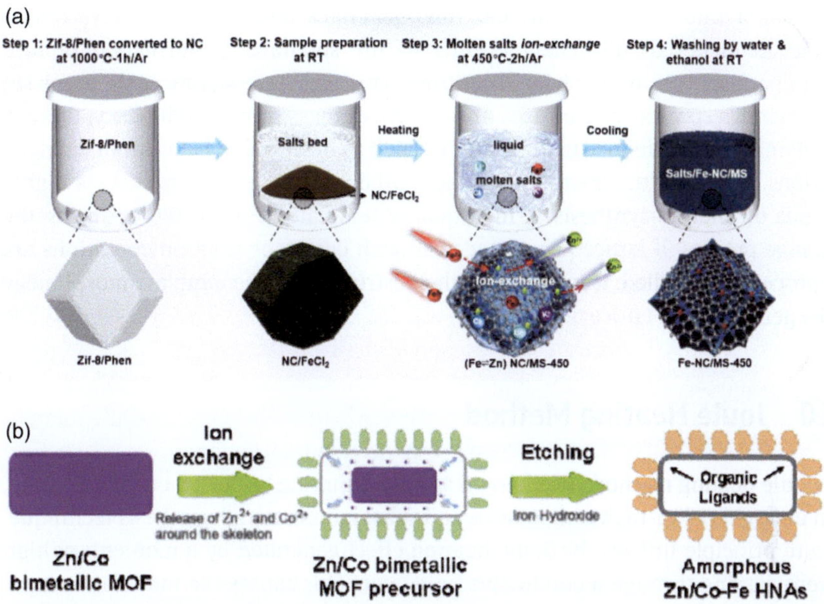

Figure 2.10 (a) Schematic diagram of the synthesis process for the Fe_y–NC/MS-T catalyst. *Source:* Reproduced from Ref. [89] / with permission of John Wiley & Sons. (b) Schematic illustration of the construction of bimetallic-MOF-derived amorphous zinc/cobalt–nickel iron hydroxide nanoarrays via an ion exchange and etching strategy. *Source:* Reproduced from Ref. [90] / with permission of John Wiley & Sons.

ZIF-8/Phen-derived nitrogen-doped carbon (NC) as the support. In a LiCl/KCl eutectic salt at low temperature (450°C), cation exchange between Fe^{2+} and Zn^{2+} enables the construction of atomically dispersed Fe–N_4 sites. This strategy effectively avoids metal agglomeration caused by Ostwald ripening during high-temperature pyrolysis, successfully synthesizing Fe–NC/MS-450 catalysts with Fe loadings as high as 3.9–4.7 wt%, which exhibit excellent ORR performance and fuel cell output performance [90]. Gu et al. developed a method based on an ion exchange-etching strategy to prepare bimetallic MOF-derived amorphous hollow nano-wall arrays for efficient OER (Figure 2.10b). This method uses Zn/Co bimetallic MOF grown on carbon cloth as the precursor. By immersing it in an ethanol–water mixed solution containing Fe^{2+}, an ion exchange reaction occurs under mild conditions, successfully constructing amorphous Zn/Co–Fe hydroxide hollow nano-wall arrays (A-Zn/Co–Fe HNAs). This structure possesses a high specific surface area, abundant active sites, and fast charge transfer capability. The optimized A-Zn/Co–Fe HNAs-120 requires an overpotential of only 226 mV at a current density of 10 mA cm^{-2} and exhibits excellent long-term stability [90].

The unique advantage of the ion exchange method lies in its exceptional morphology inheritance capability, enabling the synthesis of functional materials with complex geometric structures (such as hollow, porous, and core–shell). A typical example is the preparation of hollow structures utilizing the Kirkendall effect: when iron oxide nanospheres undergo exchange with sulfide ions, the difference in diffusion rates between different ions generates and accumulates vacancies within the material, ultimately leading to the spontaneous formation of iron sulfide nanospheres with cavity structures. This structure significantly increases the material's specific surface area and improves mass transfer efficiency, thereby enhancing its performance in applications such as lithium-ion batteries or catalytic reactions. However, this method also faces inherent challenges. Its success highly depends on the pre-synthesis of high-quality templates. Furthermore, during the exchange process, if lattice parameter mismatch occurs or reaction conditions are not properly controlled, it may lead to the destruction of the template morphology or the generation of lattice defects.

2.10 Joule Heating Method

The Joule heating method, also known as flash sintering or pulsed current heating, is an emerging ultra-high-temperature, ultrafast nanomaterial synthesis technique. Its core principle utilizes the Joule heating effect generated by a momentary high current passing through a conductive precursor. This causes the material to experience an extremely high-temperature field in a very short time (typically seconds or even milliseconds), followed by rapid cooling, thereby inducing electrochemical reactions, phase transitions, or crystallization processes, ultimately yielding the target nanomaterial. This method constitutes a radical departure from traditional furnace-based heating modes that rely on external heat sources and require prolonged high-temperature maintenance. Its heating rate can reach thousands or even tens of thousands of degrees Celsius per second, providing unique conditions for nonequilibrium synthesis and the capture of metastable structures.

When preparing iron, cobalt, and nickel-based compound nanomaterials, the typical implementation process of the Joule heating method is as follows: first, the target metal elements (such as salts or oxides of iron, cobalt, and nickel) are uniformly mixed with reactants (such as carbon sources, sulfur sources, and phosphorus sources) and necessary conductive additives (such as carbon black and graphene), and pressed into a self-supporting precursor block with certain conductivity. This precursor is then placed between two electrodes, and a short, high-intensity current pulse is applied via capacitor discharge or directly using a high-power source. The resistance encountered by the current within the precursor instantaneously generates immense Joule heat, causing the local temperature to rapidly soar to 1500–3000°C, thereby triggering ultrafast chemical reactions between the metal sources and reactants inside the precursor. After the reaction, due to the extremely low thermal inertia, the system cools at a very high rate, "freezing" the structure formed at high temperature, and finally yielding the target nanomaterial embedded within a conductive matrix (typically carbon).

The Joule heating method demonstrates several unique application pathways in synthesizing iron-, cobalt-, and nickel-based materials. One important direction is the direct synthesis of metal NP/carbon composites. For example, subjecting a mixture of iron salts and carbon black to Joule thermal shock can in situ generate iron NPs uniformly dispersed within a carbon matrix. This composite material exhibits excellent performance in areas such as the electrocatalytic OER and electromagnetic wave absorption. Another key application is the preparation of metal compounds, such as sulfides, phosphides, or nitrides. By introducing sulfur powder, red phosphorus, or nitrogen-containing compounds (e.g., urea) into the conductive precursor, the Joule heating process can instantaneously complete sulfidation, phosphidation, or nitridation reactions, producing nanocrystals like CoS_2, Ni_2P, or Fe_4N. The extremely high cooling rate of this method effectively suppresses particle agglomeration and overgrowth, favoring the formation of fine-sized, uniformly distributed nanocrystals. A more cutting-edge application is the synthesis of high-entropy alloy NPs. Subjecting a mixture of salts containing iron, cobalt, nickel, and other metals (e.g., chromium, manganese, and copper) to Joule thermal shock results in ultrafast heating and cooling. This rapid process prevents phase separation of the multiple metal elements, forcibly forming single-phase solid solution structured high-entropy alloy NPs, thereby opening a new pathway for exploring novel high-performance catalytic materials. Liu et al. developed an ultrafast Joule heating method and successfully synthesized an atomically dispersed Fe–N–C catalyst with axial oxygen coordination (FeN_4O/NC) for the first time (Figure 2.11a). This method employs periodic carbothermal shock treatment (100 A current, 3 seconds heating/6 seconds cooling, total 14 cycles, and maximum temperature ~1300 K) to rapidly dissociate Fe(II)-phenanthroline complexes adsorbed on nitrogen-doped carbon in an argon atmosphere. This releases individual Fe atoms and CO, which then combine with oxygen-containing groups, ultimately forming FeN_4O active sites with axial oxygen coordination on the NC support. This catalyst exhibits a half-wave potential of 0.903 V for the ORR along with excellent stability, providing a novel strategy for the precise regulation of the single-atom coordination environment [91]. Wang et al. adopted an ultrafast and energy-efficient Joule heating strategy (~10 seconds) to successfully prepare atomically dispersed $Cr—N_2{=}Fe$ bimetallic catalysts ($Cr—N_2{=}Fe$ ABCs) (Figure 2.11b). This method involves loading an ethanol solution of iron/chromium acetates and 1,10-phenanthroline onto carbon black, followed by Joule heating triggered by a 305 A current. The temperature rapidly rises to 1300°C within 3 seconds and then quickly cools, with the entire process lasting only about 10 seconds. This strategy effectively suppresses the agglomeration of metal atoms, constructing a diatomic site structure with synergistic interaction between CrN_4 and FeN_4, significantly enhancing the activity and stability of the ORR. The catalyst enabled stable operation in zinc–air batteries for over 300 hours [92].

The Joule heating method's significant advantages lie in its remarkable speed and efficiency. The entire synthesis process is typically completed within seconds, with extremely low energy consumption, offering unparalleled potential for energy savings and production efficiency compared to traditional furnace processes that take hours or even days. The extremely high heating and cooling rates keep the synthesis process far from equilibrium, enabling access to metastable crystal structures or ultrafine nanostructures that are difficult to achieve with conventional methods. Furthermore, this technique often uses bulk precursors, yielding ready-to-use, self-supporting electrode materials in a single step, thereby eliminating subsequent

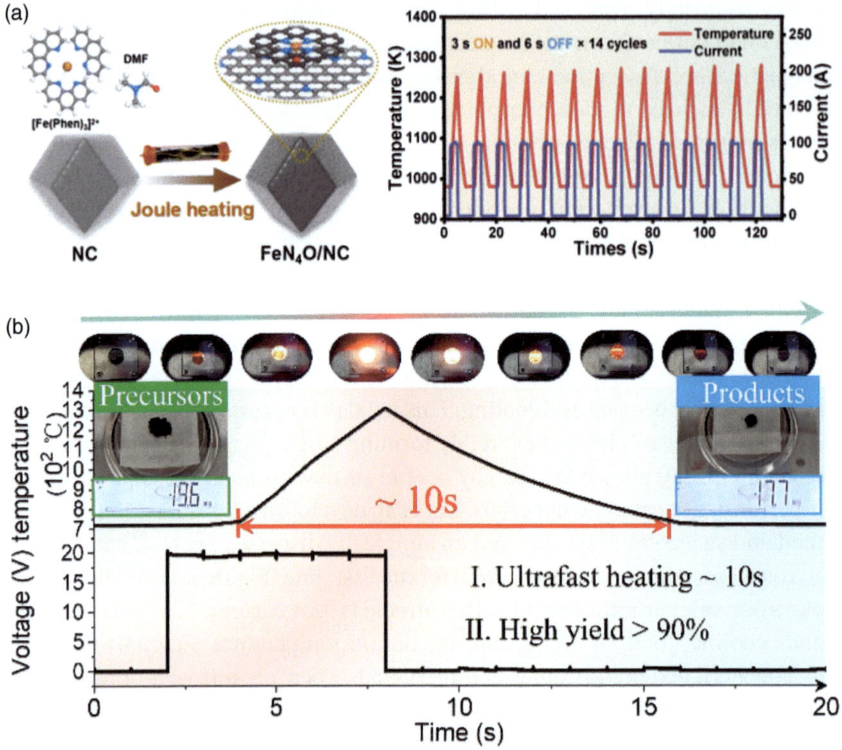

Figure 2.11 (a) Schematic diagram of the synthesis of FeN_4O/NC and the temperature–current–time profile of the periodic carbothermal shock process. *Source:* Reproduced from Ref. [91] / with permission of The Royal Society of Chemistry. (b) Changes in temperature and voltage of the catalyst during the Joule heating process. *Source:* Reproduced from Ref. [92] / with permission of Elsevier.

procedures such as coating and binding, and simplifying the device fabrication process.

However, the method also faces several challenges. Its successful implementation heavily relies on the homogeneity and conductivity of the precursor. The transient nature and extreme temperatures of the reaction process make real-time monitoring and precise control of reaction intermediates exceptionally difficult. The window for synthesis parameters is relatively narrow, and the technology is currently primarily suitable for laboratory-scale exploration, with equipment and processes for large-scale production yet to be developed.

As a disruptive synthesis platform, the Joule heating method holds significant development prospects. By designing more complex precursor structures and precisely controlling the current pulse waveform (e.g., pulse width and intensity), fine-tuning of the product's composition, crystal phase, and surface state can be achieved. Integrating this method with artificial intelligence for high-throughput screening and optimization holds promise for accelerating the discovery of new materials. It is foreseeable that the Joule heating method will continue to play a crucial role in the controllable preparation of iron-, cobalt-, and nickel-based nano-functional materials, particularly in fields like energy and catalysis, where demands on material structure and performance are increasingly stringent.

References

1 Munoz, M., de Pedro, Z.M., Casas, J.A., and Rodriguez, J.J. (2015). Preparation of magnetite-based catalysts and their application in heterogeneous Fenton oxidation—a review. *Applied Catalysis B: Environmental* 176: 249–265.

2 Wang, J.L. and Bai, Z.Y. (2017). Fe-based catalysts for heterogeneous catalytic ozonation of emerging contaminants in water and wastewater. *Chemical Engineering Journal* 312: 79–98.

3 Wang, Z., Zhang, Y., Neyts, E.C. et al. (2018). Catalyst preparation with plasmas: how does it work? *ACS Catalysis* 8: 2093–2110.

4 Wang, J.L. and Tang, J.T. (2021). Fe-based Fenton-like catalysts for water treatment: preparation, characterization and modification. *Chemosphere* 276: 130177.

5 Bukhtiyarova, M.V. (2019). A review on effect of synthesis conditions on the formation of layered double hydroxides. *Journal of Solid State Chemistry* 269: 494–506.

6 Lassoued, A., Lassoued, M.S., Dkhil, B. et al. (2018). Synthesis, photoluminescence and Magnetic properties of iron oxide (α-Fe_2O_3) nanoparticles through precipitation or hydrothermal methods. *Physica E: Low-dimensional Systems and Nanostructures* 101: 212–219.

7 Lu, A.H., Salabas, E.L., and Schuth, F. (2007). Magnetic nanoparticles: synthesis, protection, functionalization, and application. *Angewandte Chemie–International Edtion* 46: 1222–1244.

8 Dhakshinamoorthy, A., Navalon, S., Alvaro, M., and Garcia, H. (2012). Metal nanoparticles as heterogeneous Fenton catalysts. *ChemSusChem* 5: 46–64.

9 Nie, Y., Zhang, L., Li, Y.Y., and Hu, C. (2015). Enhanced Fenton-like degradation of refractory organic compounds by surface complex formation of $LaFeO_3$ and H_2O_2. *Journal of Hazardous Materials* 294: 195–200.

10 Tang, J. and Wang, J. (2017). Fe_3O_4-MWCNT magnetic nanocomposites as efficient fenton-like catalysts for degradation of sulfamethazine in aqueous solution. *ChemistrySelect* 2: 10727–10735.

11 Esteves, L.C., Oliveira, T.R., Souza, E.C. Jr., et al. (2015). A fast and environment-friendly method for determination of chemical oxygen demand by using the heterogeneous Fenton-like process (H_2O_2/$Fe_{3-x}Co_xO_4$ nanoparticles) as an oxidant. *Talanta* 135: 75–80.

12 Zhou, C., Sun, L., Zhang, A. et al. (2015). $Fe_{3-x}Cu_xO_4$ as highly active heterogeneous Fenton-like catalysts toward elemental mercury removal. *Chemosphere* 125: 16–24.

13 Gogoi, A., Navgire, M., Sarma, K.C., and Gogoi, P. (2017). Fe_3O_4 -CeO_2 metal oxide nanocomposite as a Fenton-like heterogeneous catalyst for degradation of catechol. *Chemical Engineering Journal* 311: 153–162.

14 Xu, L. and Wang, J. (2011). A heterogeneous Fenton-like system with nanoparticulate zero-valent iron for removal of 4-chloro-3-methyl phenol. *Journal of Hazardous Materials* 186: 256–264.
15 Cheng, R., Cheng, C., Liu, G.H. et al. (2015). Removing pentachlorophenol from water using a nanoscale zero-valent iron/H_2O_2 system. *Chemosphere* 141: 138–143.
16 Xu, L. and Wang, J. (2012). Fenton-like degradation of 2,4-dichlorophenol using Fe_3O_4 magnetic nanoparticles. *Applied Catalysis B: Environmental* 123–124: 117–126.
17 Hou, J., Zhang, B., Li, Z. et al. (2018). Vertically aligned oxygenated-CoS_2–MoS_2 heteronanosheet architecture from polyoxometalate for efficient and stable overall water splitting. *ACS Catalysis* 8: 4612–4621.
18 Li, P., Wang, M., Duan, X. et al. (2019). Boosting oxygen evolution of single-atomic ruthenium through electronic coupling with cobalt-iron layered double hydroxides. *Nature Communications* 10: 1711.
19 Prieto, G., Tüysüz, H., Duyckaerts, N. et al. (2016). Hollow nano- and microstructures as catalysts. *Chemical Reviews* 116: 14056–14119.
20 Sharma, R.K., Dutta, S., Sharma, S. et al. (2016). Fe_3O_4 (iron oxide)-supported nanocatalysts: synthesis, characterization and applications in coupling reactions. *Green Chemistry* 18: 3184–3209.
21 Sudha, D. and Sivakumar, P. (2015). Review on the photocatalytic activity of various composite catalysts. *Chemical Engineering and Processing Process Intensification* 97: 112–133.
22 Cai, S.-H., Chen, X.-N., Huang, M.-J. et al. (2022). Interfacial engineering of nickel/iron/ruthenium phosphides for efficient overall water splitting powered by solar energy. *Journal of Materials Chemistry A* 10: 772–778.
23 Yang, L., Huang, L., Yao, Y., and Jiao, L. (2021). In-situ construction of lattice-matching NiP_2/$NiSe_2$ heterointerfaces with electron redistribution for boosting overall water splitting. *Applied Catalysis B: Environmental* 282.
24 Bao, W., Yang, J., Yan, W. et al. (2023). Rapid in situ synthesis of sulfur-doped transition-metal oxyhydroxides by room temperature impregnation for advanced electrocatalytic splitting water. *Journal of Power Sources* 580: 233307.
25 Lin, Q., Zhao, J., Zhang, P. et al. (2024). Highly selective photocatalytic reduction of CO_2 to CH_4 on electron-rich Fe species cocatalyst under visible light irradiation. *Carbon Energy* 6: e435.
26 Wang, Z., Liu, X., Lv, M., and Meng, J. (2010). A new kind of mesoporous Fe_7Co_3/carbon nanocomposite and its application as magnetically separable adsorber. *Materials Letters* 64: 1219–1221.
27 Zhao, W., Seo, D.N., Kim, H.T., and Kim, I.J. (2010). Characterization of multi-walled carbon nanotubes (MWNTs) synthesized by CCVD using zeolite template from acetylene. *Journal of the Ceramic Society of Japan* 118: 983–988.
28 Wen, N., Xia, Y., Wang, H. et al. (2022). Large-scale synthesis of spinel $Ni_xMn_{3-x}O_4$ solid solution immobilized with iridium single atoms for efficient alkaline seawater electrolysis. *Advanced Science* 9, 2200529.

29 Mushiana, T., Khan, M., Abdullah, M.I. et al. (2022). Facile sol-gel preparation of high-entropy multielemental electrocatalysts for efficient oxidation of methanol and urea. *Nano Research* 15: 5014–5023.

30 Che, Q., Li, Q., Chen, X. et al. (2020). Assembling amorphous (Fe-Ni)Co_x-OH/Ni_3S_2 nanohybrids with S-vacancy and interfacial effects as an ultra-highly efficient electrocatalyst: inner investigation of mechanism for alkaline water-to-hydrogen/oxygen conversion. *Applied Catalysis B: Environmental* 263: 118338.

31 Ma, Y., Liu, D., Wu, H. et al. (2021). Promoting bifunctional water splitting by modification of the electronic structure at the interface of NiFe layered double hydroxide and Ag. *ACS Applied Materials & Interfaces* 13: 26055–26063.

32 Zhang, L.Y., Wang, T., Wu, H.M. et al. (2022). Nickel hydroxide array coated with NiFe alloy nanosheets for overall mixed water splitting. *Journal of Alloys and Compounds* 918: 165564.

33 Liu, C.C., Han, Y., Yao, L.B. et al. (2021). Engineering bimetallic NiFe-based hydroxides/selenides heterostructure nanosheet arrays for highly-efficient oxygen evolution reaction. *Small* 17: 2007334.

34 Yang, Y., Xie, Y.C., Yu, Z.H. et al. (2021). Self-supported NiFe-LDH@CoS_x nanosheet arrays grown on nickel foam as efficient bifunctional electrocatalysts for overall water splitting. *Chemical Engineering Journal* 419: 129512.

35 Zang, Z.H., Wang, X.W., Li, X. et al. (2021). Co_9S_8 nanosheet coupled Cu_2S nanorod heterostructure as efficient catalyst for overall water splitting. *ACS Applied Materials & Interfaces* 13: 9865–9874.

36 Kale, M.B., Borse, R.A., Mohamed, A.G.A., Wang, Y.B. (2021). Electrocatalysts by electrodeposition: recent advances, synthesis methods, and applications in energy conversion. *Advanced Functional Materials*, 31, 2101313.

37 Benetti, G., Cavaliere, E., Banfi, F., and Gavioli, L. (2020). Antimicrobial nanostructured coatings: a gas phase deposition and magnetron sputtering perspective. *Materials* 13: 784.

38 De Coster, V., Poelman, H., Dendooven, J. et al. (2020). Designing nanoparticles and nanoalloys for gas-phase catalysis with controlled surface reactivity using colloidal synthesis and atomic layer deposition. *Molecules* 25: 3735.

39 Kim, D.-S. and Hwang, N.-M. (2018). Synthesis of nanostructures using charged nanoparticles spontaneously generated in the gas phase during chemical vapor deposition. *Journal of Physics D: Applied Physics* 51: 463002.

40 Utke, I., Swiderek, P., Hoflich, K. et al. (2022). Coordination and organometallic precursors of group 10 and 11: focused electron beam induced deposition of metals and insight gained from chemical vapour deposition, atomic layer deposition, and fundamental surface and gas phase studies. *Coordination Chemistry Reviews* 458: 213851.

41 Wang, H. and Lu, J. (2018). Atomic layer deposition: a gas phase route to bottom-up precise synthesis of heterogeneous catalyst. *Acta Physico-Chimica Sinica* 34: 1334–1357.

42 Lang, Z., Song, G.-L., Liao, X. et al. (2024). Conversion of magnetron-sputtered sacrificial intermediate layer into a stable FeCo-LDH catalyst for oxygen evolution reaction. *Nano Research* 17: 4307–4313.

43 Inamdar, A.I., Chavan, H.S., Pawar, S.M. et al. (2020). NiFeCo oxide as an efficient and sustainable catalyst for the oxygen evolution reaction. *International Journal of Energy Research* 44: 1789–1797.

44 Tao, J., Zhang, Y., Wang, S. et al. (2019). Activating three-dimensional networks of Fe@Ni nanofibers via fast surface modification for efficient overall water splitting. *ACS Applied Materials & Interfaces* 11: 18342–18348.

45 Oviroh, P.O., Akbarzadeh, R., Pan, D. et al. (2019). New development of atomic layer deposition: processes, methods and applications. *Science and Technology of Advanced Materials* 20: 465–496.

46 Hou, J., Wu, Y., Zhang, B. et al. (2019). Rational design of nanoarray architectures for electrocatalytic water splitting. *Advanced Functional Materials* 29: 1808367.

47 Qi, M.M., Zachman, M.J., Li, Y.X. et al. (2025). Highly dense atomic Fe–Ni dual metal sites for efficient CO_2 to CO electrolyzers at industrial current densities. *Energy & Environmental Science* 18: 5643–5656.

48 He, B., Tan, K., Gong, Y. et al. (2020). Coupling amorphous cobalt hydroxide nanoflakes on $Sr_2Fe_{1.5}Mo_{0.5}O_{5+\delta}$ perovskite nanofibers to induce bifunctionality for water splitting. *Nanoscale* 12: 9048–9057.

49 Flahaut, E., Peigney, A., Bacsa, W.S. et al. (2004). CCVD synthesis of carbon nanotubes from (Mg,Co,Mo)O catalysts: influence of the proportions of cobalt and molybdenum. *Journal of Materials Chemistry* 14: 646–653.

50 Wang, C.C., Liu, C.P., Liu, R.M., and Liao, K.H. (2001). Characterization of co-catalyzed multiwalled carbon nanotubes by high-resolution transmission electron microscopy. *Symposium on Quantum Confined Semiconductor Nanostructures Held at the 2002 MRS Fall Meeting*, Boston, MA (2–5 December 2003), Boston, Ma, pp. 703–708.

51 Wei, H., Zheng, W., Jiang, Z., and Huang, Y. (2019). CNT coatings grown on the outer and inner surfaces of magnetic hollow carbon fibers with enhanced electromagnetic interference shielding performance. *Journal of Materials Chemistry C* 7: 14375–14383.

52 El-Gendy, A.A., Ibrahim, E.M.M., Khavrus, V.O. et al. (2009). The synthesis of carbon coated Fe, Co and Ni nanoparticles and an examination of their magnetic properties. *Carbon* 47: 2821–2828.

53 Sengupta, J., Jana, A., Singh, N.D.P. et al. (2010). Site-selective synthesis of in situ Ni-filled multi-walled carbon nanotubes using Ni(salen) as a catalyst source. *Nanotechnology* 21: 415605.

54 O'Neill, B.J., Jackson, D.H.K., Lee, J. et al. (2015). Catalyst design with atomic layer deposition. *ACS Catalysis* 5: 1804–1825.

55 Cremers, V., Puurunen, R.L., and Dendooven, J. (2019). Conformality in atomic layer deposition: current status overview of analysis and modelling. *Applied Physics Reviews* 6: 021302.

56 Johnson, R.W., Hultqvist, A., and Bent, S.F. (2014). A brief review of atomic layer deposition: from fundamentals to applications. *Materials Today* 17: 236–246.

57 Mackus, A.J.M., Merkx, M.J.M., and Kessels, W.M.M. (2019). From the bottom-up: toward area-selective atomic layer deposition with high selectivity. *Chemistry of Materials* 31: 2–12.

58 Miikkulainen, V., Leskela, M., Ritala, M., and Puurunen, R.L. (2013). Crystallinity of inorganic films grown by atomic layer deposition: overview and general trends. *Journal of Applied Physics* 113: 021301.

59 Lou, C., Yang, C., Zheng, W. et al. (2021). Atomic layer deposition of ZnO on SnO_2 nanospheres for enhanced formaldehyde detection. *Sensors and Actuators B-Chemical* 329: 129218.

60 Lu, J., Elam, J.W., and Stair, P.C. (2016). Atomic layer deposition-Sequential self-limiting surface reactions for advanced catalyst "bottom-up" synthesis. *Surface Science Reports* 71: 410–472.

61 Mackus, A.J.M., Bol, A.A., and Kessels, W.M.M. (2014). The use of atomic layer deposition in advanced nanopatterning. *Nanoscale* 6: 10941–10960.

62 Wong, M.S., Hwang, D., Alhassan, A.I. et al. (2018). High efficiency of III-nitride micro-light-emitting diodes by sidewall passivation using atomic layer deposition. *Optics Express* 26: 21324–21331.

63 Lin, C., Zhang, Y., Abdulagatov, A. et al. (2010). IEEE in ALD-enabled hermetic sealing for polymer-based wafer level packaging of MEMS. *23rd IEEE International Conference on Micro Electro Mechanical Systems (MEMS 2010)*, Hong Kong, PEOPLES R CHINA (24–28 January 2010), Hong Kong, PEOPLES R CHINA, pp. 528–531.

64 Liang, C., Zou, P., Nairan, A. et al. (2020). Exceptional performance of hierarchical Ni–Fe oxyhydroxide@NiFe alloy nanowire array electrocatalysts for large current density water splitting. *Energy & Environmental Science* 13: 86–95.

65 He, Y., Zhuang, X., Lei, C. et al. (2019). Porous carbon nanosheets: synthetic strategies and electrochemical energy related applications. *Nano Today* 24: 103–119.

66 Sun, Q., Dai, Z., Meng, X., and Xiao, F.-S. (2015). Porous polymer catalysts with hierarchical structures. *Chemical Society Reviews* 44: 6018–6034.

67 Chen, L., Yu, X., Xie, G. et al. (2021). Research progress in preparation of perovskite-type composite oxide catalyst by template method. *Journal of the Chinese Society of Rare Earths* 39: 531–542.

68 Gao, D., Zhang, M., Lu, B., and Ma, J. (2019). Recent process on preparation and application of amphiphilic Janus nanoparticles based on template method. *Fine Chemicals* 36: 1981–1988.

69 Kaplin, I.Y.O., Lokteva, E.S., Golubina, E.V., and Lunin, V.V. (2020). Template synthesis of porous ceria-based catalysts for environmental application. *Molecules* 25: 4242.

70 Kumar, L., Singh, S., Horechyy, A. et al. (2021). Block copolymer template-directed catalytic systems: recent progress and perspectives. *Membranes* 11: 318.

71 Tada, M. and Iwasawa, Y. (2003). Design of molecular-imprinting metal-complex catalysts. *Journal of Molecular Catalysis A-Chemical* 199: 115–137.

72 Guo, K., Fan, G., Gu, D. et al. (2019). Pore size expansion accelerates ammonium bisulfate decomposition for improved sulfur resistance in low-temperature NH_3-SCR. *ACS Applied Materials & Interfaces* 11: 4900–4907.

73 Park, S., Kim, T.J., Chung, Y.-M. et al. (2009). Direct synthesis of hydrogen peroxide from hydrogen and oxygen over palladium catalyst supported on SO_3H-Functionalized SBA-15. *Catalysis Letters* 130: 296–300.

74 Kim, M.J., Lee, M.W., and Lee, K.-Y. (2021). Improved catalytic wet peroxide oxidation of phenol over Pt-Fe_2O_3/SBA-15: influence of platinum species and DFT calculations. *Applied Surface Science* 541: 48409.

75 Liu, H. and Yin, H.-J. (2009). Electrical properties of carbon nanotube field-effect transistors in applied axial magnetic field. *Acta Physica Sinica* 58: 3287–3292.

76 Jang, B., Bong, S., Woo, S. et al. (2014). Facile synthesis of one-dimensional iron-oxide/carbon hybrid nanostructures as electrocatalysts for oxygen reduction reaction in alkaline media. *Journal of Nanoscience and Nanotechnology* 14: 8852–8857.

77 Tian, H., Zhang, K., Feng, X. et al. (2022). Self-supported $CoMoO_4$/NiFe-LDH core-shell nanorods grown on nickel foam for enhanced electrocatalysis of oxygen evolution. *Dalton Transactions* 51: 13762–13770.

78 Chen, Z., Liao, X., Sun, C. et al. (2021). Enhanced performance of atomically dispersed dual-site Fe-Mn electrocatalysts through cascade reaction mechanism. *Applied Catalysis B: Environmental* 288: 120021.

79 Ma, F.X., Liu, Z.Q., Zhang, G. et al. (2023). Self-sacrificing template synthesis of carbon nanosheets assembled hollow spheres with abundant active Fe–N_4O_1 moieties for electrocatalytic oxygen reduction. *Small* 19: 2207991.

80 Zhang, Y., Gao, F., Wang, D. et al. (2023). Amorphous/crystalline heterostructure transition-metal-based catalysts for high-performance water splitting. *Coordination Chemistry Reviews* 475: 214916.

81 Chen, Z.L., Ha, Y., Jia, H.X. et al. (2019). Oriented transformation of Co-LDH into 2D/3D ZIF-67 to achieve Co-N-C hybrids for efficient overall water splitting. *Advanced Energy Materials* 9: 1803918.

82 Hou, C.C., Zou, L., Wang, Y., and Xu, Q. (2020). MOF-mediated fabrication of a porous 3D superstructure of carbon nanosheets decorated with ultrafine cobalt phosphide nanoparticles for efficient electrocatalysis and zinc-air batteries. *Angewandte Chemie–International Edtion* 59: 21360–21366.

83 Sun, C., Dong, Q., Yang, J. et al. (2016). Metal–organic framework derived $CoSe_2$ nanoparticles anchored on carbon fibers as bifunctional electrocatalysts for efficient overall water splitting. *Nano Research* 9: 2234–2243.

84 Yang, J., Wang, X., Li, B. et al. (2017). Novel iron/cobalt-containing polypyrrole hydrogel-derived trifunctional electrocatalyst for self-powered overall water splitting. *Advanced Functional Materials* 27: 1606497.

85 Liu, G., Yao, R., Zhao, Y. et al. (2018). Encapsulation of Ni/Fe_3O_4 heterostructures inside onion-like N-doped carbon nanorods enables synergistic electrocatalysis for water oxidation. *Nanoscale* 10: 3997–4003.

86 Gao, T., Li, X., Chen, X. et al. (2021). Ultra-fast preparing carbon nanotube-supported trimetallic Ni, Ru, Fe heterostructures as robust bifunctional electrocatalysts for overall water splitting. *Chemical Engineering Journal* 424: 130416.

87 Cheng, Y., Guo, H., Yuan, P. et al. (2021). Self-supported bifunctional electrocatalysts with Ni nanoparticles encapsulated in vertical N-doped carbon nanotube for efficient overall water splitting. *Chemical Engineering Journal* 413: 127531.

88 Liu, Y., Luo, X., Zhou, C. et al. (2020). A modulated electronic state strategy designed to integrate active HER and OER components as hybrid heterostructures for efficient overall water splitting. *Applied Catalysis B: Environmental* 260: 118197.

89 Cheng, X.Y., Yin, S.H., Zhang, J.N. et al. (2025). Low-temperature pyrolysis: a universal route to high-loading single-atom catalysts for fuel cells. *Advanced Materials* 37: 2501707.

90 Gu, Z.X., Wei, X.L., Zhang, X.Q. et al. (2021). Bimetallic-MOF-derived amorphous zinc/cobalt-iron-based hollow nanowall arrays via ion exchange for highly efficient oxygen evolution. *Small* 17: 2104125.

91 Liu, F., Guo, X.Y., Zhong, Y. et al. (2024). Axial oxygen-bridged FeN_4O/NC produced by ultrafast Joule heating for efficient oxygen reduction reaction. *Chemical Communications* 60: 14459–14462.

92 Wang, L.L., Yang, L., Zhao, X.Y. et al. (2025). Using CrN_4 moiety to weaken the dissociation barrier of hydroxyl on adjacent single iron atom for efficient oxygen reduction. *Energy Storage Materials* 74: 103927.

3

Characterizations of Iron (Fe, Co, and Ni)-Group Metal Compound Nanomaterials

3.1 Introduction

The macroscopic properties of materials are fundamentally determined by their microstructure and surface characteristics, and advanced characterization techniques are key to establishing this "structure-property" relationship. To systematically unveil the nature of materials, this chapter focuses on eight core characterization techniques: X-ray diffraction (XRD), scanning electron microscopy (SEM), transmission electron microscopy (TEM), Brunauer–Emmett–Teller (BET)-specific surface area analysis, X-ray photoelectron spectroscopy (XPS), ultraviolet photoelectron spectroscopy (UPS), differential thermal analysis (DTA), and electron paramagnetic resonance (EPR). These techniques form a multi-scale analytical framework spanning from structure to properties: XRD is used to resolve crystal structure and phases; SEM and TEM enable high-resolution imaging of surface morphology and internal lattice structure, respectively; BET quantifies specific surface area and pore structure; XPS and UPS reveal elemental composition, chemical states, and electronic structure by detecting core-level and valence electrons, respectively; DTA tracks thermal behaviors such as phase transitions and decomposition; and EPR is specialized in detecting paramagnetic centers (e.g., free radicals and defects). This chapter aims to elaborate on the fundamental principles, key formulas, and representative application cases of the aforementioned techniques, with the goal of providing a systematic characterization methodology for cutting-edge research and development in the field of iron-group metal compound nanomaterials.

3.2 XRD

X-ray diffraction (XRD) is a fundamental characterization technique for analyzing the crystal structure of materials. Since the discovery of X-rays by Röntgen, the diffraction phenomena resulting from their interaction with matter have served as the physical basis for phase identification and crystal structure analysis. X-rays are electromagnetic waves with relatively short wavelengths (approximately 10^{-8}–10^{-10} cm), generated by bombarding a metal target with a

Iron-Group Metal Compound Nanomaterials: Preparation, Characterization, Structure, and Applications, First Edition. Jinlong Zheng.

high-energy electron beam. Their specific wavelengths (i.e., characteristic X-rays) are closely related to the elemental composition of the target. In 1912, Max von Laue proposed that crystals could act as natural three-dimensional diffraction gratings for X-rays. This insight revealed that when X-rays pass through a crystal, diffraction occurs because the spacing between interatomic planes is comparable to the X-ray wavelength, resulting in diffraction patterns with specific angular distributions and intensities. By analyzing the directions of enhanced and diminished radiation in these patterns, the atomic arrangement, lattice parameters, and phase composition of the crystal can be determined. With technological advancements, XRD has evolved into a highly automated and precise analytical method, widely applied to the structural characterization of various systems, including crystalline, amorphous, and composite materials, and even biological samples. In the study of nanomaterials, XRD can accurately identify key structural parameters such as phase composition, grain size, crystallinity, and microstrain, providing indispensable crystallographic insights for understanding the "structure-property" relationship of materials.

3.2.1 Basic Principles

The generation of X-rays typically originates from an X-ray tube. In this vacuum device, electrons emitted from the cathode are accelerated by a high-voltage electric field, forming a high-speed electron beam that strikes a metal anode target, thereby exciting X-rays. The fundamental principle of XRD analysis lies in the fact that when a monochromatic X-ray beam strikes a crystalline material, the regularly spaced atomic planes within the crystal, with interplanar distances comparable to the X-ray wavelength, cause interference among the X-rays scattered by different atomic planes. In specific directions, these scattered waves undergo constructive interference due to their identical phases, forming diffraction peaks. The spatial orientation and intensity distribution of the diffracted rays directly reflect the crystal structure of the material, meaning that substances with different structures exhibit unique diffraction patterns. To quantitatively describe the relationship between diffraction direction and crystal-structure parameters, the British physicists Bragg, father and son, proposed the renowned Bragg's law in 1913:

$$2d \sin\theta = n\lambda \tag{3.1}$$

In the formula, d is the interplanar spacing of the crystal, n is any positive integer, θ is the glancing angle, and λ is the X-ray wavelength.

3.2.2 Applications of X-Ray Diffraction Technology in Material Analysis

XRD is a key technique for characterizing the crystal structure of nanomaterials based on iron-group metal compounds. Its core applications include phase identification, unit cell parameter calculation, and crystallite size analysis, with phase analysis being particularly important. Each crystalline phase possesses a unique diffraction pattern, where the positions of diffraction lines (characterized by interplanar spacing, d-value) and their relative intensities (I) form the "fingerprint" of that phase. This fingerprint remains unchanged even when the phase is mixed with others, forming the foundation of XRD phase analysis. XRD phase analysis consists of two parts: qualitative and quantitative analysis. Qualitative analysis involves comparing the experimentally obtained d-I data set with standard powder diffraction databases (such as PDF cards) to identify the phases present in the sample. If the sample is a multiphase mixture, its diffraction pattern is a superposition of the diffraction lines from all constituent phases. Building on qualitative analysis, quantitative analysis can be performed based on the principle that the diffraction intensity of each phase is proportional to its volume fraction in the sample. Common quantitative methods include the internal standard method, external standard method, spiking method, standardless method, and whole-pattern fitting method. The internal standard and spiking methods require adding a reference phase to the sample and constructing calibration curves, but in multiphase systems, peak overlap can complicate the analysis. The external standard method necessitates the use of pure phases to create standard curves, which is often impractical. While the standardless method and whole-pattern fitting method avoid the need for complex standard series, they rely on sophisticated mathematical calculations and have certain limitations in their applicability. The physicochemical properties of nanomaterials are closely related to their particle size. Due to their large specific surface area and high surface energy, nanoparticles are prone to aggregation. As a result, conventional particle size analyzers often measure the size of aggregates rather than the primary particle size. The XRD line broadening method (Scherrer method) provides an effective means to determine the average size of nanocrystallites. This method is based on the broadening of diffraction peaks caused by crystallite refinement, and its calculation formula is:

$$D = 0.89\lambda/(\beta_{hkl}\cos\theta) \tag{3.2}$$

Among these, D represents the average crystallite size, λ represents the X-ray wavelength, β_{hkl} represents the full width at half maximum (FWHM) of the diffraction peak, and 0 represents the Bragg angle. In practical measurements, multiple low-angle diffraction lines (e.g., $2\theta \leq 50°$) are typically selected for calculation to enhance the accuracy of the results. The reliability of this method has been widely validated. For instance, Gu et al. used both the Scherrer method and transmission electron microscopy (TEM) to determine the particle sizes of nano-$CaCO_3$ and CeO_2, and the results were in close agreement, confirming the credibility of the method. In specific studies on iron-group metal compound nanomaterials, XRD plays a pivotal role. Taking Ni-doped Co_3O_4 thin films as an example [1], XRD analysis was employed to determine the phase and structural parameters of samples with different doping concentrations (0–10 wt%). As shown in Figure 3.1a, the diffraction peaks of all samples matched well with the standard card of cubic spinel-structured Co_3O_4 (JCPDS: 74-2120). The study revealed that Ni was successfully incorporated into the Co_3O_4 lattice, and as the Ni doping concentration increased, the intensity of the (311) diffraction peak enhanced, indicating improved crystal quality. To further illustrate the application of XRD in the analysis of complex nanostructures, the crystal structures of Co_3O_4 nanowires and their core-shell

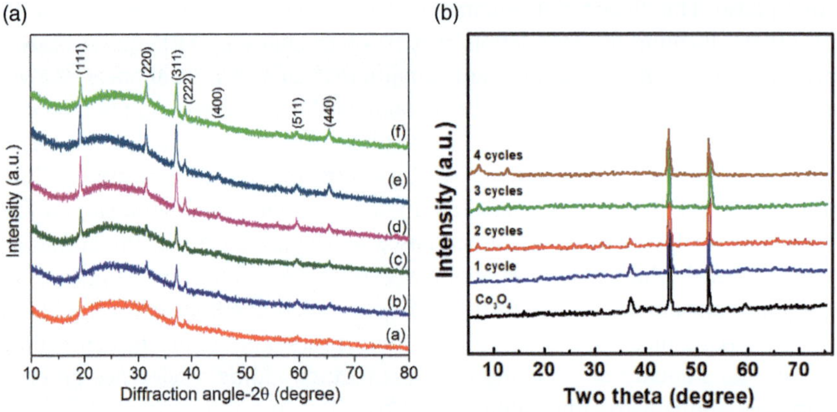

Figure 3.1 (a) XRD patterns of Co_3O_4 thin films doped with different percentages (0%, 2%, 4%, 6%, 8%, and 10%) of Ni; *Source:* Reproduced from Ref. [1] / with permission of Elsevier. (b) XRD patterns of Co_3O_4 nanowires and core-shell Co_3O_4@MOF-74 composites subjected to different solvothermal cycles; *Source:* Reproduced from Ref. [2] / with permission of Elsevier.

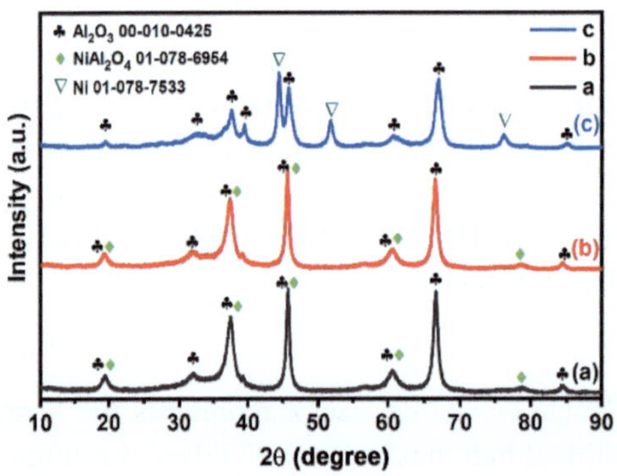

Figure 3.2 XRD patterns of the as-synthesized catalysts: (a) 10Ni90Al; (b) after reduction at 800°C; (c) after reduction at 900°C. *Source:* Reproduced from Ref. [3] / with permission of Elsevier.

structured Co_3O_4@MOF-74 composites were characterized (Figure 3.1b) [2]. The XRD patterns clearly revealed the in situ formation process of the MOF-74 shell with an increasing number of solvothermal cycles, manifested by the gradual enhancement of the characteristic diffraction peaks of MOF-74 ($2\theta = 6.6°$ and 12.7°) and the corresponding weakening of the characteristic peak of Co_3O_4 (37.1°).

In the study of catalytic materials, XRD can effectively track phase transition processes. As shown in Figure 3.2 [3], the as-synthesized catalyst exhibits characteristic peaks of Al_2O_3 and $NiAl_2O_4$. After high-temperature reduction treatment, partial $NiAl_2O_4$ peaks diminish, and distinct metallic Ni phases emerge following reduction at 900°C, which are considered the primary active sites for the ethanol reforming reaction.

Furthermore, XRD can be utilized to evaluate the crystallinity and phase dispersion of catalysts. XRD analysis of AlMoNi and AlCNT-MoNi catalysts (Figure 3.3) [4] reveals that after the introduction of active metals, the diffraction peak intensity of the γ-Al_2O_3 support significantly decreases, and no distinct sharp diffraction peaks of Ni or Mo oxides are observed, indicating high dispersion of the metal

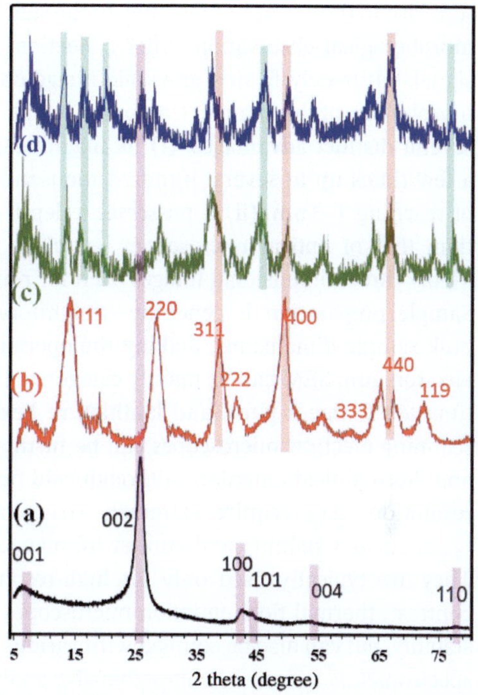

Figure 3.3 XRD patterns of the as-synthesized catalysts: (a) CNTs; (b) Al_2O_3; (c) AlMoNi catalyst; (d) AlCNT-MoNi catalyst. *Source:* Reproduced from Ref. [4] / with permission of Elsevier.

species on the support. The appearance of Mo_2C diffraction peaks, along with weak characteristic peaks of carbon nanotube (CNT) in the patterns, further demonstrates the complex composition of the materials. This high dispersion provides the catalyst with more active sites, thereby contributing to enhanced catalytic activity and stability.

3.3 SEM

Scanning electron microscopy (SEM) is an observational technique that bridges the gap between TEM and optical microscopy. Its operating principle is based on the interaction between a high-energy electron beam and the surface of a solid sample. By detecting signals such as secondary electrons and backscattered electrons generated from this interaction, high-resolution three-dimensional topographic information of the sample surface is obtained. An SEM system primarily consists of an electron optical system and a signal display system. The electron optical system includes an electron gun, electromagnetic lenses, scanning coils, and a sample chamber. During operation, the electron beam emitted by the electron gun (with a diameter of approximately 20–30 μm) is accelerated under a high voltage of 1–40 kV. Through the focusing action of the condenser and objective lenses, it is concentrated into a narrow electron probe with a diameter of only a few tens of nanometers, which is then projected onto the sample surface. The interaction between the electron beam and the sample excites various signals, including secondary electrons, backscattered electrons, and Auger electrons. Secondary electrons are highly sensitive to the surface topography of the sample and serve as the primary signal for morphological observation. After collection, amplification, and processing, these signals ultimately form a grayscale image on the display that reflects the surface morphology of the sample. Compared to optical microscopy and TEM, SEM offers several distinct advantages: (i) Its magnification is continuously adjustable from a few times up to several hundred thousand times, with high resolution capable of reaching 1–3 nm. (ii) It possesses a depth of field more than 300 times greater than that of optical microscopes, enabling the acquisition of overall clear and highly three-dimensional images even for rough and uneven sample surfaces. (iii) Sample preparation is generally straightforward, with lenient requirements for bulk sample dimensions, making the operation convenient. Based on the type of electron gun, SEM can be mainly categorized into three types: field emission guns, tungsten filament guns, and lanthanum hexaboride (LaB_6) guns. Field emission scanning electron microscopes can be further subdivided into cold field emission and thermal field emission. Although cold field emission microscopes offer higher resolution, they require extremely stringent vacuum conditions, have limited beam current stability and emitter lifespan, and necessitate periodic tip cleaning. They are typically used only for high-resolution morphological observation. In contrast, thermal field emission microscopes not only offer long-term operational stability, but can also be coupled with various accessories such as energy dispersive spectrometers to achieve comprehensive analysis of morphology, composition, and crystallographic information. Consequently, they are more widely applied in fields like materials science that require multifunctional analysis.

3.3.1 Fundamental Principles

SEM core imaging principles are based on the interaction between a high-energy electron beam and the surface of a solid sample. The operational process can be summarized as follows: the electron beam emitted by the electron gun is focused by the electromagnetic lens system into a nanoscale probe, which performs a raster scan across the sample surface point-by-point under the control of the scanning coils. The incident electrons interact with the sample, exciting various physical signals, including secondary electrons, backscattered electrons, and characteristic X-rays. These signals are collected, converted, and amplified by corresponding detectors, and then synchronously modulate the brightness of corresponding pixels on the display screen, ultimately forming a grayscale image that reflects the surface characteristics of the sample. SEM employs a unique "synchronous scanning" mechanism, where both the electron beam scanning the sample surface and the electron beam scanning the display screen are controlled by the same scan generator, achieving perfect synchronization. This design ensures a precise spatiotemporal correspondence between each "object point" on the sample surface and each "image point" on the display screen. A typical SEM image consists of nearly a million such image points, enabling high-fidelity reproduction of the sample's surface morphology, compositional distribution, and other physical property variations. Among the various imaging signals, secondary electron imaging is the most widely used and highest resolution imaging mode, due to its high sensitivity to surface topography. The specific imaging process is as follows: the accelerated (typically up to 30 keV) and focused electron beam scans the sample surface point-by-point, exciting secondary electrons. These electrons are captured by the collector, converted into light signals by a scintillator, and then transformed into electrical signals via a photomultiplier tube. These electrical signals are further amplified by a video amplifier and finally used to modulate the brightness of corresponding pixels on the display. Variations in surface topography cause changes in secondary electron yield, which manifest as brightness contrast differences in the image, forming a three-dimensional representation of the surface morphology. In addition to secondary electrons, SEM can detect various other signals. Backscattered electron contrast is related to atomic number and can be used for compositional analysis; characteristic X-rays enable qualitative and quantitative elemental analysis. Different signals have varying sampling depths, collectively forming the physical basis for SEM's multifunctional analytical capabilities.

3.3.2 Application of SEM Technology in Material Analysis

SEM can intuitively reveal the microscopic morphology and structural characteristics of materials, serving as a key technique for analyzing the surface structure, particle size, and distribution of iron-group metal compound nanomaterials. Taking the study of nickel-based oxides and their composite materials as an example [5], by comparing the SEM images of pristine NiO, Cu-doped NiO, and Cu-doped NiO/reduced graphene oxide (rGO) composites, the evolution of morphology and structure with different compositions can be clearly observed. The pristine NiO sample (Figure 3.4a) exhibits a typical nanostructure under 28,440× magnification, though its morphology is relatively irregular. The Cu-doped NiO sample (Figure 3.4b) largely retains the morphological characteristics of the pristine sample, but the size of its nanostructures is significantly reduced. This may be attributed to the introduction of doping ions inhibiting grain growth and nucleation processes, or to the "Zener effect" where grain boundary migration is dragged by solute atoms. In the Cu-doped NiO/rGO composite (Figure 3.4c), no distinct rGO nanosheet structures are observed in the SEM image. This could be due to the low content and uniform distribution of rGO in the composite, or because the rGO is covered by Cu-doped NiO nanostructures during sample preparation. In contrast, the SEM image of pristine graphene oxide (Figure 3.4d) clearly shows its highly exfoliated layered structure. The above SEM analysis results indicate that Cu doping can effectively refine NiO grains, while the introduction of rGO does not significantly alter the overall morphology, suggesting that rGO may be uniformly dispersed as a supporting matrix in the composite material.

In the study of montmorillonite (MMT)-based composite materials, SEM was employed to visually evaluate the evolution of the material's microstructure [6]. As shown in Figure 3.5a, the pristine MMT exhibits a typical layered structure. After intercalation modification, the perovskite phase ($LCCOM_{0.2}$) is uniformly distributed between the MMT layers (Figure 3.5b), confirming successful intercalation

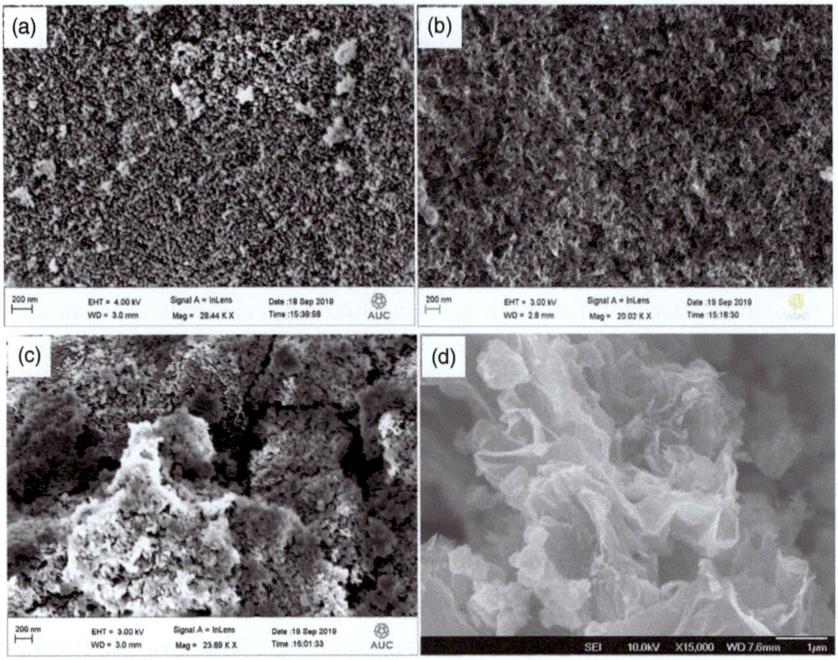

Figure 3.4 SEM images of (a) NiO; (b) Cu-doped NiO; (c) Cu-doped NiO/rGO composite; (d) pristine rGO. *Source:* Reproduced from Ref. [5] / with permission of Elsevier.

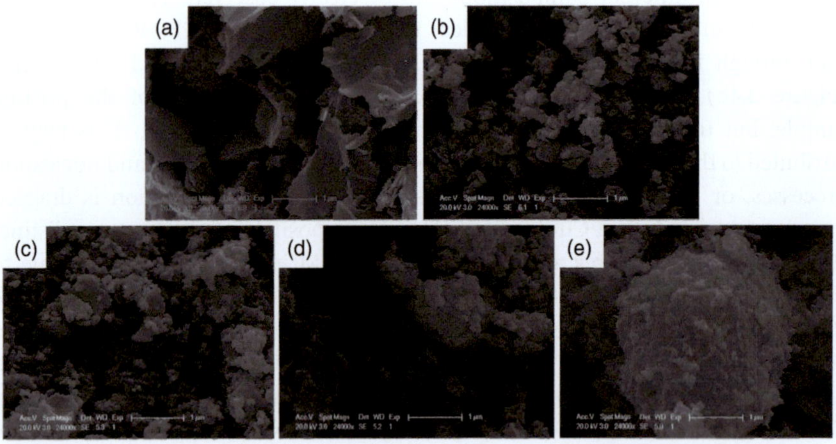

Figure 3.5 SEM images of the as-prepared catalysts: (a) MMT; (b) $LCCOM_{0.2}$; (c) $0.02Fe_2O_3/LCCOM_{0.2}$; (d) $0.04Fe_2O_3/LCCOM_{0.2}$; (e) $0.06Fe_2O_3/LCCOM_{0.2}$. *Source:* Reproduced from Ref. [6] / with permission of Elsevier.

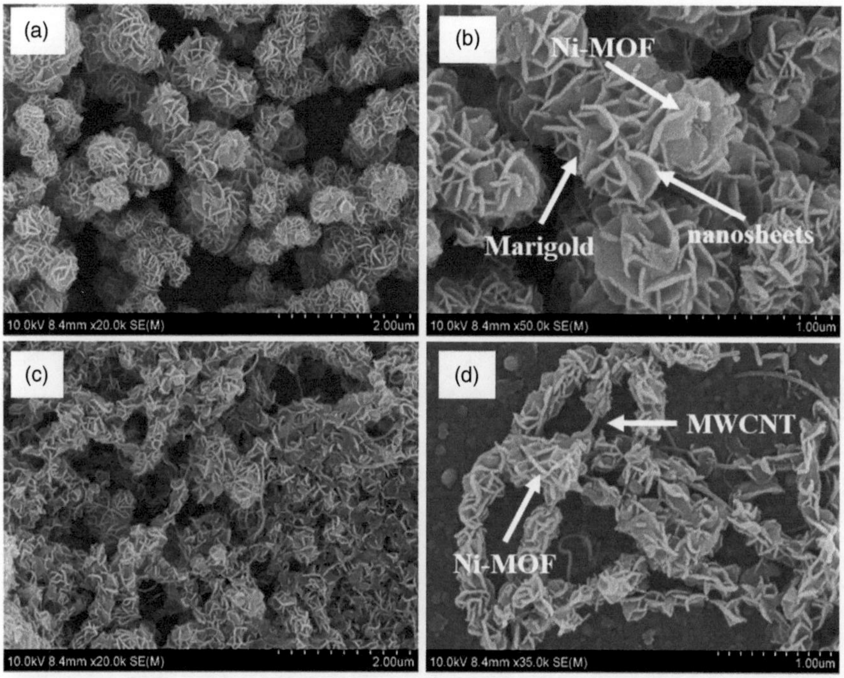

Figure 3.6 SEM images at different magnification levels: (a) and (b) pristine Ni-MOF; (c) and (d) Ni-MOF@CNT. *Source:* Reproduced from Ref. [7] / with permission of Elsevier.

compounding. When a small amount of Fe_2O_3 is introduced ($0.02Fe_2O_3/LCCOM_{0.2}$), the particles are well dispersed on the surface. At the optimal loading level ($0.04Fe_2O_3/LCCOM_{0.2}$), the distribution of Fe_2O_3 is most uniform. However, under overload conditions (Figure 3.5c–e), the Fe_2O_3 particles exhibit significant agglomeration, forming irregular blocky aggregates and causing disruption and stacking of the MMT-layered structure. This phenomenon is further corroborated by TEM results.

SEM plays a crucial role in revealing the morphology and assembly modes of metal-organic framework (MOF) materials as well. Through morphological characterization of Ni-MOF and its CNT composite (Ni-MOF@CNT) [7], it can be observed that the pristine Ni-MOF exhibits a "marigold-like" microsphere structure, formed by the self-assembly of two-dimensional nanosheets (Figure 3.6a, b). Although the Ni-MOF@CNT sample (Figure 3.6c, d) maintains a similar flower-like morphology after growth on functionalized multi-walled CNTs (MWCNTs), the size of its self-assembled microspheres is significantly reduced. Moreover, these microstructures interconnect to form a more continuous three-dimensional network structure, demonstrating the effective regulation of MOF crystal growth behavior by CNTs.

SEM images further confirm the formation of Co nanoparticle-nitrogen-doped carbon nanotubes (NCNTs) hybrids [8]. After selenization, only diffraction peaks corresponding to orthorhombic $CoSe_2$ (PDF No. 53-0449) are observed alongside the

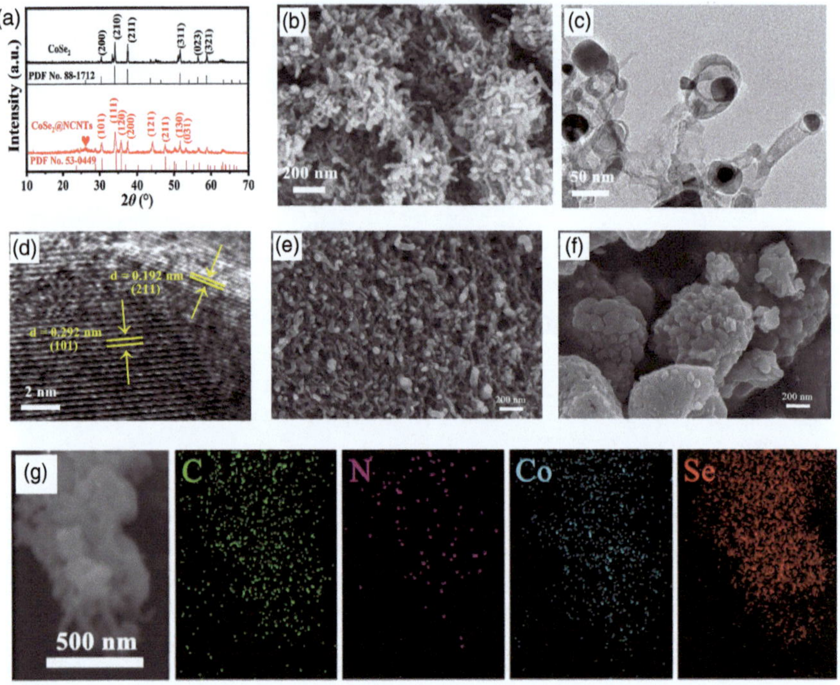

Figure 3.7 (a) XRD patterns of $CoSe_2$ and $CoSe_2$@NCNT; (b) SEM image of $CoSe_2$@NCNT; (c) TEM image of $CoSe_2$@NCNT; (d) HRTEM image taken from an individual $CoSe_2$ nanoparticle; (e) SEM image of Co@NCNTs; (f) SEM image of pure $CoSe_2$; (g) SEM and EDX elemental mapping images of C, N, Co, and Se for $CoSe_2$@NCNTs. *Source:* Reproduced from Ref. [9] / with permission of Springer Nature.

NCNTs peaks (Figure 3.7a) [9], indicating complete conversion of Co to orthorhombic $CoSe_2$. Furthermore, the morphology remains well-preserved (Figure 3.7b). TEM images (Figure 3.7c) reveal the formation of nanoparticles decorated on NCNTs. Moreover, high-resolution TEM (HRTEM) shows lattice fringes with interplanar spacings of 0.192 and 0.292 nm, corresponding to the (211) and (101) planes of orthorhombic $CoSe_2$, respectively (Figure 3.7d). For comparison, pure $CoSe_2$ particles were synthesized under the same conditions as $CoSe_2$@NCNTs but without using $Zn(NO_3)_2 \cdot 6H_2O$ and melamine. Notably, the absence of NCNTs resulted in the formation of larger cubic pyrite-type $CoSe_2$ particles (Figure 3.7e–g).

3.4 TEM

TEM is a high-resolution imaging technique that utilizes a high-energy electron beam as an illumination source, transmits it through an ultrathin specimen, and employs electromagnetic lenses for magnified imaging, thereby obtaining ultrastructural information from the sample. Its emergence broke the diffraction limit of optical microscopy, extending human observation capability from the micrometer scale down to the nanometer and even atomic scale, profoundly deepening our understanding of the microscopic world of materials. The fundamental principle of TEM imaging is based on the fact that electrons have a de Broglie wavelength much shorter than that of visible light. According to physical principles, the wavelength (λ) of the electron beam is inversely proportional to the square root of the accelerating voltage (V), meaning that shorter-wavelength electron sources can be obtained by increasing the accelerating voltage. Since Ruska invented the first transmission electron microscope in 1932, this technology, with its exceptional resolution capability (modern TEM can achieve resolutions below 0.2 nm), has become an indispensable core tool for observing submicroscopic structures (i.e., ultrastructures, typically referring to details smaller than 0.2 μm) of materials, playing a vital role in the fine structural characterization of iron-group metal compound nanomaterials.

3.4.1 Fundamental Principles

The TEM imaging system consists of an electron gun, an electromagnetic lens system, and a vacuum system. Its working principle is analogous to that of an optical microscope but employs an electron beam instead of visible light as the illumination source and uses electromagnetic fields rather than glass lenses for focusing and magnification. The imaging process begins with a high-energy electron beam emitted from the electron gun, which is converged by the condenser lens and directed onto an ultrathin sample (typically 50–100 nm). Electrons interact with atoms in the sample, resulting in elastic and inelastic scattering, thereby forming a transmitted electron beam that carries structural information of the sample. TEM imaging primarily relies on three contrast mechanisms: (i) Mass thickness contrast: Arises from differences in electron absorption and scattering across various regions of the sample. Areas with greater mass-thickness or higher atomic numbers scatter electrons more strongly, resulting in weaker transmitted electron signals and darker images, whereas areas with lower mass-thickness or atomic numbers appear brighter. This is the primary contrast source in low-magnification imaging. (ii) Diffraction contrast: When the electron beam passes through a crystalline sample, diffraction occurs under conditions satisfying Bragg's law. By selecting specific diffracted beams for imaging, microstructural features such as crystal defects, grain boundaries, and domain structures can be observed. (iii) Phase contrast: For extremely thin samples (typically <10 nm), interference occurs between transmitted and scattered electrons, forming high-resolution images. A typical TEM system includes an electron emission source, acceleration system, condenser lens, objective lens, projection lens, and detectors. The electron beam is accelerated and converged by electromagnetic lenses before irradiating the sample. During transmission, it interacts with atoms, generating transmitted and diffracted beams that contain structural information of the sample. By adjusting the imaging mode, different types of structural information can be obtained: HRTEM utilizes the interference effects of electron waves to directly reveal the projected atomic arrangement of the material, enabling precise measurement of interplanar spacings and identification of lattice defects; selected-area electron diffraction (SAED), based on Bragg's law, analyzes diffraction patterns from specific micro-regions to obtain key parameters such as crystal structure, orientation, and lattice constants. Through the integrated application of these three imaging modes, TEM enables multi-scale comprehensive analysis of iron-group metal compound nanomaterials, ranging from microscopic morphology and crystal structure to atomic arrangement, providing crucial structural insights for understanding the structure–property relationships of materials.

3.4.2 Application of TEM Technology in Material Analysis

In the study of iron-group metal compounds, TEM technology provides direct evidence for understanding the structure–property relationships of materials. Taking the Mn_2Co_1-mixed oxide with excellent catalytic performance as an example, the analytical value of TEM extends throughout structural characterization and performance mechanism investigation [10]. In the HRTEM image of the Mn_2Co_1 mixed oxide (Figure 3.8a), lattice fringes with interplanar spacings of 0.14, 0.20, 0.24, and 0.27 nm correspond to the (400), (220), (211), and (103) crystal planes of $CoMn_2O_4$, respectively, providing direct microscopic evidence for phase composition. TEM reveals the effect of doping on crystallinity through comparison. Compared with the single Mn oxide, TEM images (Figure 3.8b) show that the Mn_2Co_1-mixed oxide exhibits poor lattice fringe continuity and low crystallinity, which is consistent with the XRD results (Figure 3.8g), confirming that Co doping promotes the formation of an amorphous phase. TEM can also accurately capture defect structures related to catalytic performance. The Mn_2Co_1 mixed oxide contains a large number of lattice defects and edge dislocations (Figure 3.8c–f), with discontinuous curved grain boundaries forming abundant defects (Figure 3.8h). These defects identified by TEM (interface dislocations, lattice distortions, etc.) are accompanied by the generation of oxygen vacancies, which can enhance redox capacity, promote the migration of oxygen species, and provide active sites, demonstrating excellent catalytic activity, particularly in the low-temperature oxidation of volatile organic compounds (VOCs).

Figure 3.9a's STEM-HAADF image and Figure 3.9b's TEM image clearly reveal the fine structures of the calcined Ni/SBA-15-AE and Ni-5Mg/SBA-15-AE catalysts, respectively [11]. Leveraging the high-resolution imaging capabilities of TEM techniques, not only can the characteristic overlapping mesoporous channels of the SBA-15 support be distinctly observed, but the fibrous microstructure formed by the Ni component and the Ni–Mg-layered silicate phase generated through Ni–Mg interaction can also be precisely identified. This provides direct visual evidence for clarifying the existence forms and dispersion states of active components during the calcination process. After reduction in hydrogen at 800°C for 1 hour, TEM further demonstrates its crucial role in analyzing active components, with related characterization results collectively presented in Figure 3.9c–g. For the Ni/SBA-15-WI catalyst, TEM imaging combined with statistical analysis clearly reveals the distribution characteristics and size patterns of Ni nanoparticles: a large number of Ni nanoparticles are predominantly concentrated on the external surface of the SBA-15 support, showing evident size inhomogeneity. Precise measurements based

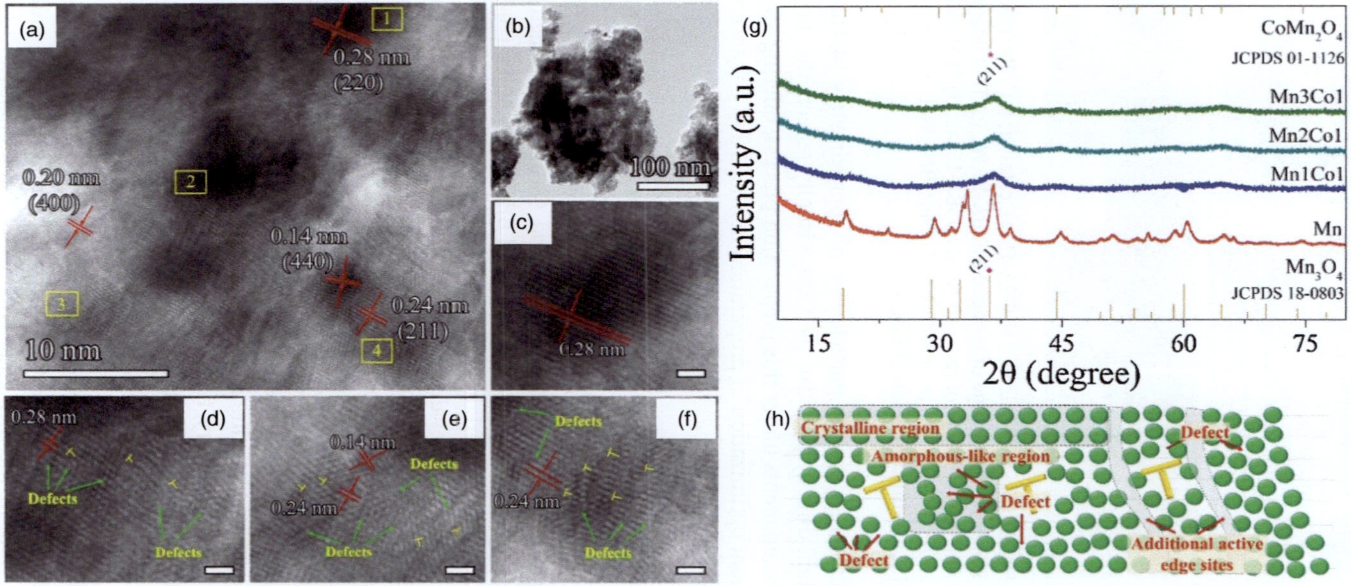

Figure 3.8 (a, b) TEM images of Mn_2Co_1-mixed oxide catalysts; (c–f) show different magnified regions from (a) with a scale bar of 2 nm. (g) XRD patterns of the as-prepared Mn_xCo_y-mixed oxides (Mn-based catalysts). (h) Schematic diagram of defect structures generated by edge dislocations. *Source:* Reproduced from Ref. [10] / with permission of Elsevier.

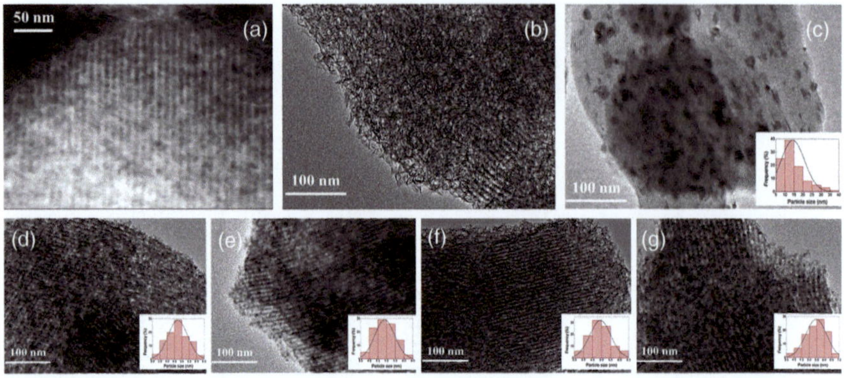

Figure 3.9 (a) STEM-HAADF image of calcined Ni/SBA-15-AE catalyst; (b) TEM image of calcined Ni-5Mg/SBA-15-AE; (c) TEM image of reduced Ni/SBA-15-WI; (d) TEM image of reduced Ni/SBA-15-AE; (e) TEM image of reduced Ni-3Mg/SBA-15-AE; (f) TEM image of reduced Ni-5Mg/SBA-15-AE; (g) TEM image of reduced Ni-7Mg/SBA-15-AE. *Source:* Reproduced from Ref. [11] / with permission of Elsevier.

on TEM images indicate an average particle size of approximately 14.4 nm. This structural information, directly provided by TEM, visually reflects the dispersion performance of Ni components under this preparation process, offering important insights for subsequent optimization of preparation methods and enhancement of active component dispersion.

3.5 BET

The BET-specific surface area measurement method (Brunauer–Emmett–Teller, abbreviated as BET testing) is based on the renowned BET theory and serves as one of the core techniques for characterizing the surface properties of porous materials. Originally developed for surface evaluation of non-microporous adsorbents, such as iron/copper catalysts, silica gel, and charcoal, the method was subsequently extended to evaluate the performance of materials like synthetic zeolites. With the emergence of novel porous materials, including mesoporous silica, porous coordination polymers (PCPs), MOFs, and covalent organic frameworks (COFs), BET testing technology has also achieved significant advancements. These materials demonstrate considerable value in fields such as gas storage and separation, catalysis, sensing, and drug delivery, further promoting the development and application of the BET method. During the evolution of adsorption theory, Langmuir first established the correlation between gas adsorption isotherms and surface area, developing a corresponding model based on the monolayer adsorption hypothesis. Meanwhile, Dubinin focused on assessing the micropore volume of microporous materials. Subsequently, researchers extended the Langmuir theory to multilayer adsorption systems, forming the BET model that more comprehensively describes actual adsorption behaviors. Although the BET theory was not initially designed for microporous adsorption, the BET region has become a universal standard for characterizing various porous materials. The International Union of Pure and Applied Chemistry (IUPAC) has recognized it as "the most widely used method for evaluating the surface area of porous and powdered materials." Since 1995, the BET method has also been adopted by the International Organization for Standardization (ISO) as the standard procedure for specific surface area determination, thus being universally regarded as one of the most crucial analytical tools in porous material characterization [12]. The BET theory, based on the multilayer adsorption model, establishes a mathematical relationship between the monolayer adsorption capacity (Vm) and multilayer adsorption capacity (V), known as the BET equation. This equation can be used to calculate the specific surface area, pore volume, and pore size distribution of materials, and analyze the morphology of nitrogen adsorption–desorption curves, playing a key role in understanding the physicochemical properties of particulate materials. Currently, the BET formula has become the most widely applied and reliable results-oriented specific surface area testing method in the industry, with the vast majority of relevant domestic and international standards established based on the BET equation.

3.5.1 Fundamental Principles

The fundamental principle of BET-specific surface area testing relies on the reversible physical adsorption of gas molecules on solid surfaces. The core concept is that the larger the specific surface area (total surface area per unit mass) of a solid material, the greater is its capacity to adsorb gas molecules. Therefore, by precisely measuring the amount of inert gas (typically nitrogen) adsorbed at a specific temperature (such as liquid nitrogen temperature, −196°C), the specific surface area can be calculated according to the BET theory. During testing, the dynamic chromatography method is generally employed: a mixed gas of nitrogen (adsorbate) and helium (carrier gas) flows through the sample, and when the sample is immersed in liquid nitrogen, adsorption occurs, producing an "adsorption peak" recorded by the detector; after removing the liquid nitrogen, the desorption of adsorbed nitrogen generates a "desorption peak." By comparing the area of the "desorption peak" with that of a "calibration peak" produced by a known volume of pure nitrogen, the amount of adsorption at that relative pressure can be accurately calculated. By measuring the adsorption amounts at a series of different relative pressures and substituting the data into the BET equation, the saturated monolayer adsorption capacity of the material can ultimately be determined, thereby calculating its specific surface area.

Adsorption isotherms graphically depict the relationship between adsorption amount and relative pressure. According to the IUPAC classification, they are mainly divided into six basic types (I to VI), whose shapes directly reveal information about the material's pore structure: Type I: Rapid rise at low pressure reaching a plateau, characteristic of microporous materials (e.g., molecular sieves), indicating micropore filling. Type II: S-shaped, characteristic of non-porous or macroporous materials, with its inflection point corresponding to the completion of monolayer adsorption, making it the most typical case for applying the BET equation. Type III: Concave throughout with no inflection point, indicating that the interaction between adsorbate molecules is stronger than that with the adsorbent. Type IV: Typical of mesoporous materials. Its initial part resembles Type II, but in the medium–high pressure range, the adsorption amount increases sharply due to capillary condensation, often accompanied by an adsorption hysteresis loop. Type V: Similar in shape to Type III, but due to capillary condensation, the adsorption amount rises more rapidly in the medium-pressure region and also exhibits a hysteresis loop. Type VI: A stepped curve, resulting from multilayer adsorption on an ideal uniform non-porous surface, rarely seen in practical materials.

For the hysteresis loops appearing in Type IV and V isotherms, IUPAC further classifies them into four categories (H1–H4) for in-depth analysis of mesoporous structures: H1 type: Steep and parallel, indicating cylindrical mesopores with uniform pore size distribution (e.g., MCM-41). H2 type: Triangular with a steep desorption branch, usually associated with "ink-bottle" pores with nonuniform pore size distribution or poorly connected channels. H3 type: No saturation plateau at high pressure, indicating slit-shaped pores formed by the aggregation of plate-like particles. H4 type: Similar to the H3 type, but with a gentler low-pressure segment and higher adsorption amount, commonly observed in slit-pore materials containing micropores. By comprehensively analyzing the type of adsorption isotherm and the shape of the hysteresis loop, the pore type, pore size distribution, channel connectivity, and geometry of the material can be systematically inferred.

3.5.2 Application of BET Technology in Material Analysis

BET technology enables the precise determination of key structural parameters such as specific surface area, pore volume, and pore size distribution by measuring the nitrogen adsorption–desorption isotherms of materials. These parameters directly influence the adsorption and catalytic performance of materials. For example, in the analysis of N-MgO@Fe_3O_4 nanomaterials [13], BET test results showed a specific surface area of 66 $m^2\ g^{-1}$, and the presence of a Type IV isotherm with a hysteresis loop confirmed the existence of a mesoporous structure. This explains the material's ability to facilitate reactant molecule transport and enhance photocatalytic activity. In another study on $TiZrO_x$-based catalysts, BET isotherms (Type IV with H1 hysteresis loops) indicated that all materials possessed highly ordered tubular mesoporous structures (Figure 3.10a). By comparing the BET data of Co_3O_4@$TiZrO_x$, $CoFeO_x$@$TiZrO_x$, and their precursor $TiZrO_x$ (Figure 3.10b, c) [14], it was found that the specific surface area decreased after loading active components (e.g., from 191.8 to 123.9 $m^2\ g^{-1}$), confirming the successful loading of active substances within the pores. The slight changes in pore size revealed the influence of different preparation methods on the pore structure, which, in turn, correlated with the optimization of catalytic combustion performance. The analysis of δ-MnO_2 and Fe/δ-MnO_2 catalysts [15] also utilized BET technology to clarify their mesoporous characteristics (Figure 3.10d). The results showed that iron doping significantly increased the specific surface area and pore volume of the material, enhancing the contact and adsorption capacity of gaseous pollutants on the surface of spherical nanomaterials, thereby improving degradation efficiency. Furthermore, in the analysis of NiAl-LDH/γ-Fe_2O_3-CdNiFe-LDH composite materials used for wastewater treatment (Figure 3.10e, f) [16], BET testing not only revealed a micro-mesoporous composite structure represented by Type IV isotherms and H4 hysteresis loops but also, through systematic changes in specific surface area (91–212 $m^2\ g^{-1}$) and pore size distribution (2–100 nm), clarified the critical role of ultrathin NiAl-LDH coating in forming slit-like mesopores and increasing active sites. This established an effective correlation between the material's structure and its high-efficiency pollutant removal capability.

Analysis of the structural characteristics of Ni/Zr and a series of Ni-x%Ce/W-Zr catalysts through N_2 adsorption–desorption isotherms (BET testing) demonstrates that BET technology can effectively reveal the significant impact of compositional changes on material texture. As shown in Figure 3.11 [17], the catalysts exhibit Type IV isotherms, consistent with the characteristics of typical mesoporous materials according to the IUPAC classification. Specific BET surface area and pore size distribution data indicate that after introducing WO_3 into the Ni/ZrO_2 catalyst, its specific surface area and total pore volume significantly increase, while the average pore size decreases, proving that WO_3 doping effectively improves the support structure. More importantly, the hysteresis loop observed in the BET tests shifts from Type H4 for the Ni/Zr catalyst to Type H1 for Ni/WZr. This change, clearly revealed by BET analysis, reflects that the introduction of WO_3 transforms the pore structure from slit-like to cylindrical mesopores with a more uniform pore size distribution, thereby facilitating the dispersion of the active Ni component and enhancing its anti-sintering performance. In contrast, although the promotion effect of CeO_2 nearly doubles the pore volume, it does not cause a significant change in the hysteresis loop type, indicating that while it can optimize pore volume, its impact on the overall geometric morphology of the mesoporous structure is limited.

3.6 XPS

X-ray photoelectron spectroscopy (XPS) is a quantitative analytical technique used to determine the elemental composition, chemical states, and electronic states of material surfaces. This technique involves irradiating the sample with X-rays and measuring the kinetic energy and quantity of photoelectrons emitted from the top 1–10 nm of the material surface, thereby obtaining the photoelectron spectrum. XPS must be conducted under ultra-high vacuum conditions and can detect all elements except hydrogen and helium (atomic number $\geq$3), with a detection limit of a few thousandths for most elements, which may reach parts per million under specific conditions. Its primary applications include determining surface elemental composition and empirical formulas, identifying impurities, ascertaining the chemical and electronic states of each element, and analyzing the uniformity of surface element distribution.

3.6.1 Fundamental Principles

XPS operates on the fundamental principle of the photoelectric effect: when a sample is irradiated with X-rays of sufficient energy, inner-shell or valence electrons

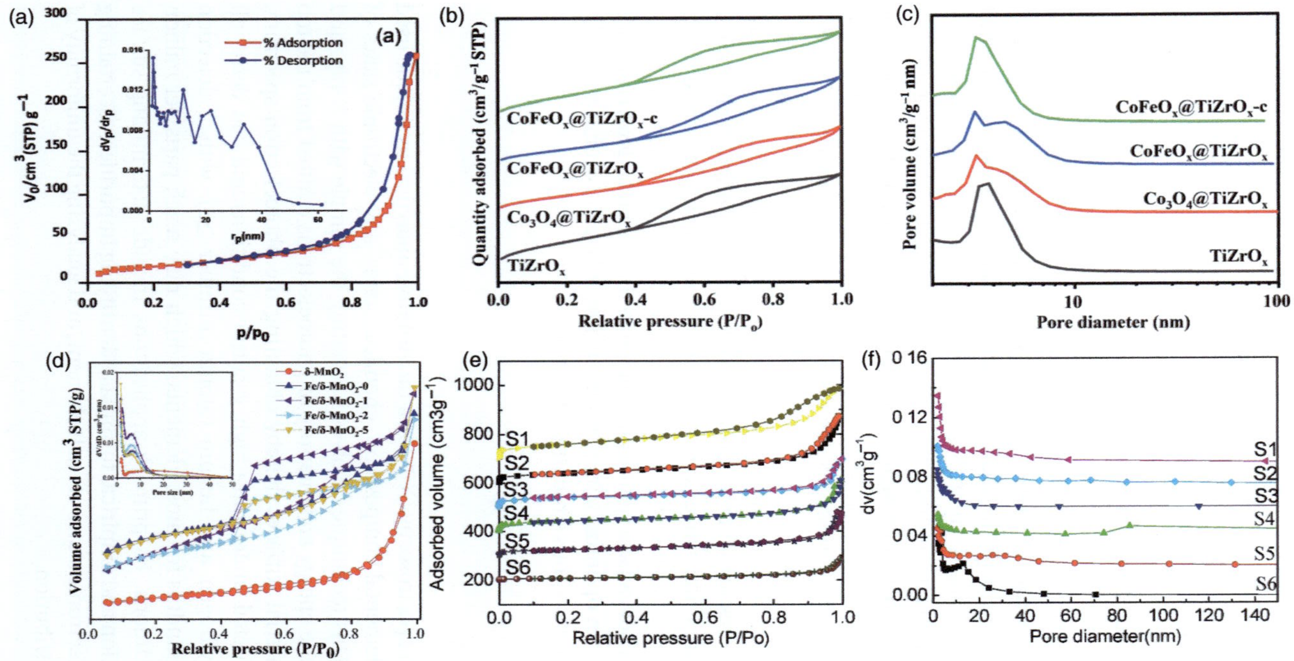

Figure 3.10 (a) BET analysis and corresponding BJH plot of the synthesized N-MgO@Fe_3O_4. *Source:* Reproduced from Ref. [13] / with permission of Elsevier / CC BY 4.0. (b) N_2 isotherms and (c) pore size distribution curves of Co_3O_4@$TiZrO_x$, $CoFeO_x$@$TiZrO_x$ and $CoFeO_x$@$TiZrO_{x\text{-}c}$; *Source:* Reproduced from Ref. [14] / with permission of Elsevier. (d) N_2 adsorption–desorption isotherms and pore size distribution measured for δ-MnO_2 and Fe/δ-MnO_2-X (X = 0, 1, 2, and 5) catalysts; *Source:* Reproduced from Ref. [15] / with permission of Elsevier. (e) Nitrogen adsorption–desorption isotherms; (f) Barrett–Joyner–Halenda (BJH) pore size distribution curves for different Ni^{2+}-Al^{3+}-LDH/γ-Fe_2O_3-Cd^{2+}-Ni^{2+}-Fe^{3+}-LDH products (S1–S6); *Source:* Reproduced from Ref. [16] / with permission of Elsevier.

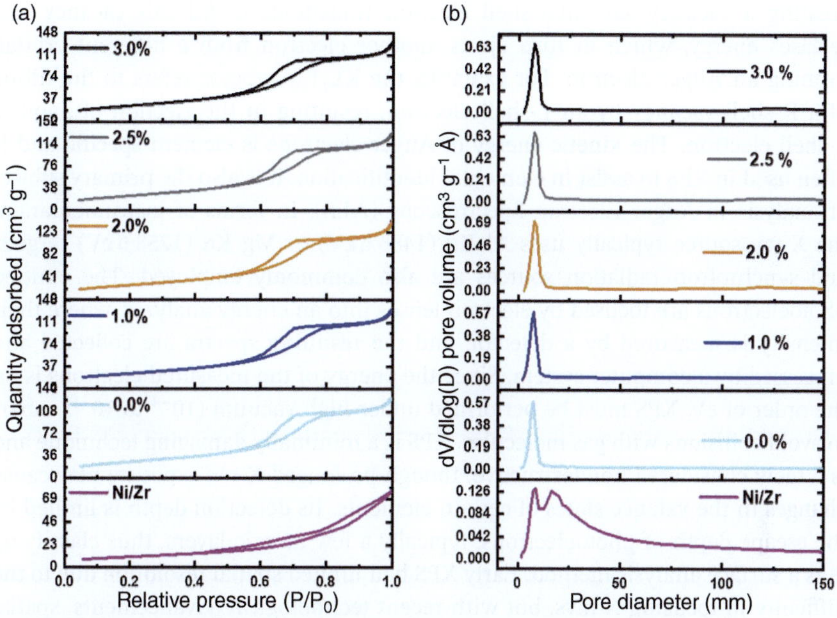

Figure 3.11 (a) N_2 adsorption isotherms of Ni/Zr and Ni-xCe/W-Zr catalysts; (b) corresponding pore size distribution. *Source:* Reproduced from Ref. [17] / with permission of Elsevier.

are emitted as photoelectrons. By measuring the energy distribution of these photoelectrons and plotting a spectrum with binding energy (or kinetic energy) on the horizontal axis and relative intensity on the vertical axis, information about the elemental and chemical states of the sample can be obtained. Electrons in atoms reside in specific orbitals with defined energy levels. When X-rays (energy $h\nu$) irradiate a solid sample (M), they can cause ionization of atoms or molecules, emitting photoelectrons with kinetic energy E_k and generating ions M^+. This process can be represented as:

$$M + h\nu \rightarrow M^+ + e^- \tag{3.3}$$

The emitted electron is referred to as a photoelectron. If the energy of this electron exceeds the vacuum level, it can overcome the surface potential barrier and escape from the material to become a free electron. This process adheres to the energy conservation equation of the Einstein photoelectric emission process:

$$E_k = h\nu - E_B \tag{3.4}$$

where E_B represents the electron binding energy. In practical analysis, using the Fermi level as the reference (where binding energy is zero), the elemental species can be identified by measuring the binding energy. Since the binding energy of an element undergoes slight variations (chemical shifts) depending on its chemical environment, this allows for the determination of the element's chemical valence and bonding state. During the X-ray-induced ionization of core energy levels, the emission of Auger electrons also occurs: after an inner-shell electron is ejected, creating a vacancy, an outer-shell electron transitions to fill this vacancy and releases energy, which in turn ejects another electron from a different orbital, forming an Auger electron. For example, the KL_1L_{23} process refers to the filling of a K-shell vacancy by an L-shell electron, resulting in the ejection of another L-shell electron. The kinetic energy of Auger electrons is element specific and is often used in XPS to assist in elemental identification; it is also the primary subject of analysis in Auger Electron Spectroscopy (AES). In terms of instrumentation, the X-ray source typically uses Al Kα (1486.6 eV) or Mg Kα (1253.6 eV) targets, and synchrotron radiation sources are also commonly employed. The emitted photoelectrons are focused by electron lenses into an energy analyzer, where their intensity is measured by a detector, and the resulting spectra are collected and processed by a computer system. Since the energy of the measured electrons is on the order of eV, XPS must be performed under high vacuum (10^{-8} to 10^{-11} mbar) to avoid collisions with gas molecules. XPS is a minimally damaging technique and is largely considered nondestructive, though prolonged X-ray exposure may cause changes in the valence states of certain elements. Its detection depth is limited by the escape depth of photoelectrons, typically a few atomic layers, thus classifying it as a surface analysis method. Early XPS had limited spatial resolution due to the difficulty in focusing X-rays, but with recent technological advancements, spatial resolution has improved to the micrometer scale, enabling elemental mapping and providing spatial distribution information of chemical states.

3.6.2 Application of XPS Technology in Material Analysis

XPS technology's core advantage lies in its ability to precisely analyze the elemental composition and chemical valence states in the near-surface region of materials (typically 1–10 nm), thereby revealing the chemical state of material surfaces and their roles in reaction processes. Over the past two decades, this technique has been widely applied in research areas such as inorganic/organic component identification, surface chemistry, mineral materials, and functional thin films, enabling qualitative, semi-quantitative, and valence state analysis of surface elements in solid samples. In the study of Co@IC/MoC@PC nanocage catalysts (Figure 3.12a) [18], XPS played a crucial role. The XPS survey spectrum (Figure 3.12b) confirmed the presence of Co, Mo, C, N, and trace oxygen elements on the material surface. Through fitting the Mo 3d high-resolution spectra (Figure 3.12c, d), XPS not only identified multiple oxidation states of Mo ranging from +2 to +6 but also clearly distinguished Mo^{2+}/Mo^{3+} in the MoC component, as well as MoO_2 (Mo^{4+}) and MoO_3 (Mo^{6+}) resulting from surface oxidation. In the Co 2p spectra (Figure 3.12d), XPS successfully identified metallic Co, cobalt oxides, and the critical Co–N structure, the latter being recognized as an important site for enhancing oxygen reduction reaction (ORR) catalytic activity. Furthermore, through analysis of the C 1s and N 1s spectra (Figure 3.12e, f), XPS further revealed the presence of Mo—C bonds in the carbon matrix, various carbon–nitrogen functional groups (such as C=N and C—N), and different nitrogen doping configurations (pyridinic N, graphitic N, etc.), clarifying the role of these nitrogen species in improving electronic conductivity and synergistically enhancing catalytic performance.

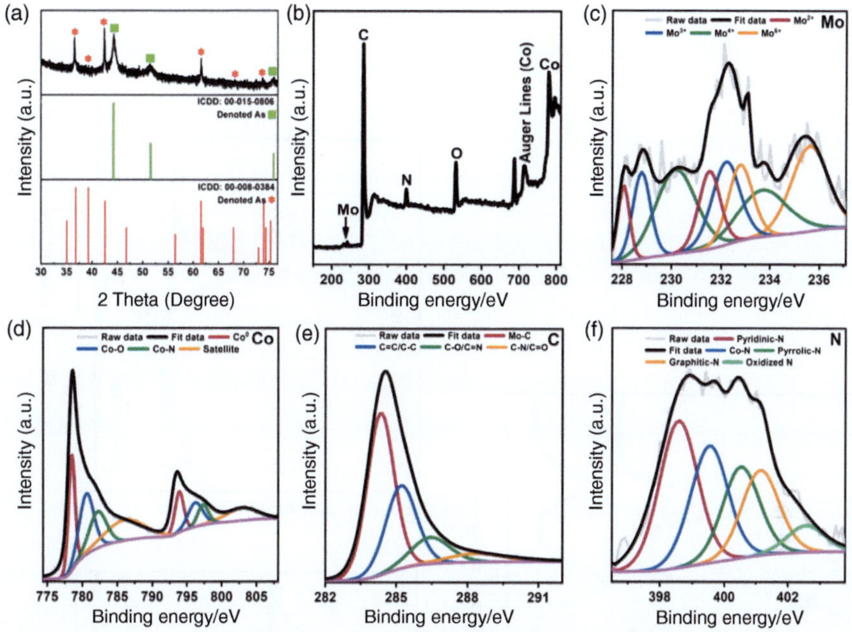

Figure 3.12 (a) X-ray diffraction (XRD) pattern of Co@IC/MoC@PC derived from pyrolysis of ZIF-67@PDA-Mo precursor; (b–f) XPS spectra of Co@IC/MoC@PC obtained from pyrolysis of ZIF-67@PDA-Mo precursor. *Source:* Reproduced from Ref. [18] / with permission of Elsevier.

In another study on NiS/$KNbO_3$ composites, XPS directly confirmed electron transfer at the heterojunction interface through precise measurement of chemical shifts in binding energy. Figure 3.13 displays the XPS spectra of NiS, $KNbO_3$, and 5% NiS/$KNbO_3$ composites [19]. Compared with pure $KNbO_3$, the XPS peaks of K 2p, Nb 3d, and O 1s in the composite material all shift toward higher binding energies, indicating a decrease in electron cloud density in $KNbO_3$. Conversely, the Ni 2p and S 2p peaks in NiS shift toward lower binding energy (Figure 3.13d, e), suggesting an increase in their electron cloud density. This pair of opposing shift trends, clearly captured by XPS analysis, provides strong evidence that electrons transfer from n-type $KNbO_3$ to p-type NiS upon contact, forming a heterojunction structure conducive to the separation of photogenerated charges. XPS analysis not only verified the elemental composition of the materials but, more importantly, revealed strong interactions between the components, demonstrating that the composite functions as a synergistic whole rather than a physical mixture.

3.7 UPS

Ultraviolet photoelectron spectroscopy (UPS) is an important branch of photoelectron spectroscopy technology. Since its development in the 1960s by Turner et al., it has become a key technique for studying the surface electronic states of solid materials. Compared with XPS, UPS uses lower-energy vacuum ultraviolet light as the

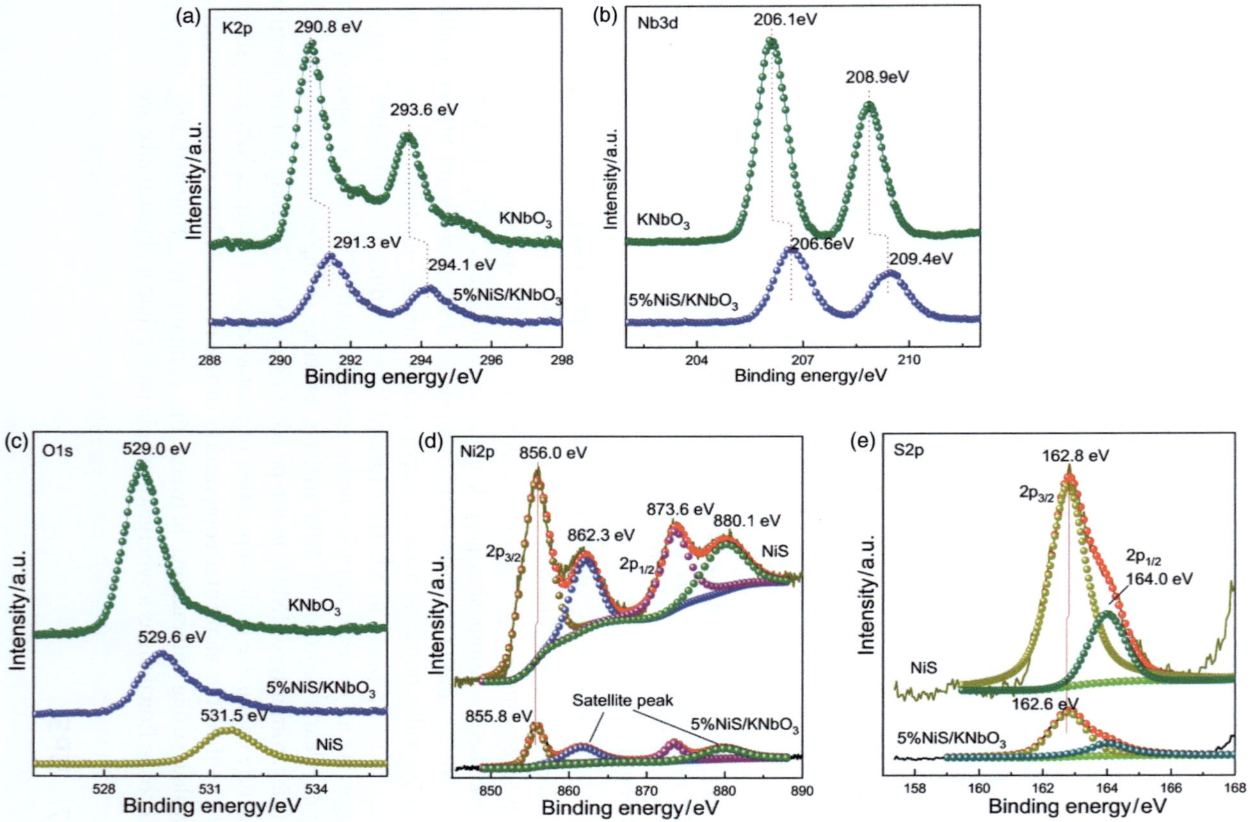

Figure 3.13 XPS spectra of NiS, $KNbO_3$, and 5% NiS/$KNbO_3$ composites: (a) K 2p; (b) Nb 3d; (c) O 1s; (d) Ni 2p; (e) S 2p. *Source:* Reproduced from Ref. [19] / with permission of Elsevier.

excitation source, enabling it to probe the fine structure of valence electron energy levels and molecular vibrational levels with high-energy resolution (10–20 meV), making it an effective method for revealing the valence electronic structure of materials. The work function is an important physical parameter of materials, highly sensitive to surface property changes and capable of reflecting conductor interface energy level alignment and carrier injection barriers, making it crucial for device performance characterization. Among various methods for measuring work function, UPS is an absolute measurement technique that can directly determine the work function of a sample. Its measurement principle is the same as that of XPS: both operate in an ultra-high vacuum environment, irradiate the sample surface with a specific energy light source to excite photoelectrons, and obtain spectra by measuring the kinetic energy distribution of these photoelectrons. By analyzing the position of the secondary electron cutoff at the low kinetic energy end of the spectrum, the work function of the sample can be determined. UPS is more commonly used than XPS for work function measurements. Due to the lower excitation energy of UPS, the emitted photoelectrons mainly originate from valence electrons, making it rarely used for elemental quantitative analysis. However, since valence electrons are directly involved in chemical bonding, UPS is particularly suitable for studying the bonding interactions in substances. Additionally, UPS can provide information on the work function of solids, the density of states (DOS) distribution, and the band structure.

3.7.1 Fundamental Principles

The fundamental principle of UPS is consistent with that of XPS, as both are based on Einstein's law of the photoelectric effect. For free molecules and atoms, the energy relationship follows the formula:

$$EK = h\upsilon - EB - \Phi_{sp} \tag{3.5}$$

where E_K represents the measured kinetic energy of the photoelectron, $h\upsilon$ represents the known incident photon energy, E_B represents the binding energy of the electron to be determined, and Φ_{sp} represents the known work function of the spectrometer (typically around 4 eV). Since the excitation sources used in UPS (such as the He I line with a photon energy of 21.22 eV) possess significantly lower energy compared to X-rays, the photoelectron signals originate exclusively from the extremely shallow surface of the sample (approximately 10 Å). These signals primarily reflect information about valence electrons near the Fermi level, making UPS an ideal tool for probing valence electron structures. An ideal UPS excitation source should generate monochromatic, high-intensity radiation. The most commonly used sources today are inert gas (e.g., helium) discharge lamps. Under ultra-high vacuum conditions (approximately 10^{-8} mbar), these lamps produce plasma via DC or microwave discharge, primarily emitting the He I resonance line (wavelength 584 Å, photon energy 21.22 eV) and the He II resonance line (wavelength 304 Å, photon energy 40.8 eV). Among these, the He I line is the most frequently used excitation source in UPS due to its excellent monochromaticity (natural line width of about 5 meV), high intensity, and low continuous background.

3.7.2 Application of UPS Technology in Material Analysis

UPS, by precisely measuring the energy distribution of valence electrons in materials, enables direct acquisition of key electronic properties such as valence band structure, work function, ionization energy, and DOS. It plays an indispensable role in elucidating the surface electronic structure of iron-group metal (Fe, Co, and Ni) compound nanomaterials and their structure–property relationships. In the study of NiO_x thin films as hole transport layers in perovskite solar cells [20], combined analysis using UPS and XPS revealed their interface modification mechanism. Through UPS testing of different sample batches over several months (Figure 3.14a, b), it was found that the average work function of as-deposited NiO_x was 4.1 eV, with an ionization energy onset of 5.2 eV (Figure 3.14c). After ultraviolet ozone (UVO) treatment, UPS measurements showed a significant increase in work function to 4.8 eV, and ionization energy also increased to 5.4 eV. Combined with valence band spectrum analysis, this confirmed the formation of intermediate gap states. These UPS data were directly correlated with the increase in surface Ni^{3+} sites, and the estimated DOS indicated an increased number of holes at the valence band (Figure 3.14d–f). Ultimately, the UPS analysis results directly linked the increase in Ni^{3+} sites with the formation of a hole-rich extraction barrier ($PbI_{2-x}Br_x$) at the interface, revealing the critical role of Ni^{3+} as catalytic sites in promoting interfacial hole extraction.

In the study of water molecule interactions with iron thin films [21], the combined use of UPS and metastable induced electron spectroscopy (MIES) (Figure 3.15a) provided dynamic evolution information of surface electronic states during the adsorption process. UPS spectra (Figure 3.15b) revealed significant Fe 3d electron emission near the Fermi level in clean iron films. As water exposure increased, this characteristic signal weakened, and an emission peak originating from the O 2p orbitals of adsorbed water emerged at 5.6 eV. Even under extremely high water exposure, UPS still detected residual metallic states, indicating that the surface reaction did not completely cover the metallic characteristics (Figure 3.15c). The mutual confirmation between UPS and MIES data clarified the electronic structure features of water dissociation, generating OH species on the iron surface, deepening the understanding of the initial oxidation and corrosion mechanisms of iron-group metals.

For the two-dimensional conductive MOF materials $M_3(HIB)_2$ (M = Ni, Cu) [22], UPS technology provided direct experimental evidence for their metallic nature. Through UPS measurements on $Ni_3(HIB)_2$ and $Cu_3(HIB)_2$ thin films prepared on highly doped silicon substrates (Figure 3.15d, e), clear electron emission (Fermi edge) was observed at the Fermi level. This directly demonstrated a nonzero density of electronic states at the Fermi level, serving as definitive evidence of metallic behavior with energy bands crossing the Fermi level. This UPS characterization result, combined with structural analysis, confirmed that these two-dimensional MOFs constructed from Ni are intrinsic metallic conductors, laying a theoretical foundation for their applications in electronic devices.

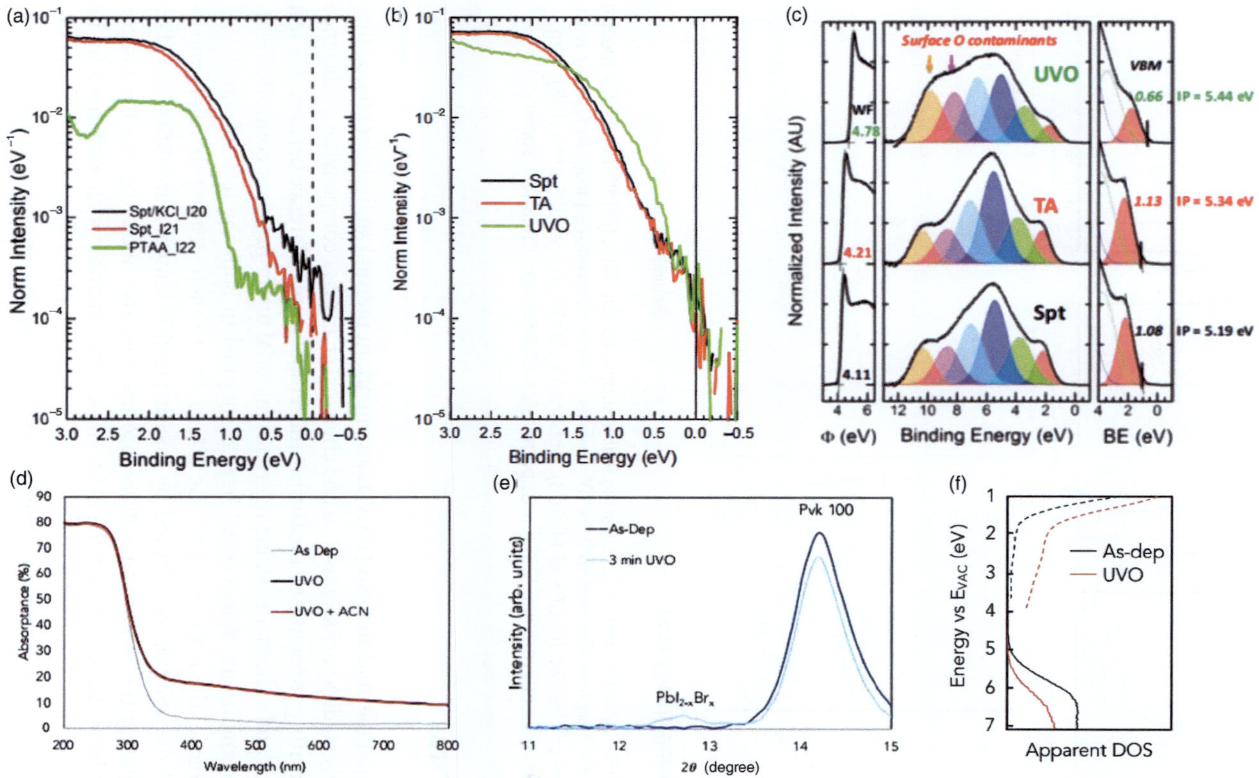

Figure 3.14 Valence band edges of sputtered $PbI_{2-x}Br_x$ on ITO: (a) at the beginning of this study and (b) several months later at the end of this study; (c) UPS spectra of as-deposited sputtered NiO_x films (Spt), after thermal annealing in air at 300°C for 1 hour (TA), and after UV-ozone cleaning for 5 minutes (UVO); (d) absorption spectra of glass slides deposited with sputtered NiO_x layers, treated with 5-minute UV-ozone (UVO), and after 5-minute UV-ozone treatment followed by water washing; (e) XRD patterns of perovskite films on NiO_x layers with and without 3-minute UVO treatment. The content of $PbI_{2-x}Br_x$ increases in perovskite deposited on UVO-treated NiO_x; (f) estimated DOS of as-deposited (as-dep) and UVO-treated NiO_x. *Source:* Reproduced from Ref. [20] / with permission of Elsevier.

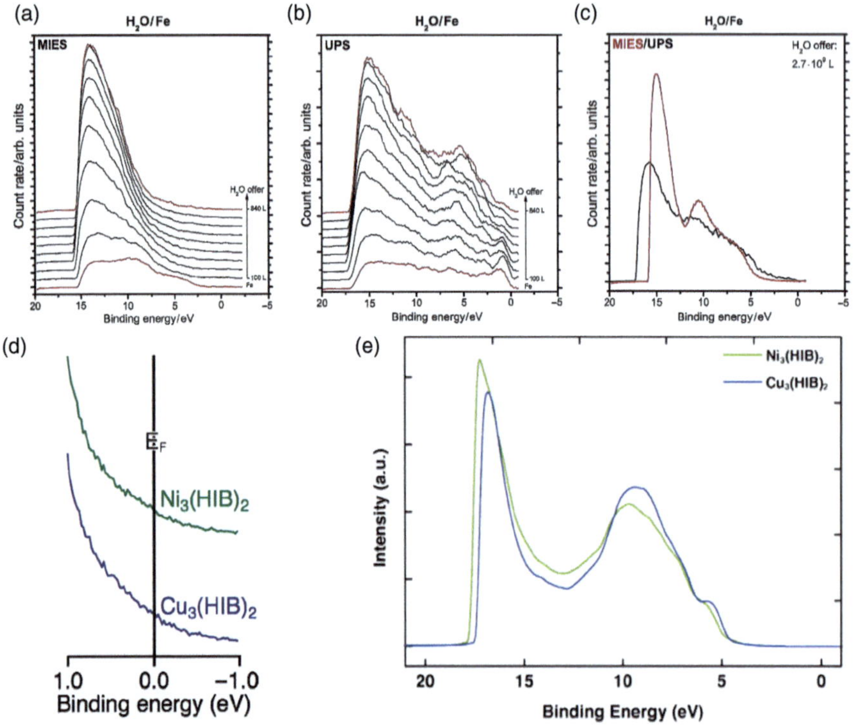

Figure 3.15 (a) MIES and (b) UPS during water supply to iron film; (c) MIES and UPS spectra provided for iron film exposed to 2.7×10^9 L water; *Source:* Reproduced from Ref. [21] / with permission of Elsevier. (d) UPS of $M_3(HIB)_2$ obtained at 300K and (e) UPS of $M_3(HIB)_2$ film; *Source:* Reproduced from Ref. [22] / with permission of American Chemical Society.

3.8 DTA

Differential thermal analysis (DTA) is an important thermal analysis technique that accurately studies various physical and chemical changes accompanying a substance during heating or cooling by measuring the temperature difference between the test substance and a thermally stable reference material under program-controlled temperature conditions. When equal heat is supplied to both the test substance and the reference, their temperature rise differs due to their distinct thermal properties, and the analysis is achieved by measuring this temperature difference. The curve plotting the temperature difference between the reference and the sample against temperature is called the DTA curve. This method is characterized by its simple operation, minimal sample requirement, and broad applicability, enabling sensitive detection of minor thermal effects resulting from processes such as phase transitions, decomposition, oxidation, or lattice structure changes at specific temperatures. Consequently, it has become an indispensable routine testing technique in numerous fields, including materials science, chemistry, mineralogy, and pharmaceutical research.

3.8.1 Fundamental Principles

DTA operates on the principle of placing the sample and a stable reference material—which undergoes no chemical or physical changes within the experimental temperature range—together in a temperature-controlled furnace, followed by programmed heating or cooling at a set rate. When the sample undergoes no changes at a given temperature, its temperature remains identical to that of the reference, resulting in a temperature difference (ΔT) of zero, and the recorder traces a smooth "baseline". However, when the sample undergoes physical changes or chemical reactions within a certain temperature range, thermal effects accompany these processes. If the reaction is endothermic, the sample temperature temporarily drops below that of the reference; if exothermic, the sample temperature temporarily rises above the reference temperature. This slight temperature difference is detected by a differential thermocouple connected to the bottoms of the sample and reference containers. The differential thermocouple converts the temperature difference into a small thermoelectric potential difference, which is then amplified and captured by the recording system. On the recorded plot—the DTA curve—endothermic and exothermic effects cause the curve to deviate from the baseline in opposite directions, producing characteristic peaks or troughs. Once the reaction concludes, the temperature difference dissipates, and the curve returns to the baseline. The DTA curve plots temperature (or time) on the horizontal axis and the temperature difference ΔT on the vertical axis. The position of a peak (or trough) on the curve corresponds to the temperature at which the thermal effect occurs; the area under the peak relates to the reaction enthalpy, while the shape of the peak is associated with the kinetic process of the reaction. Therefore, by analyzing the peak temperature, peak area, and peak shape of the DTA curve, one can not only study the properties of a substance but also identify unknown materials. To realize the above principles, a DTA instrument primarily consists of a heating furnace with temperature control, a sample holder and sample cells (containing the sample and reference), a temperature difference measurement circuit, an atmosphere control system, and a data recording and output system. To ensure accurate experimental results, the sample amount should be as small as possible, and both particle size and packing method must be uniform to minimize temperature differences between the interior and surface of the sample. Additionally, the position of the thermocouple must be precisely fixed in each experiment.

3.8.2 Application of DTA Technology in Material Analysis

DTA provides crucial information for studying the thermal stability, phase transitions, and decomposition behavior of iron-group metal (Fe, Co, and Ni) compound nanomaterials by precisely detecting thermal effects during programmed heating. In the study of cobalt oxide (CoO_x) precursors [23], the combined use of DTA and thermogravimetric analysis (TGA) revealed the decisive influence of precipitation pH on the thermal decomposition behavior of the precursors. As shown in Figure 3.16a, b, the DTA curves of precursors prepared at low pH (7.5–8.5) exhibited endothermic peaks, whereas those prepared at high pH (9.0–10.0) showed exothermic peaks. This significant difference in DTA results clearly indicated the formation of two distinct cobalt oxide precursors with markedly different thermal behaviors under varying pH conditions. Simultaneously, DTA analysis confirmed that 500°C is a suitable calcination temperature to ensure complete decomposition into stable cobalt oxides. For Fe-doped TiO_2 materials [24], DTA technology effectively discerned complex thermal transition processes. As shown in Figure 3.16c–e, the DTA curves detected an exothermic peak around 550°C, attributed to the superposition of two processes: combustion of residual carbon and the phase transition of TiO_2 from anatase to rutile. By analyzing differences in DTA peak temperatures and TGA weight loss rates among samples prepared with different additives (e.g., F-127 and K-30), it was confirmed that the additives effectively adhered to the precursor surface and influenced its thermal decomposition and phase transition pathways. In the analysis of laser-irradiated $NaFePO_4$ materials [25], combined DTA and TGA analysis (Figure 3.16f) revealed their high-temperature behavior. Within the range of 300–550°C, the DTA curve showed two exothermic peaks at 370°C and 495°C, accompanied by an increase in sample mass. DTA analysis attributed this thermal effect to the oxidation of Fe^{2+} to Fe^{3+} in the material. Additionally, an endothermic peak observed around 910°C was unequivocally identified by DTA results as the melting of $NaFePO_4$, providing important evidence for estimating the actual temperature during laser processing.

In the study of the thermal properties of PVDF polymer influenced by Co_3O_4 and its composite fillers with functionalized MWCNTs (f-MWCNTs) [26], DTA and DSC analyses indicated that the incorporation of nanofillers increased both the decomposition peak temperature and melting temperature (T_m) of PVDF (Figure 3.17a–c). DTA data revealed that as the concentration of Co_3O_4/f-MWCNTs fillers increased, the thermal decomposition peak shifted toward higher temperatures, demonstrating that the nanofillers significantly enhanced the thermal stability of the PVDF-based nanocomposites and promoted the transition of the polymer from the α-phase to the β-phase. During the evaluation of the thermal behavior of synthesized α-Fe_2O_3 (hematite) nanoparticles (C-1 NPs) [27], the DTA/DTG (Derivative Thermogravimetry) curves (Figure 3.17d, e) clearly recorded multiple thermal processes. Analysis identified an endothermic peak at 434.63°C, corresponding to the transition of the amorphous precursor into the single α-Fe_2O_3 phase. This temperature was determined as the optimal calcination temperature for forming pure-phase α-Fe_2O_3. Subsequently, a broad endothermic peak at 743.23°C marked the completion of thermal decomposition and the formation of stable α-Fe_2O_3 nanoparticles. In this context, DTA technology successfully elucidated the complete thermal decomposition pathway and phase transition sequence from the precursor to the final product.

3.9 EPR

Electron paramagnetic resonance (EPR) spectroscopy is a modern analytical technique specifically designed for detecting substances containing unpaired

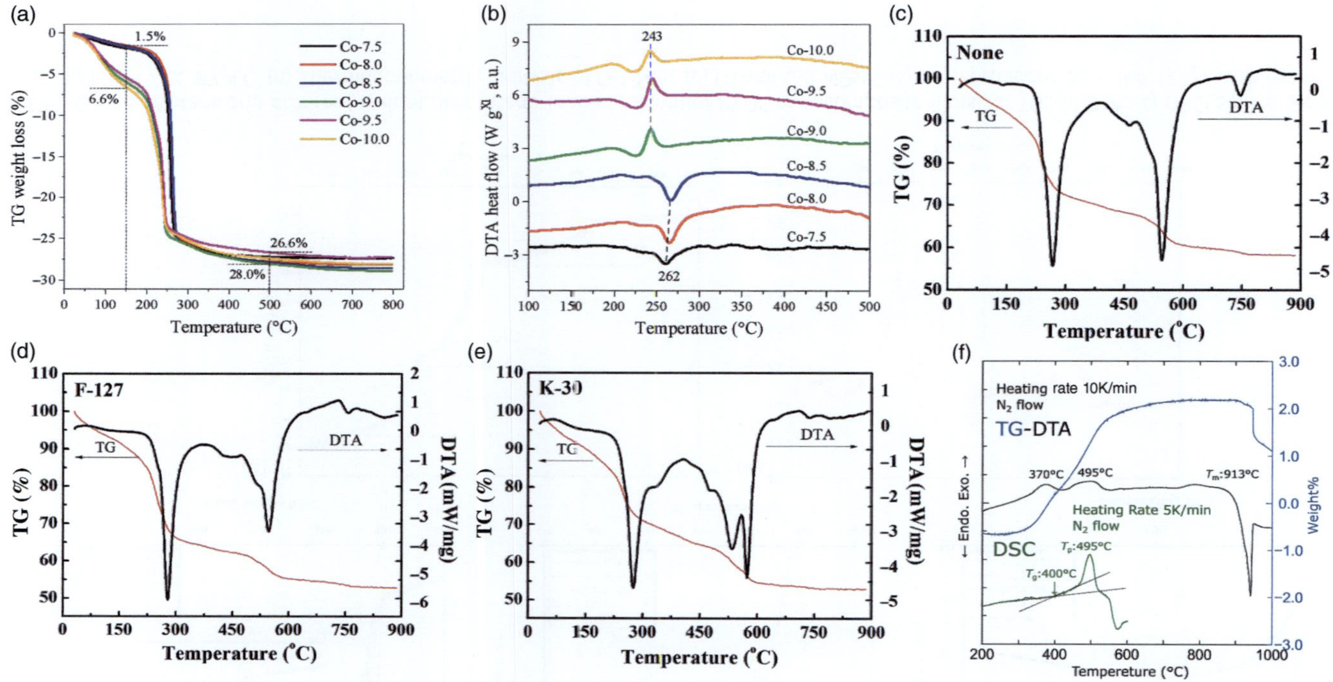

Figure 3.16 (a, b) TG/DTA curves of cobalt oxide precursors precipitated at different pH values during decomposition in flowing air (heating rate: 10°C min^{-1}); *Source:* Reproduced from Ref. [23] / with permission of Elsevier. TG-DTA curves of Fe-doped TiO_2 precursors: (c) prepared without additive; (d) prepared with F-127; (e) prepared with K-30; *Source:* Reproduced from Ref. [24] / with permission of Elsevier. (f) TG-DTA and DSC analysis of laser-induced $NaFePO_4$; *Source:* Reproduced from Ref. [25] / with permission of Elsevier.

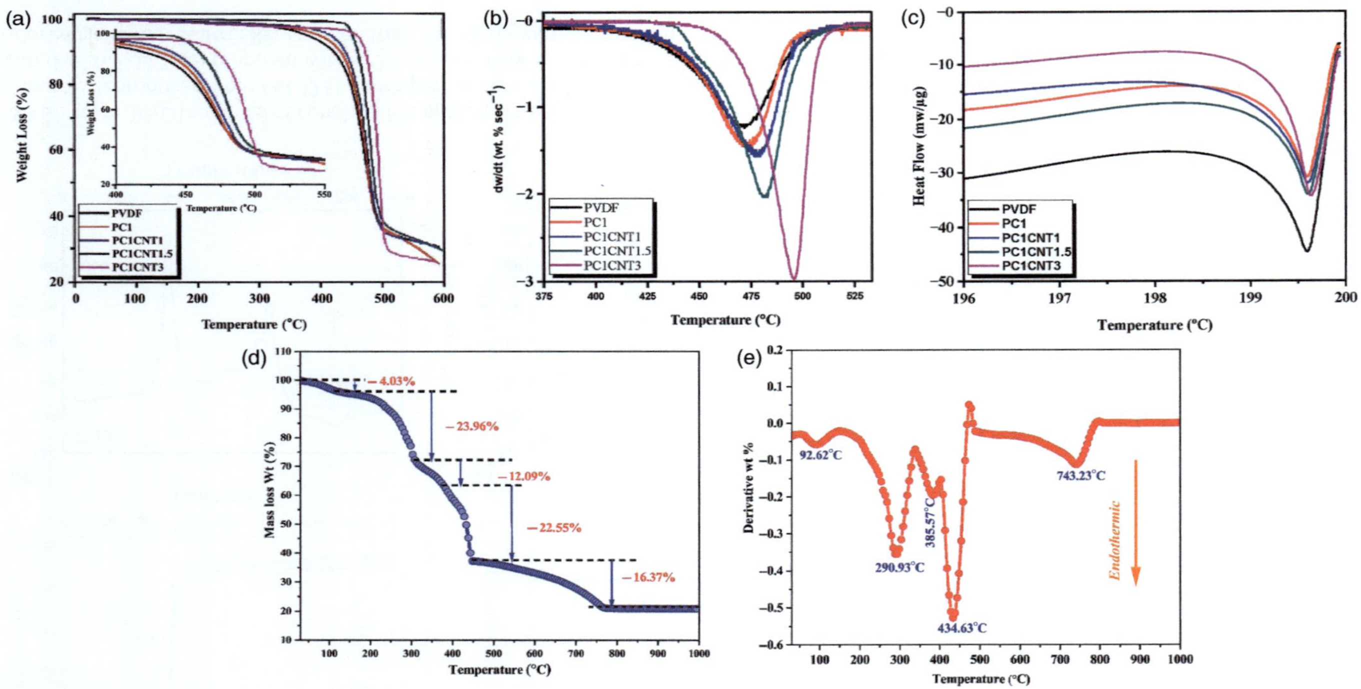

Figure 3.17 Curves of blank and different weight percentage-loaded nanofiller PVDF nanocomposite films: (a) TG; (b) DTA; (c) DSC; *Source:* Reproduced from Ref. [26] / MDPI / CC BY 4.0. (d) Thermogravimetric analysis of C-1 NPs; (e) DTA of C-1 NPs; *Source:* Reproduced from Ref. [27] / with permission of Elsevier.

electrons. This technology can nondestructively obtain structural characteristics and dynamic information during chemical or physical changes in substances, making it an important tool for characterizing the properties of high-tech materials and serving as an ideal complementary technique widely applied in physics, chemistry, and many other fields. EPR was initially used by physicists to study atomic electronic structures, crystal defects, and other issues, and was later adopted by chemists and biologists to elucidate the chemical bond characteristics of organic compounds, electron density distribution, and free radicals in biological systems.

3.9.1 Fundamental Principles

EPR spectroscopy primarily studies free radicals, diradicals or polyradicals, triplet-state molecules, transition metal ions, rare-earth ions, and lattice defects in solid materials. Its core applications include direct detection and analysis of radical intermediates, trapping of transient radicals, structural elucidation of paramagnetic ion complexes, and related monitoring in industrial processes. The paramagnetism of substances originates from the spin magnetic moment of unpaired electrons. In molecular orbitals, the magnetic moments of paired electrons cancel each other, while substances with unpaired electrons possess a permanent magnetic moment, exhibiting paramagnetism in an external magnetic field. The magnetic moment generated by electron spin can be expressed as $\mu = g_e\beta$, where β is the Bohr magneton and g_e is the g-factor (for free electrons, $g_e \approx 2.0023$). When placed in an external magnetic field H, the electron spin energy levels undergo Zeeman splitting, forming two energy states: $E = \pm\frac{1}{2}g_e\beta H$, with an energy level difference of $\Delta E = g_e\beta H$. If electromagnetic radiation with frequency $h\nu$ is applied perpendicular to H, resonance occurs when the following condition is met:

$$h\upsilon = g_e\beta H \tag{3.6}$$

Electrons between the two energy levels undergo stimulated transitions, causing some electrons in the lower energy level to absorb energy from the electromagnetic waves and transition to the higher energy level, thereby generating the EPR phenomenon. The absorption signal produced by stimulated transitions is processed by an electronic system to obtain the EPR absorption spectrum. The absorption signal recorded by an EPR spectrometer is typically a first-derivative lineshape, also referred to as a first-derivative spectrum. The g-value can be calculated using the following formula:

$$g = \frac{h\nu}{\beta H} = 0.07145\frac{v}{H(T)} \tag{3.7}$$

In this equation, the value of **H** corresponds to the peak maximum of the absorption curve, which is the magnetic field value at the midpoint between the peak and trough in the derivative curve. This allows for the calculation of the g-factor. The g-factor can then be used to infer the chemical environment of the tested element's atoms and the electronic state.

3.9.2 Application of EPR Technology in Material Analysis

EPR technology, by directly detecting unpaired electrons in materials, provides unique and crucial information for elucidating the electronic structure, defects, and reaction mechanisms of iron-group metal (Fe, Co, and Ni) compound nanomaterials. In the analysis of NiFe-based layered double hydroxide (LDH) electrocatalysts, EPR was directly employed to reveal the electronic structure and stability of their metal centers. As shown in Figure 3.18a [28], through X-band EPR tests on NiFe-3:1, NiCr-2:1, and trimetallic NiFeCr-6:2:1 LDHs, a strong signal at $g = 2.0$ was observed in the NiFe-3:1 sample, which was identified as characteristic of high-spin Fe^{3+} ($3d^5$, $S = 5/2$) in an octahedral field. In contrast, the weak signal at $g = 2.2$ in the NiCr-2:1 sample was attributed to Cr^{3+} ($3d^3$, $S = 3/2$). For the trimetallic NiFeCr-6:2:1 sample, signals from both Fe^{3+} and Cr^{3+} were identified in the EPR spectrum, but the Fe^{3+} signal intensity was significantly weaker than that in the bimetallic NiFe-3:1 sample. This directly demonstrates that Cr incorporation markedly modulates the local electronic structure of NiFe LDH. More importantly, after the oxygen evolution reaction (OER) test, the Cr^{3+} signal remained clearly visible in the EPR spectrum of the NiFeCr-6:2:1 sample, providing direct electronic-level evidence for the stability of active sites under harsh reaction conditions. In the study of $NiSe_2$-based nanomaterials, EPR served as a powerful tool for quantitatively characterizing the concentration of selenium (Se) vacancies. As shown in Figure 3.18b [29], all samples exhibited EPR signals at $g \approx 2.007$, with signal intensity directly correlated to the concentration of unpaired electrons (i.e., defect density) in the material. The signal from pristine $NiSe_2$ was extremely weak, indicating negligible intrinsic Se vacancies. In contrast, the Ar plasma-treated A-$NiSe_2$ sample showed a significantly enhanced signal, confirming the successful introduction of Se vacancies via plasma etching. Among them, the P-doped A-$NiSe_2$|P sample displayed the strongest EPR signal, attributed to the combined effects of Ar plasma etching and P-doping-induced lattice distortion, resulting in the highest concentration of Se vacancies. These EPR analysis results corroborated findings from XRD and Raman spectroscopy, collectively revealing the microscopic mechanism of optimizing the electrocatalytic performance of materials through defect engineering.

For the Mo-Co_9Se_8@NiSe/NF composite material, EPR technology was employed to systematically investigate the relationship between hydrogen peroxide (H_2O_2) etching time and the concentration of Se vacancies. As shown in Figure 3.18c [30], the unetched pristine sample exhibited only a weak EPR signal, indicating an almost negligible presence of Se vacancies. As the etching time in H_2O_2 solution increased to 40 seconds and 60 seconds, the EPR signal intensity significantly enhanced, directly demonstrating that the etching process effectively generated a large number of Se vacancies with unpaired electrons. This EPR analysis quantitatively confirmed that H_2O_2 etching is an effective strategy for introducing defects and revealed a positive correlation between defect concentration and etching time, providing a basis for precise control of material defect density. Furthermore, EPR combined with spin trapping technology played an indispensable role in elucidating the reaction mechanism of the advanced oxidation process involving $CoFe_2O_4$-based

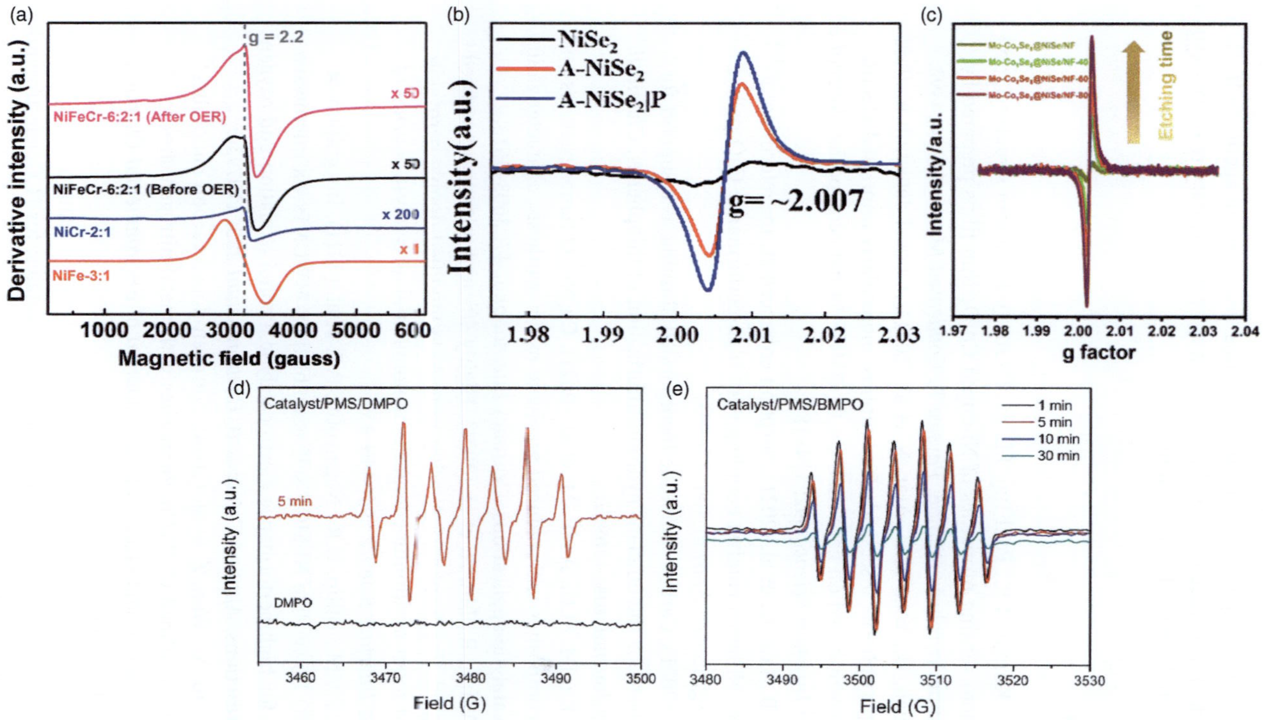

Figure 3.18 (a) X-band EPR spectra of NiFe-3:1, NiCr-2:1, and NiFeCr-6:2:1 LDH samples (before and after OER). The EPR spectra of NiCr-2:1 and NiFeCr-6:2:1 samples are magnified by factors of 200 and 50, respectively, for comparison with the NiFe-3:1 spectrum; *Source:* Reproduced from Ref. [29] / with permission of John Wiley & Sons. (b) EPR patterns of $NiSe_2$, A-$NiSe_2$, and A-$NiSe_2$|P. *Source:* Reproduced from Ref. [29] / with permission of Elsevier. (c) EPR spectra of Mo-Co_9Se_8@NiSe/NF with different etching times compared to pristine Mo-Co_9Se_8@NiSe/NF; *Source:* Reproduced from Ref. [30] / with permission of The Royal Society of Chemistry. (d, e) EPR signals of (d) DMPO and (e) BMPO adducts during the Co_3O_4@Fe_2O_3/PMS process; *Source:* Reproduced from Ref. [31] / with permission of Elsevier.

catalysts activating peroxymonosulfate (PMS). As shown in Figure 3.18d, e [31], in experiments using 5,5-dimethyl-1-pyrroline N-oxide (DMPO) as the trapping agent, a characteristic seven-line peak of 5,5-dimethyl-2-oxo-pyrroline-1-oxyl radical (DMPOX) was observed, indicating the generation of highly reactive oxygen species with strong oxidative capacity sufficient to oxidize the trapping agent itself. Here, EPR analysis clearly revealed the evolution pathway of the primary reactive oxygen species in this catalytic oxidation system.

References

1 Lavanya, S., Kumar, T.R., Poul Raj, I.L. et al. (2023). Enhanced structural, optical, and photo sensing properties of Ni-doped Co_3O_4 thin films prepared by nebulizer spray pyrolysis method. *Physica B: Condensed Matter* 649: 414492.

2 Karimi-Maleh, H., Yola, M.L., Atar, N. et al. (2021). A novel detection method for organophosphorus insecticide fenamiphos: molecularly imprinted electrochemical sensor based on core-shell Co_3O_4@MOF-74 nanocomposite. *Journal of Colloid and Interface Science* 592: 174–185.

3 Xu, Y., Lu, B., Luo, C. et al. (2021). Sorption enhanced steam reforming of ethanol over Ni-based catalyst coupling with high-performance CaO pellets. *Chemical Engineering Journal* 406: 126903.

4 Saleh, T.A. (2021). Carbon nanotube-incorporated alumina as a support for MoNi catalysts for the efficient hydrodesulfurization of thiophenes. *Chemical Engineering Journal* 404: 126987.

5 Bashir, B., Khalid, M.U., Aadil, M. et al. (2021). $Cu_xNi_{1-x}O$ nanostructures and their nanocomposites with reduced graphene oxide: synthesis, characterization, and photocatalytic applications. *Ceramics International* 47: 3603–3613.

6 Wang, Y., Wang, R., Lin, N. et al. (2021). Highly efficient microwave-assisted Fenton degradation bisphenol A using iron oxide modified double perovskite intercalated montmorillonite composite nanomaterial as catalyst. *Journal of Colloid and Interface Science* 594: 446–459.

7 Sreekanth, T.V.M., Dillip, G.R., Nagajyothi, P.C. et al. (2021). Integration of Marigold 3D flower-like Ni-MOF self-assembled on MWCNTs via microwave irradiation for high-performance electrocatalytic alcohol oxidation and oxygen evolution reactions. *Applied Catalysis B: Environmental* 285: 119793.

8 Zhang, Y., Liu, X., Wan, Z. et al. (2024). Nitrogen-doped carbon nanotubes/nanoparticles confined Co/FeCo composites with metal-nitrogen sites for efficient multifunctional electrocatalysis. *Journal of Environmental Chemical Engineering* 12: 114326.

9 Zhang, L., Liang, J., Yue, L. et al. (2022). N-doped carbon nanotubes supported $CoSe_2$ nanoparticles: a highly efficient and stable catalyst for H_2O_2 electrosynthesis in acidic media. *Nano Res* 15: 304–309.

10 Wang, P., Wang, J., An, X. et al. (2021). Generation of abundant defects in Mn-Co mixed oxides by a facile agar-gel method for highly efficient catalysis of total toluene oxidation. *Applied Catalysis B: Environmental* 282: 119560.

11 Hongmanorom, P., Ashok, J., Zhang, G. et al. (2021). Enhanced performance and selectivity of CO_2 methanation over phyllosilicate structure derived Ni-Mg/SBA-15 catalysts. *Applied Catalysis B: Environmental* 282: 119564.

12 Osterrieth, J.W.M., Rampersad, J., Madden, D. et al. (2022). How reproducible are surface areas calculated from the BET equation? *Advanced Materials* 34: e2201502.

13 Akbari, S., Moussavi, G., Decker, J. et al. (2022). Superior visible light-mediated catalytic activity of a novel N-doped, Fe_3O_4-incorporating MgO nanosheet in presence of PMS: imidacloprid degradation and implications on simultaneous bacterial inactivation. *Applied Catalysis B: Environmental* 317: 121732.

14 Wu, S., Zhao, H., Ling, W. et al. (2022). In situ self-assembly encapsulation of $CoFeO_x$ nanoparticles in ordered mesoporous $TiZrO_x$ channels for enhanced catalytic combustion of o-dichlorobenzene. *Fuel* 311: 122496.

15 Wu, Y., Guo, Q., Liu, H. et al. (2022). Effect of Fe doping on the surface properties of δ-MnO_2 nanomaterials and its decomposition of formaldehyde at room temperature. *Journal of Environmental Chemical Engineering* 10: 108277.

16 Dang, R., Ma, X.-R., Zhao, Y.-P. et al. (2023). Construction of 3D Ni^{2+}-Al^{3+}-LDH/γ-Fe_2O_3-Cd^{2+}-Ni^{2+}-Fe^{3+}-LDH structures and multistage recycling treatment of dye and fluoride-containing wastewater. *Chemical Engineering Journal* 451: 138499.

17 Al-Fatesh, A.S., Arafat, Y., Kasim, S.O. et al. (2021). In situ auto-gasification of coke deposits over a novel Ni-Ce/W-Zr catalyst by sequential generation of oxygen vacancies for remarkably stable syngas production via CO_2-reforming of methane. *Applied Catalysis B: Environmental* 280: 119445.

18 Zhang, L., Zhu, Y., Nie, Z. et al. (2021). Co/MoC nanoparticles embedded in carbon nanoboxes as robust trifunctional electrocatalysts for a Zn-air battery and water electrocatalysis. *ACS Nano* 15: 13399–13414.

19 Zhang, W., Xing, P., Zhang, J. et al. (2021). Facile preparation of novel nickel sulfide modified KNbO(3) heterojunction composite and its enhanced performance in photocatalytic nitrogen fixation. *Journal of Colloid and Interface Science* 590: 548–560.

20 Boyd, C.C., Shallcross, R.C., Moot, T. et al. (2020). Overcoming redox reactions at perovskite-nickel oxide interfaces to boost voltages in perovskite solar cells. *Joule* 4: 1759–1775.

21 Volgmann, K., Voigts, F., and Maus-Friedrichs, W. (2012). The interaction of H_2O molecules with iron films studied with MIES, UPS and XPS. *Surface Science* 606: 858–864.

22 Dou, J.-H., Sun, L., Ge, Y. et al. (2017). Signature of metallic behavior in the metal–organic frameworks M_3(hexaiminobenzene)$_2$ (M = Ni, Cu). *Journal of the American Chemical Society* 139: 13608–13611.

23 Zhang, W., Lassen, K., Descorme, C. et al. (2021). Effect of the precipitation pH on the characteristics and performance of Co_3O_4 catalysts in the total oxidation of toluene and propane. *Applied Catalysis B: Environmental* 282: 119566.

24 Wang, J., Li, X., Ren, Y. et al. (2021). The effects of additive on properties of Fe doped TiO_2 nanoparticles by modified sol-gel method. *Journal of Alloys and Compounds* 858: 157726.

25 Hiratsuka, M., Honma, T., and Komatsu, T. (2021). Vitrification of maricite $NaFePO_4$ crystal by laser irradiation and enhanced sodium ion battery performance. *Journal of Alloys and Compounds* 885: 160928.

26 Farhan, Ahmad, S., Ullah, H. et al. (2022). Investigation on crystal-structure, thermal and electrical properties of PVDF nanocomposites with cobalt oxide and functionalized multi-wall-carbon-nanotubes. *Nanomaterials (Basel)* 12: 2796.

27 Paulson, E. and Jothibas, M. (2021). Significance of thermal interfacing in hematite (α-Fe_2O_3) nanoparticles synthesized by sol-gel method and its characteristics properties. *Surfaces and Interfaces* 26: 101432.

28 Yang, Y., Dang, L., Shearer, M.J. et al. (2018). Highly active trimetallic NiFeCr layered double hydroxide electrocatalysts for oxygen evolution reaction. *Advanced Energy Materials* 8: 1703189.

29 Lin, J., Wang, H., Cao, J. et al. (2020). Engineering Se vacancies to promote the intrinsic activities of P doped $NiSe_2$ nanosheets for overall water splitting. *Journal of Colloid and Interface Science* 571: 260–266.

30 Tian, L., Chen, Z., Wang, T. et al. (2023). Mo doping and Se vacancy engineering for boosting electrocatalytic water oxidation by regulating the electronic structure of self-supported Co_9Se_8@NiSe. *Nanoscale* 15: 259–265.

31 Chen, L., Zuo, X., Yang, S. et al. (2019). Rational design and synthesis of hollow Co_3O_4@Fe_2O_3 core-shell nanostructure for the catalytic degradation of norfloxacin by coupling with peroxymonosulfate. *Chemical Engineering Journal* 359: 373–384.

Part II

Structural Characteristics of Iron-Group Metal Compound Nanomaterials

4

Structures of Iron (Fe, Co, and Ni)-Group Metal Compound Nanomaterials

Since the first successful synthesis of iron nanoparticles by German scientists Gleiter et al. in 1984 using the inert gas condensation method, significant progress has been made in the preparation, properties, and applications of nanomaterials. Catalysis represents an early example of nanotechnology developed through empirical approaches. Owing to their small size, nanomaterials exhibit a high surface-to-volume ratio, distinct surface bonding and electronic states compared to the bulk, and incomplete coordination of surface atoms, which collectively contribute to the superior performance of nanocatalysts [1]. In recent years, dimensionality has become a key focus in condensed matter physics and materials science. While traditional three-dimensional (3D) materials continue to evolve, low-dimensional (LD) materials have also advanced rapidly. LD materials are defined as materials with nanoscale dimensions in at least one direction. Their unique properties arise from quantum confinement effects at the nanoscale, allowing precise control over material characteristics at the atomic level. Generally, LD materials are categorized into three types: two-dimensional (2D), one-dimensional (1D), and zero-dimensional (0D). Nanocatalysts of different dimensions play vital roles in chemical processes by lowering reaction temperatures, controlling reaction time, enhancing efficiency and rate, and offering high selectivity [2–4]. In particular, the controlled synthesis and innovative design of iron-group metal nanocatalysts have greatly propelled research in catalytic materials, showing considerable potential for future industrial applications.

4.1 Zero-Dimensional Iron-Group Nanomaterials

With ongoing advances in nanotechnology, zero-dimensional (0D) nanomaterials have emerged as a class of materials with unique structures and properties, showing growing potential in both scientific research and industrial applications. These materials are defined as materials with nanoscale dimensions (typically 1–100 nm) in all three spatial directions, or material systems constructed from such nanoscale units as basic building blocks. Due to their minute size, the proportion of surface atoms increases significantly, leading to a higher surface state density and

Iron-Group Metal Compound Nanomaterials: Preparation, Characterization, Structure, and Applications, First Edition. Jinlong Zheng.

pronounced quantum confinement and size effects. These characteristics endow 0D materials with exceptional properties, such as narrow emission peaks, high photochemical stability, and size-tunable photoluminescence. As a result, they are widely used in various advanced fields, including photoelectrocatalysis, selective biomass oxidation, efficient energy conversion and storage systems, and highly sensitive sensors. Thus, 0D nanomaterials not only serve as ideal model systems for fundamental research but also provide a critical material foundation for the development of a new generation of high-performance nanodevices [5–8].

4.1.1 Zero-Dimensional Atomic Clusters and Nanoparticles

0D nanomaterials primarily include clusters, nanoparticles, and quantum dots. Metal-based electrocatalysts of varying sizes exhibit distinct catalytic behaviors in different chemical reactions. Low-cost non-precious iron-group metals (Fe, Co, and Ni) are promising nanomaterials with considerable potential. Leveraging the advantages of 0D structures, researchers are developing new compounds suitable for modern industrial production [9, 10].

Nanoclusters, recognized as a new class of chemical species, were only discovered in the 1980s. Comprising several to hundreds of atoms with sizes $\leq$1 nm, clusters represent a transitional state between individual atoms/molecules and bulk materials, often referred to as the "fifth state of matter" [11]. When the metal particle size approaches the Fermi wavelength (<1 nm), the electronic energy levels near the Fermi level shift from quasi-continuous to discrete states. This quantum size effect leads to molecule-like behaviors in metal nanoclusters, such as discrete Highest Occupied Molecular Orbital-Lowest Unoccupied Molecular Orbital (HOMO–LUMO) gaps, strong photoluminescence, quantum charging, unusual redox activity, chiral structures, and intrinsic magnetism [12–15].

Due to their high dispersion, well-defined composition, and atomically precise tunable surface and electronic structures, metal nanoclusters serve as ideal model nanocatalysts. They help elucidate the structure–activity relationship of nanocatalysts at the atomic level. Current research on metal nanoclusters spans catalysis, chemical sensing, electronics, bio-labeling, and biomedicine. In catalysis, iron-based catalysts have been isolated with excellent oxygen reduction reaction (ORR) performance, durability, and methanol tolerance, making them promising alternatives to Pt-based catalysts. However, research on iron-group metal nanoclusters remains limited. Research efforts have predominantly focused on clusters of Au, Ag, and Cu. However, monometallic and alloy clusters of Pt, Pd, Ir, Ni, Co, and Fe have received comparatively little attention. Furthermore, while ligands are essential for stabilizing nanoclusters during synthesis, their effect on electrocatalytic activity remains unclear [16, 17]. Nanoparticles, also known as ultrafine particles or nanoscale powders, generally have dimensions below 100 nm [18]. Research on 0D iron-group nanomaterials continues to gain attention with the ongoing development of nanotechnology. Experimental studies primarily focus on their structure, morphology, and magnetic properties, exploring applications in catalysis, magnetic materials, biomedicine, and energy.

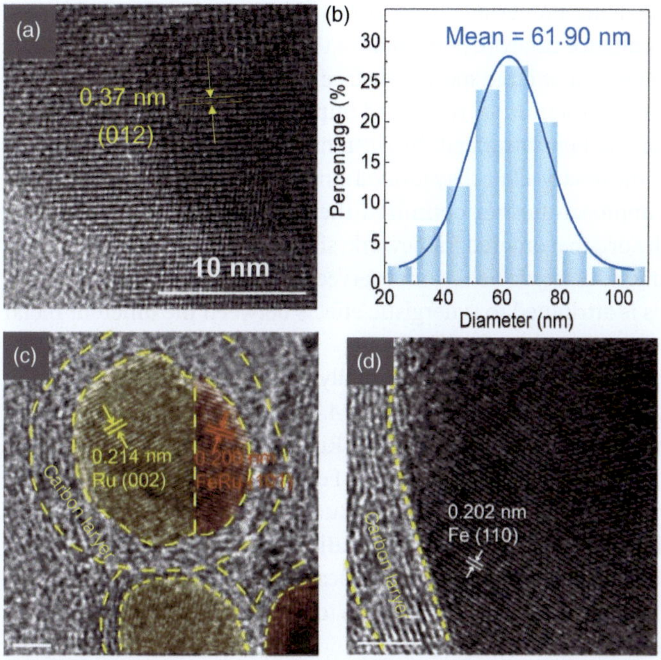

Figure 4.1 (a, b) Fe_2O_3 nanoparticles. *Source:* Reproduced from Ref. [20] / with permission of Elsevier. (c, d) Ru-FeRu nanoparticles. *Source:* Reproduced from Ref. [21] / with permission of Elsevier.

The synthesis of magnetite nanoparticles with controlled dimensions has long been of scientific and technological significance. As an important class of semimetallic materials, their dispersed particulates are widely used in ferrofluids for applications such as rotary shaft seals, oscillation damping, and position sensing. Sun and his research team synthesized Fe_3O_4 nanoparticles of various sizes with an inverse spinel structure, achieving controlled size synthesis of magnetite nanoparticles [19]. Similarly, different morphologies and sizes of iron-group metal nanoparticles can be obtained by varying synthesis methods and process parameters. Pan et al. employed a high-temperature shock strategy to realize size-controlled synthesis of Fe_2O_3 nanoparticles, while concurrently fabricating a robust core-shell architecture [20]. As shown in Figure 4.1a, b, the prepared Fe_2O_3@C nanoparticles not only possess a well-defined structure comprising a crystalline Fe_2O_3 core and a carbon shell but also exhibit a narrow size distribution precisely controlled at 61.90 ± 15.09 nm. This dual control over particle structure and size demonstrates the superiority of the synthesis method and provides a key material foundation for high-performance heterogeneous electro-Fenton catalysts.

Another research team prepared FeCoNi nanoparticles using a hydrothermal synthesis method, regulating their morphology and size by controlling the reaction temperature and time [22]. The results indicated that at 200 °C, the FeCoNi nanoparticles exhibited a spherical morphology with high magnetism. When the temperature was increased to 240 °C, the particle morphology changed to cubic, accompanied by a reduction in magnetic properties. In electrocatalysis and biomass oxidation, metal nanoparticles enhance catalytic performance through synergistic interactions between multiple metals. Their large specific surface area increases molecular collision probability, facilitating activation, reuse, and improved catalytic efficiency. Feng et al. reported carbon-shell-encapsulated Ru-FeRu nanoparticles [21]. By introducing a third metal into the binary alloy, they formed structurally ordered ternary metal nanoparticles. This approach further optimized the catalyst's electronic structure while maintaining the original ordered framework, significantly enhancing catalytic performance. The unique catalytic behavior observed in these nano-electrocatalysts with metal interfaces is attributed to synergistic effects between the different metal components.

The existence of this heterostructure was visually confirmed by high-resolution transmission electron microscopy analysis. Figure 4.1c clearly shows the interfacial structure where Ru and FeRu phases coexist in the Ru-FeRu@C/NC catalyst. Lattice spacings of 0.200 and 0.214 nm correspond to the FeRu (101) plane and the metallic Ru (002) plane, respectively. In contrast, the Ru@C/NC sample in Figure 4.1d exhibits only a single Ru phase structure, with a lattice spacing of 0.236 nm belonging to the hexagonal Ru (100) plane. This significant difference underscores the importance of developing iron-group nanoparticles to advance materials science and broaden the applications of nanomaterials.

4.1.2 Zero-Dimensional Quantum Dot Materials

Nanotechnology has played a vital role in scientific research over recent decades, with various nanomaterials now widely applied in fields ranging from catalysis to biomedicine. Following the discovery of new physical phenomena in semiconductor nanocrystals, numerous potential applications of quantum dots (QDs) have been identified. Research on QDs is still in a developmental stage, with new types continuously being developed and investigated. Due to their tunable structures, good dispersibility, remarkable optical properties, and favorable biocompatibility, QDs show great potential in biosensing and small-molecule detection [23–25]. From a chemical perspective, further progress is needed in QD synthesis to maintain and enhance their desired chemical and optoelectronic properties with high reproducibility. This requires the use of low-cost synthesis methods and precursors to scale up laboratory results for commercial applications. Moreover, a deeper understanding of QD surface characteristics, atomic arrangement, and metastable properties is essential for future advancements. From a regulatory standpoint, emphasis should be placed on developing high-quality QDs that avoid the use of heavy metals such as cadmium, lead, and mercury.

Iron-group metal-based QDs are widely used as cocatalysts due to their low cost, low toxicity, and favorable light absorption coefficients. For example, Zhang et al. used $Ni(OH)_2$ QDs as a stable cocatalyst to modify α-Fe_2O_3 for enhanced photoelectrochemical (PEC) water splitting. The deposition of cocatalysts has been shown to be an effective strategy for improving the efficiency of photoanodes in PEC water splitting. The $Ni(OH)_2$ QDs exhibited excellent stability under high current density during water oxidation, attributed to their ultrasmall size. This allows rapid hole transfer from α-Fe_2O_3 to the QDs, forming active sites for water oxidation and facilitating hole transfer to water molecules [26]. Further (photo)electrochemical analysis revealed that the $Ni(OH)_2$ QDs not only provide abundant active sites but also passivate surface states on α-Fe_2O_3, suppressing charge recombination and significantly enhancing water oxidation kinetics. In another study, Qiao et al. synthesized a novel catalyst—ultrasmall $NiFe_2O_4$ QDs uniformly dispersed on high-surface-area carbon nanotubes (CNTs)—targeting both the ORR and oxygen evolution reaction (OER). The formation mechanism of the QDs was found to depend on the Fe/Ni ratio, the role of the QDs, and the strong coupling between CNTs and $NiFe_2O_4$ in accelerating oxygen electrochemistry. A reaction mechanism for both ORR and OER occurring at the hybrid $NiFe_2O_4$ (QDs)/CNT interface was proposed for the first time. Benefiting from these advantages, the $NiFe_2O_4$ (QDs)/CNT hybrid exhibited remarkable electrochemical activity and stability, outperforming state-of-the-art noble metal-based catalysts [27].

4.2 1D Iron-Group Nanomaterials

4.2.1 Overview of 1D Nanomaterials

In one-dimensional (1D) nanomaterials, electrons move freely only along one non-nanoscale direction, following a linear path. In contrast, electrons in two-dimensional (2D) nanomaterials can move freely in two non-nanoscale dimensions. 1D nanomaterials exhibit directional electron/ion transport, high mechanical stress tolerance, and short axial pathways for electron and ion conduction. Compared with 0D nanoparticles, 1D nanomaterials offer significant inherent advantages, such as large pore volume, high surface area, structural stability, and tunable pore size, which endow them with unique physicochemical properties. These characteristics facilitate access to active sites, improve mass transport/diffusion, and enable applications across various fields [28, 29].

Since Iijima and colleagues in Japan discovered CNTs in 1991, 1D nanostructures—such as nanowires (NWs), nanorods, nanobelts, and nanotubes (NTs)—have attracted widespread interest among researchers [30]. Due to their distinct structural features and unusual physical and chemical properties, 1D nanomaterials hold significant value for fundamental research and potential applications. They have drawn attention from scientists in physics, chemistry, materials science, and biology, leading to extensive studies [31]. As foundational structures for other LD nanomaterials and key components in nanoelectronic devices and microsensors, the synthesis and application of 1D nanomaterials have become a leading research topic in materials science.

Over the past decade, numerous 1D inorganic nanomaterials have been synthesized and investigated. However, progress has been relatively slower compared with QDs, mainly due to challenges in achieving precise control over size, morphology, purity, and chemical composition. As research interest in 1D nanomaterials continues to grow, many 1D metal nanomaterials—including various iron-group-based structures—have been successfully synthesized. Several major types of iron-group 1D nanomaterials will be introduced in the following section.

Modern research shows that magnetic nanomaterials, with dimensions comparable to the electron coherence length or the wavelength of light, often exhibit unique physical effects not found in bulk materials: (i) Size effect (small-size effect): When the size of magnetic nanomaterials is similar to or smaller than characteristic physical lengths—such as the de Broglie wavelength of electrons, the wavelength of light, magnetic exchange length, domain wall width, or superconducting coherence length—distinctive properties emerge. These effects include single-domain confinement and superparamagnetism. For instance, as magnetic particle size decreases below 20 nm, coercivity can increase up to a thousandfold. However, further reduction leads to a drop in coercivity to zero, resulting in superparamagnetic behavior. (ii) Quantum size effect: When metal or semiconductor materials are scaled down to the nanoscale in one or more dimensions, approaching or falling below a critical size, carrier motion becomes confined in certain directions. This changes the electronic states near the Fermi level from quasi-continuous to discrete, altering magnetic and electrical properties—such as giving rise to the giant magnetoimpedance effect. (iii) Macroscopic quantum tunneling effect: This refers to the phenomenon where macroscopic quantities—such as magnetization, magnetic flux in quantum coherent devices, and electric charge—can traverse energy barriers in macroscopic systems. This effect determines the time limits for magnetic data storage and the size limits for microelectronic devices [32–35].

4.2.2 Types of 1D Iron-Group Metallic Materials

1D NTs are a class of nanomaterials with promising applications, offering numerous advantages due to their unique structural characteristics. These NTs exhibit excellent mechanical properties, including high strength and stiffness, enabling them to withstand significant forces while maintaining structural stability under bending or stretching. Their remarkable electron transport properties make them highly suitable for applications in sensors and electronic devices. Electrons are confined quantum-mechanically along the radial direction and can only move freely along the tube axis, resulting in unique band structures and outstanding electrical conductivity. This characteristic is highly significant for the advancement of nanoelectronics and nanodevices [36, 37]. Furthermore, due to their structural specificity, 1D NTs possess a relatively large surface-area-to-volume ratio, providing significant potential for applications in catalysis, adsorption, and sensing. For instance, they can serve as efficient catalyst supports, offering more active sites and increased reactive surface area. Additionally, the properties of NTs can be tuned based on their material composition and dimensions, highlighting their versatile application prospects.

In summary, the combination of excellent strength and stiffness, unique electronic properties, high specific surface area, multifunctionality, and the potential for low-cost controlled synthesis makes 1D NTs highly promising in nanoscience and nanotechnology. Currently, iron-group metal-based 1D NTs are extensively studied for catalytic applications. Leveraging their inherent advantages, 1D

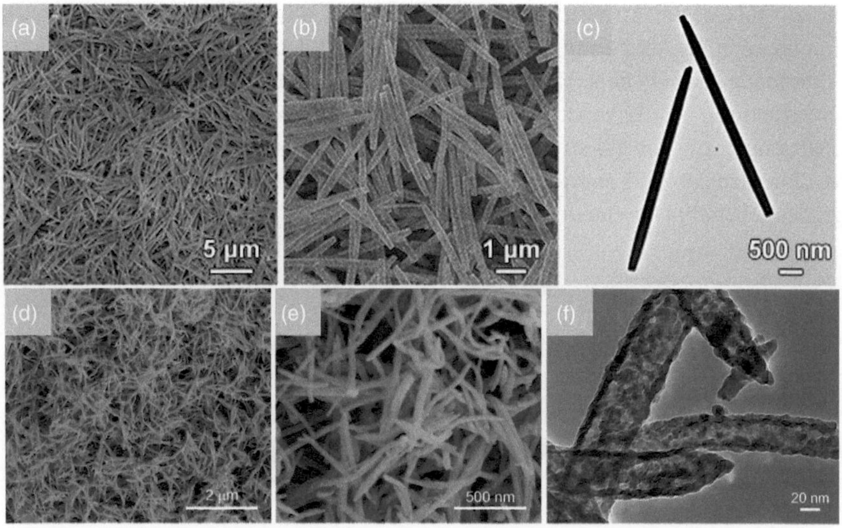

Figure 4.2 (a–c) FeMOC MRs. *Source:* Ref. [38] / John Wiley & Sons/CC BY 4.0. (d–f) $CoFeO_x$ NTs. *Source:* Reproduced from Ref. [39] / with permission of Elsevier.

NTs are often used as precursors. For example, Lou and his team developed a simple self-templating strategy to synthesize hierarchical cobalt–iron phosphide microtubes. Using iron-based metal–organic compound (FeMOC) microrods as precursors, the resulting electrocatalyst demonstrated excellent performance for both the OER and hydrogen evolution reaction (HER) in alkaline electrolytes [38]. The large surface area of hollow nano/microtubes provides abundant active sites for redox reactions, while their thin-shell topology facilitates charge transfer and electrolyte penetration, enhancing the kinetics of water-splitting reactions. Moreover, using hollow NTs as precursors to construct complex hollow structures with hierarchical morphologies can further increase the surface area and reduce charge/mass transfer resistance, offering new opportunities for designing highly active electrocatalysts (Figure 4.2a–c).

In another study, Huang and colleagues proposed a MOF-templated method to form hollow nanostructures, aiming to improve the HER catalytic performance of FeP. Benefiting from the synergistic effects of the desired morphology and atomic-level electronic structure engineering, the synthesized Co–Fe–P NTs exhibited superior catalytic performance. The high surface area of the tubular structure, combined with the synergy induced by Co substitution, provided the Co–Fe–P catalyst with abundant active sites, short mass diffusion pathways, low charge transfer resistance, and an electronic configuration favorable for hydrogen evolution at low overpotentials. Furthermore, the robust Co–Fe–P NTs demonstrated excellent long-term stability and durability (Figure 4.2d–f) [39]. Zheng Jinlong's team also prepared Fe_2O_3 NT precursors using a rapid and scalable hydrothermal method. These Fe_2O_3 NTs have proven to be excellent OER catalysts, showing great promise for replacing noble metal oxide catalysts like IrO_2 or RuO_2. Additionally, their porous structure and strong magnetism make Fe_2O_3 NTs suitable for coupling with other compounds, rendering them ideal for catalyst development. Electrocatalytic performance strongly depends on morphology and nanostructure. Among various nanostructures, NTs have attracted increasing attention due to their unique functional properties. The 1D structure with internal void space offers a high surface area, low density, and short pathways for mass and charge transport, providing significant benefits for electrocatalysis.

4.2.3 NWs

NWs are defined as 1D nanostructures where one dimension significantly exceeds 100 nm, while the other two dimensions remain below 100 nm. They can exist as crystalline or self-assembled structures and are not limited by material composition; even organic or polymeric materials can form NWs through self-assembly. Typically appearing as straight or curved solid 1D materials, common examples include silver NWs, copper NWs, and silicon carbide (SiC) NWs [40, 41]. Research on iron-group NW arrays primarily focuses on their catalytic properties and nanomagnetic applications. Among these, magnetic studies of iron-group metal NWs have attracted the most extensive attention. The preparation and magnetic investigation of nanomagnetic materials remain important in nanoscience and technology, with ferromagnetic metal NW arrays receiving particular interest. Due to their high aspect ratio, the shape anisotropy in NWs often dominates over magnetocrystalline anisotropy. When the easy magnetization axis of the crystal structure aligns with the shape anisotropy direction, the magnetic properties of the 1D nanostructure are significantly enhanced.

Structurally, 1D cylindrical magnetic nanostructures include NWs, NTs, and multilayer NWs. Although the most effective and hard magnetic materials contain rare-earth elements, their scarcity poses limitations. FeCo alloy NW arrays are considered promising alternatives for high-frequency applications as hard magnetic materials. Their controllable and strong effective anisotropy can shift ferromagnetic resonance into the microwave frequency range without requiring an external magnetic field. Bulk FeNi alloy is typically a soft magnetic material; however, with increased shape anisotropy in high-aspect-ratio nanostructures, it exhibits semi-hard magnetic behavior [42]. Figure 4.3 presents hysteresis loops of FeCo and FeNi alloy NWs under different structural parameters, revealing significant dependencies of magnetic properties on composition and diameter. In Figure 4.3a, b, FeCo NWs display high squareness ratios under parallel field orientation, indicating that the easy magnetization axis is primarily aligned along the NW length. In contrast, Figure 4.3c, d shows that Fe28.3Ni71.7 NWs with diameters below 100 nm exhibit higher coercivity under perpendicular fields than parallel fields. This behavior diverges from both the curling mode and coherent rotation models of magnetization reversal, possibly due to complex magnetic anisotropy distributions caused by transverse grains. As the diameter increases, both coercivity and squareness ratio decrease, suggesting the formation of multi-domain structures and increased grain size. These findings demonstrate that precise control of composition and geometric

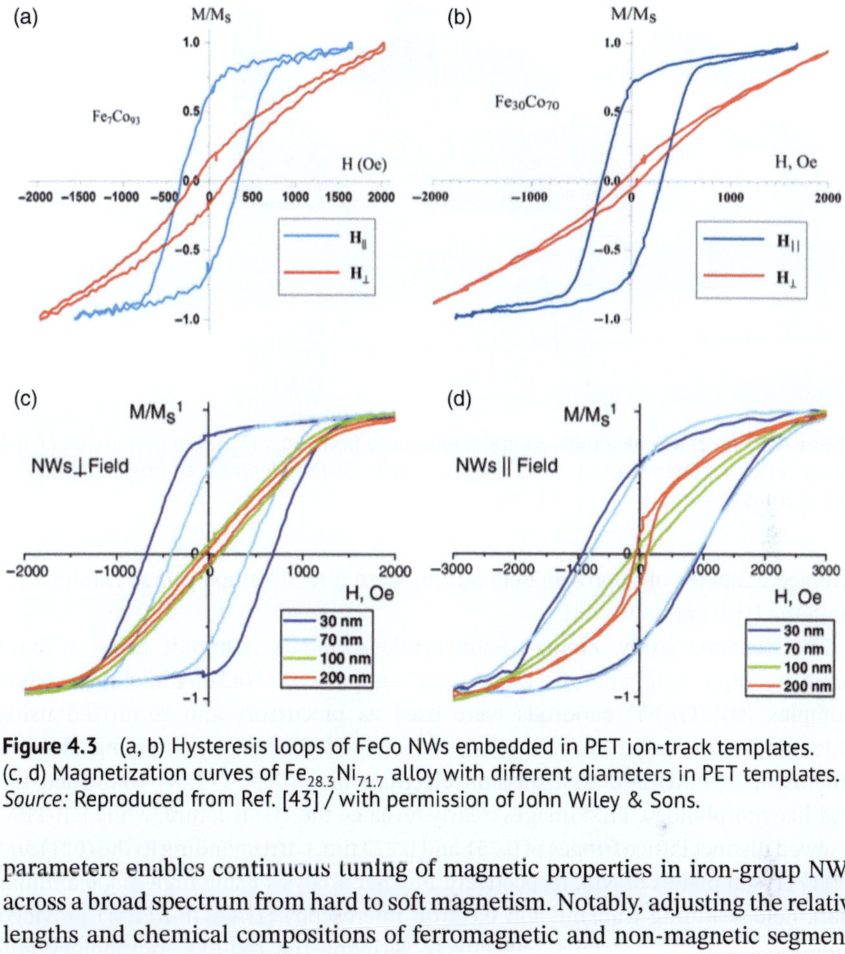

Figure 4.3 (a, b) Hysteresis loops of FeCo NWs embedded in PET ion-track templates. (c, d) Magnetization curves of $Fe_{28.3}Ni_{71.7}$ alloy with different diameters in PET templates. *Source:* Reproduced from Ref. [43] / with permission of John Wiley & Sons.

parameters enables continuous tuning of magnetic properties in iron-group NWs, across a broad spectrum from hard to soft magnetism. Notably, adjusting the relative lengths and chemical compositions of ferromagnetic and non-magnetic segments allows controlled sequencing of soft and semi hard magnetic layers within a single NW. This capability is crucial for developing novel nanoscale spin valve systems [43].

4.2.4 Nanorods and Nanobelts

Nanorods sharesimilarities with NWs but are characterized by dimensions that do not exceed 100 nm in all three spatial directions. They are typically crystalline, although self-assembled organic or polymeric structures with similar morphology are also classified as nanorods. Generally, nanorods are short, straight, 1D cylindrical solid nanomaterials [44]. Nanobelts, on the other hand, are long, thin, ribbon-like structures with rectangular cross-sections. Their thickness is on the nanoscale, while their width can extend to several hundred nanometers. Unlike the near-circular cross-sections of NTs or NWs, nanobelts exhibit a quadrangular shape, with typical width-to-thickness ratios in the range of tens of nanometers [45]. Zhang et al. prepared NiO nanorod arrays in situ on carbon fiber paper (CFP) using a simple cation exchange method. The resulting NiO nanorods, with an

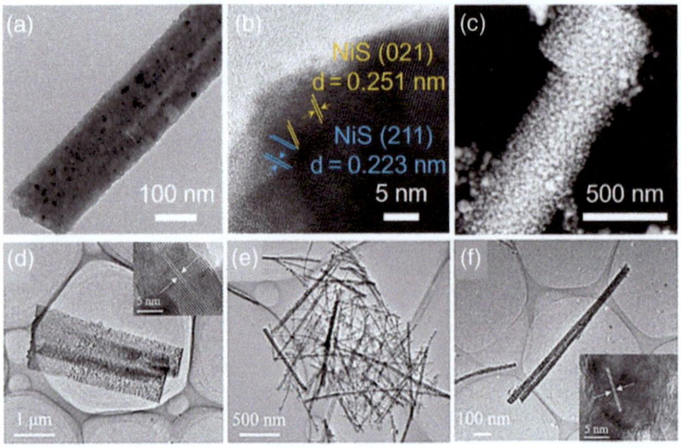

Figure 4.4 (a–c) NiS nanorods. *Source:* Reproduced from Ref. [47] / with permission of The Royal Society of Chemistry. (d–f) $NiCo_2O_4$ nanobelts. *Source:* Reproduced from Ref. [48] / with permission of John Wiley & Sons.

average diameter of approximately 80 nm, were directly employed as catalysts for both the HER and OER [46].

In a separate study, Zhang's team synthesized NiS nanorods using a "hard template–high temperature calcination" approach. Nickel–dimethylglyoxime complex ($(DMG)_2Ni$) nanorods were used as precursors and sulfurized using thiourea as the sulfur source. Structural characterization by scanning electron microscope (SEM) and transmission electron microscope (TEM) confirmed the rod-like morphology. TEM images clearly revealed the 1D structure, while HR-TEM showed distinct lattice fringes of 0.251 and 0.223 nm, corresponding to the (021) and (211) crystal planes of NiS, respectively. Further analysis using high-angle annular dark-field scanning transmission electron microscopy (HAADF-STEM) provided additional structural details, collectively verifying the rod-like morphology and crystalline characteristics, as shown in Figure 4.4a–c [47]. Figure 4.4d–f displays the morphological features of Fe-doped $NiCo_2O_4$ (Fe–$NiCo_2O_4$), clearly illustrating the structural characteristics typical of 1D nanobelts. The combination of Fe doping and the unique nanobelt structure endowed the Fe–$NiCo_2O_4$ sample with excellent OER catalytic performance.

Morphological control represents a key strategy for enhancing material properties. Nanomaterials generally possess a large specific surface area, which not only increases the effective area for chemical reactions but also provides more active sites. Furthermore, nanostructures with specific morphologies can shorten the migration distance of holes to the material surface. 1D nanostructures such as nanobelts and nanorods, with their small characteristic dimensions, can match the hole diffusion length, thereby reducing the recombination loss of electron–hole pairs at bulk grain boundaries. This unique structure facilitates the directional migration of photogenerated electrons and holes along the tube wall to the material substrate and surface, respectively, thereby significantly improving the overall catalytic performance [48].

4.3 2D Iron-Group Nanomaterials

4.3.1 Overview of 2D Materials

2D nanomaterials are defined as materials in which electrons can move freely within a plane but are confined to the nanoscale (1–100 nm) in the third dimension. A landmark achievement was made in 2004 when Geim's team at the University of Manchester successfully isolated single-atom-layer carbon crystals—graphene—from graphite using mechanical micro-stress exfoliation [49, 50]. This breakthrough immediately ignited global academic interest and spurred systematic research into 2D materials. Since the discovery of graphene, numerous novel 2D materials have been successively identified and investigated. Over the past decade, research on 2D materials for catalysis has advanced rapidly. These materials are now widely employed in traditional heterogeneous catalysis, electrocatalysis, and photocatalysis, demonstrating unique and irreplaceable catalytic characteristics in key energy-related processes such as oxygen activation, water splitting, and carbon dioxide reduction. Their distinctive 2D planar geometry and exceptionally high specific surface area set them apart from bulk materials. These structural advantages not only provide excellent electron transport efficiency, charge separation capability, and abundant surface reaction sites but also lead to unique properties in conductivity, adsorption, and catalysis. Consequently, 2D materials find applications across diverse fields, including electronic information technology, advanced healthcare, environmental remediation, aerospace engineering, and consumer products [49, 51–54].

Building on graphene as a starting point, research has expanded to identify four main categories of 2D nanomaterials: the graphene-like family, 2D transition metal dichalcogenides, 2D oxides, and other layered structures. Formally known as 2D atomic crystal materials, they can theoretically extend infinitely within the plane, providing an immense surface area. Charge carrier migration and heat diffusion are confined to the 2D plane, resulting in remarkable features such as atomic-scale thickness, high carrier mobility, linear energy spectra, and exceptional mechanical strength. These properties establish them as a potentially revolutionary class of materials, with each type offering distinct advantages [55–57]. To date, numerous catalysts based on transition metal nanomaterials have been developed and effectively applied across various catalytic systems. Although noble metals like Au, Ru, Pt, and Rh exhibit excellent catalytic performance, their high cost and scarcity have motivated researchers to explore non-precious alternatives. Transition metals such as Fe, Co, and Ni are abundant in the Earth's crust and more affordable, providing a practical foundation for industrial-scale applications. Compounds and nanomaterials derived from these metals are cost-effective, making them economically attractive with broad potential for further development.

Owing to their high specific surface area and tunable structures, 2D materials have been extensively utilized in catalysis since their emergence. They are particularly promising for reactions such as the HER, OER, ORR, carbon dioxide reduction reaction (CO_2RR), nitrogen reduction reaction (NRR), and pollutant degradation [58–60].

4.3.2 Iron-Group 2D Nanomaterials

A growing number of 2D nanomaterials are currently being developed and extensively investigated. Synthesis methods are broadly categorized based on whether the starting crystals are layered or non-layered. 2D metallic materials with non-layered structures primarily consist of pure metals and their alloys. Their synthesis can be achieved through both chemical and mechanical approaches. Chemical methods typically involve spatially confined synthesis, template-directed synthesis, surfactant/ligand-assisted energy-controlled synthesis, hydrothermal/solvothermal synthesis, seed-mediated growth, and other specialized routes. Mechanical methods comprise various compression techniques, such as large-scale hot pressing and repeated size reduction processes [61]. In contrast to non-layered 2D metals, layered materials are mainly represented by metallenes. The range of synthetic methods for these materials is less diverse than for non-layered structures, with primary techniques focusing on molecular beam epitaxy (MBE) and various exfoliation approaches. Additionally, when metallenes are produced in non-pure forms, separation techniques are employed to isolate the synthesized metallenes without damage. The diversity of 2D metallic materials produced through these various methods has been intensively studied to fully exploit their potential across different fields. Leveraging their unique properties—such as predominantly exposed active metal sites, rapid kinetics for energy storage (fast diffusion of electrons and ions), and surface plasmon resonance (SPR)—applications have been developed in catalysis, electronic/battery devices, and various SPR-related technologies [62–64].

Significant research efforts are focused on the prospective development of 2D iron-group (iron, cobalt, and nickel) compound nanomaterials. Structures like nanosheets, nanoplates, and nanobelts are widely used in catalysis, as illustrated in Figure 4.5 [65, 66]. These 2D metallic nanostructures possess distinctive electronic and physicochemical properties, including a high aspect ratio, large specific surface area, high density of surface-unsaturated atoms, and high electron mobility. These unique characteristics provide them with significant advantages over their bulk counterparts in surface-related applications, such as electrochemical water splitting, the ORR, and the CO_2RR.

2D nanomaterials are attractive candidates for energy conversion and storage devices due to their large surface area, which enables effective enrichment and exposure of active sites [67, 68]. In the field of water-splitting catalysis, various non-precious metal-based catalytic materials have been extensively developed. These materials include carbon materials, metals, metal oxides/hydroxides, sulfides, and phosphides, as well as nitrides and selenides, meeting the needs of different catalytic scenarios. As a typical 2D material, metal-based layered double hydroxides possess notable advantages: low raw material cost, simple preparation methods, and excellent electrocatalytic performance for both the hydrogen evolution reaction and OER. Their chemical composition and electronic structure can be readily tuned through metal ion doping or interlayer anion exchange, demonstrating significant potential for replacing precious metal-based catalysts in efficient water splitting [69, 70].

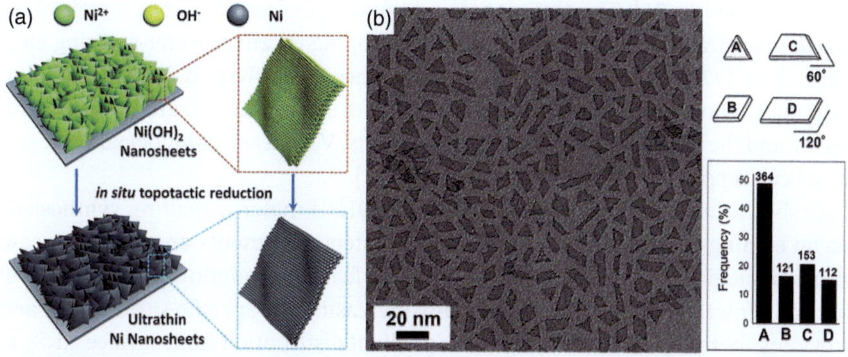

Figure 4.5 Representative morphology of nanoplates. (a) In situ topotactic reduction method for fabrication of the atomically layered Ni nanosheet array. *Source:* Reproduced from Ref. [65] / with permission of John Wiley & Sons. (b) TEM image of the top view of the prepared nanoplates and shape distribution diagram by investigating 750 plates. *Source:* Reproduced from Ref. [66] / with permission of American Chemical Society.

Layered double hydroxides (LDHs) are a class of layered compounds consisting of stacked sheets of edge-sharing MO_6 octahedra. The general formula is $[M^{2+}{}_{1-x}M^{3+}{}_x(OH)_2]^{v+}$ (where M is typically a first-row transition metal), representing positively charged metal hydroxide layers, with charge-balancing interlayer anions and water molecules $[A^{n-}{}_x/n{\cdot}H_2O]$. LDH materials have emerged as highly promising earth-abundant transition metal-based electrocatalysts, owing to their tunable interlayer anions [71]. In summary, 2D materials—particularly those based on non-precious metals—are considered ideal for water splitting due to their cost-effectiveness, favorable electrocatalytic performance, and straightforward synthesis. To further enhance the catalytic properties of LDHs, several engineering strategies have been developed, including heteroatom doping, vacancy creation, and morphological control [72, 73].

Introducing heteroatoms into the crystal structure of Ni/Co/Fe-based LDHs can form new chemical bonds and alter the electron cloud distribution of host atoms, thereby generating new high-activity catalytic sites tailored for water splitting. Doping often induces local crystal lattice distortion, creating numerous oxygen vacancies. These vacancies increase the density of electronic states near the Fermi level, improve electrical conductivity, and expose more active sites to the electrolyte. The combined effects enhance electron transfer efficiency and increase the proportion of catalytic sites interacting with reaction intermediates. Furthermore, oxygen vacancies and heteroatoms can synergistically modulate the electronic structure of active sites, optimizing the adsorption energy of reaction intermediates. Heteroatom doping includes both precious metals (Pt, Ru, and Ir) and non-precious metals. Precious metal doping helps form highly intrinsic catalytic sites with optimized intermediate adsorption energies, leading to low overpotentials in OER and HER [74]. Numerous non-precious metal-doped LDHs—such as those incorporating Fe, V, Cr, W, Mo, or Ce—also exhibit bifunctional electrocatalytic performance for both OER and HER. Examples include W-$Ni(OH)_2$/NiOOH, Mo-NiCo-LDH, Fe-NiCo-LDH, and

Ta-FeNi-LDH [75, 76]. The abundance and tunability of non-precious metals make this doping strategy highly promising for developing excellent water-splitting electrocatalysts [77]. The electrocatalytic performance of catalysts is closely related to their morphology, which influences the number of active sites, electron transfer processes, and the diffusion of reactants and products. Various morphological strategies have been reported to enhance LDH activity [71].

As shown in Figure 4.6a, the Ni_3S_4/NiFe-LDH/IF heterostructure, synthesized via an etching-driven gradient engineering strategy, represents an exemplary case of morphological optimization. This structure features nanoflowers composed of vertically interlayered sheets, significantly increasing the specific surface area and providing more exposed active sites. The gradient distribution from surface Ni_3S_4 to inner NiFe-LDH establishes continuous electron transport pathways and optimizes reactant penetration and product desorption, contributing to its excellent performance in OER, overall water splitting, and anion exchange membrane water electrolysis [77]. Due to weak interlayer interactions, anions can be easily intercalated into LDH layers, adjusting the interlayer spacing and enhancing the electrochemically active surface area. Larger interlayer spacing facilitates the formation of LDH nanosheets with high specific surface area and active site exposure density, and can promote the formation of anion defects, further improving electrocatalytic performance. When the interlayer spacing exceeds a certain threshold, it can also promote the exfoliation of LDHs into ultrathin nanosheets.

Traditional LDHs have a layered stacking structure that shields many active sites and hinders electron and mass transfer during electrocatalysis. An exfoliation strategy can be employed to separate them into smaller, ultrathin nanosheets, significantly enhancing performance. These exfoliated monolayers provide a large surface area and more electrochemically active sites, facilitating reactions with reactants, as illustrated in Figure 4.6b, c [78, 79]. Creating vacancies is now considered an essential method for enhancing the initial electrocatalytic performance of Ni/Co-based LDHs. Vacancy regulation typically involves creating cation vacancies or anion vacancies. The formation of oxygen vacancies implies the generation of numerous unsaturated atoms, which can significantly improve the surface adsorption energy of anions from the electrolyte and reduce the activation energy of reaction steps. The presence of metal vacancies provides abundant unsaturated coordination atoms that modulate the electronic configuration around active metal sites, thereby promoting HER/OER activity, as depicted in Figure 4.6d [80].

4.3.3 Iron-Group Metal–Organic Frameworks

Iron-group metal–organic frameworks represent another significant class of 2D materials in catalysis. Metal organic frameworks (MOFs) are emerging porous crystalline materials composed of metal ions or clusters coordinated with organic ligands. They are widely used in chemical sensing, gas adsorption, and energy conversion/storage due to their distinct advantages, including high porosity, structural diversity, and tunable composition. By assembling suitable metal nodes with various linkers, MOFs can be tailored to form matrices with controlled pore

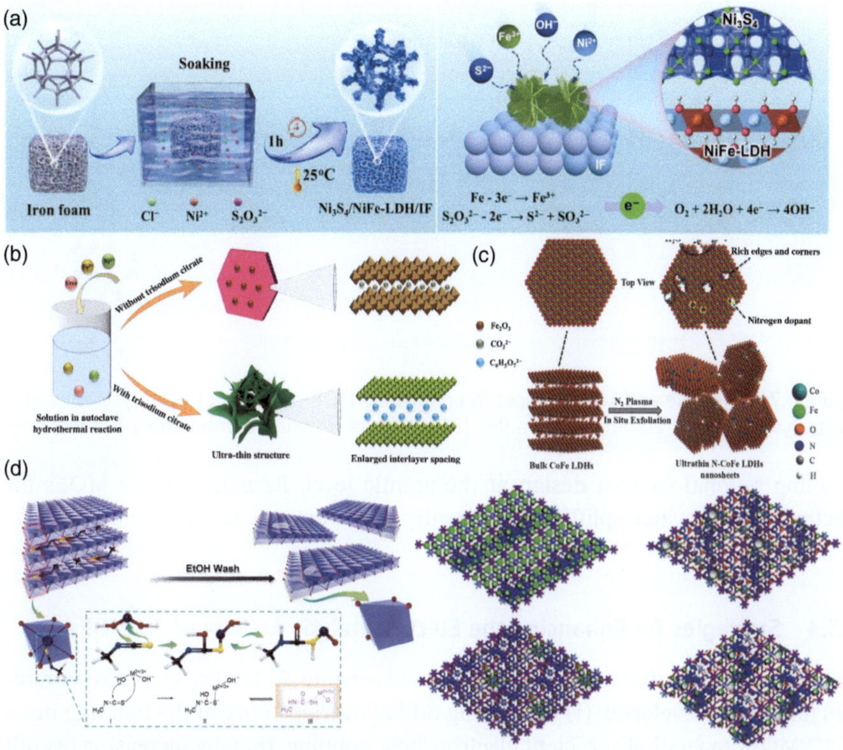

Figure 4.6 Schematic illustrations of preparation processes for Ni(Co, Fe)-based LDHs: (a) Synthesis of Ni_3S_4/NiFe-LDH/IF. *Source:* Reproduced from Ref. [77] / with permission of John Wiley & Sons. (b) Formation of NiFe LDH without and with sodium citrate. *Source:* Reproduced from Ref. [78] / with permission of Elsevier. (c) Exfoliation of bulk CoFe-LDHs into ultrathin CoFe-LDH nanosheets via N_2 plasma treatment. *Source:* Reproduced from Ref. [79] / with permission of John Wiley & Sons. (d) Mechanism for generating multiple vacancies and corresponding electronic structure of defective NiFe LDH. *Source:* Reproduced from Ref. [80] / with permission of Elsevier.

sizes, morphologies, and functionalities, featuring uniformly distributed active centers. From an electrochemical perspective, MOFs can be viewed as spatial architectures in which discrete functional units are strategically positioned to facilitate bond-breaking and bond-forming reactions.

Active sites and electrical conductivity are generally recognized as key factors determining the electrochemical performance of HER and OER catalysts. Compared to their bulk counterparts, 2D MOFs have emerged as promising catalytic materials due to their unique characteristics: expanded surface area, rapid mass transport, and enhanced conductivity. Constructed from single or few atomic layers, 2D MOFs form layered materials with high lateral-to-thickness aspect ratios. Since electrocatalytic processes occur at surfaces/interfaces, the presence of defects or coordinatively unsaturated sites in 2D MOFs can significantly promote catalytic activity. Furthermore, the well-defined periodicity of their surface structures provides an ideal platform for investigating precise structure–property relationships,

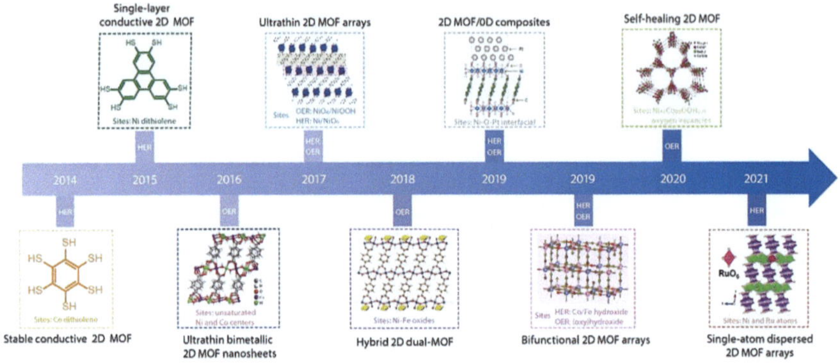

Figure 4.7 Timeline and key advances in pristine 2D MOF-based electrocatalysts for water electrolysis. *Source:* Reproduced from Ref. [81] / with permission of John Wiley & Sons.

enabling rational catalyst design at the atomic level. Research on 2D MOFs for electrochemical water splitting is currently in a phase of rapid development, as illustrated in Figure 4.7 [81].

4.3.4 Strategies for Enhancing the Electrocatalytic Activity of 2D MOFs

To improve the electrocatalytic performance of pristine 2D MOFs, several key strategies have been developed: (i) Enhancing orbital overlap between the building units of 2D MOFs to facilitate efficient electron/hole hopping, thereby increasing overall electrical conductivity. (ii) Reducing the thickness of 2D MOF nanosheets to expose more accessible electrochemically active sites. (iii) Constructing composite materials by integrating 2D MOFs with other functional nanomaterials to create a synergistically enhanced catalytic system. (iv) Growing 2D MOF arrays in situ on conductive substrates to achieve superior electron and ion transport properties [82, 83].

The plasma-enhanced electrocatalytic system involving the Au nanorods (AuNRs)/NiCo-MOF composite illustrates a clear structure–activity relationship. Under light excitation, AuNRs act as photosensitizers, generating hot electrons and holes that are efficiently injected into the MOF's energy bands. TEM analysis confirms the structural basis for this mechanism, showing that AuNRs are uniformly dispersed and embedded within the 2D MOF nanosheets, thereby forming ideal heterojunction interfaces. This synergistic interaction enhances overall charge transport efficiency (Strategy 1) while modulating the electronic structure of active sites, promoting the formation of highly active Ni^{+} species and significantly improving the kinetics of the OER, as shown in Figure 4.8a [84]. In the realm of conductive MOFs, 2D conductive cobalt dithiolene MOFs have demonstrated high efficiency and stability as HER electrocatalysts in acidic media. Subsequently, other 2D conductive MOFs based on specific π-conjugated or redox-active organic linkers have been proposed, including monolayer 2D conductive nickel dithiolene MOF nanosheets with thicknesses of 0.7–0.9 nm [86]. Recent research increasingly focuses on thinning intrinsically insulating yet electrochemically active MOFs to monolayer or few-layer structures to fully unlock their catalytic potential.

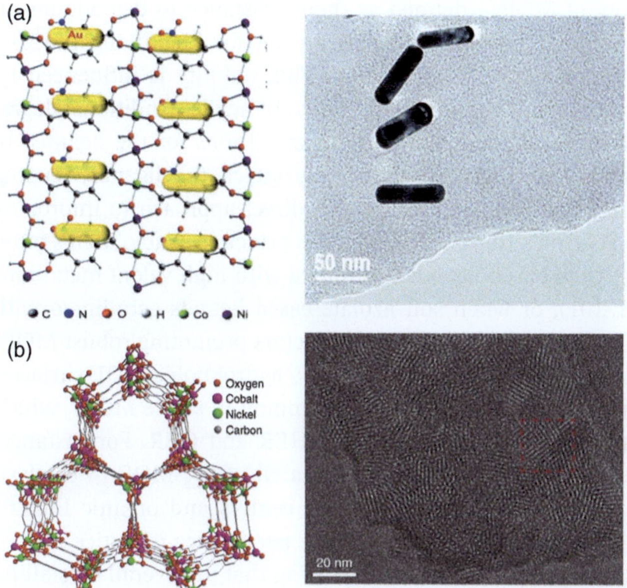

Figure 4.8 (a) Schematic structure and TEM image of AuNRs/NiCo-MOF. *Source:* Reproduced from Ref. [84] / with permission of Elsevier. (b) Crystal structure and HRTEM image of $Ni_{0.5}Co_{0.5}$-MOF-74. *Source:* Reproduced from Ref. [85] / with permission of Springer Nature.

The dimensionality of MOFs is largely governed by the coordination geometry and connectivity of their building blocks. For instance, Co^{2+} ions can adopt either tetrahedral or octahedral coordination geometries, depending on the steric demands of the coordinating linkers. Connection of building blocks along one direction yields 1D chain structures, whereas connection in two directions produces 2D-layered structures that can stack into multilayers via weak interactions. Compared to traditional bulk MOFs, 2D MOFs offer significant advantages for overall water splitting. Their increased surface area and abundant accessible unsaturated metal sites promote interaction with reactant molecules. Their nanoscale thickness shortens diffusion distances and enhances mass transport, thereby increasing reaction rates. While bulk MOFs often suffer from poor conductivity, 2D MOFs can form vacancies that act as shallow donors, increasing charge carrier concentration and enhancing charge transfer. Defects introduced during the synthesis of 2D MOFs can help improve intrinsic catalytic activity. Some MOFs exhibit high stability under water-splitting conditions, maintaining their crystal structure, composition, and morphology after catalytic reactions. Metal sites within these stable 2D MOFs are often identified as the active centers. For example, Wang and colleagues reported highly efficient and robust NiRu-MOF nanosheet arrays based on H_2BDC linkers for the electrochemical HER [87]. The diversity of organic linkers and metal building blocks in MOFs provides a customizable platform for designing tailored electrocatalysts. Figure 4.8b shows the crystal structures of AuNRs/NiCo-MOF and $Ni_{0.5}Co_{0.5}$-MOF-74 [85].

The chemical stability of MOFs—defined as their resistance to degradation in operational environments—is a fundamental parameter for identifying intrinsic catalytic sites and ensuring practical application. This stability is influenced by the strength of the metal–ligand coordination bonds. In acidic conditions, degradation often results from proton competition with metal ions for the ligands. In alkaline conditions, hydroxide ions can chemically attack the ligands, leading to decomposition. Enhancing bond robustness is a direct approach to improving stability. According to Pearson's Hard/Soft Acid–Base principle, stable MOFs often form when hard carboxylate-based ligands coordinate with high-valent metal ions (e.g., MIL-53 and MIL-101), or when soft azolate-based ligands coordinate with soft, low-valent metal ions (e.g., ZIFs). Additional factors promoting robust MOFs include high connectivity of metal nodes and linkers, hydrophobic MOF surfaces, and linker rigidity. Carboxylate-based linkers are common in stable MOFs, which can serve as highly active electrocatalysts for the HER and OER. For instance, Figure 4.8b displays the crystal structure model of a $Ni_{0.5}Co_{0.5}$-MOF-74 catalyst, featuring 1D hexagonal channels formed by metal centers and organic linkers. After OER testing, HR-TEM imaging confirmed the persistence of lattice fringes corresponding to the (110) plane of MOF-74, indicating that the overall crystalline framework was largely preserved despite potential local reconstruction of metal sites, demonstrating significant structural stability. Furthermore, iron-, cobalt-, and nickel-based MOFs constructed from terephthalic acid or its functionalized derivatives also show promising catalytic performance and structural durability. Therefore, the rational selection of metal species and ligands with stable active sites is crucial for designing efficient and stable MOF-based electrocatalysts.

4.4 3D Iron-Group Nanomaterials

3D nanostructures are composite materials formed by integrating one or more fundamental structural units from 0D, 1D, and 2D nanomaterials [88]. Iron-group metals are widely used in the preparation of nanocomposites. As the second most common metal element, iron has spurred the development of various synthesis methods and novel nanomaterials for optimizing water electrolysis performance, showing very broad application prospects. Since the early 20th century, metallic nickel and its oxides have been found to exhibit high electrocatalytic activity for the OER in alkaline solutions. Subsequent engineering efforts have led to a series of nickel-based bifunctional electrocatalysts for overall water splitting. Similarly, cobalt offers moderate hydrogen adsorption energy alongside high conductivity. While most catalysts corrode easily in acidic environments, making bifunctional catalyst design challenging under such conditions, cobalt-based sulfides and phosphides still demonstrate good catalytic activity in acidic electrolytes, showing excellent driving capability for catalytic reactions.

Given that single-component electrocatalysts often show limited performance, constructing 3D composite materials using iron-group metals, combined with structural engineering strategies—including heteroatom doping, strain engineering, interface/hybrid engineering, and single-atom construction—can modify the electronic structure of catalysts and continuously optimize their catalytic activity [89–91]. Combining two or more catalytic materials to form heterostructures has become an effective strategy for enhancing catalyst performance. The following section focuses on the primary advantages of heterostructures for achieving efficient catalysis.

4.4.1 Construction Strategies for 3D Heterostructures

Heterostructures are primarily formed through: (i) epitaxial growth of specific components; (ii) ion exchange using specific templates; and (iii) in situ segregation of active phases. Well-designed 3D iron-group compound heterostructures can preserve the intrinsic properties of each component while maximizing synergistic effects. Among various optimization strategies, surface/interface engineering plays a crucial role in tuning the properties of 3D mixed heterostructures. Since electrocatalytic reactions occur on catalyst surfaces or at interfaces, the precise design of surface and interface characteristics—guided by theoretical calculations and implemented via chemical modification or physical treatment—can significantly enhance overall catalytic performance. Specifically, interface engineering modifies the local coordination environment and electronic structure of active sites, potentially creating new highly active centers. Moreover, the interface itself can serve as a unique synergistic active site, promoting multiple steps such as reactant adsorption, intermediate transformation, and product desorption, thereby accelerating electrocatalytic reaction kinetics.

To achieve and verify such unique interface structures rich in active sites, researchers have developed various synthesis routes. For instance, epitaxial growth can form clean interfaces with lattice coherence; heterogeneous coating techniques enable core–shell structures or surface functionalization; and hybridization methods facilitate 3D interpenetrating networks [92–94]. These methods provide a solid technical foundation for precisely constructing heterointerfaces and analyzing their structure–activity relationships. Among strategies for building hybrid nanostructures, epitaxial growth is a typical method for in situ preparation of hybrid nanostructures with desired crystal phases and exposed interfaces. A small lattice mismatch between the epitaxial layer and the substrate crystal is crucial, as it reduces interfacial energy and is a core prerequisite for high-quality epitaxial growth, directly affecting the crystallinity and interfacial stability of the nanostructure. Currently, liquid-phase and vapor-phase epitaxy are the most widely used techniques, allowing precise control over structural morphology and chemical composition, thus providing flexible preparation routes for optimizing material properties.

Additionally, exposing metals to air to induce surface self-oxidation offers a simple and efficient epitaxial growth strategy, enabling spontaneous formation of hybrid structures with specific interfacial features without complex equipment. Successful epitaxial growth depends on good lattice matching and elemental similarity: the former ensures directional continuity of crystal growth, while the latter reduces elemental interdiffusion and defect formation at the interface, collectively ensuring

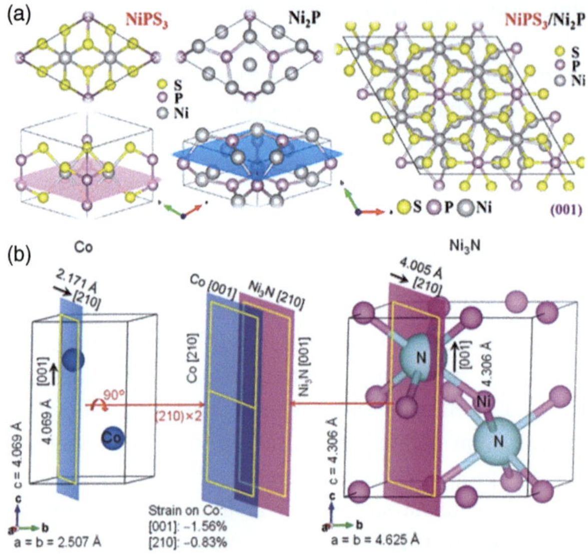

Figure 4.9 Schematic illustration of lattice matching. (a) Schematic models to illustrate the lattice matching between $NiPS_3$ and Ni_2P. *Source:* Reproduced from Ref. [96] / with permission of American Chemical Society. (b) Schematic illustration of lattice matching of metallic Co and Ni_3N. *Source:* Reproduced from Ref. [97] / with permission of John Wiley & Sons.

structural integrity and functionality [95]. Yan et al. constructed a $NiPS_3/Ni_2P$ heterojunction based on the good lattice matching between $NiPS_3$ and Ni_2P crystal structures to further optimize the material's architecture (Figure 4.9a). This epitaxially grown heterojunction significantly enhances the electronic interaction between $NiPS_3$ and Ni_2P. Changes in atomic coordination on $NiPS_3$ (110) affect the adsorption free energy. Experiments confirm that in situ connection with epitaxial Ni_2P is an effective strategy for markedly improving the overall water-splitting electrocatalytic activity of $NiPS_3$. The well-matched lattice parameters between $NiPS_3$ and Ni_2P form the foundation for constructing this epitaxial heterojunction. Theoretically, density functional theory (DFT) calculations verify lower energy barriers and a favorable built-in electric field at the epitaxial interface due to enhanced electronic interaction, consistent with electrochemical test results. These findings highlight the importance of rationally designing heterostructures with inwardly grown epitaxial interfaces to maximize the electrocatalytic activity of 2D layered ternary thiophosphates.

Furthermore, electrocatalytic performance can be enhanced by fine-tuning the interfacial electronic structure. Hong et al. developed a novel atomic epitaxial growth method to design a Co–Ni_3N NW array, where the nano-confinement effect at the interface significantly improved catalytic performance (Figure 4.9b). The epitaxial structure of Co–Ni_3N at the nanoscale promotes electron transfer between two different domains across the epitaxial interface, leading to substantially enhanced catalytic activity for both HER and OER [96, 97]. The nano-confinement effect occurring during the in situ atomic epitaxial growth of two compatible materials demonstrates an effective pathway for achieving high-performance electrocatalysis and energy storage.

The heterogeneous coating strategy optimizes the catalytic performance of 3D composite materials by constructing core–shell nanostructures, offering multi-dimensional control. The enhancement mechanism primarily stems from two aspects. At the electronic structure level, interfacial interactions between the core and shell (such as electron transfer or strain effects induced by lattice mismatch) optimize the adsorption energy of active sites for reaction intermediates, thereby significantly improving intrinsic activity. At the structural stability level, the outer shell provides a "confinement effect" on the core, effectively shielding it from corrosion, poisoning, or sintering in reactive environments, ensuring structural integrity and sustained catalytic activity during long-term operation.

In the literature, this unique nanostructure is commonly referred to as a "core–shell" structure. While often visualized as concentric spheres (core–shell particles), these structures are not limited to particles. They can also manifest as 1D nanorods or NWs, 2D nanosheets, or 3D frameworks encapsulated by shell materials. The specific components of the core and shell vary depending on application requirements. For example, in the Co_3O_4@NiP_x core–shell electrocatalyst system, multidimensional characterization and theoretical calculations collectively paint a complete picture of its structure–activity relationship [98]. SEM and TEM images clearly reveal the core–shell morphology, where the structural distinction between the core and shell provides the basis for performance tuning. The SAED pattern further confirms its polycrystalline nature. Analysis of the Ni 2p orbital in XPS spectra, comparing peak positions and intensities of samples like Co_3O_4@NiP_x and NiP_x, combined with DFT calculations, confirms directional electron transfer from the Co_3O_4 core to the NiP_x shell. This significantly enhances the electronegativity of Ni sites, optimizing their adsorption free energies for HER/OER intermediates (ΔGH^* as low as 0.07 eV). The free energy diagram further quantifies the energy changes during the HER, showing the reaction energy barrier for Co_3O_4@NiP_x is significantly lower than those of the control samples, like Co_3O_4@NiO and Co_3O_4. Overall, this suite of characterizations comprehensively illustrates the advantages of the Co_3O_4@NiP_x core–shell heterostructure, directly demonstrating its significant role in enhancing intrinsic catalytic activity.

Beyond electrocatalysis, the heterogeneous coating strategy also excels in photo-Fenton catalysis. A representative example is the Fe_3O_4@Au/COF core–shell catalyst. The SEM image clearly shows the tightly integrated core–shell morphology comprising an Fe_3O_4 core, an intermediate layer of Au nanoparticles, and a COF shell. Energy dispersive spectrometer (EDS) mapping and corresponding elemental distribution profiles visually confirm the uniform distribution of Fe, Au, C, N, and O within the composite, strongly supporting the successful construction of this core–shell structure. The Fe_3O_4 core provides magnetism and basic catalytic sites, while the Au nanoparticles act as a "bridge." Their SPR effect synergizes with the π-conjugated porous structure of the COF shell, significantly enhancing the separation efficiency of photogenerated electron–hole pairs and thereby substantially improving the degradation performance in photo-Fenton catalysis. Regarding structural stability, the COF shell provides a confinement effect, protecting the Fe_3O_4 core and Au nanoparticles from corrosion, poisoning, or sintering, ensuring structural integrity and sustained activity during long-term operation. Experimental results show that after five cycles in photo-Fenton reactions, the Fe_3O_4@Au/COF catalyst maintained its SERS signal intensity and pollutant degradation efficiency without significant decay, fully validating the protective role of the core–shell structure for the core components in complex reactive environments and highlighting the superior advantages of the heterogeneous coating strategy for multidimensional catalytic performance tuning [99].

The universality and powerful functionality of core–shell and heterostructure strategies are further demonstrated across a broader range of material systems. Moving beyond classical core–shell models, constructing strongly coupled heterointerfaces has become a frontier in electrocatalyst design. For instance, Zhang et al. demonstrated a simple surface engineering strategy for synthesizing an ORR electrocatalyst by coating polydopamine onto a Co-based MOF followed by auxiliary pyrolysis. This process resulted in Co nanoparticles encapsulated by several N-doped graphitic carbon layers. The formed interfacial Co–N–C structure acts as an active site, favoring interfacial charge transfer and endowing the synthesized catalyst with desirable ORR performance [100].

Similarly, Yang et al. synthesized a hierarchical compound, Ni_xFe_{1-x}-AHNAs. This nanomaterial consists of a crystalline-core phase and surface amorphous nanodomains. The distinct composition of the crystalline and amorphous phases can be clearly observed in Figure 4.10g–I [101]. This unique crystalline-core/amorphous-shell heterojunction effectively modulates the electronic structure and exposes abundant active sites, exhibiting a synergistically enhanced catalytic effect.

From precisely tuned core–shell geometries to multicomponent heterointerface coupling, the essence of this strategy lies in the fine tailoring of material surface and interface structures. Through directed regulation of the interfacial electronic structure and microenvironment at the nanoscale or even atomic scale, synergistic optimization of the catalyst's intrinsic activity, structural stability, and reaction mass-transfer efficiency has been successfully achieved. This strategy not only provides diverse solutions for overcoming the activity–stability trade-off in specific catalytic reactions but, more importantly, establishes a systematic methodology connecting microscopic structural design to macroscopic performance control. It lays a solid theoretical and practical foundation for the rational design of a new generation of high-performance, durable catalytic materials, showing profound application prospects.

4.4.2 Structural Advantages of Iron-Group 3D Compounds

Heterostructures comprising iron-group metal compounds combined with other metallic or nonmetallic compounds significantly enhance electron transport during charging and discharging processes. Electrode materials prepared using these as active substances exhibit high theoretical specific capacitance, high electronic

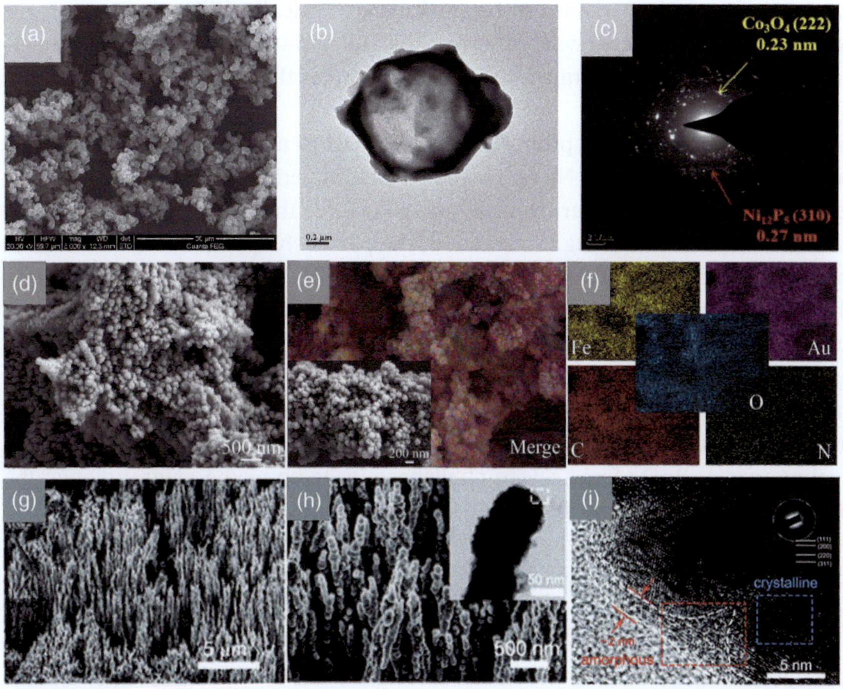

Figure 4.10 (a) SEM image of Co_3O_4@NiP_x. (b) TEM image of Co_3O_4@NiP_x. (c) SAED image of Co_3O_4@NiP_x. *Source:* Reproduced from Ref. [98] / with permission of Elsevier. (d) SEM image of Fe_3O_4@Au/COF materials. (e, f) The EDS mapping image and corresponding element mappings of Fe, Au, C, N, and O of Fe_3O_4@Au/COF. *Source:* Reproduced from Ref. [99] / with permission of Elsevier. (g–i) Hierarchical $NixFe_{1-x}$-AHNAs compound. *Source:* Reproduced from Ref. [101] / with permission of The Royal Society of Chemistry.

conductivity, and rich valence state transitions in alkaline aqueous electrolytes. Consequently, constructing electrode materials with heterostructures is regarded as an effective strategy for obtaining high-performance catalysts.

The structural advantages of heterostructure electrocatalysts include synergistic effects, strain effects, electronic interactions, chemical doping, and nanostructure engineering [102]. Among these, the synergistic effect induced by heterointerfaces and multicomponent materials is a distinctive feature not found in single-component materials. Typically, heterostructure materials consist of two or more different components that play distinct roles in the electrocatalytic reaction. Specifically, with the formation of interfaces between different components, the physical and chemical properties of the material, such as wettability and electronic structure, may change. Coupling active components with conductive substrates like carbon shells, nickel foam, or MXenes can significantly enhance electronic conductivity. The interfacial bonding in these heterostructures can act as a bridge for electron transfer. Beyond conductivity, properties such as hydrophilicity, anti-gas shielding, active site density, electrochemical stability, mechanical properties, and mass transfer efficiency can also be modulated. Typically, different components exhibit specific activities toward different fundamental steps of the electrochemical reaction, generating a synergistic effect that promotes the overall reaction kinetics [103–105].

For example, Guan et al. proposed using FeOOH as a cocatalyst to enhance the electrochemical activity of Ni_3N. The composite material demonstrated excellent performance, achieving overpotentials of 244 mV for the OER and 67 mV for the HER at 10 mA cm^{-2}. During the reaction, the positively charged FeOOH tends to bond with O atoms, whereas the negatively charged Ni_3N tends to adsorb H atoms. Therefore, the synergy between FeOOH and Ni_3N can activate water molecules and accelerate the cleavage of the H—OH bond, as illustrated in Figure 4.11a. Building on this heterostructure design concept, efforts have been made to achieve similar strain and electronic synergy in iron-group catalytic materials to enhance their intrinsic catalytic activity. Unlike iron-group materials that perform excellently in alkaline media, RuO_2 faces severe challenges in harsh acidic environments: the inherent strong covalency of the Ru—O bond can easily trigger the lattice oxygen oxidation mechanism, leading to oxygen vacancy formation, structural collapse, and dissolution of active sites, thereby severely limiting its application in proton exchange membrane electrolyzers.

$LiCoO_2$ was strategically chosen as a support to construct a $RuO_2/LiCoO_2$ heterojunction to investigate the structural origin of the catalyst's high activity. This material was selected because of its known dynamic electrochemical delithiation. In-depth characterization of its microstructure and electronic states was conducted. HRTEM images revealed a clear heterojunction interface between the RuO_2 nanoparticles and the $LiCoO_2$ support. This intimate lattice contact introduced significant lattice strain into RuO_2. Further X-ray absorption spectroscopy analysis confirmed that this strain effect led to a shortening of the Ru—O bond length and a negative shift in the Ru absorption edge, indicating increased electron density at the Ru sites and an upward shift of the d-band center. This optimized electronic structure effectively tuned the adsorption energy of reaction intermediates on the active sites, significantly enhancing the intrinsic activity for the OER, as shown in Figure 4.11b [106].

Heterostructure materials typically undergo inevitable electronic structure changes at the interface within each phase, which can be viewed as electronic interaction. Surface electron modulation is related to charge transfer across the heterointerface, governed by the band structure alignment of the heterostructure. Based on the work function differences between different components, the heterostructure can bend the band structure of semiconductors. Electron transfer occurs across the interface until the work functions equalize. Recently, Liu et al. rationally designed an atomic-scale Co_3O_4/CeO_2 interface, which demonstrated significantly improved OER activity with a low overpotential of 270 mV and a high turnover frequency of 0.25 s^{-1}. This heterostructure possesses a notable 2D electron gas and abundant oxygen vacancies, contributing to its excellent OER performance. Electron energy loss spectroscopy indicated the presence of free electrons at the heterointerface, with a carrier density reaching ~3.8×10^{14} cm^{-2}, resulting in high

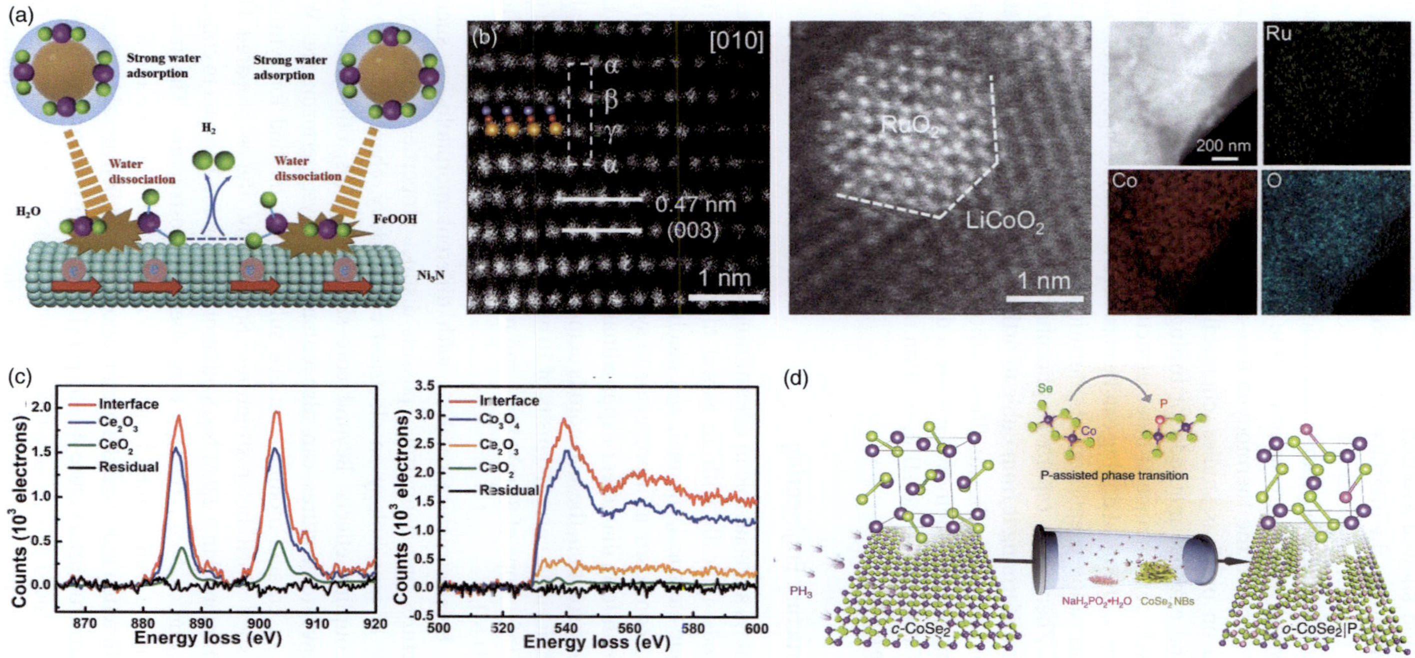

Figure 4.11 (a) Schematic of FeOOH-promoted electrolysis. *Source:* Reproduced from Ref. [104] / with permission of Elsevier. (b) $RuO_2/LiCoO_2$ heterojunction. *Source:* Reproduced from Ref. [106] / with permission of Springer Nature. (c) Charge density distribution. *Source:* Reproduced from Ref. [107] / with permission of John Wiley & Sons. (d) Synthesis schematic of o-$CoSe_2$|P. *Source:* Ref. [108] / Springer Nature / CC BY 4.0.

electron mobility and conductivity. Analyses of the density of states and charge density distribution showed electron transfer from Co to Ce, leading to excess electron accumulation at the interface, as depicted in Figure 4.11c [107].

Theoretically, the rate of electrocatalytic reactions is hindered by kinetic reaction barriers and the thermodynamic adsorption/desorption energies of reaction intermediates. Modifying electrocatalysts at the molecular level has been reported to substantially improve the thermodynamics and/or kinetics of electrocatalytic reactions. Based on previous studies, chemical doping is considered one of the most rational and effective methods for tuning electronic structures to enhance the electrocatalytic performance of heterostructure materials [109]. Depending on the doping element, chemical doping can be classified into three types: noble metal doping, early transition metal/non-3d transition metal doping, and non-metal element doping. Zheng et al. reported a phosphorus-doping-induced phase transition in $CoSe_2$, transforming it from cubic to orthorhombic phase [108]. The study showed that in 1 M KOH, the modified orthorhombic $CoSe_2$ with appropriate phosphorus doping (8 wt%) delivered a current density of 10 mA cm^{-2} at a very low overpotential of 104 mV for the HER, with an onset potential as low as 31 mV, as shown in Figure 4.11d.

4.4.3 Nanostructure Engineering

Nanostructure plays a crucial role in determining electrocatalytic performance. For instance, nanorods, typically with an aspect ratio (length/diameter) of less than 10, exhibit unique electrical, thermal, and mechanical properties. Li et al. utilized a MoO_3@ZIF-67 composite as a precursor to synthesize 1D MoO_2-$Co_2Mo_3O_8$@C nanorods with a uniform diameter of approximately 200 nm. When used as an OER catalyst in alkaline solution, these nanorods demonstrated favorable electrocatalytic performance, achieving a current density of 10 mA cm^{-2} at an overpotential of only 320 mV, with a Tafel slope of 88 mV dec^{-1} and relatively good durability, as shown in Figure 4.12a [110].

Furthermore, heterostructure catalysts with different morphologies, such as core–shell structures and nanoarrays, possess distinct properties, primarily due to the influence of morphology on electrical conductivity, charge transfer rate, and resultant coupling effects. Beyond core–shell structures, other nanostructures like hollow architectures can also enhance catalyst performance. Wang et al. reported an ethanol-modified surface sulfurization method for synthesizing nano-core–shell CoNi heterostructures. Specifically, the as-prepared CoNi hydroxide was immersed in a 2.0 M Na_2S ethanol solution for 24 hours to construct a $Co_{1.8}Ni(OH)_{5.6}$@$Co_{1.8}NiS_{0.4}(OH)_{4\cdot 8}$ core–shell heterostructure. The resulting sample maintained a well-defined hexagonal plate-like morphology, and selected area electron diffraction indicated the preservation of the bulk hydroxide properties within the heterostructure (Figure 4.12b) [111]. The high performance is largely attributed to the reduced particle size and increased surface atom ratio, which make the structure more active for O_2 evolution and facilitate ion diffusion through the 1D nanostructure.

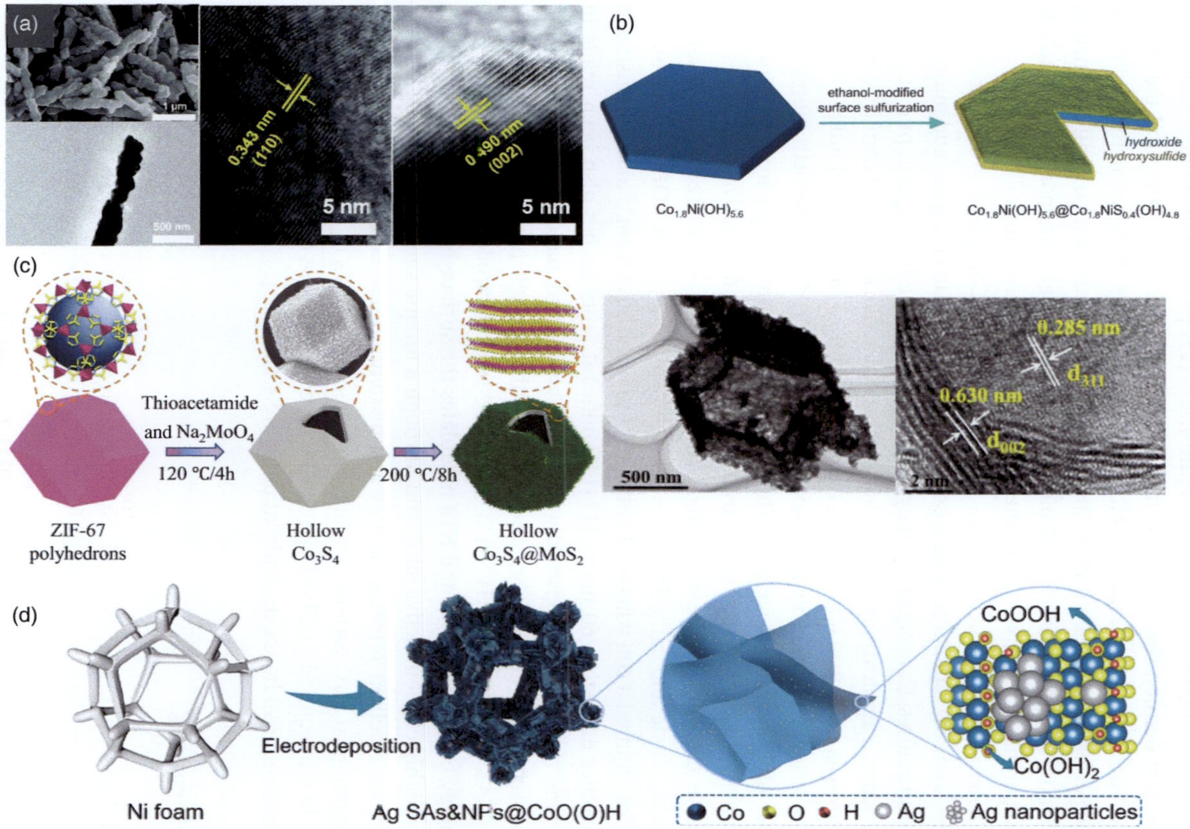

Figure 4.12 Nanostructure engineering: (a) MoO_2-$Co_2Mo_3O_8$@C. *Source:* Reproduced from Ref. [110] / with permission of The Royal Society of Chemistry. (b) $Co_{1.8}Ni(OH)_{5.6}@Co_{1.8}NiS_{0.4}(OH)_{4.8}$. *Source:* Reproduced from Ref. [111] / with permission of John Wiley & Sons. (c) $Co_3S_4@MoS_2$ nanostructure. *Source:* Reproduced from Ref. [112] / with permission of American Chemical Society. (d) Synthesis schematic of Ag SAs&NPs@CoO(O)H. *Source:* Reproduced from Ref. [113] / with permission of John Wiley & Sons.

Template-assisted strategies are primarily used for synthesizing hollow structures in catalysts, including hard templates and self-sacrificing templates. In an interesting study by Guo et al., ZIF-67 was used as a morphological template and Co as a precursor. A perforated Co_3S_4@MoS_2 heterostructure was then prepared via a two-step hydrothermal process (Figure 4.12c) [112]. Due to the hollow nanostructure's low density, shell permeability, abundant electroactive sites, and the synergistic effect between Co_3S_4 and MoS_2, the Co_3S_4@MoS_2 heterostructure exhibited good performance for both HER and OER. Nanostructures with overall dimensions exceeding 100 nm are termed 3D nanostructures. Typical 3D nanomaterials include mesoporous structures, core–shell structures, micro/nanospheres, hierarchical/heterogeneous structures, and hollow structures. Compared to LD nanostructures, some 3D nanostructures possess higher specific surface areas and multi-level pore structures. Consequently, they provide significant advantages in practical electrochemical applications, such as highly exposed active sites, good contact with the electrolyte, enhanced ion/electron transfer and diffusion, and rapid bubble release.

For example, Zhang et al. proposed hollow CeO_x/CoP heterostructures via a novel MOF-templated method. The abundant heterointerfaces in CeO_x/CoP can modulate the surface electronic structure, create numerous active defect sites, and achieve excellent intrinsic activity. When this hollow-structured catalyst served as both anode and cathode for overall water splitting in 1 M KOH electrolyte, the device required a cell voltage of only 1.63 V to reach 10 mA cm^{-2} and demonstrated durability exceeding 60 hours [114]. Significant progress has also been made in systems modified synergistically by single atoms and nanoparticles. Guo et al. proposed a CoO(O)H catalyst co-modified with Ag single atoms and nanoparticles (Ag SAs&NPs@CoO(O)H) via a facile one-step electrodeposition method. The synergy between Ag SAs and NPs modulates the surface electronic structure, creates highly efficient active sites, and, through a unique site-switching mechanism, achieves excellent intrinsic activity and dissolution resistance. When paired with a commercial Pt/C cathode for alkaline water electrolysis in 1 M KOH electrolyte, the device required a cell voltage of 1.9 V to reach 1000 mA cm^{-2} and showed stable operation for 90 hours, as depicted in Figure 4.12d [113].

References

1 Hendrich, C.M., Sekine, K., Koshikawa, T. et al. (2021). Homogeneous and heterogeneous gold catalysis for materials science. *Chemical Reviews* 121: 9113–9163.

2 Debnath, P.C. and Yeom, D.-I. (2021). Ultrafast fiber lasers with low-dimensional saturable absorbers: status and prospects. *Sensors* 21: 3676.

3 Liu, W., Yin, R., Xu, X. et al. (2019). Structural engineering of low-dimensional metal-organic frameworks: synthesis, properties, and applications. *Advanced Science* 6: 1802373.

4 Shao, C., Yu, X., Yang, N. et al. (2017). A review of thermal transport in low-dimensional materials under external perturbation: effect of strain, substrate, and clustering. *Nanoscale and Microscale Thermophysical Engineering* 21: 201–236.

5 Cheng, P., Chen, X., Gao, H. et al. (2021). Different dimensional nanoadditives for thermal conductivity enhancement of phase change materials: fundamentals and applications. *Nano Energy* 85: 105948.

6 Li, M. and Xia, Z. (2021). Recent progress of zero-dimensional luminescent metal halides. *Chemical Society Reviews* 50: 2626–2662.

7 Lin, H., Zhou, C., Tian, Y. et al. (2018). Low-dimensional organometal halide perovskites. *ACS Energy Letters* 3: 54–62.

8 Liu, S., Fang, X., Lu, B., and Yan, D. (2020). Wide range zero-thermal-quenching ultralong phosphorescence from zero-dimensional metal halide hybrids. *Nature Communications* 11: 4649.

9 Jana, J., Ngo, Y.-L.T., Chung, J.S., and Hur, S.H. (2020). Contribution of carbon dot nanoparticles in electrocatalysis: development in energy conversion process. *Journal of Electrochemical Science and Technology* 11: 220–237.

10 Li, S., Zhang, Y., and Huang, H. (2022). Black phosphorus-based heterostructures for photocatalysis and photoelectrochemical water splitting. *Journal of Energy Chemistry* 67: 745–779.

11 Shang, L., Xu, J., and Nienhaus, G.U. (2019). Recent advances in synthesizing metal nanocluster-based nanocomposites for application in sensing, imaging and catalysis. *Nano Today* 28: 100767.

12 Antoine, R., Broyer, M., and Dugourd, P. (2023). Metal nanoclusters: from fundamental aspects to electronic properties and optical applications. *Science and Technology of Advanced Materials* 24: 2222546.

13 Cook, A.W. and Hayton, T.W. (2018). Case studies in nanocluster synthesis and characterization: challenges and opportunities. *Accounts of Chemical Research* 51: 2456–2464.

14 Ehara, M. and Priyakumar, U.D. (2019). Gold-palladium nanocluster catalysts for homocoupling: electronic structure and interface dynamics. *Chemical Record* 19: 947–959.

15 Ren, C., Shu, T., Du, X. et al. (2022). Luminescent sensors based on the assembly of coinage metal nanoclusters. *Chemosensors* 10: 253.

16 Oshima, T., Wang, Y., Lu, D. et al. (2019). Photocatalytic overall water splitting on Pt nanocluster-intercalated, restacked $KCa_2Nb_3O_{10}$ nanosheets: the promotional effect of co-existing ions. *Nanoscale Advances* 1: 189–194.

17 Tao, L., Huang, M., Guo, S. et al. (2019). Surface modification of $NiCo_2Te_4$ nanoclusters: a highly efficient electrocatalyst for overall water-splitting in neutral solution. *Applied Catalysis B: Environmental* 254: 424–431.

18 Ijaz, I., Gilani, E., Nazir, A., and Bukhari, A. (2020). Detail review on chemical, physical and green synthesis, classification, characterizations and applications of nanoparticles. *Green Chemistry Letters and Reviews* 13: 59–81.

19 Sun, S.H. and Zeng, H. (2002). Size-controlled synthesis of magnetite nanoparticles. *Journal of the American Chemical Society* 124: 8204–8205.

20 Pan, S., Liu, C., Li, Y. et al. (2022). Ultrafast self-assembly Fe_2O_3 nanoparticles confined in carbon layers toward robust heterogeneous electro-Fenton reaction. *Electrochimica Acta* 433: 141262.

21 Feng, W., Feng, Y., Chen, J. et al. (2022). Interfacial electronic engineering of Ru/FeRu nanoparticles as efficient trifunctional electrocatalyst for overall water splitting and Zn-air battery. *Chemical Engineering Journal* 437: 135456.

22 Bazioti, C., Lovvik, O.M., Poulia, A. et al. (2022). Probing the structural evolution and its impact on magnetic properties of $FeCoNi(AlMn)_{(x)}$ high-entropy alloy at the nanoscale. *Journal of Alloys and Compounds* 910: 164724.

23 Reiter, D.E., Kuhn, T., and Axt, V.M. (2019). Distinctive characteristics of carrier-phonon interactions in optically driven semiconductor quantum dots. *Advances in Physics: X* 4: UNSP 1655478.

24 Sothmann, B., Sanchez, R., and Jordan, A.N. (2015). Thermoelectric energy harvesting with quantum dots. *Nanotechnology* 26: 032001.

25 Zheng, Z., Ji, H., Yu, P., and Wang, Z. (2016). Recent progress towards quantum dot solar cells with enhanced optical absorption. *Nanoscale Research Letters* 11: 266.

26 Rong, J., Wang, Z., Lv, J. et al. (2021). $Ni(OH)_2$ quantum dots as a stable cocatalyst modified alpha-Fe_2O_3 for enhanced photoelectrochemical water-splitting. *Chinese Journal of Catalysis* 42: 1999–2009.

27 Xu, N., Zhang, Y., Zhang, T. et al. (2019). Efficient quantum dots anchored nanocomposite for highly active ORR/OER electrocatalyst of advanced metal-air batteries. *Nano Energy* 57: 176–185.

28 Zaiyin, H. and Chunfang, C. (2006). Research and progress in direct-oxidized growth of quasi- one-dimensional oxide nanomaterials. *Materials Review* 20: 52–55,59.

29 Zhou, J., Wang, T., Chen, Y., and Zhang, Y. (2008). Progresses in the hydrothermal synthesis of one-dimensional nanomaterials. *Chemistry* 71: 510–517.

30 Wencheng, L.I.U., wei, C.A.I., and Xianglong, M. (2006). Development of quasi one-dimensional nanomaterials. *Journal of Functional Materials* 37: 515–518,523.

31 Li, Y., Yang, X.-Y., Feng, Y. et al. (2012). One-dimensional metal oxide nanotubes, nanowires, nanoribbons, and nanorods: synthesis, characterizations, properties and applications. *Critical Reviews in Solid State and Materials Sciences* 37: 1–74.

32 Jariwala, D., Sangwan, V.K., Lauhon, L.J. et al. (2013). Carbon nanomaterials for electronics, optoelectronics, photovoltaics, and sensing. *Chemical Society Reviews* 42: 2824–2860.

33 Lee, K., Mazare, A., and Schmuki, P. (2014). One-dimensional titanium dioxide nanomaterials: nanotubes. *Chemical Reviews* 114: 9385–9454.

34 Yao, S. and Zhu, Y. (2015). Nanomaterial-enabled stretchable conductors: strategies, materials and devices. *Advanced Materials* 27: 1480–1511.

35 Zhang, Z., Ouyang, Y., Cheng, Y. et al. (2020). Size-dependent phononic thermal transport in low-dimensional nanomaterials. *Physics Reports-Review Section of Physics Letters* 860: 1–26.

36 Faraji, M., Yousefi, M., Yousefzadeh, S. et al. (2019). Two-dimensional materials in semiconductor photoelectrocatalytic systems for water splitting. *Energy & Environmental Science* 12: 59–95.

37 Peng, S., Jin, G., Li, L. et al. (2016). Multi-functional electrospun nanofibres for advances in tissue regeneration, energy conversion & storage, and water treatment. *Chemical Society Reviews* 45: 1225–1241.

38 Zhang, P., Lu, X.F., Nai, J. et al. (2019). Construction of hierarchical Co-Fe oxyphosphide microtubes for electrocatalytic overall water splitting. *Advanced Science* 6: 1900576.

39 Chen, J., Liu, J., Xie, J.-Q. et al. (2019). Co-Fe-P nanotubes electrocatalysts derived from metal-organic frameworks for efficient hydrogen evolution reaction under wide pH range. *Nano Energy* 56: 225–233.

40 Ganji, S. (2019). Nanowire growths, and mechanisms of these growths for developing novel nanomaterials. *Journal of Nanoscience and Nanotechnology* 19: 1849–1874.

41 Tuo, K., Mao, L., Li, S. et al. (2019). Research progress in nanowire—like anode materials for lithium ion batteries. *Modern Chemical Industry* 39: 57–61.

42 Wei, X.-W., Wu, K.-L., Zhu, G.-X. et al. (2012). Ultrasonic-assisted surfactant-free synthesis of highly magnetized FeCo alloy nanocrystallite from ferric and cobalt salt. *Journal of Alloys and Compounds* 539: 21–25.

43 Panina, L.V., Zagorskiy, D.L., Shymskaya, A. et al. (2022). 1D nanomaterials in Fe-group metals obtained by synthesis in the pores of polymer templates: correlation of structure, magnetic, and transport properties. *Physica Status Solidi a-Applications and Materials Science* 219: 2100538.

44 Lohse, S.E. and Murphy, C.J. (2013). The quest for shape control: a history of gold nanorod synthesis. *Chemistry of Materials* 25: 1250–1261.

45 Yang, J., Son, J.S., Yu, J.H. et al. (2013). Advances in the colloidal synthesis of two-dimensional semiconductor nanoribbons. *Chemistry of Materials* 25: 1190–1198.

46 Zhang, T., Wu, M.-Y., Yan, D.-Y. et al. (2018). Engineering oxygen vacancy on NiO nanorod arrays for alkaline hydrogen evolution. *Nano Energy* 43: 103–109.

47 Zhang, X., Wang, J., Zong, K. et al. (2024). Porous nickel sulfide nanorods serve as a multifunctional electrocatalyst for the hydrogen evolution reaction, urea electrooxidation reaction, and nitrate reduction reaction. *Journal of Materials Chemistry A* 12: 25519–25529.

48 Wei, P., Yang, Y., Kang, H. et al. (2020). Controllable synthesis of Fe-doped $NiCo_{(2)}O_{(4)}$nanobelts as superior catalysts for oxygen evolution reaction. *Chemistry—a European Journal* 26: 13725–13729.

49 Wang, Y., Wang, L., Zhang, X. et al. (2021). Two-dimensional nanomaterials with engineered bandgap: synthesis, properties, applications. *Nano Today* 37: 101059.

50 Yuan, R., Chen, L., and Wu, C. (2022). Heat conduction behavior of two-dimensional nanomaterials and their interface regulation. *Acta Chimica Sinica* 80: 839–847.

51 Cai, S. and Yang, R. (2020). Two-dimensional nanomaterials with enzyme-like properties for biomedical applications. *Frontiers in Chemistry* 8: 565940.

52 Gao, L., Song, Z., Sun, Z. et al. (2018). Application and development of novel two-dimensional nanomaterials in electrochemistry. *Chinese Journal of Applied Chemistry* 35: 247–258.

53 Shams, M., Mansukhani, N., Hersam, M.C.C. et al. (2023). Environmentally sustainable implementations of two-dimensional nanomaterials. *Frontiers in Chemistry* 11: 1132233.

54 Zhang, W., Qian, L., Jiang, N. et al. (2021). Two-dimensional nanomaterials for biosensing. *Journal of Analytical Science* 37: 485–492.

55 Fang, W., Huang, L., Zaman, S. et al. (2020). Recent progress on two-dimensional electrocatalysis. *Chemical Research in Chinese Universities* 36: 611–621.

56 Huang, X., Zeng, Z., and Zhang, H. (2013). Metal dichalcogenide nanosheets: preparation, properties and applications. *Chemical Society Reviews* 42: 1934–1946.

57 Iqbal, A., Hong, J., Ko, T.Y., and Koo, C.M. (2021). Improving oxidation stability of 2D MXenes: synthesis, storage media, and conditions. *Nano Convergence* 8: 9.

58 Jun, B.-M., Kim, S., Heo, J. et al. (2019). Review of MXenes as new nanomaterials for energy storage/delivery and selected environmental applications. *Nano Research* 12: 471–487.

59 Khan, A.H., Ghosh, S., Pradhan, B. et al. (2017). Two-dimensional (2D) nanomaterials towards electrochemical nanoarchitectonics in energy-related applications. *Bulletin of the Chemical Society of Japan* 90: 627–648.

60 Liu, B. and Zhou, K. (2019). Recent progress on graphene-analogous 2D nanomaterials: properties, modeling and applications. *Progress in Materials Science* 100: 99–169.

61 Tan, C., Cao, X., Wu, X.-J. et al. (2017). Recent advances in ultrathin two-dimensional nanomaterials. *Chemical Reviews* 117: 6225–6331.

62 Tan, C., Liu, Z., Huang, W., and Zhang, H. (2015). Non-volatile resistive memory devices based on solution-processed ultrathin two-dimensional nanomaterials. *Chemical Society Reviews* 44: 2615–2628.

63 Yang, W., Gan, L., Li, H., and Zhai, T. (2016). Two-dimensional layered nanomaterials for gas-sensing applications. *Inorganic Chemistry Frontiers* 3: 433–451.

64 Zhu, C., Du, D., and Lin, Y. (2017). Graphene-like 2D nanomaterial-based biointerfaces for biosensing applications. *Biosensors & Bioelectronics* 89: 43–55.

65 Kuang, Y., Feng, G., Li, P. et al. (2016). Single-crystalline ultrathin nickel nanosheets array from in situ topotactic reduction for active and stable electrocatalysis. *Angewandte Chemie, International Edition* 55: 693–697.

66 Jang, K., Kim, H.J., and Son, S.U. (2010). Low-temperature synthesis of ultrathin rhodium nanoplates via molecular orbital symmetry interaction between rhodium precursors. *Chemistry of Materials* 22: 1273–1275.

67 Bi, Z.-X., Guo, R.-T., Hu, X. et al. (2022). Research progress on photocatalytic reduction of CO_2 based on LDH materials. *Nanoscale* 14: 3367–3386.

68 Bodhankar, P.M., Sarawade, P.B., Singh, G. et al. (2021). Recent advances in highly active nanostructured NiFe LDH catalyst for electrochemical water splitting. *Journal of Materials Chemistry A* 9: 3180–3208.

69 Arif, M., Yasin, G., Shakeel, M. et al. (2018). Coupling of bifunctional CoMn-layered double hydroxide@graphitic C_3N_4 nanohybrids towards efficient photoelectrochemical overall water splitting. *Chemistry—An Asian Journal* 13: 1045–1052.

70 Chen, S., Zhang, Y., Zhong, H., and Cao, Z. (2022). Trimetallic CoFeCr-LDH@MoS_2 as a highly efficient bifunctional electrocatalyst for overall water splitting. *Colloids and Surfaces a-Physicochemical and Engineering Aspects* 655: 130146.

71 Li, C., Bao, Y., Liu, E. et al. (2023). Recent advances of modified Ni (Co, Fe)-based LDH 2D materials for water splitting. *Molecules* 28: 1475.

72 Li, X.-P., Zheng, L.-R., Liu, S.-J. et al. (2022). Heterostructures of NiFe LDH hierarchically assembled on MoS_2 nanosheets as high-efficiency electrocatalysts for overall water splitting. *Chinese Chemical Letters* 33: 4761–4765.

73 Liu, W., Bao, J., Guan, M. et al. (2017). Nickel-cobalt-layered double hydroxide nanosheet arrays on Ni foam as a bifunctional electrocatalyst for overall water splitting. *Dalton Transactions* 46: 8372–8376.

74 Wang, T., Wang, W., Shao, W. et al. (2021). Synthesis and electronic modulation of nanostructured layered double hydroxides for efficient electrochemical oxygen evolution. *ChemSusChem* 14: 5112–5134.

75 Meng, X. and Deng, D. (2017). Two-dimensional materials for electrocatalytic water splitting. *Chinese Science Bulletin* 62: 3154–3174.

76 Hu, Z., Xu, Y., Xiang, W. et al. (2022). Amorphous Ni-P-S@FeOOH/CC catalyst for high oxygen evolution activity: preparation, characterization and modeling. *Chemical Engineering Science* 258: 117761.

77 Liu, Y., Guo, J., Liu, X. et al. (2025). Corrosion-driven Ni_3S_4 gradient in NiFe-LDH enables durable industrial-scale water electrolysis. *Angewandte Chemie International Edition* 137: e202516894.

78 Xu, H., Zhang, W.-D., Liu, J. et al. (2022). Intercalation-induced partial exfoliation of NiFe LDHs with abundant active edge sites for highly enhanced oxygen evolution reaction. *Journal of Colloid and Interface Science* 607: 1353–1361.

79 Wang, Y., Xie, C., Zhang, Z. et al. (2018). In situ exfoliated, N-doped, and edge-rich ultrathin layered double hydroxides nanosheets for oxygen evolution reaction. *Advanced Functional Materials* 28: 1703363.

80 Wang, Y., Tao, S., Lin, H. et al. (2021). Atomically targeting NiFe LDH to create multivacancies for OER catalysis with a small organic anchor. *Nano Energy* 81: 105606.

81 Wang, C.-P., Lin, Y.-X., Cui, L. et al. (2023). 2D metal-organic frameworks as competent electrocatalysts for water splitting. *Small* 19: 2207342.

82 Borja, L. (2017). 2D electrocatalytic MOF sets efficiency record for water splitting. *MRS Bulletin* 42: 550–551.

83 Ge, K., Zhang, Y., Zhao, Y. et al. (2022). Room temperature preparation of two-dimensional black phosphorus@metal organic framework heterojunctions and their efficient overall water-splitting electrocatalytic reactions. *ACS Applied Materials & Interfaces* 14: 31502–31509.

84 Xiao, L., Wang, Z., and Guan, J. (2022). 2D MOFs and their derivatives for electrocatalytic applications: recent advances and new challenges. *Coordination Chemistry Reviews* 472: 214777.

85 Zhao, S., Tan, C., He, C.-T. et al. (2020). Structural transformation of highly active metal-organic framework electrocatalysts during the oxygen evolution reaction. *Nature Energy* 5: 881–890.

86 Feng, C., Li, Z., Wang, J. et al. (2019). Synthesis of metal-organic framework-derived cobalt disulfide with high-performance oxygen reduction reaction catalytic properties. *Journal of Electroanalytical Chemistry* 840: 27–34.

87 Xu, Y., Yu, S., Ren, T. et al. (2020). Hydrophilic/aerophobic hydrogen-evolving electrode: NiRu-based metal-organic framework nanosheets in situ grown on conductive substrates. *ACS Applied Materials & Interfaces* 12: 34728–34735.

88 Chang, H., Liang, Z., Wang, L., and Wang, C. (2022). Research progress in improving the oxygen evolution reaction by adjusting the 3d electronic structure of transition metal catalysts. *Nanoscale* 14: 5639–5656.

89 Anantharaj, S. and Aravindan, V. (2020). Developments and perspectives in 3D transition-metal-based electrocatalysts for neutral and near-neutral water electrolysis. *Advanced Energy Materials* 10: 1902666.

90 Chaudhari, N.K., Jin, H., Kim, B., and Lee, K. (2017). Nanostructured materials on 3D nickel foam as electrocatalysts for water splitting. *Nanoscale* 9: 12231–12247.

91 Chen, R., Yan, Z.-H., and Kong, X.-J. (2020). Recent advances in first-row transition metal clusters for photocatalytic water splitting. *Chemphotochem* 4: 157–167.

92 Bai, S., Yin, W., Wang, L. et al. (2016). Surface and interface design in cocatalysts for photocatalytic water splitting and CO_2 reduction. *RSC Advances* 6: S7446–S7463.

93 Dai, Y., Yu, J., Cheng, C. et al. (2020). Engineering the interfaces in water-splitting photoelectrodes—an overview of the technique development. *Journal of Materials Chemistry A* 8: 6984–7002.

94 Hu, Y., Gao, C., and Xiong, Y. (2018). Surface and interface design for photocatalytic water splitting. *Dalton Transactions* 47: 12035–12040.

95 Su, T., Shao, Q., Qin, Z. et al. (2018). Role of interfaces in two-dimensional photocatalyst for water splitting. *ACS Catalysis* 8: 2253–2276.

96 Lian, Q., Zhong, L., Du, C. et al. (2019). Interfacing epitaxial dinickel phosphide to 2D nickel thiophosphate nanosheets for boosting electrocatalytic water splitting. *ACS Nano* 13: 7975–7984.

97 Zhu, C., Wang, A.-L., Xiao, W. et al. (2018). In situ grown epitaxial heterojunction exhibits high-performance electrocatalytic water splitting. *Advanced Materials* 30: e1705516.

98 An, B., Bian, R., Dong, J. et al. (2024). Theoretical and experimental investigation on electrostatic field dynamics of Co_3O_4@NiP_x electrocatalyst with core shell structure in overall water splitting reactions. *Chemical Engineering Journal* 485: 149903.

99 Xie, Y., Yi, C., Zhou, Q. et al. (2025). Magnetic organic frame material Fe_3O_4@Au/COF for sensitive SERS detection and photo-Fenton degradation of tetracycline. *Journal of Environmental Chemical Engineering* 13: 119472.

100 Liu, W., Zhang, L., Yan, W. et al. (2016). Single-atom dispersed Co-N-C catalyst: structure identification and performance for hydrogenative coupling of nitroarenes. *Chemical Science* 7: 5758–5764.

101 Liang, C., Zou, P., Nairan, A. et al. (2020). Exceptional performance of hierarchical Ni-Fe oxyhydroxide@NiFe alloy nanowire array electrocatalysts for large current density water splitting. *Energy & Environmental Science* 13: 86–95.

102 Radhakrishnan, J., Ratna, S., and Biswas, K. (2022). Metal oxide/2D layered TMDs composites for H_2 evolution reaction via photocatalytic water splitting—a mini review. *Inorganic Chemistry Communications* 145: 109971.

103 Xiao, Y., Guo, X., Yang, N., and Zhang, F. (2021). Heterostructured MOFs photocatalysts for water splitting to produce hydrogen. *Journal of Energy Chemistry* 58: 508–522.

104 Zheng, D., Yu, L., Liu, W. et al. (2021). Structural advantages and enhancement strategies of heterostructure water-splitting electrocatalysts. *Cell Reports Physical Science* 2: 100443.

105 Zhong, H., Feng, Y., and Alonso-Vante, N. (2022). Heterostructures based on transition metal chalcogenides and layered double hydroxides for enhanced water splitting. *Current Opinion in Electrochemistry* 34: 101016.

106 Wang, L., Hung, S.-F., Zhao, S. et al. (2025). Modulating the covalency of Ru-O bonds by dynamic reconstruction for efficient acidic oxygen evolution. *Nature Communications* 16: 3502.

107 Liu, Y., Ma, C., Zhang, Q. et al. (2019). 2D electron gas and oxygen vacancy induced high oxygen evolution performances for advanced Co_3O_4/CeO_2 nanohybrids. *Advanced Materials* 31: e1900062.

108 Zheng, Y.-R., Wu, P., Gao, M.-R. et al. (2018). Doping-induced structural phase transition in cobalt diselenide enables enhanced hydrogen evolution catalysis. *Nature Communications* 9: 2533.

109 Wang, Y., Li, X., Zhang, M. et al. (2020). Lattice-strain engineering of homogeneous $NiS_{(0.5)}Se_{(0.5)}$core-shell nanostructure as a highly efficient and robust electrocatalyst for overall water splitting. *Advanced Materials* 32: e2000231.

110 Li, Y., Xu, H., Huang, H. et al. (2018). One-dimensional MoO_2-$Co_2Mo_3O_8$@C nanorods: a novel and highly efficient oxygen evolution reaction catalyst derived from metal-organic framework composites. *Chemical Communications* 54: 2739–2742.

111 Wang, B., Tang, C., Wang, H.-F. et al. (2019). A nanosized CoNi hydroxide@hydroxysulfide core-shell heterostructure for enhanced oxygen evolution. *Advanced Materials* 31: e1805658.

112 Guo, Y., Tang, J., Qian, H. et al. (2017). One-pot synthesis of zeolitic imidazolate framework 67-derived hollow Co_3S_4@MoS_2 heterostructures as efficient bifunctional catalysts. *Chemistry of Materials* 29: 5566–5573.

113 Guo, X., Zhang, H., Xia, W. et al. (2024). Constructing Ag single atoms and nanoparticles co-decorated CoO(O)H as highly active electrocatalyst for oxygen evolution reaction under large current density. *Advanced Functional Materials* 34: 2316539.

114 Zhang, T., Wu, X., Fan, Y. et al. (2020). Hollow CeO_x/CoP heterostructures using two-dimensional Co-MOF as template for efficient and stable electrocatalytic water splitting. *ChemNanoMat* 6: 1119–1126.

Part III

Diverse Applications of These Nanomaterials

5

Applications of Iron (Fe, Co, and Ni)-Group Metal Compound Nanomaterials in Rechargeable Batteries

Rechargeable batteries serve as the cornerstone of the electrochemical energy storage market and play a pivotal role in building a new energy system and promoting energy conservation and carbon reduction. Although lithium-ion batteries have been widely adopted in consumer electronics and electric vehicles, their performance limitations are becoming increasingly evident as energy storage demands grow more diverse, underscoring the urgent need to develop new rechargeable battery systems. Among various key materials, iron (Fe, Co, and Ni)-group metal compound nanomaterials have demonstrated great potential in enhancing battery performance due to their unique electronic structures and electrochemical properties. Current research has expanded into multiple systems, including lithium-ion batteries (LIBs), sodium/potassium-ion batteries (SIBs/KIBs), lithium-sulfur batteries (Li–S), and metal–air batteries. However, critical scientific challenges—such as understanding electrode reaction mechanisms, charge transfer kinetics, material structure design, and interface regulation—still require systematic and in-depth investigation at the material level to advance the practical application of next-generation high-performance batteries.

5.1 LIB

The increasing depletion of fossil fuels, coupled with stringent control of carbon dioxide emissions, has created urgent demands for the sustainability of energy systems, significantly accelerating the development of alternative energy sources such as solar, nuclear, wind, tidal, and bioenergy [1]. Among various renewable energy technologies, lithium-ion batteries (LIBs), which operate based on the exchange of lithium ions between electrodes, offered promising solutions to the energy crisis and related environmental challenges during their early commercialization phase. Over the past few decades, LIBs have become the primary power source for portable electronic devices and are deeply integrated into nearly every aspect of daily life [2, 3]. However, as the application scope of LIBs expands into next-generation high-power and high-energy-density energy storage systems, they face significant challenges in further enhancing their performance, particularly in terms of energy

Iron-Group Metal Compound Nanomaterials: Preparation, Characterization, Structure, and Applications, First Edition. Jinlong Zheng.

density. Against this backdrop, nanomaterials composed of iron-group metal compounds (Fe, Co, and Ni) have emerged as a research hotspot in electrode materials for rechargeable batteries due to their unique advantages. These materials are not only abundant in the Earth's crust and cost-effective, but their variable valence states and excellent electrochemical activity also make them suitable for charge storage requirements in various systems, such as lithium-ion, sodium-ion, and potassium-ion batteries (KIBs). Whether serving as cathode materials to enable efficient ion intercalation/deintercalation reactions or as anode materials to enhance electrical conductivity and cycling stability, iron-group metal compound nanomaterials provide critical support for overcoming the performance bottlenecks of rechargeable batteries. They have become a key driving force in advancing energy storage technologies toward higher efficiency and lower costs [2, 3].

As an integrated system, the performance of LIBs is profoundly influenced by the development level of their constituent materials. Commercial LIBs typically consist of a cathode, a porous separator, and a Li^+-conducting electrolyte. The cathode commonly employs lithium metal oxides such as $LiCoO_2$, while the separator, primarily made of polyolefins like polyethylene or polypropylene, facilitates Li^+ migration and isolates the electrodes. The electrolyte, usually composed of $LiPF_6$ in organic carbonates, serves as the medium for Li^+ transport [4, 5]. The charge–discharge mechanism is well-defined: during charging, Li^+ deintercalates from the cathode, migrates through the electrolyte and separator, and embeds into the anode; during discharging, Li^+ follows the reverse path back to the cathode, while electrons flow directionally through the external circuit, completing the energy conversion cycle [6]. In this process, electrode materials, as the core components for lithium storage and charge transfer, directly determine key performance metrics of LIBs, such as capacity, charging efficiency, and cycle life. Iron-group metal compound nanomaterials (Fe, Co, and Ni) are emerging as a key focus in electrode material innovation, leveraging their unique advantages. Compared to traditional $LiCoO_2$, these materials are not only more abundant in the Earth's crust and cost-effective but also offer larger specific surface areas and shorter ion diffusion paths due to their nanostructured design. Their variable valence states further enable compatibility with diverse electrochemical reactions. Whether employed as cathodes to construct high-voltage ($\approx$5 V) systems or as anodes to enhance lithium storage capacity, iron-group metal compound nanomaterials provide critical support for overcoming the energy density limitations of LIBs, serving as a key driver for performance advancement [7].

5.1.1 Cathode

Battery voltage is fundamentally determined by the difference in redox energy between the cathode and anode. This implies that to achieve higher voltages, the energy band of the cathode material should be as low as possible, while that of the anode material should be as high as possible. In other words, the cathode needs to stabilize high oxidation states of metals at lower energy bands, whereas the anode needs to stabilize low oxidation states at higher energy bands [8]. Iron-group metals (Fe, Co, and Ni), with their diverse valence states, serve as ideal candidates for achieving this goal. Their compound nanomaterials further provide extensive opportunities for voltage regulation due to their structural advantages. Thus, the

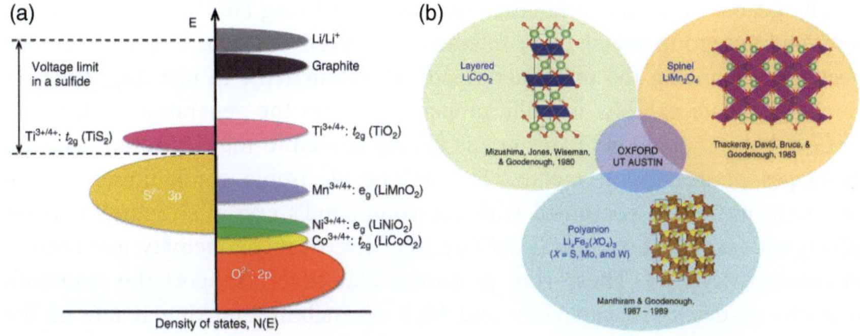

Figure 5.1 (a) In cathodes, the higher energy level of the S^{2-}: 3p band top limits the battery voltage to <2.5 V. In contrast, the lower energy level of the O^{2-}: 2p band top enables the material to access lower energy bands through higher oxidation states, thereby significantly increasing the battery voltage to approximately 4 V; (b) three types of oxide cathode materials discovered in the 1980s. *Source*: Ref. [9] / Springer Nature / CC BY 4.0.

key challenge lies in enabling metal ions such as Fe, Co, and Ni to occupy lower energy bands at higher oxidation states, thereby effectively increasing the battery voltage. After extensive fundamental research on transition metal oxides and related materials, Goodenough's team designed oxide cathodes based on a critical insight (Figure 5.1a) [9]: the energy at the top of the S^{2-} 3p band is higher than that of the O^{2-} 2p band. This difference implies that in sulfides, attempts to lower the cathode redox energy by increasing the oxidation states of metals like Fe or Co (e.g., $Co^{3+/4+}$) are limited by the higher S^{2-} 3p band top and may even lead to the oxidation of S^{2-} to molecular disulfide ions $(S_2)^{2-}$. In contrast, in oxides, the lower energy of the O^{2-} 2p band top allows for the introduction of low-energy-band high oxidation states of metals like Fe, Co, and Ni (e.g., $Co^{3+/4+}$), significantly reducing the cathode redox energy and thereby elevating the battery voltage to the 4 V level. Building on this core concept, Goodenough's research group successively discovered three major types of oxide cathode materials centered on iron-group metals (Fe, Co, and Ni) in the 1980s (Figure 5.1b). The nanostructures derived from these materials remain a key focus in the study of high-voltage cathodes for rechargeable batteries today. These three types of oxide cathode materials correspond to the three major systems of layered, spinel, and polyanion, covering different structural types and electrochemical properties, laying a core foundation for the subsequent commercial application, modification and optimization of lithium-ion battery cathode materials [10].

5.1.1.1 Layered Oxides

As a typical representative of iron-group metal (Fe, Co, and Ni) compound nanomaterials, the layered $LiCoO_2$ (Figure 5.2a) oxide cathode possesses an O3 structure [11]. Within the cubic close-packed oxide ion array, Li^+ and Co^{3+} (iron-group metal ions) are arranged in an alternating pattern [16]. The significant differences in charge and size between these ions promote a high degree of cation ordering, forming the structural basis for rapid two-dimensional ion diffusion and enhanced conductivity within the lithium layers [17]. Lithium-ion migration follows an octahedral–tetrahedral–octahedral path through shared faces, which has the lowest energy barrier, further meeting the ion transport efficiency requirements of rechargeable batteries. Simultaneously, cation ordering facilitates direct Co–Co (between iron-group metal ions) interactions within the cobalt planes, thereby effectively enhancing the material's electrical conductivity. In fact, $Li_{1-x}CoO_2$ in the charged state exhibits metallic properties due to the generation of holes in the $Co^{3+/4+}$:$t_{2g}{}^{6-x}$ band, a characteristic closely related to the electronic structure advantages of iron-group metals. The unique electronic configuration of the iron-group metal Co, combined with the compound's crystal structure, endows $LiCoO_2$ with excellent structural stability, high electrical conductivity, and lithium ion conductivity [18]. These core properties collectively support the material's rapid charge–discharge capability and high reversibility, making it one of the highest-performing cathode materials in the field of rechargeable batteries, with an operating voltage of approximately 4 V. Compared to sulfide cathodes from the 1970s, $LiCoO_2$ not only increased the operating voltage of rechargeable batteries from <2.5 to ~4 V but also eliminated the need for metallic lithium anodes, enabling pairing with lithium-free anodes such as graphite. This established $LiCoO_2$ as a core cathode material for modern LIBs [19], highlighting the critical application value of iron-group metal compound nanomaterials in rechargeable batteries. However, influenced by the electronic orbital characteristics of the iron-group metal Co, the $Co^{3+/4+}$ energy band overlaps with the O^{2-}:2p band, leading to lattice oxygen release in $LiCoO_2$ during deep charging ($(1-x) < 0.5$). This limits its practical capacity to approximately 140 mA h g^{-1}, representing one of the areas where iron-group metal compound nanomaterials require further optimization in rechargeable battery applications.

Layered $LiMO_2$ (where M is a 3d transition metal) systems have gradually become an important research direction for cathode materials in rechargeable batteries. Among them, Fe and Ni, as core iron-group metal elements, have drawn research focus due to the performance limitations of their related compounds. $LiMO_2$ systems containing Fe, due to their low octahedral site stabilization energy (OSSE), are prone to layered-to-spinel transformations or other structural distortions during charge–discharge cycles, making it difficult to meet the structural stability requirements for cathode materials. $LiNiO_2$, influenced by the electronic properties of the iron-group metal Ni, tends to undergo reduction reactions of Ni^{3+}, and lithium volatilization during synthesis hampers the formation highly ordered crystal structures, limiting its direct application in rechargeable batteries [20, 21]. To balance the dual demands of material cost and capacity in rechargeable batteries, research focus has shifted to composite systems of iron-group metals (Ni and Co) and Mn, leading to the development of iron-group metal composite compound nanomaterials $LiNi_{1-y-z}Mn_yCo_zO_2$ (NMC) by substituting part of Co with Mn and Ni [22]. As a typical iron-group metal composite nanomaterial, NMC features synergistic functionality of the two iron-group metals Ni and Co with Mn. Mn^{4+} primarily provides structural stability (without participating in electrochemical reactions), with its $Mn^{3+/4+}$ energy band located above the top of the O^{2-}:2p band, reducing the likelihood of the oxygen loss and offering high chemical stability, though its low OSSE makes it susceptible to cation migration and structural transformations. The iron-group metal Co has high OSSE, endowing the material with

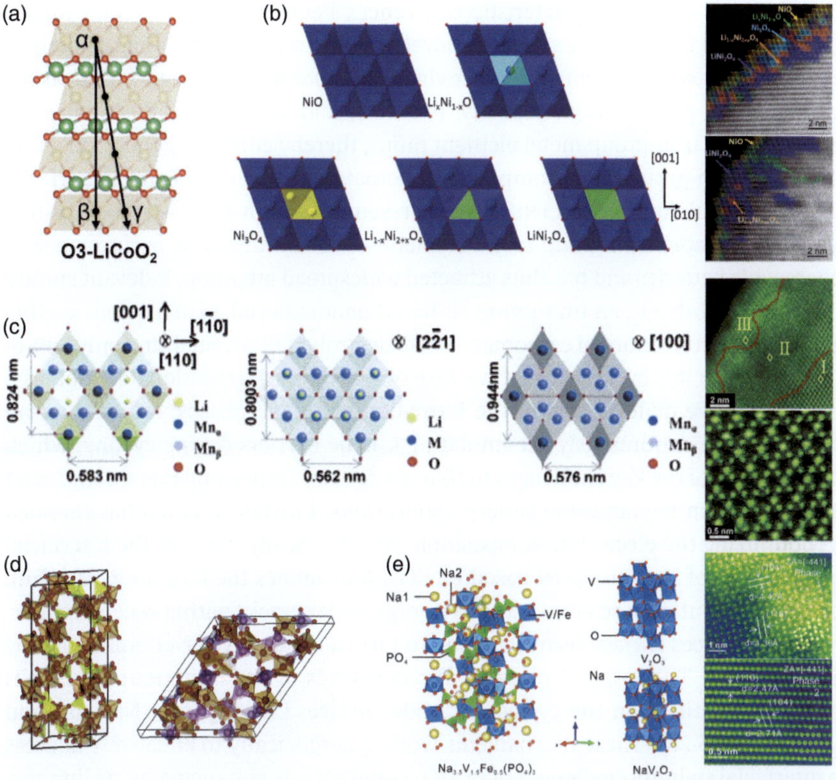

Figure 5.2 (a) Structural schematic of O3-type $LiCoO_2$ viewed along the [001] direction. *Source:* Reproduced from Ref. [11] / with permission of Springer Nature. (b) Atomic models of five degradation phases (all models are viewed along the [100] direction of the $LiNiO_2$ trigonal structure, equivalent to the [110] direction of the cubic structure) and HAADF-STEM characterization of the cycled $LiNiO_2$ electrode. *Source:* Reproduced from Ref. [12] / with permission of John Wiley & Sons. (c) Crystal structure of $LiMn_2O_4$ viewed along the [1$\bar{1}$0] cubic direction; structural schematic of layered cathode material $LiCoO_2$ along the [2$\bar{2}$1] trigonal direction; schematic of tetragonal Mn_3O_4 structure viewed along the [100] tetragonal direction; and HAADF-STEM image of spinel lithium manganese oxide along the [110] tetragonal direction, where Region I shows high consistency with the $LiMn_2O_4$ spinel structure. *Source:* Reproduced from Ref. [13] / with permission of John Wiley & Sons. (d) Crystal structures of $Fe_2(SO_4)_3$ with NASICON-type rhombohedral structure and $Fe_2(MoO_4)_3$ with monoclinic structure. *Source:* Reproduced from Ref. [14] / with permission of American Chemical Society. (e) Schematic crystal structures of $Na_{3.5}V_{1.5}Fe_{0.5}(PO_4)_3$; magnified atomic-resolution HAADF-STEM images of Phase 1 and Phase 2 taken along the [−441] zone axis. *Source:* Reproduced from Ref. [15] / with permission of The Royal Society of Chemistry.

excellent structural stability, and it exhibits metallic conductivity upon charging, ensuring efficient electron transport, but the $Co^{3+/4+}$ energy band overlaps with the top of the O^{2-}:2p band, posing a risk of oxygen evolution. The properties of the iron-group metal Ni lie between the two, with the $Ni^{3+/4+}$ energy band close to, but not significantly overlapping, the O^{2-}:2p band, allowing charging to Ni^{4+} without causing oxygen loss, and moderate OSSE providing good structural stability

[23, 24]. Based on the characteristic differences between the iron-group metals Ni and Co, the current research and development trend of NMC nanomaterials focuses on increasing Ni content and reducing Co content, aiming to achieve the comprehensive goals of high capacity, low cost, and structural stability through optimization of iron-group metal element ratios, thereby enhancing the application potential of iron-group metal compound nanomaterials in rechargeable batteries.

The interfacial stability of $LiNiO_2$—a representative high-nickel cathode system with Ni as the core iron-group metal element—directly affects its performance in rechargeable batteries and has thus attracted widespread attention. Relevant studies [12] show that $LiNiO_2$, an iron-group Ni-based nanomaterial, exhibits poor cycling stability in the conventional carbonate-based electrolyte LP57, retaining only 28% of its capacity after 400 cycles at a high voltage of 4.4 V. This degradation does not originate from active material loss or the formation of fatigue phases; rather, the core issue lies in the continuously accumulating kinetic barriers during cycling, which represents one of the key challenges to be overcome for iron-group metal compound nanomaterials in rechargeable battery applications. Further research has revealed that optimizing the electrolyte composition can specifically improve the interfacial characteristics of iron-group Ni-based $LiNiO_2$. It promotes the formation of a thin, inorganic-rich interphase layer, reduces intergranular cracks within secondary particles, and induces surface reconstruction to form a conductive layer dominated by spinel $LiNi_2O_4$. Three spinel phases (all compounds of the iron-group metal Ni) have been identified on the cycled electrode surface: $LiNi_2O_4$, $Li_{1-x}Ni_{2+x}O_4$, and Ni_3O_4 (Figure 5.2b). These mechanisms work synergistically to effectively address the interfacial stability challenges of $LiNiO_2$, significantly enhancing its cycling performance. Under the same testing conditions, the capacity retention rate after 400 deep cycles increases to 84%, providing important technical support for the efficient application of iron-group metal compound nanomaterials in rechargeable batteries.

5.1.1.2 Spinel Oxides

Spinel $LiMn_2O_4$ (Figure 5.2c), as an iron-group metal compound nanomaterial (Mn shares similar electronic structure and chemical properties with Fe-group metals, both belonging to the transition metal system and often included in the research scope of iron-group related compounds), features a structure derived from lithium insertion into the magnetite (Fe_3O_4, a typical iron-group metal Fe compound) type framework, making it one of the most promising cathode materials in rechargeable batteries. In this material, $Mn^{3+/4+}$ occupies the 16d octahedral sites, while Li^+ occupies the 8a tetrahedral sites. The stable $[Mn_2]_{16d}O_4$ framework forms three-dimensional ionic diffusion channels through shared octahedral edges, allowing Li^+ to migrate between 8a tetrahedral sites via adjacent vacant 16c octahedra with the lowest energy barrier [13]—a structural feature closely related to the crystalline characteristics of iron-group metal compounds. Similar to iron-group metal Co-based $LiCoO_2$, $(Li_{1-x})_{8a}[Mn_2]_{16d}O_4$ contains mixed-valence high-spin $Mn^{3+/4+}$: the shared MnO_6 octahedra with $t_{2g}{}^3e_g{}^{1-x}$ configuration exhibit direct Mn–Mn interactions at the edges that promote good hopping electron conduction, though the material still behaves as a small-polaron semiconductor during charge/discharge processes [25]. Its three-dimensional stable structure, high electronic conductivity, and lithium-ion conductivity collectively endow the material with rapid charge/discharge capability and high reversibility, meeting the core performance requirements for electrode materials in rechargeable batteries. $Li_{1-x}Mn_2O_4$ operates at around 4 V with a practical capacity <130 mA h g^{-1}, since only one Li^+ can be reversibly extracted per two Mn ions on average [26]. Additional Li^+ can be inserted into vacant 16c octahedral sites at 3 V to form $[Li_2]_{16c}[Mn_2]_{16d}O_4$, while the original 8a-site Li^+ spontaneously migrates to vacant 16c octahedral sites, resulting in a 1 V voltage drop [27]—demonstrating the significant influence of site energy in solids on voltage. However, the Jahn–Teller distortion induced by Mn^{3+}: $t_{2g}{}^3e_g{}^1$ in $[Li_2]_{16c}[Mn_2]_{16d}O_4$ leads to cubic→tetragonal phase transition accompanied by large volume changes and two-phase reactions, making the capacity in the 3 V region practically unusable [28], which represents a key issue requiring optimization for iron-group-related transition metal compound nanomaterials in rechargeable battery applications.

Spinel $LiMn_2O_4$ offers significantly lower cost compared to iron-group Co-based $LiCoO_2$ (with Mn being two orders of magnitude cheaper than Co), yet it exhibits critical performance shortcomings in rechargeable battery applications: Mn^{3+} is prone to disproportionation reactions in the presence of trace H^+, generating Mn^{4+} and Mn^{2+}, where the dissolved Mn^{2+} can poison the graphite anode, severely limiting battery cycle life [29, 30]. Partial substitution of Mn with Li (e.g., 5 at.%) can suppress long-range Mn–Mn interactions and reduce Mn dissolution, thereby enhancing the material's cycling stability in rechargeable batteries [31] and providing an optimization strategy for adapting such transition metal compounds to battery systems dominated by iron-group metal compounds. Within the spinel LiM_2O_4 system, the only currently known stable M elements are Ti, V, and Mn, primarily because conventional high-temperature synthesis struggles to stabilize the highly oxidized $M^{3+/4+}$ states [32]. Among these, $LiTi_2O_4$ operates at a low voltage (~1.5 V), while LiV_2O_4 has a relatively low voltage (~3 V) and structural instability, making both unsuitable for meeting the performance requirements of electrode materials in rechargeable batteries. Regarding the preparation of iron-group metal (Co, Ni)-based spinel compounds, studies have attempted to chemically extract 50% Li from iron-group Co-based $LiCoO_2$ and Ni-based $LiNiO_2$, followed by medium-temperature calcination to convert them into spinel phases $Li_{0.5}CoO_2$ and $Li_{0.5}NiO_2$. However, due to the instability of high-valence Co/Ni states, the final products were mixtures of spinel-like phases and Co_3O_4/NiO (both being iron-group metal compounds) with poor crystallinity and electrochemical performance [33–36], limiting the direct application of iron-group metal spinel compounds.

5.1.1.3 Polyanionic Oxides

Building upon early research on the preparation of low-valent Mo^{4+} oxides through hydrogen reduction of $Ln_2(MoO_4)_3$ (where Ln represents lanthanide elements or Y), researchers have further applied the same hydrogen reduction strategy to synthesize polyanionic structural nanomaterials based on iron-group metal Fe, such as $Fe_2(MoO_4)_3$ and $Fe_2(WO_4)_3$ [37], thereby enriching the material systems of iron-group metal compounds for rechargeable batteries. These iron-based polyanionic nanomaterials exhibit excellent lithium-ion storage performance: each formula unit can reversibly intercalate/deintercalate two lithium ions, forming $Li_2Fe_2(MoO_4)_3$ or $Li_2Fe_2(WO_4)_3$, both demonstrating a discharge plateau around 3 V. This voltage is significantly higher than that of simple iron-group Fe oxides like Fe_2O_3 or Fe_3O_4 (<2.5 V). Despite utilizing the same $Fe^{2+}/^{3+}$ redox couple, the unique advantages of the polyanionic structure substantially enhance the voltage, better aligning with the voltage performance requirements for electrode materials in rechargeable batteries. Further studies revealed that another iron-group Fe-based polyanionic nanomaterial, $Fe_2(SO_4)_3$, which shares the same structure as $Fe_2(MoO_4)_3$, achieves a further increased discharge voltage of 3.6 V [38], fully demonstrating the voltage advantages of iron-group Fe-based polyanionic compounds in rechargeable batteries and providing important references for performance optimization and application expansion of such materials.

A comparison of the voltages of $Fe_2(XO_4)_3$ (X = Mo, W, S) with Fe_2O_3 reveals that the highly charged counter cations (Mo^{6+}, W^{6+}, and S^{6+}) significantly modulate the energy level of the $Fe^{2+/3+}$ redox couple by altering the nature of the Fe—O bond [14]. In the $Fe_2(XO_4)_3$ structure (Figure 5.2d), FeO_6 octahedra and XO_4 tetrahedra are connected via corner-sharing, forming a three-dimensional –O–Fe–O–X–O–Fe–O– network. The relatively strong covalent bonds in Mo—O and W—O pairs weaken the covalency of the Fe—O bond through an inductive effect, thereby lowering the energy of the $Fe^{2+/3+}$ redox couple and increasing the operational voltage to 3.0 V in $Fe_2(MoO_4)_3$ and $Fe_2(WO_4)_3$. The even stronger covalency of the S—O bond further reduces the redox energy, resulting in a higher voltage of 3.6 V for $Fe_2(SO_4)_3$. Following the commercialization of the $LiCoO_2$/graphite LIB in 1991, and building on prior knowledge of molybdates, sulfates, and phosphates, researchers further explored lithiated phosphates as cathodes. This effort culminated in 1997 with the identification of olivine-structured $LiFePO_4$ as an excellent cathode material [39].

To further enhance the overall electrochemical performance of polyanionic systems in iron-group metal compound nanomaterials, researchers have introduced multiphase integration strategies into the design of cathode materials for rechargeable batteries. Zhao et al. [15] proposed a composite system scheme involving iron-group metal Fe doping, designing a polyanionic-layered oxide integrated cathode embedded in a porous carbon framework (Figure 5.2e). The core of this design is a symbiotic heterostructure (NVFP–VO) composed of iron-group Fe-modified sodium-rich polyanionic phase $Na_{3.5}V_{1.5}Fe_{0.5}(PO_4)_3$ (NVFP) and layered oxide V_2O_3, representing a typical application case of iron-group metal composite compound nanomaterials. The performance advantages of this heterostructure are closely related to the doping of iron-group Fe and the composite system design. As a core iron-group metal element, Fe doping not only reduces the bandgap of $Na_3V_2(PO_4)_3$ but also activates the V^{4+}/V^{5+} redox couple, simultaneously enhancing the material's electronic conductivity and specific capacity. The heterostructure itself accelerates ion transport and reaction kinetics, meeting the efficient charge–discharge requirements of rechargeable batteries. Meanwhile, the low-strain V_2O_3 phase acts as a structural stabilizer, alleviating volume fluctuations and stress gradients in NVFP, while its activation process can contribute additional capacity. Benefiting from the synergistic effects of iron-group Fe-doping modification and multiphase integration, this cathode material demonstrates exceptional electrochemical performance: it achieves a reversible capacity exceeding $130\,mA\,h\,g^{-1}$ at 0.1C, and maintains a capacity retention rate of 72.6% after over 100,000 cycles at a high rate of 100C, representing the longest cycle life reported to date for polyanionic iron-group metal composite compound cathodes. This breakthrough provides innovative ideas and technical support for high-performance applications of iron-group metal compound nanomaterials in rechargeable batteries.

Overall, the three types of oxide cathode materials dominated by iron-group metal compound nanomaterials—layered, spinel, and polyanionic structures—each possess distinct technical characteristics and development bottlenecks, collectively forming the core system of cathode materials for rechargeable batteries. Among them, iron-group metal (Co, Ni, and Fe)-based layered materials exhibit diverse systems and, along with iron-group-related spinel structures, offer advantages such as high electronic conductivity and dense packing, making them widely used mainstream types in current rechargeable batteries. Polyanionic materials, with cost-effective metals like iron-group Fe as core components, demonstrate excellent thermal stability and safety due to their strong covalent bonding frameworks [40], enabling sustainable material design. However, their inherent poor conductivity and low volumetric energy density limit their application in scenarios requiring high energy density, such as portable devices and electric vehicles. Regarding specific iron-group metal compound nanomaterials, the practical application of spinel structures is still dominated by Mn based materials, but the manganese dissolution leads to capacity decay and negatively impacts anode performance [41]. Although this can be partially mitigated through lithium-site doping, the fundamental problem remains unresolved. Meanwhile, $LiMn_{1.5}Ni_{0.5}O_4$, which contains the iron-group metal Ni, offers a high-voltage platform of ~4.7 V, but its application is hindered by the slow development of high-voltage electrolytes [42], making practical implementation challenging. The contribution of iron-group metal-based cathode material systems to enhancing the energy density of rechargeable batteries is nearing saturation. Future breakthroughs in LIB energy density will rely more on innovative design and performance optimization of iron-group metal compound nanomaterials in anode systems, further expanding the application boundaries of such.

5.1.2 Anode

While lithium metal anodes offer exceptional theoretical specific capacity ($3860\,mA\,h\,g^{-1}$), low density, and the most negative electrode potential, and were historically used in primary lithium batteries, they face multiple challenges in rechargeable batteries. These include severe volume fluctuations during cycling, electrode pulverization tendency, poor interfacial stability, and lithium dendrite formation during charging that may penetrate separators, causing short circuits or even thermal runaway [43], which have long limited their practical application. With increasing energy density demands in rechargeable batteries, lithium metal anodes

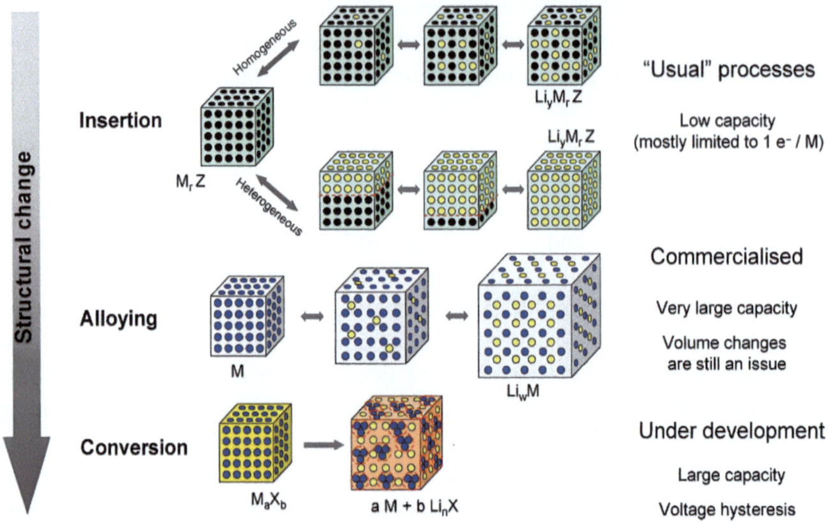

Figure 5.3 Schematic illustration of different reaction mechanisms observed in lithium battery electrode materials. Black circles: voids in the crystal structure; blue circles: metal atoms; yellow circles: lithium. *Source:* Reproduced from Ref. [46] / with permission of The Royal Society of Chemistry.

have regained attention due to their theoretical capacity comparable to silicon (~4200 mA h g^{-1}) and lower density, yet the fundamental technical bottlenecks of dendrite growth and cyclic volume changes remain unresolved [44], creating potential application opportunities for iron-group metal compound nanomaterials. Given the inherent limitations of lithium metal anodes, research focus on anode materials for rechargeable batteries over the past decades has shifted toward developing new systems with higher specific capacity than graphite and better safety profiles than lithium metal. Iron-group metal compound nanomaterials have emerged as a core research direction in this field due to their unique electronic structures and physicochemical properties. Related studies predominantly concentrate on structural design and application-oriented synthesis methods of these nanomaterials [45], with their widely recognized lithium storage mechanisms encompassing three types: intercalation reactions, conversion reactions, and alloying reactions (Figure 5.3), providing crucial support for breakthroughs in rechargeable battery anode performance. Simultaneously, the manufacturing quality of lithium metal carriers (requiring strict inert atmosphere processing, with carrier uniformity, pore structure, specific surface area, and surface properties significantly influencing electrode performance) also offers optimization space for applications of iron-group metal compound nanomaterials as carriers or modification agents.

5.1.2.1 Intercalation Reaction

The intercalation reaction represents the most classical lithium storage mechanism in LIBs. It refers to the reversible insertion of lithium ions as mobile guests into the lattice vacancies of a host material, a process also termed a "topotactic reaction."

An ideal host crystal must possess structural features, such as one-dimensional channels, two-dimensional planes, or three-dimensional networks, to allow free insertion/extraction of lithium ions while maintaining the essential integrity of the host framework. This provides the electrode with excellent cycling stability [47]. Anode materials based on the intercalation mechanism primarily include carbon-based materials, titanium-based oxides, and spinel-type compounds [48, 49]. Among these, artificial graphite, benefiting from mature processing technology and stable performance, dominates the current commercial anode market. Its lithium storage mechanism is based on the reversible intercalation and deintercalation of lithium ions between graphite layers, forming LiC_6 intercalation compounds with a theoretical specific capacity of 372 mA h g^{-1}. The conventional intercalation reaction can be represented as: $6C + xLi^+ + xe^- \leftrightarrow LiC_6$ [50].

Natural graphite can be classified into flake graphite and amorphous graphite (microcrystalline graphite) based on crystallinity [51]. Flake graphite exhibits high crystallinity (>98%) and anisotropy, delivering an initial Coulombic efficiency of 90–93% and a reversible capacity of 340–370 mA h g^{-1} at 0.1C, approaching its theoretical limit. However, its layered structure is prone to exfoliation during cycling, compromising cycle stability. Jiang et al. synthesized a microcrystalline graphite/NiO hexagonal nanosheet composite via a hydrothermal method (Figure 5.4a). The microcrystalline graphite acted as a conductive framework, enhancing overall stability and enabling the material to retain a high capacity of 1111.6 mA h g^{-1} after 45 cycles [52]. Nevertheless, limited by its lower crystallinity and inferior electrochemical performance, microcrystalline graphite has received less research attention and application compared to flake graphite.

Among synthetic carbon materials, mesocarbon microbeads (MCMB) have attracted widespread attention due to their unique spherical structure. MCMBs are formed through liquid-phase polycondensation of pitch-based precursors, with commercial variants typically having diameters of 5–40 μm. They are characterized by a smooth surface and high tap density [55]. Their lithium storage performance is significantly influenced by the degree of graphitization and crystalline texture. Chang et al. [53] prepared five types of MCMBs with distinct crystalline alignment structures (Figure 5.4b). Their study found that MCMBs with a "longitude-defective alignment" exhibited the best overall performance. Although not the highest in capacity, they demonstrated remarkable cycling stability: after graphitization at 3000°C, the capacity retention rate reached 98.9% over 20 cycles, effectively balancing capacity and cycle life requirements. The spherical morphology of MCMBs facilitates dense electrode packing, while their relatively small specific surface area helps suppress side reactions. The radially aligned carbon layers provide fast diffusion channels for lithium ions, endowing the material with good rate capability. However, during long-term cycling, the edge carbon layers in MCMBs are prone to structural exfoliation due to repeated lithium ion insertion/extraction, leading to capacity fade. Studies have shown that surface coating modifications can effectively enhance their structural stability [56].

Graphitized carbon fibers represent another class of high-performance carbon anodes, typically produced from polyacrylonitrile, phenolic resin, or pitch-based

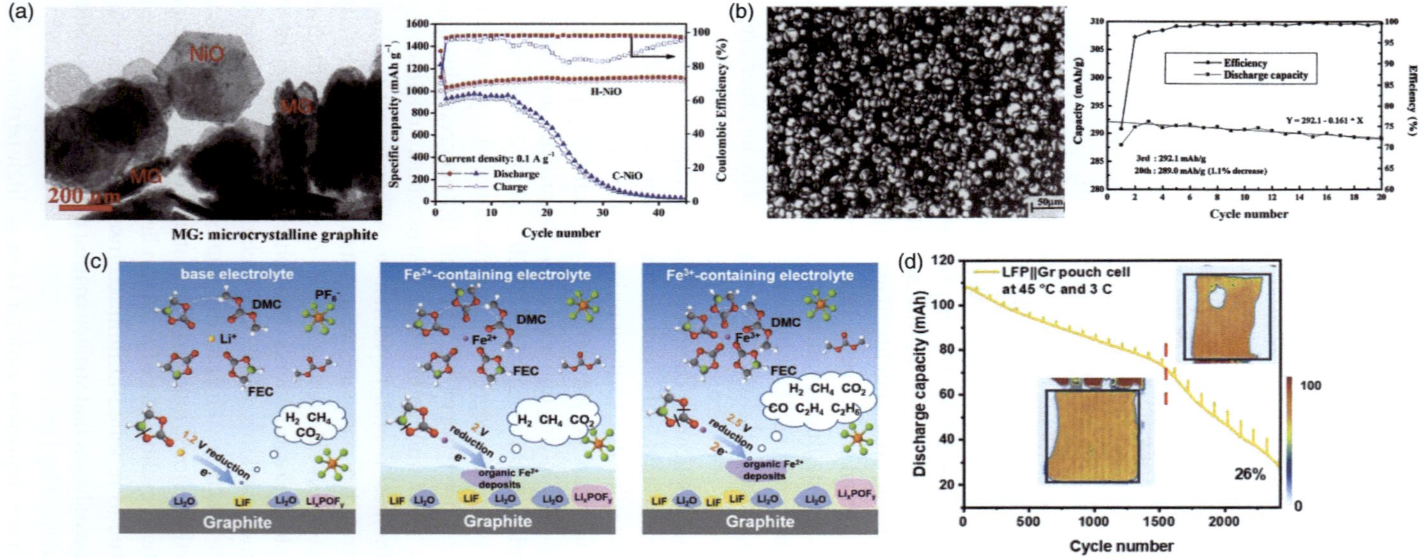

Figure 5.4 (a) Hexagonal sheet-like NiO supported on microcrystalline graphite and its capacity decay profile. *Source:* Reproduced from Ref. [52] / with permission of Elsevier. (b) Optical characteristics of longitude-defective aligned MCMB soaked in pitch and the corresponding capacity fading curve. *Source:* Reproduced from Ref. [53] / with permission of Elsevier. (c) Schematic illustration of the reactivity of Fe^{2+}/Fe^{3+} in the base electrolyte and their catalytic derivation of SEI structure. (d) Long-term cycling performance and capacity retention of LFP||Gr pouch cells. *Source:* Reproduced from Ref. [54] / with permission of John Wiley & Sons.

fibers via high-temperature conversion. These materials feature a characteristic coaxial layered structure with diameters of approximately 200–500 nm and highly oriented graphitic layers. Their lithium-ion diffusion coefficient is significantly higher (by about an order of magnitude) than that of conventional graphite, and the initial Coulombic efficiency can reach up to 97%. However, their reversible capacity is generally lower than that of graphite-based anodes [57]. Notably, the performance of graphite-based anodes—a core component of LIBs—is highly susceptible to interference from impurity ions within the battery system, prompting systematic investigation in related research. Wang et al. [54] studied the impact of Fe^{3+} on battery performance in lithium iron phosphate (LFP) || graphite cells, revealing that the adverse effects induced by Fe^{3+} are far more severe than those caused by Fe^{2+}(Figure 5.4c, d). While Fe^{2+} primarily promotes conventional electrolyte reduction reactions, increasing the generation of H_2, CH_4, and CO_2, Fe^{3+} in contrast promotes radical recombination and catalyzes two-electron reduction reactions, triggering side reactions that release CO, C_2H_4, and C_2H_6. This research not only elucidates the fundamental mechanisms of impurity ion effects on graphite-based anodes but also provides important references for performance optimization of carbon anodes, including graphitized carbon fibers. Although such materials exhibit excellent ion transport properties, their current research and application scale in LIB anodes remain relatively limited due to complex preparation processes and high costs. In addition to graphite-based materials, hard carbon and lithium titanate ($Li_4Ti_5O_{12}$) are also important intercalation-type anodes. Hard carbon, a non-graphitizable carbon material typically produced by pyrolysis of polymers, possesses a porous structure and stores lithium mainly through reversible adsorption/desorption in micropores and surface adsorption, delivering a reversible specific capacity of 300–500 mA h g^{-1}. However, its relatively high average redox potential (approximately 1 V vs. Li^+/Li) and the absence of a distinct voltage plateau reduce the energy density of full cells [58]. Recent research has focused on carbon source selection, process optimization, and compositing with high-capacity materials [59].

Although silicon–carbon (Si/C) composite anodes currently account for less than 1% of the market share, their exceptional theoretical specific capacity positions them as a critical development direction for next-generation high-energy-density LIBs. The silicon–carbon anode system (SiO_x/C, $0 \leq x \leq 2$) effectively combines the high capacity of silicon with the structural stability of carbon by integrating active silicon into a carbon matrix. Elementary silicon offers a theoretical specific capacity of up to 4200 mA h g^{-1} and a volumetric capacity of 9786 mA h cm^{-3}, both superior to those of lithium metal. However, its substantial volume expansion (up to 320%) during lithiation severely limits practical application. By compositing with carbon materials, the volume effects can be effectively buffered while enhancing the electronic conductivity of the electrode, thereby significantly improving cycling performance [60]. Currently, commercial SiO_x/C anodes typically achieve mass-specific capacities in the range of 400–650 mA h g^{-1}, about 1.2–2.0 times that of graphite anodes. Under laboratory conditions, capacities can reach 500–1500 mA h g^{-1} with a cycle life exceeding 1000 cycles at 1C rate. As battery manufacturers worldwide accelerate their industrialization efforts, this system is progressively transitioning from the laboratory stage to large-scale application. In terms of structural design, commonly used carbon frameworks include graphite, MCMB, carbon nanotubes, and graphene, among other forms. Based on the carbon coating configuration, they are mainly categorized into continuous and discontinuous types [61, 62]. Continuous coating (e.g., core–shell structures) effectively suppresses silicon volume expansion, reduces continuous SEI film growth during cycling, and helps improve the initial Coulombic efficiency. Discontinuous coatings (e.g., uniformly dispersed structures) facilitate electrolyte penetration and thereby enhance rate performance, although this is usually achieved at the cost of lower initial efficiency. It should be noted that when non-graphitic carbons such as acetylene black, pyrolytic carbon, or carbon nanotubes are used as hosts, their inherently higher average operating voltage (about 1 V) and lack of a distinct voltage plateau can lead to an elevated overall voltage plateau in the silicon composite anode, consequently reducing the operating voltage and power density of the full cell. Current challenges for SiO_x/C anodes include high costs (approximately 10 times that of graphite anodes), insufficient practical capacity, and cycling stability issues. The research focus is primarily on developing new processes to reduce costs and enhance overall performance. Furthermore, incorporating non-carbon hosts such as transition metal compounds, MXene, and polymers for composite modification has proven effective in improving capacity and cycling stability. Continued optimization of material structures and preparation processes is expected to gradually overcome existing barriers and promote their large-scale adoption in the mass market [63–65].

5.1.2.2 Conversion Reaction

The conversion reaction represents another important lithium storage mechanism distinct from intercalation, primarily occurring in transition metal compounds that lack typical insertion sites. Such materials can generally be denoted as M_aX_b (where M is a metal such as Mn, Fe, Co, Ni, or Cu; X is an anion such as O, S, F, P, or N), and the reaction follows the pathway [66]: $M_aX_b + (b \times c)Li^+ + (b \times c)e^- \leftrightarrow aM^0 + bLi_cX$. During the reaction, the material is reduced to metallic M^0 and forms lithium salt Li_cX, with a theoretical specific capacity typically ranging from 700 to 1000 mA h g^{-1}. Compared with commercial silicon–carbon anodes, conversion-type materials offer advantages in cost control and generally exhibit volume changes of less than 250%, demonstrating favorable structural reversibility and cycling stability. Yang et al. [67] employed an electrodeposition combined with an in-situ phosphorization strategy to directly grow CoP nanosheet arrays on carbon cloth (Figure 5.5a), constructing a binder-free electrode. This structure establishes intimate interfacial contact between CoP and the carbon cloth, effectively enhancing electron conduction. The three-dimensional network not only provides buffer space for volume expansion but also shortens the ion/electron transport pathways. As a result, the CoP@CC electrode maintained a high discharge capacity of 919.6 mA h g^{-1} after 200 cycles at a current density of 0.5 A g^{-1} (Figure 5.5b).

Structural design plays a critical role in determining the performance of conversion-type materials. During cycling, active particles are prone to component segregation and contact loss due to repeated volume expansion and contraction.

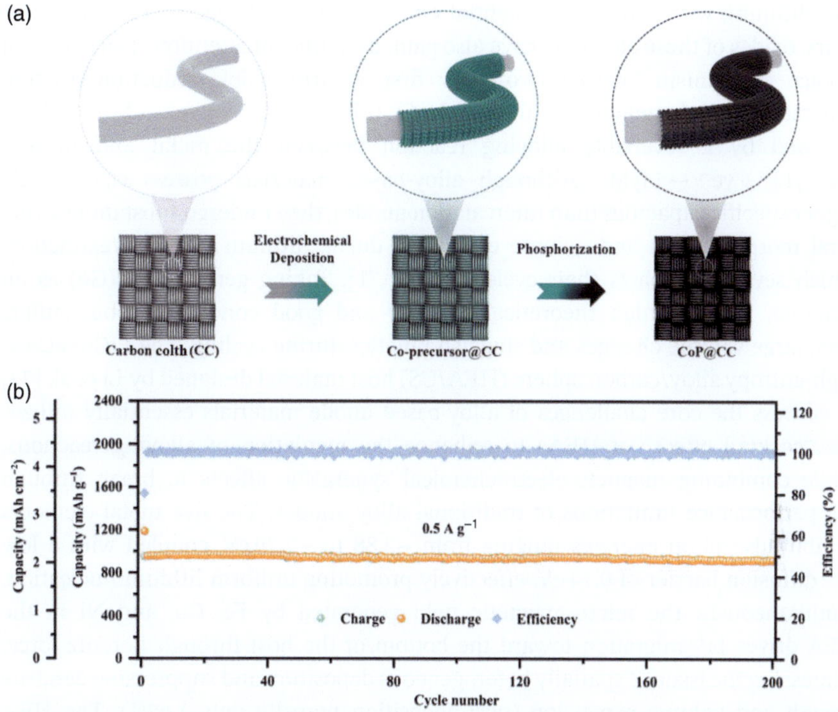

Figure 5.5 (a) Schematic illustration of the preparation process for the CoP@CC sample. (b) Long-term cycling performance of the NCM523||CoP@CC full cell over 200 cycles at a current density of 0.5 A g^{-1}. *Source:* Reproduced from Ref. [67] / with permission of John Wiley & Sons.

Encapsulating the nano-active phase within conductive frameworks such as carbon materials, polymers, or MXene can significantly enhance specific capacity and cycling stability. Oxide compositing has also been shown to provide certain improvements [68]. However, such materials generally suffer from high redox potentials, resulting in lower operating voltages of the assembled full cells. Their voltage compatibility with existing electronic devices still requires further optimization. Liu et al. [69] characterized the conversion reaction process of the spinel-structured $(LiFeZnNiCoMn)_3O_4$ when used as an anode in LIBs (LIBs). The test results showed that as the diffraction peaks of the spinel phase disappeared, diffraction rings corresponding to Fe, Co, Ni metals, and Li_2O gradually emerged—this phenomenon signifies the occurrence of the conversion reaction. Furthermore, after full charging, the diffraction rings of Fe, Co, Ni, and Li_2O gradually vanished, and only the diffraction signals of the parent spinel phase and rock-salt phase could be detected, confirming the high reversibility of the conversion reaction in this electrode.

5.1.2.3 Alloying Reaction

The alloying reaction primarily refers to the lithium storage process in which group IV and V elements (such as Si, Ge, Sn, and Sb) form intermetallic compounds with lithium, with a reaction potential typically below 1 V (vs. Li^+/Li). In recent years, oxides of these elements have also gained significant attention. Their lithium storage mechanism involves two steps: first, an irreversible reduction reaction generates the elemental metal and Li_2O ($MOx + 2xLi^+ + 2xe^- \rightarrow M^0 + xLi_2O$), followed by a reversible alloying reaction between the metal and lithium ($M + yLi^+ + ye^- \leftrightarrow LiyM$). Although alloy-based materials possess significantly higher specific capacities than intercalation anodes, they undergo substantial structural reorganization and volume expansion during lithium insertion/extraction, which severely impacts their cycle life [70, 71]. Taking germanium (Ge) as an example, it offers high theoretical capacity and good conductivity but suffers from large volume changes and sluggish kinetics during cycling. The PtCuFeCoNi high-entropy alloy/carbon sphere (HEA/CS) host material designed by Li et al. [72] to address the core challenges of alloy-based anode materials essentially utilizes the "cocktail effect" of HEAs to enhance the regulation of alloying reactions, while combining magneto-electrochemical synergistic effects to break through the performance limitations of traditional alloy anodes. The five metal elements exhibit adsorption energies ranging from −3.88 to −2.39 eV, coupled with a low Li^+ diffusion barrier of 0.14 eV, effectively promoting uniform lithium nucleation. Simultaneously, the micro-magnetic field generated by Fe, Co, and Ni in the HEA drives Li^+ migration toward the bottom of the host through Lorentz force, addressing the issue of spatially heterogeneous deposition and suppressing dendrite growth and volume expansion (post-deposition porosity only 1.63%). The HEA also selectively adsorbs $TFSI^-$ anions, inducing the formation of an LiF-rich SEI film and accelerating interfacial reaction kinetics. Symmetric cells based on this material achieved stable cycling for 4200 hours under 1 and 1 mA h cm^{-2} conditions, while pouch cells assembled with NCM811 demonstrated an energy density of 410 Wh kg^{-1} with 83% capacity retention after 120 cycles, laying the foundation for the commercialization of high-energy-density lithium metal batteries. This research not only provides a theoretical basis and technical pathways for the application of HEAs in alloy anodes but also establishes a new paradigm through the innovative concept of magneto-electrochemical synergistic regulation to address common issues such as volume expansion and sluggish kinetics in alloy materials, holding significant importance for advancing the commercialization of high-energy-density lithium metal batteries.

5.2 SIBs

LIBs have become the dominant technology in the secondary battery market due to their high energy density and long cycle life, with iron-group (Fe, Co, and Ni) metal compound nanomaterials serving as core cathode materials (such as Co-based $LiCoO_2$, Ni-based NMC, and Fe-based polyanionic compounds) that provide critical support for their performance breakthroughs. However, as lithium resources become increasingly scarce and prices fluctuate, developing sustainable new energy storage systems has become a shared priority for both academia and industry [73], which also drives the expansion of application scenarios for iron-group metal compound nanomaterials into alternative systems such as sodium-ion batteries (SIBs). Sodium is abundant in the Earth's crust (ranking 6th in abundance) and shares similar chemical properties with lithium, making room-temperature SIBs an ideal alternative for large-scale energy storage. Iron-group metal (Fe, Co, and Ni) compound nanomaterials, with their tunable electronic structures, stable crystal structures, and diverse modification potential, have become core material directions compatible with SIBs. Although SIBs operate on similar principles to LIBs, the larger mass and radius of sodium ions result in relatively lower mass and volumetric energy densities. Furthermore, the transport behavior, structural evolution, and interfacial characteristics of sodium ions in electrode materials differ significantly from those of lithium ions, posing specific requirements for the structural design (such as layered and polyanionic types) and performance optimization of iron-group metal compound nanomaterials. Limited by the intrinsic properties of sodium ions, SIBs struggle to directly compete with LIBs in fields like new energy vehicle power batteries. However, their unique potential in resource cost, safety, and cycle life highly aligns with the sustainable advantages of iron-group metal compounds (particularly low-cost Fe-based materials) [74]. Therefore, developing iron-group metal compound nanomaterial systems tailored to the characteristics of SIBs is crucial. In terms of industrialization, SIBs can leverage the mature manufacturing processes and equipment of LIBs to accelerate the commercial application of iron-group metal compound nanomaterials in this system; meanwhile, the role of iron-group metal compounds in enhancing the high-low temperature performance and safety of SIBs still requires further exploration.

5.2.1 Cathode Materials

Cathode materials are critical in determining the energy density, power characteristics, and cycling stability of SIBs. Current research primarily focuses on layered oxides, polyanionic compounds, Prussian blue analogues (PBA), and metal–organic frameworks (MOFs). Layered oxides, with the general formula Na_xTMO_2 (where TM represents transition metals), have been extensively studied due to their simple synthesis, high capacity, and environmental friendliness. Based on the coordination environment of sodium ions and the stacking mode of oxygen layers, they can be classified into O3, P3, P2, O2, and other types [75] (Figure 5.6a). Among these, P2 and O3 are two common configurations. During charge–discharge processes, O3 and P2 phases tend to transform into P3 and O2 phases, respectively. Such phase transitions can lead to increased polarization and structural stress, thereby affecting cycling stability. To enhance performance, researchers often employ strategies such as doping, coating, or constructing composite structures.

O3-type-layered oxide cathodes, while offering high specific capacity, face two inherent challenges that limit their practical application: first, the materials are air-sensitive and prone to spontaneous lattice sodium extraction when exposed to air; second, they exhibit sluggish sodium-ion diffusion kinetics, primarily due to the high energy barrier associated with tetrahedral site hopping during

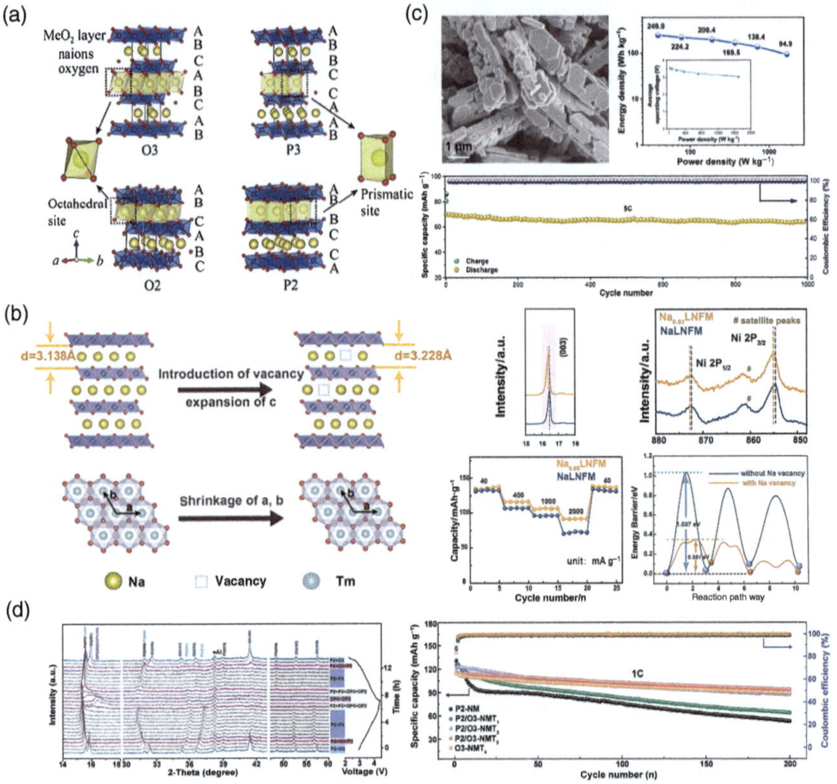

Figure 5.6 (a) Crystal structure schematics of O3-, P3-, O2-, and P2-type-layered oxide cathodes. *Source:* Reproduced from Ref. [75] / with permission of Elsevier. (b) Schematic of parameter changes after Na vacancy introduction, comparison of (003) lattice plane angles and XPS spectra, along with the rate performance of samples and migration energy barriers of sodium ions. *Source:* Reproduced from Ref. [76] / with permission of John Wiley & Sons. (c) Microstructure of NNMMO-HNA, corresponding power density plot, and cycling stability. *Source:* Reproduced from Ref. [77] / with permission of Elsevier. (d) Operando XRD patterns during the first charge/discharge at 0.1C in the voltage range of 2.2–4.4 V, and cycling performance of P2/O3-NMT3. *Source:* Reproduced from Ref. [78] / with permission of Elsevier.

migration. More critically, enhancing air stability typically requires strengthening the lattice's binding to sodium, which often further impedes sodium-ion diffusion and degrades kinetic performance. This inherent trade-off makes dual modification of O3-type materials particularly challenging. Yuan et al. [76] addressed this issue by introducing sodium vacancies into $Na_{0.93}Li_{0.12}Ni_{0.25}Fe_{0.15}Mn_{0.48}O_2$ through a simple strategy (Figure 5.6b). The sodium vacancies reduce the charge density on transition metal ions, enhancing the material's oxidation resistance and ensuring stable lattice sodium retention in $Na_{0.93}Li_{0.12}Ni_{0.25}Fe_{0.15}Mn_{0.48}O_2$ when exposed to air. Additionally, this approach creates more Na^+ diffusion sites and expands the interlayer spacing of sodium layers, resulting in a reduction of the energy barrier from approximately 1000–300 meV. The modified material maintains 70.8% capacity retention at a high current density of 2000 mA g^{-1}.

P2-type $Na_{0.67}Ni_{0.33}Mn_{0.67}O_2$, as a typical layered transition metal oxide cathode material, is regarded as a highly promising cathode system for SIBs due to its high energy density characteristics. However, significant capacity decay during cycling and poor rate performance at high current densities have severely hindered its practical application. Peng et al. [77] reported that P_2-$Na_{0.67}Ni_{0.23}Mg_{0.1}Mn_{0.67}O_2$, through Mg^{2+} doping, effectively suppresses the P2–O2 phase transition and reduces the activation energy for interfacial charge transfer. The material features a hierarchical one-dimensional nanostructure composed of nanoplate subunits, which increased the capacity retention from 74.8% to 92.2% after 150 cycles. In a half-cell configuration, it maintained 90.9% capacity retention over 1000 cycles at 5C. More importantly, compared with previously reported state-of-the-art values for layered oxide cathodes, the full cell achieved the highest average operating voltage of 3.56 V and a high energy density of 249.9 Wh kg^{-1} (Figure 5.6c).

In the development of layered cathode materials for SIBs, constructing P2/O3 biphasic composite structures has been demonstrated as an effective strategy for enhancing sodium storage performance. However, the specific mechanism of the biphasic synergy, particularly the structural evolution at high voltages (>4.2 V vs. Na^+/Na), remains unclear. To address this issue, Yu et al. [78] successfully prepared a high-voltage stable composite layered oxide—$Na_{0.85}Ni_{0.4}Mn_{0.33}Ti_{0.33}O_2$—with a specific P2:O3 phase ratio (24.8%: 75.2%) by controlling the Ti substitution level in $Na_{0.85}Ni_{0.34}Mn_{0.66}$-$xTi_xO_2$. When charged to a high voltage of 4.4 V, this biphasic cathode effectively suppressed the formation of the detrimental O2 phase and underwent a fully reversible structural evolution from P2/O3 → P2/P3 → OP4/OP2. The study further revealed an "interlocking effect" at the phase boundaries, which significantly alleviated structural strain and lattice volume changes under high voltage, thereby enhancing the structural stability of the material. Benefiting from these characteristics, the prepared P2/O3 composite cathode exhibited a high capacity retention of 80.6% after 200 cycles at 1C. More notably, a full cell assembled with a hard carbon anode achieved a high energy density of 294.6 Wh kg^{-1} (Figure 5.6d).

Building on the industrial success of phosphate-based materials (e.g., $LiFePO_4$) in LIBs, polyanionic compounds—including phosphates, fluorophosphates, and pyrophosphates—have also attracted extensive research as cathode materials for SIBs due to their stable structures and favorable thermodynamic properties [79]. These materials generally exhibit high operating voltages and can be represented by the general formula $Na_xM_y[(XO_m)^{n-}]_z$, where M is an electroactive transition metal, and X is a nonmetallic element such as P, S, or Si. Compared to layered oxide cathodes, polyanionic materials demonstrate superior thermal stability and safety. However, their intrinsic low electronic conductivity limits their charge–discharge performance at high current rates. To address this, strategies such as coating or doping are commonly employed to enhance conductivity and improve electrochemical performance [80]. Among various polyanionic materials, Na_3V_2 $(PO_4)_3$ with a

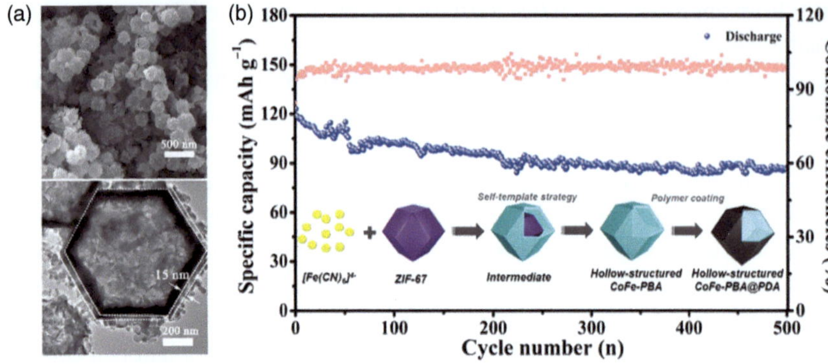

Figure 5.7 (a) Micromorphology of polydopamine-coated cobalt–iron Prussian blue analogue, and (b) long-term cycling stability (0.1 A g^{-1}). *Source:* Reproduced from Ref. [82] / with permission of American Chemical Society.

NASICON (sodium superionic conductor) structure has garnered significant attention for its high voltage and high theoretical specific capacity. Its crystal framework consists of PO_4 tetrahedra and VO_6 octahedra connected at their vertices, providing three types of sodium ion sites. However, only two sodium ions can be extracted during charging, corresponding to a theoretical specific capacity of 118 mA h g^{-1}. Unmodified $Na_3V_2(PO_4)_3$ suffers from poor conductivity and unsatisfactory cycling stability.

PBA cathode materials possess a typical perovskite-like face-centered cubic structure with the general formula $A_xM[Fe(CN)_6]\gamma{\cdot}zH_2O$ ($0 < x < 2$, $0 < y < 1$), where A represents alkali metals such as Li, Na, or K, and M denotes transition metals like Fe, Mn, Co, Ni, or Cu. In this crystal framework, alkali metal ions occupy the body-center sites, Fe is coordinated to C, and the transition metal M is coordinated to N [81]. Huang et al. developed a simple self-templating method to successfully prepare hollow-structured cobalt-iron PBAs (CoFe-PBA) [82]. The hollow structure effectively buffers volume changes during sodium ion extraction/insertion and shortens the ion diffusion pathways. After further constructing a thin polydopamine (PDA) coating on the material surface, the resulting CoFe-PBA@PDA composite exhibited a high discharge capacity of 123.1 mA h g^{-1} at a current density of 0.1 A g^{-1}, with a capacity retention rate of 71.5% after 500 cycles (Figure 5.7a, b).

5.2.2 Anode Materials

The similarity in working principles between SIBs and LIBs has led to extensive exploration of carbon-based materials, titanium-based compounds, and transition metal compounds as anode systems for SIBs [83]. However, conventional graphite anodes are unsuitable for reversible sodium ion insertion/extraction due to their small interlayer spacing, shifting research focus toward amorphous carbon materials such as soft carbon and hard carbon [84]. Among non-carbon materials, titanium-based oxides have attracted considerable attention due to the stability of Ti^{4+} in air and their suitable operating potential (the Ti^{4+}/Ti^{3+} redox couple lies

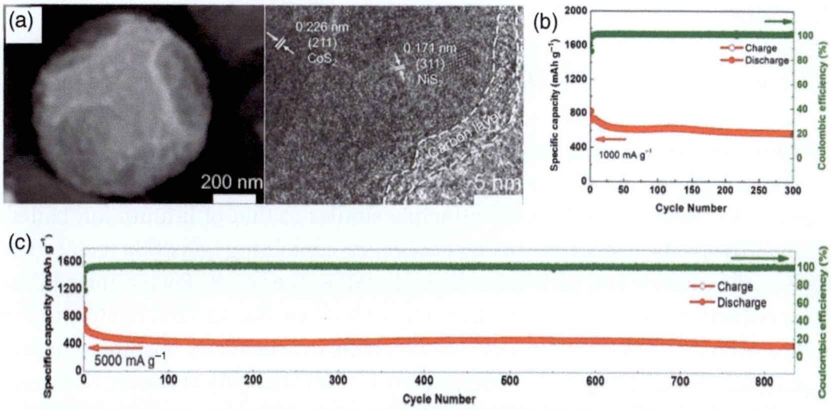

Figure 5.8 (a) Microstructure of NCSC. (b) Cycling performance of NCSC at a current density of 1000 mA g^{-1}. (c) Cycling performance of NCSC at a current density of 5000 mA g^{-1}. *Source:* Reproduced from Ref. [87] / with permission of Elsevier.

within the 0–2 V range). Different crystal structures of titanium-based oxides exhibit distinct sodium storage potentials, providing avenues for material design [85].

Transition metal compounds, particularly those of the iron-group metals, are regarded as promising anode materials for SIBs due to their excellent physicochemical properties and unique structural features. However, these materials commonly suffer from severe volume expansion during cycling, leading to rapid capacity decay and poor cycling stability [86]. To address this challenge, researchers have focused on enhancing their electrochemical performance through nanostructural design and synthesis process innovation. Deng et al. constructed concave-structured NiS_2@CoS_2 nanocubes (NCSC) through chemical etching and sulfurization of a Ni–Co coordination polymer precursor [87]. The material exhibited a high specific capacity of 848 mA h g^{-1} at 0.1 A g^{-1} and maintained a stable capacity of 572 mA h g^{-1} after 830 cycles at a high current density of 5 A g^{-1} (Figure 5.8a–c). The performance improvement is mainly attributed to the introduction of a dual-carbon framework, which significantly enhances the material's conductivity and promotes sodium-ion diffusion. Meanwhile, the well-designed concave structure and surface carbon layer effectively buffer volume changes during cycling, inhibit electrode structure collapse, and thereby improve the transport efficiency of sodium ions and electrons.

While SIBs currently lag behind LIBs in terms of capacity, cycle life, and rate performance, significant progress has been made in recent years in improving their power density and energy density. The continuous innovation of novel anode materials such as iron-group metal compounds has laid the foundation for further breakthroughs in the performance of SIBs. In the future, with ongoing optimization of material design and deeper mechanistic understanding, SIBs are expected to gradually approach the performance level of LIBs and play an important role in specific application scenarios.

5.3 KIBs

With growing concerns about the sustainability of lithium resources, new electrochemical energy storage systems, particularly SIBs, have been extensively developed. Meanwhile, potassium—the seventh most abundant element in the Earth's crust and, like sodium, an alkali metal—has gradually gained attention as a promising charge carrier. Based on a working principle similar to that of lithium-ion batteries/SIBs, KIBs can be constructed using potassium-containing transition metal compounds as the cathode and carbonaceous materials as the anode. Potassium exhibits a lower reduction potential than sodium (K: −2.93 V vs. Na: −2.71 V, relative to the standard hydrogen electrode), which enables KIBs to potentially operate at higher voltages, thereby achieving higher energy density than SIBs [88]. However, the larger ionic radius of potassium requires electrode materials with more spacious ion channels to enable efficient and reversible intercalation/deintercalation. For this reason, structures with open frameworks, such as metal–MOFs and PBAs, have been widely explored as cathode materials [89]. Zhang et al. [90] have provided a breakthrough practice for the application of PBAs in KIBs. The team designed and synthesized a quaternary medium-entropy PBA (Q-ME-PBAs) composed of Fe, Ni, Co, and Cu as a cathode material through medium-entropy lattice engineering. Its open framework structure and three-dimensional lattice distortion capability effectively accommodate the insertion/extraction of potassium ions, significantly reducing volume changes and diffusion barriers. Experimental results show that the material maintains over 90% capacity after 300 cycles at 0.25C, achieves 74.7% capacity retention after 10,000 cycles (over 147 days) at 3.75C, and full cells paired with hard carbon or graphite anodes demonstrate excellent cycling stability and rate performance. This confirms that structurally optimized PBAs can effectively address the technical challenges posed by the large ionic radius of potassium ions, providing an important reference for the development of high-performance cathode materials for KIBs.

In alkali metal-ion battery systems, graphite—a well-established anode material for LIBs—fails to function directly in SIBs due to its insufficient sodium storage capability. In contrast, hard carbon has become the mainstream choice for sodium-ion battery anodes owing to its high capacity and good cycling stability. Notably, the limited energy density of graphite in LIBs primarily stems from the significant voltage variation during the initial stage of lithium intercalation, which lacks a flat voltage plateau [91]. This characteristic has inspired researchers to explore graphite for potassium-ion battery systems. Compared to the performance of hard carbon in SIBs, graphite exhibits a lower operating voltage in KIBs, offering higher energy density advantages. It can achieve a reversible capacity exceeding 270 mA h g^{-1}, demonstrating promising application potential. Beyond carbon materials, a range of iron-group compounds capable of intercalation, conversion, and alloying reactions have been extensively studied as anodes for KIBs [92]. Adekoya et al. constructed a Co_3O_4@nitrogen-doped carbon (N–C) composite through interface engineering (Figure 5.9a, b). By leveraging the strong potassium ion adsorption, high conductivity, and reduced diffusion barriers provided by the nitrogen-doped carbon, the electrochemical performance of Co_3O_4 was effectively optimized [93].

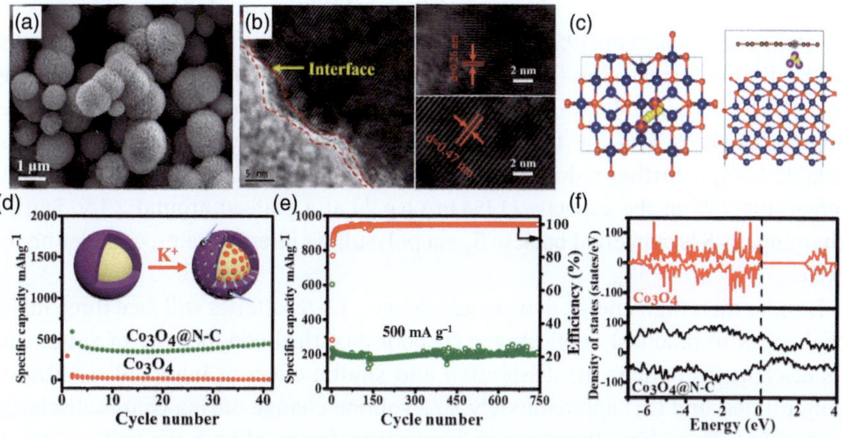

Figure 5.9 Material characterization of the prepared cobaltosic oxide-based nitrogen-doped carbon composite (Co_3O_4@N–C): (a) scanning electron microscopy image and (b) high-resolution transmission electron microscopy image; (c) potassium migration pathways in Co_3O_4 and (e) Co_3O_4@N–C. Color scheme: Co: blue; O: red; C: brown; N: light blue; K: purple. (d) Correspondingly calculated potassium diffusion barriers in Co_3O_4 and Co_3O_4@N–C. (e) Cycling performance of Co_3O_4@N–C at 50 mA g^{-1}. (f) Long-term cycling performance at a current density of 500 mA g^{-1}. *Source:* Reproduced from Ref. [93] / with permission of American Chemical Society.

This composite structure not only increases the interlayer spacing of Co_3O_4 but also significantly reduces the diffusion resistance of potassium ions (Figure 5.9c, d) through the nitrogen-doped carbon layer. It enhances overall conductivity while mitigating structural damage to the electrode during cycling. Benefiting from these advantages, Co_3O_4@N-C retained a high capacity of 448.7 mA h g^{-1} after 40 cycles at a current density of 50 mA g^{-1}, and maintained a stable capacity of 213 mA h g^{-1} after 740 cycles even under a high current density of 500 mA g^{-1} (Figure 5.9e, f).

5.4 Li–S

Lithium–sulfur (Li–S) batteries have emerged as a strong candidate for next-generation high-energy-density storage systems due to the high theoretical specific capacities of their constituent elements. The lithium metal anode boasts a theoretical specific capacity of 3860 mA h g^{-1}, while the elemental sulfur cathode is not only abundant and environmentally benign but also offers a theoretical capacity of 1675 mA h g^{-1}. As the sulfur cathode stores lithium via multistep conversion reactions rather than conventional intercalation, Li–S batteries achieve a theoretical energy density of up to 2500 Wh kg^{-1} at an average operating voltage of 2.15 V, significantly surpassing existing LIB systems [94]. Li–S batteries operate based on the reversible redox reaction between lithium metal and cyclic S_8. During discharge, lithium is oxidized at the anode to Li^+ ions and electrons, which migrate through the electrolyte and external circuit to the cathode, respectively. At the cathode, sulfur is reduced, ultimately forming lithium sulfide (Li_2S). The process reverses during charging [95]. The actual reaction pathway is more complex: in the initial discharge stage, S_8 is lithiated to form soluble long-chain lithium polysulfides (e.g., Li_2S_8, Li_2S_6, and Li_2S_4). This stage occurs at around 2.3 V and contributes approximately 25% of the theoretical capacity (418 mA h g^{-1}). Subsequently, the soluble Li_2S_4 is further reduced to solid short-chain Li_2S_2 and Li_2S, delivering the remaining 75% of the capacity (1254 mA h g^{-1}) at a plateau around 2.1 V. During charging, Li_2S is converted back to S_8 via polysulfide intermediates, completing the cycle.

Despite their high energy density advantages, Li–S batteries still face three major challenges in practical applications: (i) poor electrical conductivity of sulfur and its discharge products; (ii) dissolution and shuttle effect of intermediate lithium polysulfides; and (iii) approximately 80% volume change during charge/discharge cycles [96]. To address these issues, synergistic design of both the sulfur cathode and lithium anode is required to enhance ion/electron transport efficiency, suppress polysulfide migration, and provide sufficient space and structural flexibility to accommodate volume changes. One of the most effective strategies currently is to uniformly disperse sulfur within a conductive matrix capable of physical confinement or chemical adsorption. Iron-group metal compounds have shown promising potential in this direction. Yuan et al. developed a polyhedral porous $CoSe_2$/C nanocomposite as a sulfur host (Figure 5.10a–d) [97]. In this material, the porous structure effectively mitigates the volume expansion of sulfur during cycling, while the highly conductive $CoSe_2$ not only improves reaction kinetics but also exerts strong adsorption on polysulfides due to its Lewis acidic surface, significantly suppressing the shuttle effect. As a result, the $CoSe_2$/C–S composite electrode maintained a reversible capacity of 363 mA h g^{-1} after 700 cycles at a rate of 2C (Figure 5.10e).

5.5 Metal–Oxygen

Metal–air batteries represent a novel electrochemical energy storage system based on metal oxidation and oxygen reduction reactions (ORR), with theoretical energy densities 3–30 times higher than those of commercial LIBs. A typical metal–air battery consists of four core components: an air electrode, a metal electrode, an electrolyte, and an optional separator. The air electrode directly uses oxygen from ambient air as the reactant, eliminating the need for built-in oxidizers. It is generally composed of a catalyst layer for facilitating the ORR and a gas diffusion layer to promote gas transport. Commonly used metal electrodes include reactive metals such as zinc, lithium, aluminum, magnesium, and sodium. For highly reactive metals like lithium and sodium, nonaqueous aprotic electrolytes are often employed to ensure stability, whereas systems based on zinc, aluminum, and magnesium can operate stably in alkaline aqueous electrolytes [98]. In some configurations, a separator is used to isolate different electrolytes, regulate mass transport, and suppress short circuits caused by dendrite formation. Among various metal–air batteries, lithium–air and zinc–air

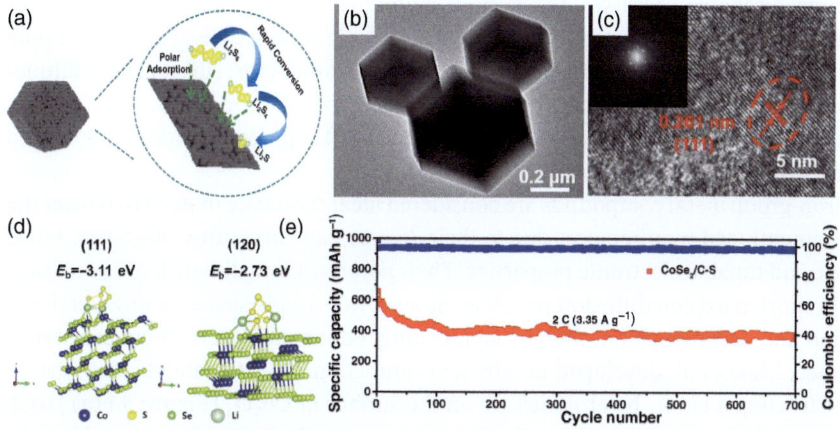

Figure 5.10 (a) Proposed reaction mechanism between $CoSe_2$/C and polysulfides. (b, c) TEM images of the $CoSe_2$/C nanocomposite. (d) Schematic diagram of the interaction between Li_2S_6 and the $CoSe_2$ nanocomposite. (e) Long-term cycling performance and Coulombic efficiency of the $CoSe_2$/C–S composite in a half-cell after activation. *Source:* Reproduced from Ref. [97] / with permission of Elsevier.

batteries have been the most extensively studied, while other systems also exhibit distinct application potential.

The electrode reaction pathways in rechargeable metal–air batteries depend on the metal type and electrolyte composition [99]. In aqueous alkaline electrolytes, typical reactions proceed as follows: Metal electrode: $M = M^{n+} + ne^-$; Air electrode: $O_2 + 4e^- + 2H_2O = 4OH^-$. Here, M represents metals such as Zn, Al, Mg, or Fe, and n represents their ionic valence. The metal ions generated during discharge may further react with OH^- to form corresponding hydroxides or oxides. In nonaqueous metal–air batteries using lithium, sodium, or potassium as the anode, the air electrode reactions involve the combination of oxygen with metal ions to form superoxides or peroxides: Metal electrode: $M = M^+ + e^-$; Air electrode: $xM^+ + O_2 + xe^- = MxO_2$ (x = 1 or 2). The oxygen molecule is first reduced to a superoxide anion (O_{2-}), which then combines with metal ions to form M_xO_2 products. Since lithium, sodium, and potassium metals are highly sensitive to moisture and carbon dioxide in air, such batteries are typically operated in a pure oxygen environment and are correspondingly referred to as Li–O_2, Na–O_2, and K–O_2 batteries.

5.5.1 Li–O_2

In recent years, lithium–oxygen (Li–O_2) batteries have attracted extensive attention due to their exceptional gravimetric energy density advantages. A typical Li–O_2 battery consists of a lithium metal anode, a porous oxygen cathode, and an electrolyte containing lithium salts. During discharge, oxygen diffuses to the cathode interface and undergoes a reduction reaction with lithium ions, sequentially forming lithium superoxide (LiO_2) and further converting to lithium peroxide (Li_2O_2). The reverse decomposition reaction occurs during charging. To achieve efficient and reversible charge–discharge processes, the cathode materials in Li–O_2 batteries must possess the following key characteristics: (i) good electronic conductivity; (ii) rational pore structure to facilitate the transport of oxygen and lithium ions; (iii) excellent bifunctional catalytic activity for both the ORR and oxygen evolution reaction (OER); and (iv) a stable electrode–electrolyte interface capable of suppressing side reactions and preventing the coverage of active sites [98].

Iron-group metal compounds are considered ideal candidate materials to meet the aforementioned requirements due to their diverse nanostructures, abundant active sites, and tunable electronic properties. Their porous or low-dimensional structures help shorten oxygen diffusion pathways and provide ample space for product deposition and decomposition, thereby significantly enhancing electrochemical performance. Zhao et al. developed an efficient bifunctional electrocatalytic cathode by constructing a Fe_2O_3 heterostructure on the surface of Co_3O_4 (Figure 5.11a) [100]. Compared to single-component Fe_2O_3 or Co_3O_4, the Co_3O_4/Fe_2O_3 composite cathode exhibited a higher discharge capacity (reaching 14,847.8 mA h g^{-1}) and maintained excellent capacity performance even at high current densities. Furthermore, the cycling stability of the composite electrode was significantly improved, achieving up to 280 stable cycles (Figure 5.11b, c). Density functional theory (DFT) calculations revealed that the strong electronic coupling effect at the heterointerface effectively modulates the electronic structure of Co_3O_4, significantly reducing the adsorption energy barriers for O_2 and the intermediate product LiO_2, thereby effectively lowering the reaction overpotential (Figure 5.11g, h).

5.5.2 Na–O_2

As a typical analogue of lithium–air batteries, sodium–air batteries (commonly referred to as Na–O_2 batteries) represent a promising energy storage strategy and a competitive alternative. Although Na–O_2 batteries have a lower theoretical energy density than Li–O_2 systems (1105 Wh kg^{-1} based on NaO_2), they offer advantages such as higher abundance, lower charging overpotential (<0.2 V), and better reversibility [101]. Like other metal–air batteries, sodium–air batteries also consist of four typical components: an air electrode, a metal anode, an aprotic/aqueous electrolyte, and a separator. Currently, two main configurations are under investigation: aprotic Na–O_2 batteries and hybrid sodium–air batteries (HSAB). Both involve the dissolution/deposition of Na at the anode during discharge/charging, with corresponding ORR/OER reactions at the cathode, similar to the reactions in lithium–air batteries. However, Na–O_2 batteries differ significantly from Li–O_2 batteries in their discharge mechanism. Compared to Li^+, Na^+ has a larger ionic radius and lower charge density, which stabilizes O_{2-}. Disproportionation does not occur spontaneously, while Na_2O_2 has a higher Gibbs free energy of formation (−449.7 kJ mol^{-1}) than NaO_2 (−436.5 kJ mol^{-1}) [102]. This one-electron transfer reaction endows Na–O_2 chemistry with low overpotential and high reversibility. Under prolonged testing, NaO_2 typically converts to Na_2O_2, accompanied

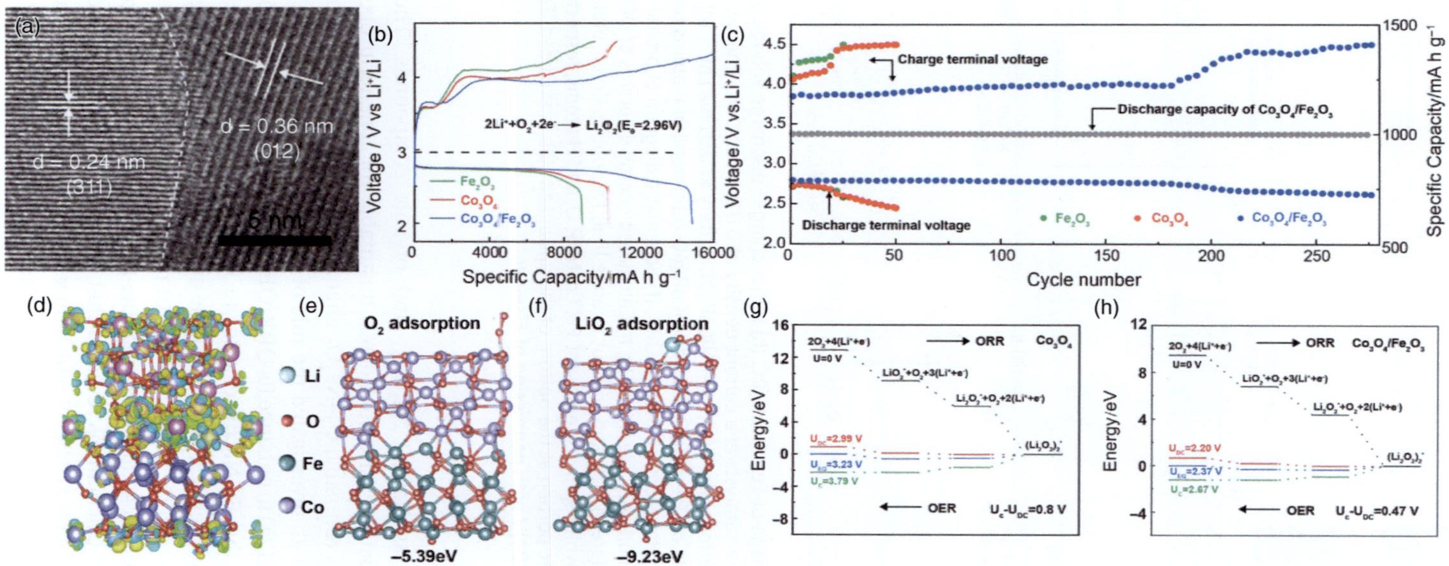

Figure 5.11 (a) High-resolution transmission electron microscopy image of Co_3O_4/Fe_2O_3. (b) Specific capacity during the first discharge–charge cycle. (c) Terminal discharge/charge voltage versus capacity of the Co_3O_4/Fe_2O_3 cathode. (d) Charge density difference at the Co_3O_4/Fe_2O_3 interface. Adsorption configurations of (e) O_2 and (f) LiO_2 on $Co_3O_4@Fe_2O_3$. Free energy calculations during discharge/charge processes for (g) Co_3O_4 and (h) Co_3O_4/Fe_2O_3. *Source:* Reproduced from Ref. [100] / with permission of John Wiley & Sons.

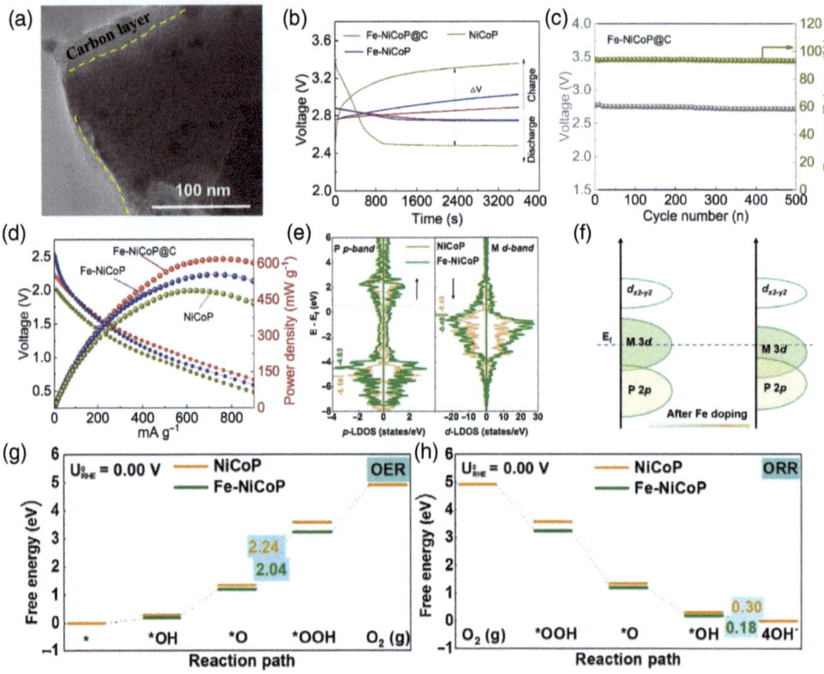

Figure 5.12 (a) TEM image of Fe–NiCoP@C. (b) Initial charge–discharge voltage profiles of batteries with different catalysts at a current density of 0.01 mA cm^{-2}. (c) Energy efficiency and cycling performance of the Fe–NiCoP@C battery at 0.01 mA cm^{-2}. (d) Power density of HSABs with NiCoP, Fe–NiCoP, and Fe–NiCoP@C. (e) p-band of P atoms related to OER and d-band of all metal atoms related to ORR. (f) Schematic diagram of orbital variations in metal and nonmetal elements. (g, h) Calculated energy diagrams of OER/ORR on NiCoP surfaces with (green line) and without (orange line) Fe doping. Green, orange, dark red, pink, red, and white spheres represent Fe, Co, Ni, P, O, and H atoms, respectively. For clarity, the symmetric portion at the bottom of the optimized slab is excluded. *Source:* Reproduced from Ref. [103] / with permission of Elsevier.

by byproducts such as NaOH, Na_2CO_3, and organic compounds. Therefore, NaO_2-based Na–O_2 batteries are considered unsuitable for long-term operation, which ideally requires reversible O_2–Na_2O_2 chemistry without the disproportionation of NaO_2. Kang et al. investigated the role of iron dopants in Fe-doped NiCoP (Fe–NiCoP) (Figure 5.12a) to enhance its ORR/OER kinetics [103]. The carbon layer effectively stabilized the Fe-NiCoP–electrolyte interface and improved electron conduction and electrochemical activity. As the air cathode in an HSAB, the HSAB exhibited a discharge voltage of 2.74 V at 0.01 mA cm^{-2}, a round-trip efficiency of 93.26% at the 500th cycle, and a state-of-the-art power density of 621 mW g^{-1} (Figure 5.12b–d). DFT calculations revealed that Fe dopants effectively modulated the electronic conductivity of NiCoP and reduced the binding energy of reaction intermediates (Figure 5.12e, f), thereby optimizing the rate-determining steps for both ORR and OER (Figure 5.12g, h).

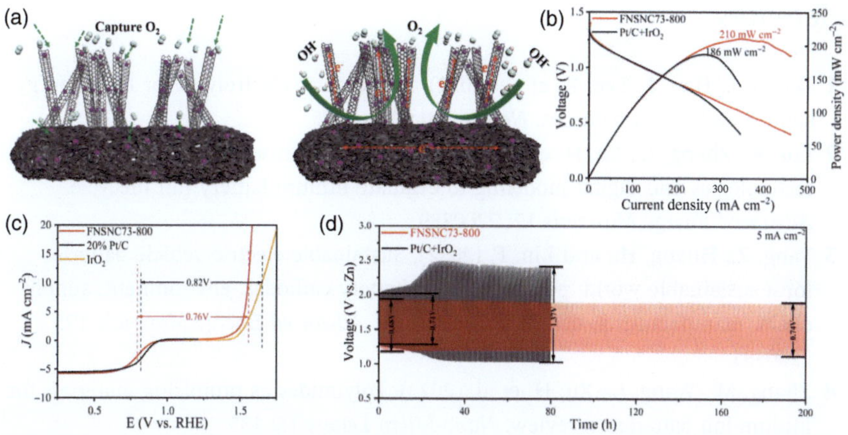

Figure 5.13 (a) Schematic diagram of the ORR and OER mechanisms in FNSNC73-800. (b) Discharge polarization curves and corresponding power density profiles. (c) Potential gap between FNSNC73-800 and Pt/C + IrO_2. (d) Galvanostatic cycling of rechargeable zinc–air batteries using FNSNC73-800 or Pt/C + IrO_2. *Source:* Reproduced from Ref. [107] / with permission of Elsevier.

5.5.3 Zn–O_2

Unlike the aforementioned metal–air batteries that require aprotic electrolytes, zinc–air batteries typically operate in aqueous electrolyte systems, offering prominent advantages such as low cost, high safety, and environmental friendliness, which have attracted extensive attention in recent years [104]. In a typical alkaline zinc–air battery, the zinc anode loses electrons during discharge and reacts with OH^- to form soluble $Zn(OH)_4{}^{2-}$. When $Zn(OH)_2$ in the electrolyte reaches saturation, the zincate ions ($Zn(OH)_4{}^{2-}$) further decompose into ZnO. On the cathode side, oxygen gains electrons and is directly reduced to OH^-. The overall reaction of the zinc–air battery is: $2Zn + O_2 \rightarrow 2ZnO$ [105]. Despite their significant competitiveness in cost and safety, the commercialization of zinc–air batteries still faces two major challenges: first, the sluggish kinetics of the ORR and OER lead to high charge–discharge overpotentials and limited energy efficiency; second, the zinc anode in alkaline environments is prone to issues such as dendrite growth, hydrogen evolution side reactions, and carbonate deposition, which severely impact cycling life [106].

To enhance the overall electrode kinetics, research has focused on developing efficient and stable iron-group compound-based cathode catalysts. Wu et al. prepared a three-dimensional self-supporting FeNi alloy/porous carbon composite catalyst through synergistic modulation of Fe^{3+} and Ni^{2+} [107]. In this material, doping with nitrogen and sulfur heteroatoms introduced abundant defects and active sites, which synergistically enhanced the bifunctional ORR/OER catalytic activity together with the FeNi alloy phase (Figure 5.13a), exhibiting a narrow voltage gap of 0.76 V (Figure 5.13b). The zinc–air battery assembled with this catalyst achieved a high peak power density of 210 mW cm^{-2} and maintained a low overpotential of 0.74 V during a 200-hour long-term charge–discharge test, demonstrating excellent cycling stability and energy conversion efficiency (Figure 5.13c, d).

References

1 Lee, M.J., Han, J., Lee, K. et al. (2022). Elastomeric electrolytes for high-energy solid-state lithium batteries. *Nature* 601: 217–222.

2 Liu, X., Zhang, L., Yu, H. et al. (2022). Bridging multiscale characterization technologies and digital modeling to evaluate lithium battery full lifecycle. *Advanced Energy Materials* 12: 2200889.

3 Yang, Z., Huang, H., and Lin, F. (2022). Sustainable electric vehicle batteries for a sustainable world: perspectives on battery cathodes, environment, supply chain, manufacturing, life cycle, and policy. *Advanced Energy Materials* 12: 2200383.

4 Zhang, M., Wang, L., Xu, H. et al. (2023). Polyimides as promising materials for lithium-ion batteries: a review. *Nano-Micro Letters* 15: 135.

5 Cavers, H., Molaiyan, P., Abdollahifar, M. et al. (2022). Perspectives on improving the safety and sustainability of high voltage lithium-ion batteries through the electrolyte and separator region. *Advanced Energy Materials* 12: 2200147.

6 Kong, H., Guan, Y., Wang, J. et al. (2022). An intercalation–conversion hybrid mechanism enables covalent organic frameworks with superior Li-ion storage. *Journal of Materials Chemistry A* 10: 20866–20873.

7 Shi, Y., Yin, P., Li, J. et al. (2023). Ultra-high rate capability of in-situ anchoring FeF_3 cathode onto double-enhanced conductive Fe/graphitic carbon for high energy density lithium-ion batteries. *Nano Energy* 108: 108181.

8 Liu, J., Yue, M., Wang, S. et al. (2022). A review of performance attenuation and mitigation strategies of lithium-ion batteries. *Advanced Functional Materials* 32: 2107769.

9 Manthiram, A. (2020). A reflection on lithium-ion battery cathode chemistry. *Nature Communications* 11: 1550.

10 Xie, J. and Lu, Y.-C. (2020). A retrospective on lithium-ion batteries. *Nature Communications* 11: 2499.

11 Wang, L., Hung, S.-F., Zhao, S. et al. (2025). Modulating the covalency of Ru-O bonds by dynamic reconstruction for efficient acidic oxygen evolution. *Nature Communications* 16: 3502.

12 Su, L., Jarvis, K., Charalambous, H. et al. (2023). Stabilizing high-nickel cathodes with high-voltage electrolytes. *Advanced Functional Materials* 33: 2213675.

13 Lee, S., Yoon, G., Jeong, M. et al. (2015). Hierarchical surface atomic structure of a manganese-based spinel cathode for lithium-ion batteries. *Angewandte Chemie International Edition* 54: 1153–1158.

14 Shirakawa, J., Nakayama, M., Wakihara, M., and Uchimoto, Y. (2007). Changes in electronic structure upon lithium insertion into $Fe_2(SO_4)_3$ and $Fe_2(MoO_4)_3$ investigated by X-ray absorption spectroscopy. *The Journal of Physical Chemistry B* 111: 1424–1430.

15 Zou, Z., Mu, Y., Han, M. et al. (2025). Integrated polyanion-layered oxide cathodes enabling 100000 cycle life for sodium-ion batteries. *Energy & Environmental Science* 18: 2216–2230.

16 Li, S., Sun, Y., Gao, A. et al. (2022). Sustainable $LiCoO_2$ by collective glide of CoO_6 slabs upon charge/discharge. *Proceedings of the National Academy of Sciences* 119: e2120060119.

17 Wang, J., Liang, Z., Zhao, Y. et al. (2022). Direct conversion of degraded $LiCoO_2$ cathode materials into high-performance $LiCoO_2$: a closed-loop green recycling strategy for spent lithium-ion batteries. *Energy Storage Materials* 45: 768–776.

18 Chen, J., Lin, Z., Xiang, W. et al. (2022). Investigation of degradation mechanism of $LiCoO_2$/graphite batteries with multiscale characterization. *Electrochimica Acta* 436: 141374.

19 Shen, B., Xu, X., Liu, W. et al. (2021). Realizing an excellent cycle and rate performance of $LiCoO_2$ at 4.55 V by Li ionic conductor surface modification. *ACS Applied Energy Materials* 4: 12917–12926.

20 Zhang, X., Zhang, Y., Liu, J. et al. (2021). Syntheses, challenges and modifications of single-crystal cathodes for lithium-ion battery. *Journal of Energy Chemistry* 63: 217–229.

21 Choi, J.U., Voronina, N., Sun, Y.-K., and Myung, S.-T. (2020). Recent progress and perspective of advanced high-energy co-less Ni-rich cathodes for Li-ion batteries: yesterday, today, and tomorrow. *Advanced Energy Materials* 10: 2002027.

22 Cui, S., Wei, Y., Liu, T. et al. (2016). Optimized temperature effect of Li-ion diffusion with layer distance in $Li(Ni_xMn_yCo_z)O_2$ cathode materials for high performance Li-ion battery. *Advanced Energy Materials* 6: 1501309.

23 Kim, M., Zhu, J., Li, L. et al. (2020). Understanding reactivities of Ni-Rich $Li[Ni_xMn_yCo_{1-x-y}]O_2$ single-crystal cathode materials. *ACS Applied Energy Materials* 3: 12238–12245.

24 Zheng, J., Liu, T., Hu, Z. et al. (2016). Tuning of thermal stability in layered $Li(Ni_xMn_yCo_z)O_2$. *Journal of the American Chemical Society* 138: 13326–13334.

25 Hwang, B.J., Tsai, Y.W., Santhanam, R. et al. (2003). In situ X-Ray absorption spectroscopy studies of nanosized $LiAl_{0.15}Mn_{1.85}O_4$ cathode material in an aqueous solution. *Journal of the Electrochemical Society* 150: A335.

26 Lee, E., Lee, K.-R., and Lee, B.-J. (2017). Interatomic potential of Li–Mn–O and molecular dynamics simulations on Li diffusion in spinel Li_1–xMn_2O_4. *The Journal of Physical Chemistry C* 121: 13008–13017.

27 Horne, C.R., Bergmann, U., Grush, M.M. et al. (2000). Electronic structure of chemically-prepared $Li_xMn_2O_4$ determined by Mn X-ray absorption and emission spectroscopies. *The Journal of Physical Chemistry B* 104: 9587–9596.

28 Chung, K.Y. and Kim, K.-B. (2004). Investigations into capacity fading as a result of a Jahn–Teller distortion in 4V $LiMn_2O_4$ thin film electrodes. *Electrochimica Acta* 49: 3327–3337.

29 Terada, Y., Nishiwaki, Y., Nakai, I., and Nishikawa, F. (2001). Study of Mn dissolution from $LiMn_2O_4$ spinel electrodes using in situ total reflection X-ray fluorescence analysis and fluorescence XAFS technique. *Journal of Power Sources* 97-98: 420–422.

30 Zhan, C., Lu, J., Jeremy Kropf, A. et al. (2013). Mn(II) deposition on anodes and its effects on capacity fade in spinel lithium manganate–carbon systems. *Nature Communications* 4: 2437.

31 Yang, Y., Xie, C., Ruffo, R. et al. (2009). Single nanorod devices for battery diagnostics: a case study on $LiMn_2O_4$. *Nano Letters* 9: 4109–4114.

32 Bhattacharya, J. and Wolverton, C. (2013). Relative stability of normal vs. inverse spinel for 3d transition metal oxides as lithium intercalation cathodes. *Physical Chemistry Chemical Physics* 15: 6486–6498.

33 Choi, S. and Manthiram, A. (2006). Synthesis and electrochemical properties of spinel $LiCo_2O_4$ cathodes. *Journal of the Electrochemical Society* 149: A162–A166.

34 Lee, K.-K., Yoon, W.-S., Kim, K.-B. et al. (2001). Thermal behavior and the decomposition mechanism of electrochemically delithiated $Li_{1-x}NiO_2$. *Journal of Power Sources* 97-98: 321–325.

35 Gardner, G., Al-Sharab, J., Danilovic, N. et al. (2016). Structural basis for differing electrocatalytic water oxidation by the cubic, layered and spinel forms of lithium cobalt oxides. *Energy & Environmental Science* 9: 184–192.

36 Mukherjee, P., Lu, P., Faenza, N. et al. (2021). Atomic structure of surface-densified phases in Ni-rich layered compounds. *ACS Applied Materials & Interfaces* 13: 17478–17486.

37 Manthiram, A. and Goodenough, J.B. (1987). Lithium insertion into $Fe_2(MO_4)_3$ frameworks: comparison of M = W with M = Mo. *Journal of Solid State Chemistry* 71: 349–360.

38 Manthiram, A. and Goodenough, J.B. (1989). Lithium insertion into $Fe_2(SO_4)_3$ frameworks. *Journal of Power Sources* 26: 403–408.

39 Li, Z., Zhang, D., and Yang, F. (2009). Developments of lithium-ion batteries and challenges of $LiFePO_4$ as one promising cathode material. *Journal of Materials Science* 44: 2435–2443.

40 Mohamed, N. and Allam, N.K. (2020). Recent advances in the design of cathode materials for Li-ion batteries. *RSC Advances* 10: 21662–21685.

41 Hou, X., Liu, X., Wang, H. et al. (2023). Specific countermeasures to intrinsic capacity decline issues and future direction of $LiMn_2O_4$ cathode. *Energy Storage Materials* 57: 577–606.

42 Seok, E., Kim, M., Lee, S. et al. (2023). Electrostatic interaction-driven inorganic coating layer toward improving battery performance for 5 V class high-voltage cathode in secondary batteries. *Chemical Engineering Journal* 453: 139737.

43 Ghanbarzadeh, B., Khatibi, A., Asadi, A., and Shokri, B. (2022). Regulating lithium-ion flux in the solid electrolyte interphase layer to prevent lithium dendrite growth on lithium metal anode. *Journal of Energy Storage* 47: 103668.

44 Wang, Q., Sun, D., Zhou, X. et al. (2020). Amide-based interface layer with high toughness in situ building on the li metal anode. *ACS Applied Materials & Interfaces* 12: 25826–25831.

45 Bi, J., Du, Z., Sun, J. et al. (2023). On the road to the frontiers of lithium-ion batteries: a review and outlook of graphene anodes. *Advanced Materials* 35: 2210734.

46 Palacín, M.R. (2009). Recent advances in rechargeable battery materials: a chemist's perspective. *Chemical Society Reviews* 38: 2565–2575.

47 Chen, X., Loaiza, L.C., Monconduit, L., and Seznec, V. (2021). 2D silicon–germanium-layered materials as anodes for Li-ion batteries. *ACS Applied Energy Materials* 4: 12552–12561.

48 Kitta, M. and Kohyama, M. (2017). Nanoscale controlled Li-insertion reaction induced by scanning electron-beam irradiation in a $Li_4Ti_5O_{12}$ electrode material for lithium-ion batteries. *Physical Chemistry Chemical Physics* 19: 11581–11587.

49 Shen, X., Tian, Z., Fan, R. et al. (2018). Research progress on silicon/carbon composite anode materials for lithium-ion battery. *Journal of Energy Chemistry* 27: 1067–1090.

50 Xu, J., Jeon, I.-Y., Ma, J. et al. (2017). Understanding of the capacity contribution of carbon in phosphorus-carbon composites for high-performance anodes in lithium ion batteries. *Nano Research* 10: 1268–1281.

51 Zhao, L., Ding, B., Qin, X.-Y. et al. (2022). Revisiting the roles of natural graphite in ongoing lithium-ion batteries. *Advanced Materials* 34: e2106704.

52 Jiang, J., Liu, J., Chen, Y. et al. (2020). Preparation of hierarchical hexagonal nanoplates NiO composite with microcrystalline graphite for highly reversible lithium storage. *Journal of Alloys and Compounds* 815: 152333.

53 Chang, Y.-C., Sohn, H.-J., Ku, C.-H. et al. (1999). Anodic performances of mesocarbon microbeads (MCMB) prepared from synthetic naphthalene isotropic pitch. *Carbon* 37: 1285–1297.

54 Zhang, J., Weng, S., Zhong, C. et al. (2025). Worse interference of Fe^{3+} than Fe^{2+} on degrading the interphase and performance of $LiFePO_4$||graphite battery. *Advanced Materials* 37: e13736.

55 Liu, C.-Y., Yang, Y., Yao, M., and Fang, H.-T. (2019). Lithium difluoro (oxalate) borate improving the zero-volt storage performance of lithium-ion batteries by offering anode SEI film tolerance to high potentials. *Energy Storage Materials* 18: 148–154.

56 Zou, Z. and Jiang, C. (2019). Nitrogen-doped amorphous carbon coated mesocarbon microbeads as excellent high rate Li storage anode materials. *Journal Of Materials Science & Technology* 35: 644–650.

57 Ramos, A., Cameán, I., Cuesta, N., and García, A.B. (2014). Graphitized stacked-cup carbon nanofibers as anode materials for lithium-ion batteries. *Electrochimica Acta* 146: 769–775.

58 Xie, L., Tang, C., Bi, Z. et al. (2021). Hard carbon anodes for next-generation Li-ion batteries: review and perspective. *Advanced Energy Materials* 11: 2101650.

59 Wang, J., Chen, X., Liu, X. et al. (2016). Capacity-increasing robust porous SiO_2/Si/graphene/C microspheres as an anode for Li-ion batteries. *RSC Advances* 6: 45077–45084.

60 Zhu, M., Yang, J., Yu, Z. et al. (2017). Novel hybrid Si nanocrystals embedded in a conductive SiO_x@C matrix from one single precursor as a high performance anode material for lithium-ion batteries. *Journal of Materials Chemistry A* 5: 7026–7034.

61 Zhang, Q., Han, S., Tian, F. et al. (2021). Molten salt derived graphene-like carbon nanosheets wrapped SiO_x/carbon submicrospheres with enhanced lithium storage. *Chinese Journal of Chemistry* 39: 1233–1239.

62 Zheng, Y., Kong, X., Usman, I. et al. (2020). Rational design of the pea-pod structure of SiO_x/C nanofibers as a high-performance anode for lithium ion batteries. *Inorganic Chemistry Frontiers* 7: 1762–1769.

63 Zhang, Y., Mu, Z., Lai, J. et al. (2019). MXene/Si@SiO_x@C layer-by-layer superstructure with autoadjustable function for superior stable lithium storage. *ACS Nano* 13: 2167–2175.

64 Ge, J., Guo, T., Shen, H. et al. (2023). Regulating charge/ion transfer kinetic behavior via embedded TiOX in hierarchical TiO_X/SiO_X@C with optimal expansion stress for lithium storage. *Journal of Power Sources* 555: 232357.

65 Guo, X., Li, W., Zhang, Q. et al. (2022). Ultrasmall metal (Fe, Co, Ni) nanoparticles strengthen silicon oxide embedded nitrogen-doped carbon superstructures for long-cycle-life Li-ion-battery anodes. *Chemical Engineering Journal* 432: 134413.

66 Yu, S.-H., Lee, S.H., Lee, D.J. et al. (2016). Conversion reaction-based oxide nanomaterials for lithium ion battery anodes. *Small* 12: 2146–2172.

67 Yang, Y., Xia, J., Guan, X. et al. (2022). In situ growth of CoP nanosheet arrays on carbon cloth as binder-free electrode for high-performance flexible lithium-ion batteries. *Small* 18: e2204970.

68 Rehman, W.u., Jiang, Z., Qu, Z. et al. (2023). Preparation of Mn_2SnO_4 wrapped with N-doped reduced graphene oxide as a stable anode material for lithium–ion storage. *Journal of Alloys and Compounds* 939: 168829.

69 Liu, X., Yu, Y., Li, K. et al. (2024). Intergrating hollow multishelled structure and high entropy engineering toward enhanced mechano-electrochemical properties in lithium battery. *Advanced Materials* 36: e2312583.

70 Yoon, J.-M., Lee, Y.-H., and Park, C.-M. (2023). Optimized Ga-based nanocomposite for superior Li-ion battery anodes. *Materials Today Energy* 35: 101327.

71 Li, C., Tu, S., Ai, X. et al. (2023). Stress-regulation design of lithium alloy electrode toward stable battery cycling. *Energy & Environmental Science* 6: e12267.

72 Jin, Q., Han, F., Zhang, L. et al. (2025). Magneto-electrochemical synergistic lithiophilic high-entropy alloy enables durable lithium metal anode. *Energy Storage Materials* 83: 104717.

73 Zhu, A., Bian, X., Han, W. et al. (2023). The application of deep eutectic solvents in lithium-ion battery recycling: a comprehensive review. *Resources, Conservation and Recycling* 188: 106690.

74 Kumar, A. and Nagmani; Puravankara, S. (2022). Symmetric sodium-ion batteries—materials, mechanisms, and prospects. *Materials Today Energy* 29: 101115.

75 Delmas, C., Fouassier, C., and Hagenmuller, P. (1980). Structural classification and properties of the layered oxides. *Physica B+C* 99: 81–85.

76 Yuan, X.-G., Guo, Y.-J., Gan, L. et al. (2022). A universal strategy toward air-stable and high-rate O_3 layered oxide cathodes for Na-ion batteries. *Advanced Functional Materials* 32: 2111466.

77 Peng, B., Sun, Z., Zhao, L. et al. (2021). Dual-manipulation on P_2-$Na_{0.67}Ni_{0.33}Mn_{0.67}O_2$ layered cathode toward sodium-ion full cell with record operating voltage beyond 3.5 V. *Energy Storage Materials* 35: 620–629.

78 Yu, L., Cheng, Z., Xu, K. et al. (2022). Interlocking biphasic chemistry for high-voltage P_2/O_3 sodium layered oxide cathode. *Energy Storage Materials* 50: 730–739.

79 Liu, Y., Cui, X., Liu, Y., and Xia, Y. (2023). Perspective on iron-based phosphate cathode for commercial sodium-ion cells. *Small* 19: e2302972.

80 Xiong, F., Li, J., Zuo, C. et al. (2023). Mg-doped $Na_4Fe_3(PO_4)_2(P_2O_7)$/C composite with enhanced intercalation pseudocapacitance for ultra-stable and high-rate sodium-ion storage. *Advanced Functional Materials* 33: 2211257.

81 Peng, J., Zhang, W., Liu, Q. et al. (2022). Prussian blue analogues for sodium-ion batteries: past, present, and future. *Advanced Materials* 34: e2108384.

82 Huang, T., Niu, Y., Yang, Q. et al. (2021). Self-template synthesis of prussian blue analogue hollow polyhedrons as superior sodium storage cathodes. *ACS Applied Materials & Interfaces* 13: 37187–37193.

83 Yang, X. and Rogach, A.L. (2020). Anodes and sodium-free cathodes in sodium ion batteries. *Advanced Energy Materials* 10: 2000288.

84 Zhao, L.-F., Hu, Z., Lai, W.-H. et al. (2021). Hard carbon anodes: fundamental understanding and commercial perspectives for Na-ion batteries beyond Li-ion and K-ion counterparts. *Advanced Energy Materials* 11: 2002704.

85 Guo, S., Yi, J., Sun, Y., and Zhou, H. (2016). Recent advances in titanium-based electrode materials for stationary sodium-ion batteries. *Energy & Environmental Science* 9: 2978–3006.

86 Gabriel, E., Hou, D., Lee, E., and Xiong, H. (2022). Multiphase layered transition metal oxide positive electrodes for sodium ion batteries. *Energy Science & Engineering* 10: 1672–1705.

87 Deng, Q., Liu, X., Li, Z. et al. (2023). Cobalt-nickel bimetallic sulfide (NiS_2/CoS_2) based dual-carbon framework for super sodium ion storage. *Journal of Colloid and Interface Science* 633: 480–488.

88 Wei, C., Tao, Y., Fei, H. et al. (2020). Recent advances and perspectives in stable and dendrite-free potassium metal anodes. *Energy Storage Materials* 30: 206–227.

89 Ye, Z., Jiang, Y., Li, L. et al. (2021). Rational design of MOF-based materials for next-generation rechargeable batteries. *Nano-Micro Letters* 13: 203.

90 Wang, Y., Zhou, S., Li, N. et al. (2025). Long-durable potassium ion batteries enabled by medium-entropy lattice engineering on Prussian blue analogues cathodes. *Advanced Energy Materials* 15: 2405007.

91 Wu, X., Chen, Y., Xing, Z. et al. (2019). Advanced carbon-based anodes for potassium-ion batteries. *Advanced Energy Materials* 9: 1900343.

92 Yan, D., Xie, M., Shao, Y. et al. (2021). Fast and durable anodes for sodium-/potassium-ion hybrid capacitors: tailoring self-adaptive nanocages inside hybrid fibers with high alignment. *Journal of Materials Chemistry A* 9: 13986–13995.

93 Adekoya, D., Chen, H., Hoh, H.Y. et al. (2020). Hierarchical Co_3O_4@N-doped carbon composite as an advanced anode material for ultrastable potassium storage. *ACS Nano* 14: 5027–5035.

94 Deng, R., Wang, M., Yu, H. et al. (2022). Recent advances and applications toward emerging lithium–sulfur batteries: working principles and opportunities. *Energy & Environmental Science* 5: 777–799.

95 Li, H., Xue, W., Wang, L., and Liu, T. (2021). Two competing reactions of sulfurized polyacrylonitrile produce high-performance lithium–sulfur batteries. *ACS Applied Materials & Interfaces* 13: 25002–25009.

96 Hou, L.-P., Zhang, X.-Q., Li, B.-Q., and Zhang, Q. (2021). Challenges and promises of lithium metal anode by soluble polysulfides in practical lithium–sulfur batteries. *Materials Today* 45: 62–76.

97 Yuan, B., Hua, D., Gu, X. et al. (2020). Polar, catalytic, and conductive $CoSe_2$/C frameworks for performance enhanced S cathode in Li–S batteries. *Journal of Energy Chemistry* 48: 128–135.

98 Mei, J., Liao, T., Liang, J. et al. (2020). Toward promising cathode catalysts for nonlithium metal–oxygen batteries. *Advanced Energy Materials* 10: 1901997.

99 Yousaf, M., Naseer, U., Li, Y. et al. (2021). A mechanistic study of electrode materials for rechargeable batteries beyond lithium ions by in situ transmission electron microscopy. *Energy & Environmental Science* 14: 2670–2707.

100 Zhao, Y., Tang, W., Liu, W. et al. (2023). Interfacial Engineering of Co_3O_4/Fe_2O_3 Nano-Heterostructure Toward Superior Li-O_2 Batteries. *Small* 19: e2205532.

101 Bi, X., Wang, R., Amine, K., and Lu, J. (2019). A critical review on superoxide-based sodium–oxygen batteries. *Small Methods* 3: 1800247.

102 Yadegari, H., Sun, Q., and Sun, X. (2016). Sodium-oxygen batteries: a comparative review from chemical and electrochemical fundamentals to future perspective. *Advanced Materials* 28: 7065–7093.

103 Kang, Y., Wang, S., Zhu, S. et al. (2021). Iron-modulated nickel cobalt phosphide embedded in carbon to boost power density of hybrid sodium–air battery. *Applied Catalysis B: Environmental* 285: 119786.

104 Lv, X.-W., Wang, Z., Lai, Z. et al. (2024). Rechargeable zinc–air batteries: advances, challenges, and prospects. *Small* 20: 2306396.

105 Liu, X., Zhang, G., Wang, L., and Fu, H. (2021). Structural design strategy and active site regulation of high-efficient bifunctional oxygen reaction electrocatalysts for Zn–Air battery. *Small* 17: e2006766.

106 Liu, Q., Wang, L., and Fu, H. (2023). Research progress on the construction of synergistic electrocatalytic ORR/OER self-supporting cathodes for zinc–air batteries. *Journal of Materials Chemistry A* 11: 4400–4427.

107 Wu, R., Wang, X., Ge, L. et al. (2023). N, S co-doped carbon with embedment of FeNi alloy as bifunctional oxygen electrocatalysts for rechargeable Zinc-air batteries. *Carbon* 202: 141–149.

6

Applications of Iron (Fe, Co, and Ni)-Group Metal Compound Nanomaterials in Supercapacitors

6.1 Overview of Supercapacitors

6.1.1 Energy Crisis and the Importance of Energy Storage Technology

The combined effect of accelerating global industrialization and population growth has led to a continuous rise in humanity's demand for fossil fuels. This trend not only causes the rapid depletion of nonrenewable resources such as oil, coal, and natural gas but also triggers severe environmental pollution problems. Frequent extreme weather events pose direct and serious threats to both ecosystems and human health. In terms of human health, the increased incidence of respiratory diseases and rising cases of deaths due to extremely high temperatures are closely linked to environmental degradation caused by the excessive use of fossil fuels.

Against this backdrop, the development of renewable energy sources such as wind energy, solar energy, tidal energy, and biomass energy has become the core path to alleviate the energy crisis and environmental pressures. In recent years, the global renewable energy industry has developed rapidly, with the installed capacities of photovoltaic power generation, wind power generation, and other technologies continuing to grow. However, such renewable energy sources are significantly constrained by natural conditions, exhibiting prominent intermittency and instability. Solar power generation relies on light intensity, resulting in huge differences in power generation between day and night; wind power generation is affected by wind speed, leading to frequent fluctuations in instantaneous power; tidal energy is limited by tidal cycles, with limited stable power generation hours per day. This characteristic means that the direct grid connection of renewable energy sources will severely impact grid frequency stability and voltage quality, and may even cause power outages and other accidents, greatly restricting their large-scale commercial application. Therefore, the large-scale application of renewable energy sources highly depends on efficient, stable, and low-cost energy storage technologies as "energy buffers" and "dispatch regulators."

As the core link connecting the production end and terminal application of renewable energy, energy storage technology's core value lies in realizing the spatial-temporal transfer of energy: storing electrical energy during periods of

Iron-Group Metal Compound Nanomaterials: Preparation, Characterization, Structure, and Applications, First Edition. Jinlong Zheng.

energy surplus and releasing it during periods of energy shortage, thereby ensuring the continuity and stability of energy supply. Current mainstream energy storage technologies can be divided into three categories: mechanical energy storage, electrochemical energy storage, and electromagnetic energy storage. Among them, mechanical energy storage, such as pumped hydro storage and compressed air energy storage, although relatively mature in technology, has obvious shortcomings: pumped hydro storage is limited by geographical conditions, requiring specific terrain to build reservoirs, with long construction cycles and high investment costs; compressed air energy storage has high requirements for sealing technology and low energy conversion efficiency; while electromagnetic energy storage, such as superconducting magnetic energy storage, is costly and difficult to apply on a large scale. In contrast, electrochemical energy storage technology has the advantages of fast response speed, flexible site selection, and high modularity. As an important member of the electrochemical energy storage family, supercapacitors (SCs) have become a key support with their unique performance. SCs have fast charging and discharging speeds, enabling rapid suppression of grid power fluctuations; high power density, which can meet the instantaneous high-power demands, such as braking energy recovery of new energy vehicles and frequency modulation of smart grids; and long cycle life, significantly reducing long-term operation and maintenance costs. In the field of new energy vehicles, SCs efficiently recover braking energy and improve energy utilization efficiency; in the field of smart grids, SCs can quickly respond to frequency deviations and ensure stable grid operation. It is these irreplaceable characteristics that make SCs play an increasingly important role in promoting the transformation of the energy structure from fossil energy dominance to renewable energy dominance, becoming one of the core technologies to ensure the safe and efficient operation of the energy system.

6.1.2 Definition, Characteristics, and Advantages of Supercapacitors

SCs, also known as electrochemical capacitors, are electrochemical energy storage devices based on charge storage at the electrode/electrolyte interface or fast reversible Faraday reactions [1]. Compared with traditional energy storage devices (such as lithium-ion batteries and lead-acid batteries), the core advantages of SCs are reflected in the following five aspects, highlighting their irreplaceable application value in many fields:

(1) Fast charging and discharging rate, suitable for fast energy response scenarios. SCs can achieve charging and discharging in seconds or even milliseconds, far superior to the hour-level charging speed of lithium-ion batteries. This characteristic makes them an ideal choice for peak power output and fast energy recovery scenarios. For example, in the research of SCs based on iron–cobalt–nickel sulfides, devices using $NiCo_2S_4$ mesoporous microaggregates as electrode materials can still achieve fast-charging and discharging responses at a high current density of $10\,A\,g^{-1}$, with a discharge time of only a few seconds, which can efficiently capture the instantaneous power fluctuations generated by new energy power generation systems (such as wind power and photovoltaic) [2].

In addition, in the emergency power supply scenario of microelectronic devices, SCs can complete charging in a few seconds and quickly release energy to provide instantaneous power support for sensors, microcontrollers, etc.

(2) High power density, meeting the demand for instantaneous high power. The power density of SCs can usually reach 10–100 kW kg^{-1}, which is 10–100 times that of lithium-ion batteries. It can easily meet the instantaneous energy demand of high-power equipment. In the braking energy recovery test of new energy vehicles, it can instantly absorb the peak power generated during braking, and the energy recovery efficiency is significantly higher than that of traditional battery systems.

(3) Long cycle life, reducing long-term use costs. The number of charge–discharge cycles of SCs can reach more than 10^5 times, far exceeding that of lithium-ion batteries. This characteristic significantly reduces the replacement frequency and maintenance costs during the long-term use of equipment. For example, using the CoO@Ni–Co–S hierarchical nanostructure as the SC material, after 10,000 cycles at a current density of 10 A g^{-1}, the capacity retention rate is still over 84% without obvious attenuation [3]. This long-life characteristic makes it highly competitive in scenarios requiring long-term stable operation, such as industrial energy management and smart grids.

(4) Wide operating temperature range, adapting to extreme environments. SCs can operate stably between −40°C and 70°C and can adapt to extreme environmental conditions such as low-temperature cold and high-temperature dryness. They also have high safety without the common explosion or leakage risks of lithium-ion batteries. In low-temperature performance tests, SCs can still maintain more than 80% of the capacity at room temperature in a −40°C environment, while the capacity attenuation of lithium-ion batteries usually exceeds 50% at the same temperature [4]; in a high-temperature environment (70°C), the SC based on $NiCo_2O_4$@MnO_2 hollow nanotubes still has a capacity retention rate of 71.3% with a coulombic efficiency of 95.9% after 5000 cycles [5]. This advantage allows it to be applied to energy storage systems in extreme environments, such as plateaus, polar regions, and deserts.

(5) Environmentally friendly, in line with the concept of sustainable development. Most electrode materials of SCs (such as iron–cobalt–nickel-based compounds) do not contain precious metals such as Ru and Pt and can be prepared through green synthesis processes such as the hydrothermal method and coprecipitation method. The raw materials are widely available and low in toxicity, which is in line with the concept of "green chemistry" and sustainable development. In addition, the recovery rate of iron–cobalt–nickel-based compound electrode materials is high, and metal elements such as Ni, Co, and Fe can be recovered through the acid dissolution–precipitation method, effectively reducing resource waste and environmental pollution.

These characteristics give SCs broad application prospects in fields such as braking energy recovery of new energy vehicles, peak shaving of smart grids, emergency power supplies for portable electronic devices, and startup of industrial high-power equipment. For example, in the field of new energy vehicles, SCs can form a hybrid energy storage system with lithium-ion batteries. The former is responsible for braking energy recovery and instantaneous power output during startup and acceleration, while the latter is responsible for long-range power supply, significantly improving the overall energy utilization efficiency of the vehicle; in the field of smart grids, SCs can act as "peak shaving buffers" to suppress power fluctuations caused by the grid connection of wind power and photovoltaics and ensure stable grid frequency.

6.1.3 Types of Supercapacitors

According to the differences in charge storage mechanisms and electrode structures, SCs are mainly divided into three categories: electric double-layer capacitors (EDLC), pseudocapacitors, and battery-type capacitors [6]. Figure 6.1 shows the differences in charge storage mechanisms and electrochemical responses between electric double-layer capacitance, pseudocapacitance, and battery-type materials through cyclic voltammetry (CV) and galvanostatic charging/discharging (GCD) curve comparison.

6.1.3.1 Electric Double-Layer Capacitors (EDLC)

The charge storage of EDLCs is based on the electrostatic adsorption at the electrode–electrolyte interface, which does not involve Faraday redox reactions and belongs to a physical energy storage process. During charging, anions and cations in the electrolyte migrate to the positive and negative electrodes, respectively, forming a closely arranged electric double layer on the electrode surface; during discharging, ions return to the electrolyte to realize charge release [8]. The energy storage process of EDLCs is only based on non-Faraday activity, so it can respond instantaneously to potential changes, and its capacitance and current are independent of potential.

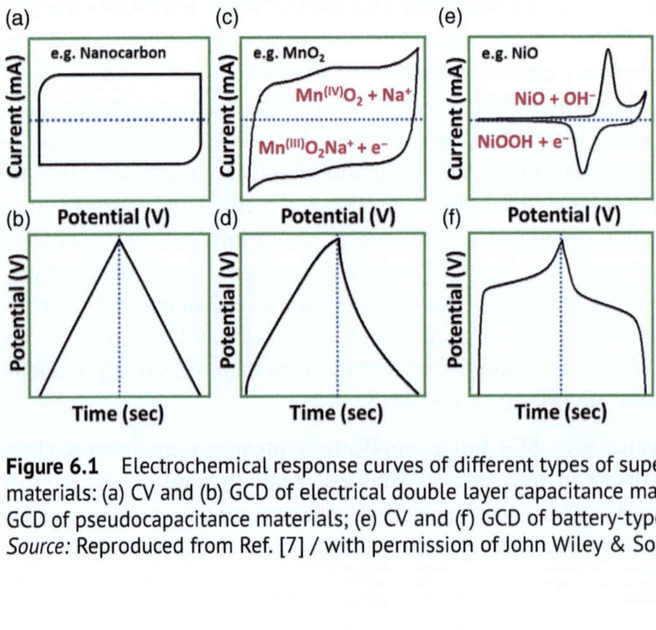

Figure 6.1 Electrochemical response curves of different types of supercapacitive electrode materials: (a) CV and (b) GCD of electrical double layer capacitance materials; (c) CV and (d) GCD of pseudocapacitance materials; (e) CV and (f) GCD of battery-type materials. *Source:* Reproduced from Ref. [7] / with permission of John Wiley & Sons.

Therefore, the electrochemical characteristics of SC materials are manifested as rectangular CV curves and GCD (Figure 6.1a, b).

The electrode materials are usually carbon materials with high specific surface area, such as activated carbon (AC), graphene, and carbon nanotubes, whose specific surface area can reach up to 1000–3000 $m^2 g^{-1}$, providing sufficient sites for ion adsorption. For example, the EDLC with AC as the electrode has a power density of 5000 $W kg^{-1}$ in 6 M KOH electrolyte, and the capacity retention rate is still over 90% after 100,000 cycles [9]. This type of device has the characteristics of excellent cycle stability and good charge–discharge reversibility, but the specific capacitance is relatively low (usually <250 $F g^{-1}$), which limits the improvement of energy density.

6.1.3.2 Pseudocapacitors

The charge storage process of pseudocapacitors is different from that of EDLCs. The charge storage of pseudocapacitors originates from fast reversible Faraday redox reactions occurring on or near the surface of electrode materials, involving ion adsorption/desorption, intercalation/deintercalation, or oxidation state changes. Since charge storage involves redox processes, this means that the electrode material behaves like a battery in terms of reaction kinetics, but its electrochemical behavior shows a quasi-rectangular CV and a nearly linear GCD profile similar to EDLCs (Figure 6.1c, d). This is because the Faraday redox process is very fast and not limited by solid-state transport. Usually, the capacitance and energy of pseudocapacitive materials are higher than those of EDLC materials. Due to the slow reaction kinetics of pseudocapacitive materials and volume expansion during repeated charge–discharge processes, the cycle performance of pseudocapacitors is much lower than that of EDLCs.

Pseudocapacitive mechanisms are divided into three different types: (i) Underpotential deposition (UPD), which refers to the deposition of a monolayer of metal ions on the surface of other metals above their redox potential. (ii) Pseudocapacitance based on redox reactions, which is a mechanism for charge storage through fast Faraday processes. In this process, ions undergo reversible adsorption or redox reactions on or near the surface of the electrode material, which is common in SC systems using aqueous electrolytes (such as sulfuric acid, potassium hydroxide, sodium sulfate, sodium hydroxide, potassium chloride, and ammonium chloride). Although its energy storage essence involves Faraday reactions, the extremely fast reaction rate makes its electrochemical response highly similar to that of EDLC materials: the CV curve is close to rectangular, and the galvanostatic charge–discharge curve is approximately triangular. Therefore, the charge storage of this type of pseudocapacitor combines two mechanisms: electric double-layer formation and surface redox. Typical materials with redox pseudocapacitive behavior include transition metal oxides, such as RuO_2, MnO_2, Mn_3O_4, and MnOOH [10]. (iii) Pseudocapacitance based on intercalation/deintercalation, which originates from the process of electrolyte ions intercalating into the layered structure of redox-active electrode materials, followed by fast Faraday reactions. Unlike the surface reaction mechanism, this type of charge storage is not limited to the electrode surface but extends to the bulk phase of the material, thereby achieving higher capacity. This mechanism is mainly found in nonaqueous electrolyte systems. Typical intercalated pseudocapacitive materials include transition metal oxides (MoO_3, Nb_2O_5, TiO_2, WO_3, V_2O_5, etc.) and transition metal sulfides (MoS_2, TiS_2, VS_2, etc.) [11]. These materials usually have two-dimensional (2D) layered or three-dimensional (3D) open structures, which are conducive to the rapid transport and intercalation of electrolyte ions. A key feature of such materials is that their ion intercalation/deintercalation processes do not undergo significant phase transitions. It should be important to note that materials that undergo phase transitions during cycling were often mistakenly classified as pseudocapacitive materials in the past. However, electrochemical research has now clearly distinguished such materials from traditional pseudocapacitors and defined them as "battery-type materials" [12].

6.1.3.3 Battery-Type Supercapacitors

The charge storage mechanism of battery-type materials is essentially different from that of capacitive electrodes, and the process is more similar to that of traditional secondary batteries. In battery-type SCs, charge storage relies on diffusion-controlled Faraday redox reactions, accompanied by phase transitions of the electrode material. Therefore, such materials show obvious redox peaks in CV curves, and typical voltage plateaus in GCD tests (Figure 6.1e, f). Typical battery-type materials include transition metal compounds based on oxides/hydroxides, phosphides, and sulfides/selenides (Cu, Co, Ni, etc.). Such materials usually have high specific capacity, but their slow reaction kinetics and phase transitions during charge–discharge often lead to poor rate performance [13].

Battery-type electrode materials undergo significant phase transitions when reacting with electrolyte ions. Although their multielectron transfer processes usually bring higher charge storage capacity than intercalation-type materials, the slow electrochemical reaction kinetics during repeated intercalation/deintercalation processes limit their rate performance and cycle stability. Typical materials include iron oxides, nickel/cobalt-based oxides/hydroxides, layered double hydroxides (LDHs), and binary metal sulfides [14].

6.2 Advantages and Disadvantages of Iron–Cobalt–Nickel Metal Compounds as Electrode Materials for Supercapacitors

6.2.1 Advantages

6.2.1.1 Abundant Resources and Low Cost

Iron, cobalt, and nickel are abundant transition metal elements in the Earth's crust. Among them, iron has a crustal abundance of approximately 5.6%, making it the fourth most abundant element; the abundances of nickel and cobalt are both significantly higher than those of precious metals (such as ruthenium and platinum). In terms of cost, the ore resources of these elements are widely distributed. After long-term industrial development, the mining and refining technologies have become highly mature, directly leading to a substantial reduction in raw material costs. Taking iron as an example, its unit price is only 1/2000 that of the precious metal ruthenium, demonstrating an extremely prominent cost advantage.

Compared to precious metal-based electrode materials (such as RuO_2), the preparation of iron–cobalt–nickel-based compounds does not require expensive raw materials, and the synthesis processes are simple. They can be produced on a large scale using low-cost methods such as hydrothermal, solvothermal, and coprecipitation methods. For instance, using a simple hydrothermal method with common chemicals like nickel salts, cobalt salts, and urea as raw materials, NiCo-LDH nanosheet arrays can be grown on nickel foam substrates [15]. The low cost and abundant resources lay the foundation for the industrial application of iron–cobalt–nickel-based compounds.

6.2.1.2 High Theoretical Specific Capacity and Abundant Redox Chemistry

Iron, cobalt, and nickel elements possess multiple variable valence states (Fe^{2+}/Fe^{3+}, $Co^{2+}/Co^{3+}/Co^{4+}$, and Ni^{2+}/Ni^{3+}). This characteristic enables them to participate in multiple reversible redox reactions, providing abundant active sites for charge storage. Consequently, they exhibit a theoretical specific capacity far exceeding that of traditional carbon materials ($<250\ F\ g^{-1}$) [16].

Notably, iron–cobalt–nickel-based transition metal sulfides and phosphides exhibit superior electrochemical performance. Due to the lower electronegativity of sulfur/phosphorus atoms compared to oxygen atoms, their electronic structure is more flexible, leading to significantly enhanced electrical conductivity and redox activity. For example, metal–organic framework (MOF)-derived $CuCo_2S_4$ nanomaterials achieve a specific capacity of $400.10\ C\ g^{-1}$, with a capacity retention rate of 89.41% after 5500 cycles at a current density of $10\ A\ g^{-1}$ [17]. These materials' high specific capacity characteristics provide core support for enhancing the energy density of SCs.

6.2.1.3 The Synergistic Effect of Iron, Cobalt, and Nickel Elements

By constructing binary or ternary metal compounds, a significant synergistic effect can be generated among iron (Fe), cobalt (Co), and nickel (Ni) elements. This effect optimizes electrochemical performance from three aspects: electronic structure, active sites, and reaction kinetics, enabling the comprehensive performance of multicomponent compounds to surpass that of single-metal compounds. In terms of electronic structure modulation, the electronegativity differences among different metal ions lead to charge redistribution at the interface, reducing the energy barrier for redox reactions. For instance, in Zn–Co–Ga trimetallic phosphide nanosheets (ZCGP), the introduction of Zn and Ga induces the splitting of Co's 3d orbitals, altering the electron cloud density distribution of Co's d orbitals. This causes the originally concentrated electron cloud to disperse into different sublevels due to orbital splitting, thereby lowering the electron transfer barrier and enhancing electrical conductivity [18].

In terms of active site amplification, the introduction of multiple metals leads to the formation of abundant heterogeneous structures on the material surface, with each metal ion serving as an independent active center for reactions. For example, in NiFeCo-S@NiFeCo-TH heterogeneous ultrathin nanosheets, Ni, Fe, and Co ions provide different redox couples, synergistically contributing to the capacity, resulting in a specific capacity significantly higher than that of single-metal sulfides [19].

In terms of reaction kinetics optimization, multi-metal compounds exhibit shorter ion diffusion paths and lower charge transfer resistance. For example, in Cu_2NiSnS_4 materials, the presence of nickel improves charge transfer and ion diffusion, thereby enhancing charge storage performance [20]. This synergistic effect enables the comprehensive performance of binary/ternary iron–cobalt–nickel-based compounds to far exceed that of single-metal compounds.

6.2.2 Disadvantages

6.2.2.1 Poor Intrinsic Conductivity

Although iron–cobalt–nickel-based compounds possess the advantage of high specific capacity, most materials suffer from insufficient intrinsic conductivity, with hydroxides and some oxides exhibiting particularly prominent issues. Specifically, the conductivity of $Ni(OH)_2$ is below 10^{-8} S cm^{-1}, while that of Fe_2O_3 is only approximately 10^{-9} S cm^{-1}. Such low conductivity severely impedes electron transport efficiency, slows redox reaction kinetics, and ultimately leads to poor rate performance of the materials [21].

Even relatively conductive sulfides and phosphides still exhibit inferior electron transport capabilities compared to metals or carbon materials like graphene. For instance, although the conductivity of $NiCo_2S_4$ is higher than that of $NiCo_2O_4$, polarization phenomena still occur during high-rate charging and discharging, limiting its application in high-power scenarios [22]. To enhance conductivity, modification through methods such as composite material construction, elemental doping, or defect engineering is required.

6.2.2.2 Insufficient Cycling Stability

During the charge–discharge cycles of iron–cobalt–nickel-based compounds, ion intercalation and deintercalation can induce volume changes in the material, with some materials exhibiting volume variations of 20–50%. Such significant volume fluctuations can lead to active material detachment and electrode structure collapse, resulting in continuous capacity decay. For example, NiS electrodes are prone to particle agglomeration and structural degradation during cycling, demonstrating a capacity retention rate of only 59.6% after 2000 cycles [23].

Furthermore, interface compatibility issues between the electrode and electrolyte can further exacerbate the problem of insufficient cycling stability. In aqueous electrolytes, iron–cobalt–nickel-based compounds are prone to dissolution and corrosion, leading to increased interface impedance. In organic electrolytes, challenges such as poor wettability and slow ion diffusion rates exist. Additionally, side reactions at the interface, including electrolyte decomposition and metal ion dissolution, further accelerate capacity decay.

6.3 Electrochemical Performance Evaluation and Energy Storage Mechanism

6.3.1 Performance Testing Methods and Evaluation Metrics

6.3.1.1 Three-Electrode Testing System

The three-electrode testing system is a fundamental setup for evaluating the electrochemical performance of electrode materials, consisting of a working electrode, a counter electrode, and a reference electrode. Among these, the working electrode is the iron–cobalt–nickel-based compound electrode under test, serving as the core site for redox reactions; the counter electrode, typically a platinum sheet or graphite, only provides a current loop and does not participate in the main electrochemical reactions; the reference electrode (e.g., Hg/HgO, Ag/AgCl, and saturated calomel electrode) acts as a stable potential benchmark, ensuring the accuracy of potential measurements.

This system enables precise testing of the electrochemical performance of single electrode materials, such as specific capacity, rate capability, and cycling stability, providing fundamental data for material optimization. For example, testing the CV curves and GCD curves of an electrode in a three-electrode system can clarify its redox behavior and capacitive characteristics [27]. It should be noted that practical application devices adopt a two-electrode system, consisting of a positive electrode, a negative electrode, an electrolyte, and a separator, which can directly reflect the actual operational performance of the device.

6.3.1.2 Cyclic Voltammetry (CV) and Energy Storage Mechanism Analysis

CV is a core technique for studying the electrochemical behavior of electrode materials. By linearly scanning the electrode potential and recording the current–potential relationship curve (voltammogram), it enables the determination of energy storage mechanisms, assessment of electrochemical reversibility, and analysis of redox reaction characteristics [28].

From the perspective of energy storage mechanisms, the CV curve of an EDLC exhibits a rectangular shape with no distinct redox peaks, corresponding to the fast reversibility of electrostatic adsorption. In contrast, the CV curve of a pseudocapacitor shows prominent redox peaks, originating from Faradaic reactions. The CV curves of iron–cobalt–nickel-based compounds typically display significant redox peaks, indicating that their pseudocapacitive behavior primarily stems from Faradaic redox reactions (Figure 6.2a). For example, the CV curve of $Co_{1.29}Ni_{1.71}O_4$@$CoNi_2S_4$ heterostructured nanosheets exhibits multiple pairs of redox peaks, reflecting the synergistic redox effects of multiple metals [29].

By analyzing the relationship between peak current (i) and scan rate (v) ($i = av^b$), the charge storage type can be distinguished: when $b = 1$, it indicates capacitive contribution; when $b = 0.5$, it signifies diffusion-controlled contribution. For example, the b-value of the CuNiS electrode is 0.63 ($0.5 < b < 1$), indicating that its charge storage involves both surface capacitive and diffusion-controlled processes [30].

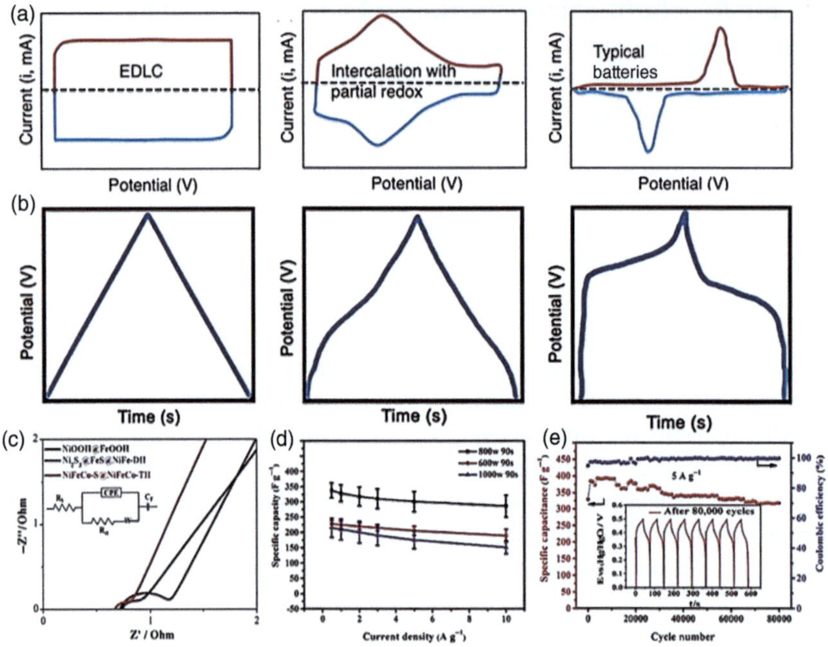

Figure 6.2 (a) Schematic of cyclic voltammograms for capacitor-like, intercalation-type, and battery-like materials. *Source:* Reproduced from Ref. [24] / with permission of American Chemical Society. (b) Schematic of galvanostatic charge/discharge profiles for capacitor-like, intercalation-type, and battery-like materials. *Source:* Reproduced from Ref. [25] / with permission of The Royal Society of Chemistry. (c) Nyquist plots of different materials. *Source:* Reproduced from Ref. [19] / with permission of Elsevier. (d) Specific capacitance of $CoV_2O_6{\cdot}2H_2O$ nanosheet. (e) cycling stability and variation of coulombic efficiency of $CoV_2O_6{\cdot}2H_2O$ at 5 A g^{-1}. *Source:* Reproduced from Ref. [26] / with permission of Elsevier.

6.3.1.3 Galvanostatic Charge–Discharge (GCD) and Specific Capacity Calculation

GCD testing involves charging and discharging the electrode at a constant current while recording the voltage change over time. This method allows for the calculation of specific capacity and evaluation of charge–discharge symmetry and rate performance. The GCD curve of an EDLC exhibits a symmetrical triangular shape with voltage changing linearly over time, whereas the GCD curve of a pseudocapacitor displays voltage plateaus corresponding to the potential range of redox reactions (Figure 6.2b) [25].

The areal-specific capacity (Q) in a three-electrode system or asymmetric SC (ASC) is calculated using the following equation [31]:

$$Q = \frac{I\Delta t}{A} \tag{6.1}$$

where I—discharge current, A.

Δt—discharge time, s.

A—area of the active material electrode, cm^2.

The areal-specific capacitance (C_A) and gravimetric-specific capacitance (C_m) are calculated using the following equations:

$$C_A = \frac{I\Delta t}{A\Delta V} \tag{6.2}$$

$$C_m = \frac{I\Delta t}{m\Delta V} \tag{6.3}$$

where I—discharge current, A.

Δt—discharge time, s.

A—area of the active material electrode, cm^2.

m—mass of active material, g.

ΔV—voltage difference during charge/discharge, V.

6.3.1.4 Electrochemical Impedance Spectroscopy (EIS) and Kinetic Analysis

EIS measures the impedance response of an electrode at different frequencies to analyze charge transfer resistance, ion diffusion resistance, and electrode/electrolyte interface characteristics, commonly represented by a Nyquist plot. The Nyquist plot consists of a semicircle in the high-frequency region and a linear segment in the low-frequency region: the intersection with the real axis in the high-frequency region represents the solution resistance (R_s), including electrolyte resistance and contact resistance between the electrode and current collector; the diameter of the semicircle represents the charge transfer resistance (R_{ct}), reflecting the difficulty of redox reactions on the electrode surface; the slope of the linear segment in the low-frequency region represents the Warburg impedance (Z_ω), which is related to the diffusion rate of electrolyte ions.

EIS analysis of iron–cobalt–nickel-based compounds can directly reflect their electrochemical kinetic performance. For example, the Nyquist plot of NiFeCo-S@NiFeCo-TH heterogeneous ultrathin nanosheets shows a small semicircle diameter and a steep low-frequency linear slope, indicating low charge transfer resistance and a fast ion diffusion rate. The Rct value is approximately 0.15 Ω, demonstrating excellent electron transport efficiency (Figure 6.2c) [19].

6.3.1.5 Energy Density, Power Density, and Cycle Life

Energy density (E) and power density (P) are key metrics for evaluating the practical value of SC devices (Figure 6.2d), which is calculated using the following formulas [31]:

$$E = \frac{1}{2} C_m \Delta V^2 \tag{6.4}$$

$$P = \frac{E}{\Delta t} \tag{6.5}$$

where C_m—gravimetric-specific capacitance in a two-electrode system, F g^{-1}.

Δt—discharge time in a two-electrode system, s.

ΔV—voltage window during charge/discharge in a two-electrode system, V.

Energy density reflects the energy storage capability of the device, while power density reflects the rate of energy output. These two parameters often require balancing through material design and device optimization.

Cycle life, evaluated by the capacitance retention rate after prolonged GCD cycling tests, is a critical indicator of SC durability. Iron–cobalt–nickel-based compound devices typically exhibit cycle lives of 5000–10,000 cycles, with some optimized devices approaching 100,000 cycles. For example, $CoV_2O_6 \cdot 2H_2O$ nanosheets demonstrate a capacity retention rate of 96.4% after 80,000 cycles at a current density of 5 A g^{-1} (Figure 6.2e) [26].

6.3.2 In-Depth Analysis of Charge Storage Mechanisms

6.3.2.1 Surface-Controlled and Diffusion-Controlled Pseudocapacitive Behaviors

The charge storage behaviors of Fe–Co–Ni-based compounds can be categorized into surface-controlled pseudocapacitance and diffusion-controlled battery-type behavior. The ratio of these two mechanisms directly affects the rate capability and power density of the materials. Surface-controlled pseudocapacitance originates from fast redox reactions on or near the electrode surface, featuring high reaction kinetics and weak dependence on scan rate, with a b-value close to 1. In contrast, diffusion-controlled battery-type behavior arises from ion intercalation/deintercalation reactions within the bulk phase of the material, characterized by slow reaction rates limited by diffusion, with a b-value close to 0.5 [30].

The contribution ratio of the two mechanisms can be quantified by analyzing CV curves at different scan rates. For example, the $CuCo_2O_4$@NiMn LDH electrode exhibits a diffusion-controlled contribution of ~27.65% and a surface-controlled contribution of ~72.35% at a low scan rate (0.1 mV s^{-1}). When the scan rate increases to 1 mV s^{-1}, the surface-controlled contribution rises to 97.03%, indicating that surface redox reactions dominate charge storage at high rates (Figure 6.3a, b) [32]. Nanostructure design (e.g., ultrathin nanosheets and porous structures) can shorten ion diffusion pathways, increase surface active sites, and enhance the proportion of surface-controlled pseudocapacitance, thereby improving rate capability.

6.3.2.2 Ion Intercalation/Deintercalation Reaction Mechanism

The charge storage of Fe–Co–Ni-based compounds is mainly based on redox reactions involving the intercalation/deintercalation of electrolyte ions (e.g., OH^-, Li^+, and Na^+), with reaction mechanisms varying among different material types.

For hydroxides (e.g., $Ni(OH)_2$ and $Co(OH)_2$), the redox reaction occurring in alkaline electrolytes is as follows:

$$Ni(OH)_2 + OH^- \leftrightarrow NiOOH + H_2O + e^- \tag{6.6}$$

$$Co(OH)_2 + OH^- \leftrightarrow CoOOH + H_2O + e^- \tag{6.7}$$

$$CoOOH + OH^- \leftrightarrow CoO_2 + H_2O + e^- \tag{6.8}$$

For spinel-type oxides (e.g., $FeCo_2O_4$ and $NiCo_2O_4$), they exhibit more complex reaction mechanisms, involving the synergistic redox reactions of multiple metal ions:

$$FeCo_2O_4 + H_2O + OH^- \leftrightarrow FeOOH + 2CoOOH + e^- \tag{6.9}$$

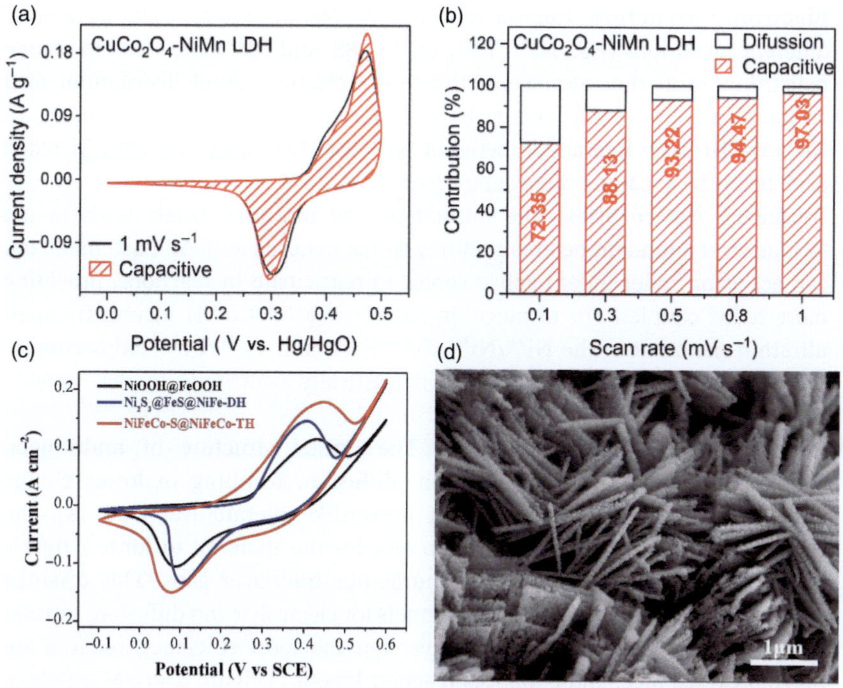

Figure 6.3 (a) Cyclic voltammetry profiles were collected at a scan rate of 1 mV s^{-1} for the $CuCo_2O_4$@NiMn LDH sample. (b) Capacitive- and diffusion-controlled contributions at different scan rates. *Source:* Ref. [32] / MDPI / CC BY 4.0. (c) CV curves of the NiOOH@FeOOH//NIF, Ni_2S_3@FeS@NiFe-DH//NIF, and NiFeCo-S@NiFeCo-TH//NIF samples at a scan rate of 10 mV s^{-1}. *Source:* Reproduced from Ref. [19] / with permission of Elsevier. (d) SEM image of $CuCo_2S_4$-6 hour. *Source:* Reproduced from Ref. [33] / MDPI / CC BY 4.0.

$$CoOOH + OH^- \leftrightarrow CoO_2 + H_2 + e^- \tag{6.10}$$

$$FeOOH + H_2O \leftrightarrow Fe(OH)_3 \leftrightarrow FeO_4^{2-} + 3e^- \tag{6.11}$$

For sulfides (e.g., $CuCo_2S_4$ and $NiCo_2S_4$), oxidation occurs on the surface during electrochemical processes, forming active metal oxide/hydroxide layers; the actual redox reactions take place between the active layers and the electrolyte:

$$CuCo_2S_4 + OH^- \leftrightarrow CuS_{4-2x}OH + 2CoS_xOH + e^- \tag{6.12}$$

$$CoS_xOH + OH^- \leftrightarrow CoS_xO + H_2O + e^- \tag{6.13}$$

$$CuS_{4-2x}OH + OH^- \leftrightarrow CuS_{4-2x}O + H_2O + e^- \tag{6.14}$$

All these reactions involve the intercalation/deintercalation of ions and the transfer of electrons. The reversibility and rate of the reactions determine the specific capacitance, rate capability, and cycling stability of the materials [34].

6.3.2.3 Synergistic Redox Mechanism of Fe, Co, and Ni

The synergistic effect of Fe, Co, and Ni elements is the core of the high performance of binary/ternary metal compounds, and their synergistic redox mechanism is mainly reflected in three aspects:

1) **Electronic Structure Regulation:** Differences in electronegativity among different metal ions (e.g., Ni^{2+}: 1.91, Co^{3+}: 1.88, and Fe^{3+}: 1.83) lead to charge redistribution at the interface, optimize the electron cloud distribution of d orbitals, and reduce the energy barrier of redox reactions. For example, in NiCoAl-LDH, the interaction between Ni^{2+} and Co^{3+} causes a shift in redox peak potentials, making the reaction more facile [35].
2) **Increased Active Sites:** The introduction of multiple metals leads to the formation of abundant heterostructures on the material surface. Each metal ion can act as an independent active center to participate in reactions, providing more redox couples. For instance, in NiFeCo-S@NiFeCo-TH heterostructured ultrathin nanosheets, the Ni^{2+}/Ni^{3+}, Fe^{2+}/Fe^{3+}, and Co^{2+}/Co^{3+} redox couples provided by Ni, Fe, and Co ions synergistically contribute to the capacity (Figure 6.3c) [19].
3) **Reaction Kinetics Optimization:** The crystal structure of multi-metal compounds is more favorable for ion diffusion, resulting in lower charge transfer resistance. For example, in flower-like hierarchical $CuCo_2S_4$, the synergistic effect between Cu and Co enables the material to form a tightly bonded hierarchical structure with numerous interlayer gaps. This oriented structure provides more interlayer channels for electrolyte ion diffusion, exposes more reaction sites, improves interlayer ion transport efficiency, reduces ion transport resistance, and enhances reaction kinetics (Figure 6.3d). Meanwhile, the introduction of Cu regulates the bond energy of Co—S bonds, optimizes the electron cloud distribution of S atoms, and facilitates the adsorption of OH^- and its participation in redox reactions [33].

6.3.3 Advanced Characterization and Theoretical Calculation

6.3.3.1 Application of In Situ/Ex Situ Characterization Technologies

In situ characterization technology can real-time capture the structural evolution, compositional changes, and charge transfer behaviors of electrode materials during the charge–discharge process, providing direct evidence for revealing energy storage mechanisms.

In situ X-ray diffraction (in situ XRD) can track the dynamic changes of crystal structures during the charge–discharge process. For example, during the cycling of the NiFe-LDH@CoNi-LDH (O_V-NFL@CNL) electrode, in situ XRD observations revealed the dynamic shift of the diffraction angle of the (003) crystal plane with ion intercalation/deintercalation: OH^- intercalation in the early stage of charging causes a low-angle shift of the diffraction angle and an expansion of the interlayer spacing; excessive OH^- intercalation when the voltage approaches 0.5 V triggers hydroxylation reconstruction of the lamellae and in situ phase transition; the hysteresis of OH^- desorption in the early stage of discharge leads to a negative shift of the diffraction angle, and the diffraction angle returns to its initial state when discharged to 0 V, confirming that the material achieves highly reversible energy storage behavior through elastic lattice deformation (Figure 6.4a, b) [36].

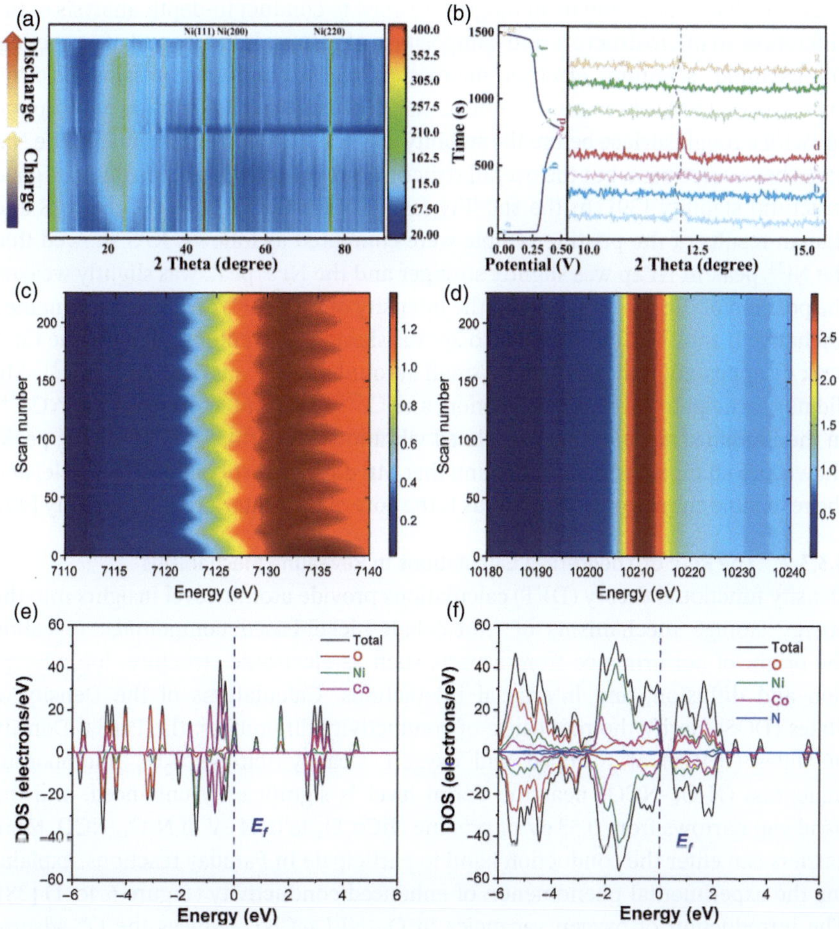

Figure 6.4 (a) In situ XRD pattern and (b) partial enlargement of O_v-NFL@CNL. *Source:* Reproduced from Ref. [36] / with permission of Elsevier. (c) Evolution of the operando XANES spectra for the FWO electrode at Fe K-edge and (d) Evolution of the operando XANES spectra for the FWO electrode at the W L3-edge. *Source:* Reproduced from Ref. [37] / with permission of John Wiley & Sons. (e) Density of state of NCO and (f) Density of state of N-O_v-NCO. *Source:* Reproduced from Ref. [38] / with permission of Elsevier.

In situ X-ray absorption spectroscopy (in situ XAS) technology (including X-ray absorption near-edge structure [XANES] and extended X-ray absorption fine structure [EXAFS]), combined with electrochemical tests (CV and GCD), enables the dynamic tracking of "valence state changes—local structure evolution" of elements during charge storage. For instance, dynamic tracking using in situ XAS clarified that the charge storage mechanism of $FeWO_4$ is the reversible redox reaction of Fe^{2+}/Fe^{3+} ($Fe^{2+} + OH^- \leftrightarrow Fe^{3+}(OH^-) + e^-$), where Fe is the only active site; in contrast, W^{6+} does not participate in electron transfer throughout the process, and only constructs a stable crystal framework through WO_4^{2-} groups to inhibit Fe ion dissolution and electrode structure collapse (Figure 6.4c, d) [37].

Ex situ characterization technologies are used to conduct in-depth analysis of the differences in microstructure and composition of materials before and after cycling. For example, scanning electron microscopy (SEM) and transmission electron microscopy (TEM) observations showed that NiS exhibited a microflower morphology with a rough surface before the stability test; after the cycle stability test, the NiS nanosheets were bent, but the overall structure did not collapse, confirming its good structural stability [39]. In the stability test of $NiMnCoO_4$ (NMCO), the XPS and Raman results of the pristine sample were compared as follows: XPS showed that the Ni^{3+} peak in Ni 2p was slightly stronger and the Ni^{2+} peak was slightly weaker, the position of the Mn 2p peak and the intensity of the Mn^{2+}/Mn^{3+} peaks remained unchanged, and the Co^{3+} peak in Co 2p was significantly enhanced while the Co^{2+} peak disappeared, indicating that a small amount of Ni^{2+} was oxidized to Ni^{3+}, Mn did not participate in the redox reaction, and Co^{2+} was completely oxidized to Co^{3+}. In the Raman spectrum, the Mn-related vibrations (superimposed on the E_g peak) showed no change, further confirming that Mn did not participate in the cycle, and the reversible change of Co—O bonds is the core of the material's high stability [40].

6.3.3.2 The Role of Theoretical Calculations in Revealing Mechanisms

Density functional theory (DFT) calculations provide atomic-level insights into the energy storage mechanisms of iron–cobalt–nickel-based compounds, revealing the origin of performance from aspects such as electronic structure, ion adsorption and diffusion, and interfacial interactions. Calculations of the Density of States (DOS) clarify the root cause of conductivity differences: the Partial Density of States (PDOS) of N-doped and oxygen vacancy-rich $NiCo_2O_4$ microporous nanograss (N-O_V-NCO) near the Fermi level is significantly enhanced, and the band gap narrows from 0.53 eV of pristine $NiCo_2O_4$ to 0.24 eV of N-O_V-NCO. More carriers can enter the conduction band to participate in Faraday reactions, explaining the experimental phenomenon of enhanced conductivity (Figure 6.4e, f) [38]. The introduction of oxygen vacancies in O_V-NFL@CNL reduces the O* adsorption energy from −1.77 eV to −6.80 eV, and local charge rearrangement forms highly active adsorption sites, lowering the ion diffusion barrier. Calculating the charge distribution at the heterojunction interface can reveal the essence of the synergistic effect. For example, the differential charge density map at the $Co_{1.29}Ni_{1.71}O_4$@$CoNi_2S_4$ interface shows that charge transfers from $Co_{1.29}Ni_{1.71}O_4$ to $CoNi_2S_4$, forming a built-in electric field that promotes electron transport and improves electrochemical activity [29].

6.4 Classification, Structure, and Characteristics of Iron-, Cobalt-, and Nickel-Based Compounds

6.4.1 Monometallic Compounds

6.4.1.1 Oxides

Fe, Co, and Ni-based monometallic oxides are among the earliest electrode materials studied for SCs. These materials have become research hotspots due to their comprehensive advantages, including typical crystal structures, well-defined redox characteristics, high theoretical specific capacitance, abundant resources, and environmental friendliness. However, they all face the common core bottleneck of poor intrinsic conductivity, which severely limits the full demonstration of their actual electrochemical performance. To overcome this limitation, researchers are committed to targeted optimization of each system through nanostructure and composite strategies.

Magnetite (Fe_3O_4) is regarded as an ideal SC electrode material due to its high conductivity and excellent theoretical specific capacitance ($2299\,F\,g^{-1}$). These superior electrochemical properties stem from its inverse spinel crystal structure (space group Fd-3m), in which oxygen ions form a face-centered cubic close-packed arrangement, with Fe^{2+} and Fe^{3+} occupying octahedral and tetrahedral interstitial sites, respectively, enabling rapid charge transport through an electron hopping mechanism. The capacitive behavior of Fe_3O_4 was first investigated and reported by Wu et al. in 2003, where Fe_3O_4 in Na_2SO_3 electrolyte delivered a relatively high capacitance of $510\,F\,g^{-1}$ with an operating potential of 1.2 V. [41] To address issues such as rapid capacity decay during cycling and easy agglomeration of nanoparticles in pure Fe_3O_4, research has shifted toward carbon composite materials. By encapsulating ultrafine Fe_3O_4 nanoparticles within hollow porous carbon capsules, the composite material not only maintained a high specific capacitance ($466\,F\,g^{-1}$) but also achieved outstanding stability with 92.4% capacity retention after 5000 cycles, demonstrating the great potential of the carbon composite strategy [42].

Cobalt oxide (Co_3O_4) has a spinel structure, with Co^{2+} occupying tetrahedral sites and Co^{3+} occupying octahedral sites, and a theoretical specific capacitance of $3560\,F\,g^{-1}$. However, its low electrical conductivity limits its rate performance. A prominent advantage of Co_3O_4 lies in the high tunability of its nanostructure and morphology (such as nanowires and mesoporous spheres). Leveraging this characteristic, constructing conductive networks by compositing with carbon materials like graphene can effectively synergistically enhance both rate performance and cycling stability, resulting in overall performance far surpassing that of the single component [43, 44]. A CuO–Co_3O_4 (1:1) heterostructure derived from bimetallic MOFs exhibited a specific capacitance as high as $1564.4\,F\,g^{-1}$ at $1\,A\,g^{-1}$, significantly superior to the individual components CuO and Co_3O_4. The ASC constructed from this material showed a capacity retention rate of 91.2% after 10,000 cycles and demonstrated higher charge transfer efficiency [45]. Nickel oxide (NiO) has a rock-salt structure and is considered a promising electrode material due to its simple preparation and low cost. However, its practical application is often limited by poor intrinsic conductivity and significant volume expansion during cycling. In NiO/Ni/reduced graphene oxide (rGO) nanocomposites, the synergistic effect between the NiO/Ni heterostructure and rGO effectively enhances the conductivity and structural stability of the material, resulting in excellent specific capacity and cycling performance. For example, this composite material delivered a specific capacity as high as $649.22\,C\,g^{-1}$ at $3\,A\,g^{-1}$ and maintained 81% capacity retention after 5000 cycles at $20\,A\,g^{-1}$ [46].

6.4.1.2 Hydroxides

Fe-, Co-, and Ni-based hydroxides mostly possess layered structures, with exchangeable anions and water molecules between the layers, which facilitates ion diffusion and results in high theoretical specific capacity. Nickel hydroxide ($Ni(OH)_2$) has two crystal forms, α and β, with β-$Ni(OH)_2$ being more stable and having an interlayer spacing of about 0.46 nm. It undergoes a Ni^{2+}/Ni^{3+} redox reaction in alkaline electrolytes. For example, the issue of easy agglomeration and poor conductivity of pure $Ni(OH)_2$ material was effectively addressed by constructing rambutan-like composite microspheres through in situ growth of $Ni(OH)_2$ nanosheets on the surface of porous carbon microspheres [47]. This composite structure achieved a specific capacity of 248.9 mAh g^{-1} at 1 A g^{-1}, far exceeding the 125.7 mAh g^{-1} of pure $Ni(OH)_2$, and also showed significantly improved cycling stability, maintaining 81.1% capacity after 3000 cycles. The hybrid SC assembled with it achieved an energy density of 41.3 Wh kg^{-1} and retained 85.2% capacity after 20,000 cycles.

Cobalt hydroxide ($Co(OH)_2$) has three crystal forms: α, β, and γ. The interlayer spacing of α-$Co(OH)_2$ usually depends on the intercalated anions; it is larger (4.6 Å) than that of β-$Co(OH)_2$ without intercalation, allowing more convenient ion diffusion. Studies have shown that constructing a core–shell heterostructure by combining β-$Co(OH)_2$, which has superior electrochemical performance, with highly conductive $CoNi_2S_4$ can greatly enhance the overall performance of the material (Figure 6.5a). For instance, the areal-specific capacity of the β-$Co(OH)_2$@$CoNi_2S_4$ core–shell electrode reached as high as 588.9 μAh cm^{-2}, which is 5.7 times that of a single β-$Co(OH)_2$ electrode, and the assembled flexible device achieved an energy density of 38.07 Wh kg^{-1} and maintained 95.7% capacity after 5000 cycles (Figure 6.5b, c) [48]. Iron oxyhydroxide (FeOOH) has four crystal forms—α, β, γ, and δ—with α-FeOOH (goethite) and γ-FeOOH (lepidocrocite) being the most common. Their layered structures are beneficial for ion migration. Besides crystal phase control, constructing low-crystallinity nanostructures is another effective strategy to improve the electrochemical performance of FeOOH. Research has found that low-crystallinity FeOOH nanoparticles, due to having more active sites and faster ion migration kinetics, can achieve a specific capacity of 1066 F g^{-1}, significantly outperforming many well-crystallized FeOOH materials [50].

6.4.1.3 Other Compounds

Fe-, Co-, and Ni-based sulfides and phosphides have become research hotspots in recent years due to their higher electrical conductivity and richer redox reactions compared to oxides/hydroxides.

Iron disulfide (FeS_2) has a high theoretical specific capacity, but its low conductivity leads to practical challenges, such as insufficient conductivity and structural instability during cycling. To synergistically enhance both conductivity and structural stability, sulfidation treatment and composite structure design have become key strategies. Research shows that sulfidation of MOF-derived iron oxide precursors can successfully prepare Fe_3O_4/FeS_2 composite materials

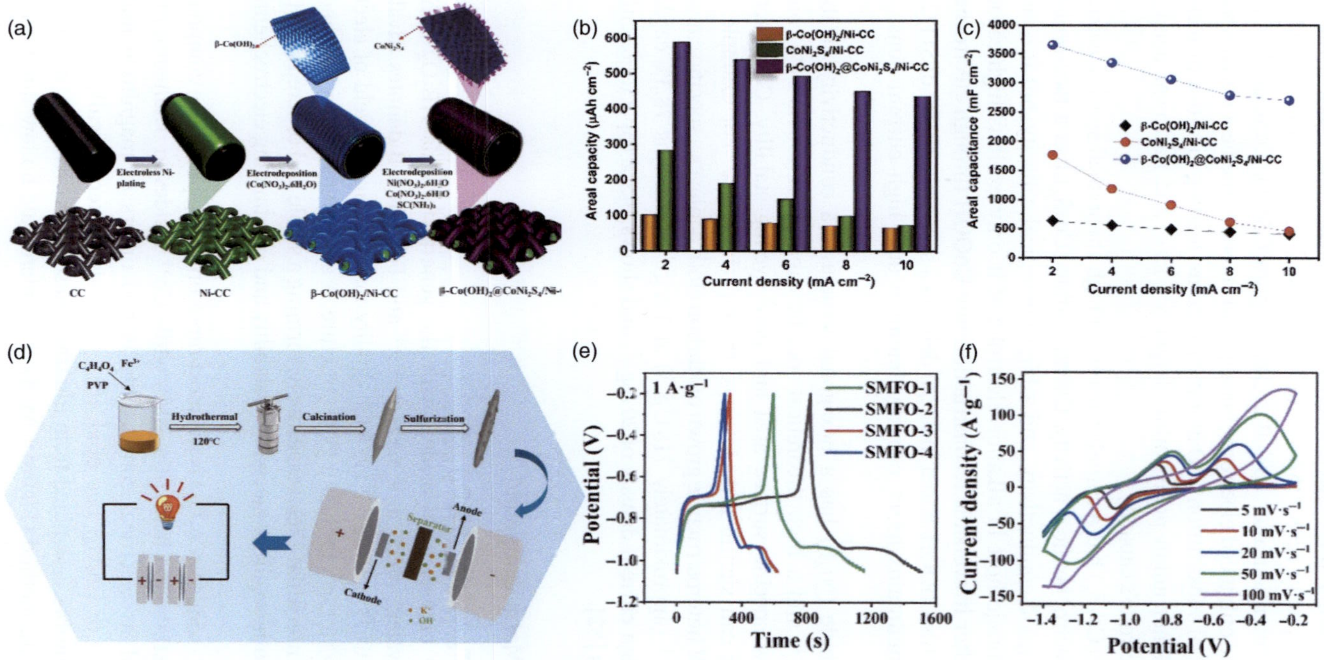

Figure 6.5 (a) Schematic diagram of the fabrication process of the β-$Co(OH)_2$@$CoNi_2S_4$/Ni-CC electrode, (b) areal capacity at different current densities, and (c) areal capacitance of the three electrodes. *Source:* Reproduced from Ref. [48] / with permission of Elsevier. (d) Schematic diagram of the synthesis and application of Fe_3O_4/FeS_2 composite material. (e) GCD curves at 1 A g^{-1}, (f) CV curves at different scan rates. *Source:* Reproduced from Ref. [49] / with permission of Elsevier.

(Figure 6.5d) [49]. The advantage of this strategy is verified by experimental data: compared to the unsulfidized oxide precursor, the optimized composite material exhibits a significantly increased conductivity of 60.4 mS cm^{-1}, and achieves a higher specific capacity (208.5 mAh g^{-1}@1 A g^{-1}) and excellent cycling stability (78.6% capacity retention after 10,000 cycles) (Figure 6.5e–f). Compared to pure FeS_2 material, the enhanced electrochemical performance of the Fe–O–S composite is due to the partial substitution of O atoms by S atoms, leading to lattice distortion. The resulting higher anion polarization, larger S^{2-} size, and narrower band gap enhance the electrode's capacity.

Cobalt sulfide (Co_3S_4) has a spinel structure, a high theoretical specific capacity, and excellent electrochemical activity. Cobalt sulfide (Co_9S_8) has a face-centered cubic structure, good conductivity, and a high theoretical capacity, but it is not a stable energy storage host in alkaline electrolytes; it only exhibits high-capacity characteristics after irreversible reconstruction into CoOOH through electrochemical activation. This CoOOH-based material delivers a specific capacity of 511 C g^{-1} at 1 A g^{-1} and maintains 65.9% capacity retention at a high current density of 50 A g^{-1} [51].

Nickel phosphide (Ni_2P) has a hexagonal structure, a conductivity of about 1.2×10^{-3} S cm^{-1}, and a high theoretical specific capacity of 1951.2 C g^{-1}, but it suffers from low specific surface area and poor cycling stability. Constructing heterostructures can significantly improve their performance. For example, the Ni_2P/NiMo-LDH heterostructure grown on carbon cloth achieves a specific capacity of 198.6 mAh g^{-1} (approximately 715 C g^{-1}) at 1 A g^{-1}, maintains 78.3% capacity retention at a high current density of 20 A g^{-1}, and shows 84.6% capacity retention after 5000 cycles [52].

6.4.2 Binary Metal Compound

6.4.2.1 Spinel-Type Oxide (AB_2O_4)

The spinel-type binary metal oxides, with the general formula AB_2O_4 (where A represents divalent metal ions such as Ni^{2+} and Fe^{2+}, and B denotes trivalent metal ions like Co^{3+} and Fe^{3+}), are high-quality materials in the field of electrochemical energy storage. In their crystal structure, A and B ions occupy tetrahedral and octahedral interstitial sites, respectively, forming a stable and ordered structure that possesses both excellent mechanical stability and outstanding electrochemical performance.

Nickel–cobalt oxide ($NiCo_2O_4$) is the most representative spinel-type compound. Ni^{2+} occupies tetrahedral interstitial sites, while Co^{3+} resides in octahedral interstitial sites, and its electrical conductivity is two orders of magnitude higher than that of NiO or Co_3O_4. Luo et al. [53] synthesized hierarchically structured tetragonal microtubes assembled from ultrathin mesoporous $NiCo_2O_4$ nanosheets via a self-template method (Figure 6.6a–d). The microscale hollow tubular main body, 2D nanosheet shell, and internal 2–5 nm mesoporous structure endow the material with the advantages of structural stability, rapid ion diffusion, and high specific surface area (Figure 6.6e). This material exhibits a specific capacitance

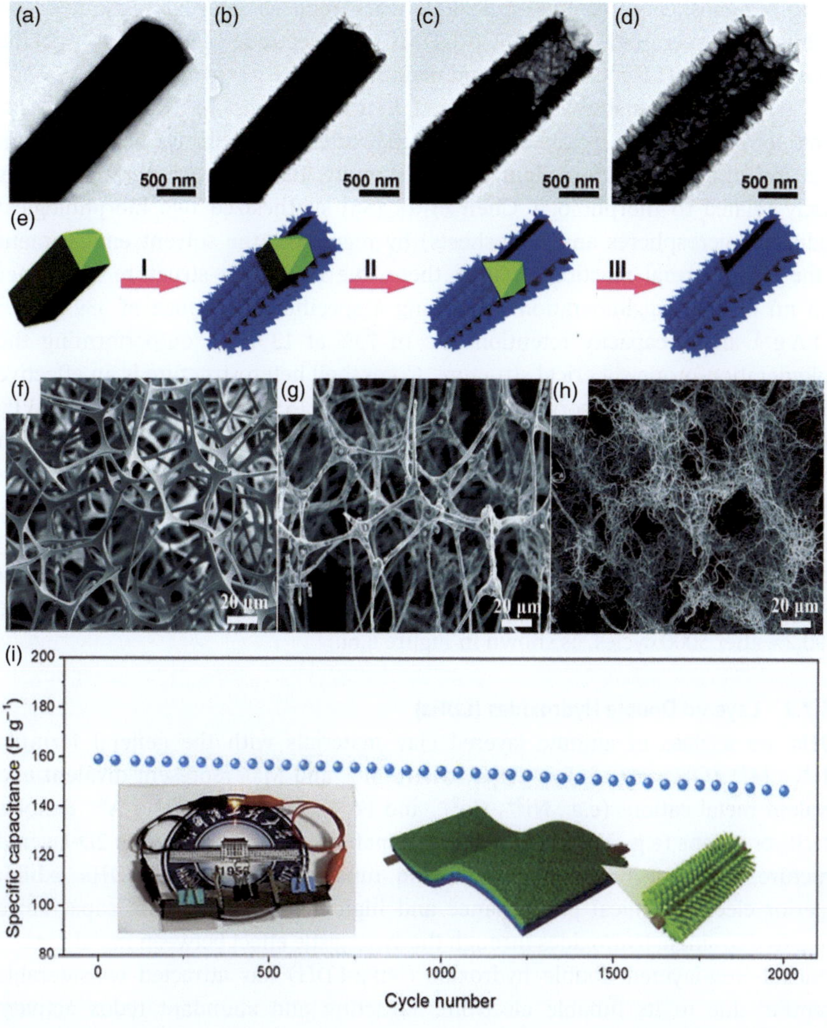

Figure 6.6 (a–d) Representative transmission electron microscopy images of the samples obtained after different reaction times. (e) Schematic illustration of the formation process of hierarchical structured NiCo-LDH tetragonal microtubes. *Source:* Reproduced from Ref. [53] / with permission of The Royal Society of Chemistry. (f–h) Field-emission scanning electron microscopy images of $NiCo_2O_4$ nanosheets on different substrates. *Source:* Reproduced from Ref. [54] / with permission of Elsevier. (i) Cycling stability of the $FeCo_2O_4$@PPy ASC device over 5000 cycles at a current density of $8\,A\,g^{-1}$. *Source:* Reproduced from Ref. [55] / with permission of American Chemical Society.

of $1387.9\,F\,g^{-1}$ at $2\,A\,g^{-1}$, a capacity retention rate of 62% at $30\,A\,g^{-1}$, and only a 10.6% capacity loss after 12,000 cycles at $10\,A\,g^{-1}$. Traditional carbonized melamine foam (CMF) is prone to shrinkage and fracture after high-temperature carbonization, which limits electrode performance. Yan et al. [54] modified CMF with branching via chemical vapor deposition (CVD) technology (Figure 6.6f–h), constructing a multi-carbonization channel MCMF framework and loading ultrathin

$NiCo_2O_4$ nanosheets. Its cross-linked network reduces internal resistance and inhibits agglomeration. The $NiCo_2O_4$/MCMF electrode achieves a specific capacitance of 1541 F g^{-1} at 1 A g^{-1} and maintains 1120 F g^{-1} even at 20 A g^{-1}.

$FeCo_2O_4$ forms a stable and open crystal structure, with Fe^{2+} and Co^{3+} occupying tetrahedral and octahedral interstitial sites, respectively, boasting both rapid ion diffusion and excellent cycling stability. Its capacitive performance is closely related to morphology. Chen et al. [56] synthesized two morphological products (microspheres and nanosheets) by regulating the solvent environment of the solvothermal reaction. Among them, the nanosheet structure is thinner with no obvious agglomeration, exhibiting a specific capacitance of 339.0 C g^{-1} at 1 A g^{-1} and a capacity retention rate of 73% at 13 A g^{-1}, outperforming the agglomeration-prone spherical structure. Core–shell heterostructure is an effective strategy to enhance SC performance. Wang et al. [55] prepared $FeCo_2O_4$@PPy core–shell nanowire electrodes via a hydrothermal method. $FeCo_2O_4$ provides high faradaic capacitance, while the polypyrrole (PPy) shell improves electrical conductivity and electrolyte wettability, realizing synergistic performance optimization. The electrode delivers a specific capacitance of 2269 F g^{-1} at 1 A g^{-1}, far exceeding the 1215.1 F g^{-1} of pure $FeCo_2O_4$. It maintains a capacity retention rate of 79.2% at 20 A g^{-1} and shows excellent long-term durability with a capacitance retention rate of 90.2% after 5000 cycles, as shown in Figure 6.6i.

6.4.2.2 Layered Double Hydroxides (LDHs)

LDHs are a class of anionic layered clay materials with the general formula $[M^{2+}{}_{1-x}M^{3+}{}_x(OH)_2]^{x+}(A^{n-})_{x/n}\cdot yH_2O$, where M^{2+} and M^{3+} represent divalent and trivalent metal cations (e.g., Ni^{2+}, Co^{2+}, and Fe^{3+}), respectively, and A^{n-} denotes interlayer anions (e.g., NO_3^- and CO_3^{2-}). Benefiting from their unique 2D-layered structure, high specific surface area, and tunable composition, LDHs exhibit superior electrochemical performance and high theoretical specific capacitance compared to single hydroxides through the synergistic effect between metal ions.

Nickel–iron-layered double hydroxide (NiFe-LDH) has attracted considerable attention due to its tunable electronic structure and abundant redox activity. The introduction of Fe^{3+} can optimize the material's electronic structure, enhancing electrical conductivity and electrochemical activity. Studies have shown that the Ni/Fe molar ratio exerts a decisive influence on the material's structure and performance. Ashok et al. [57] found that when Ni:Fe = 3:1, the material forms a porous flower-like structure, exhibiting a high specific capacitance of 1368 F g^{-1} and excellent cycling stability. This special morphology not only provides more active sites but also shortens ion transport paths, thereby significantly improving charge storage efficiency. Cobalt–iron-layered double hydroxide (CoFe-LDH) also demonstrates outstanding electrochemical performance. Zhao et al. [58] successfully prepared CoFe-LDH with a hollow core–shell structure via an ingenious template method. This structure effectively alleviates the stacking problem of traditional LDH materials, delivering high specific capacitances of 523 F g^{-1} and 489 F g^{-1} in KOH and NaCl electrolytes, respectively. Notably, LDH materials also show great potential in the field of capacitive deionization. The chloride-pre-intercalated CoFe-LDH

(CoFeCl-LDH) developed by Pan et al. [59] achieves optimal performance when the molar ratio is 3:1, with a desalination capacity of 100.2 mg g^{-1} and a desalination rate of 0.38 mg g^{-1} s^{-1}. These indicators are at the leading level among similar materials, providing a new solution for high-efficiency desalination technology.

6.4.2.3 Binary Sulfides/Phosphides

Binary metal sulfides have demonstrated significant advantages in electrochemical energy storage devices such as SCs and batteries, owing to their unique electronic structures and physicochemical properties. Compared to their oxide counterparts, sulfides possess a narrower band gap and higher intrinsic electrical conductivity. Meanwhile, binary metal phosphides have emerged as another promising class of electrode materials, exhibiting near-metallic conductivity, excellent thermal stability, and electron-rich characteristics. The introduction of phosphorus atoms can modulate the electronic structure of metal atoms, thereby enhancing the charge transfer efficiency between the electrode material and the electrolyte.

For $NiCo_2S_4$ materials, the main challenge lies in improving the cycling stability at high current densities. Wang et al. [60] addressed this by constructing a $NiCo_2S_4$/polyaniline (PANI) core–shell nanosheet array. The PANI coating improved the interfacial hydrophilicity and buffered volume changes, resulting in a specific capacitance of 1879 F g^{-1} at 1 A g^{-1} and a capacity retention of 72% even at a high rate of 20 A g^{-1}. Another strategy involves designing hierarchical structures. The Parkin team [61] synthesized dendritic $NiCo_2S_4$@$NiCo_2S_4$, which delivered a high areal capacity of 4.43 mAh cm^{-2} at 240 mA cm^{-2} and maintained approximately 93% capacity after 10,000 cycles, demonstrating exceptional structural stability. $CuCo_2S_4$ exhibits higher intrinsic conductivity due to its narrower band gap. Zhang et al. [62] composited $CuCo_2S_4$ microflowers with rGO. First-principle calculations confirmed that its bandgap (0.533 eV) is significantly smaller than that of its oxide precursor (1.207 eV). The composite electrode achieved a specific capacitance of 1280.2 F g^{-1}, effectively combining the high activity of the sulfide with the conductive network of rGO.

NiCoP is highly promising due to its metal-like conductivity and electron-rich properties, with performance heavily dependent on its microstructure. Wang et al. [63] synthesized one-dimensional (1D) hollow NiCoP nanorods via the Kirkendall effect. This structure enhanced ion transport efficiency, yielding a specific capacity of 273.4 μAh cm^{-2} at 30 mA cm^{-2}. Fu et al. [64] prepared a high-mass-loading (8.6 mg cm^{-2}) NiCo–P nano-wall structure on nickel foam using a fluoride-free synthesis method. The 3D interconnected network facilitates electrolyte infiltration, enabling the electrode to deliver a high specific capacitance of 1861 F g^{-1} at 1 A g^{-1}, along with good rate performance (57.5% retention at 10 A g^{-1}) and cycling stability (85.6% capacity retention after 8000 cycles).

6.4.3 Ternary Metal Compounds and Multicomponent Systems

To further optimize the electronic structure and electrochemical performance, researchers have constructed ternary metal compounds by introducing a third metal element (Fe, Mn, Al, Zn, etc.). The synergistic effect of multiple metal elements is utilized to increase the number of active sites, improve conductivity, and enhance structural stability.

6.4.3.1 Ternary Oxide/Hydroxide

Nickel–iron–cobalt oxide $NiFeCoO_4$ has its electronic structure optimized and abundant active sites through the synergistic effect of three metal ions, with a theoretical specific capacity of approximately 1200 F g^{-1}. Shi et al. used $[NiFe(CN)_6]^-$ intercalated NiCo single-bond LDH as a precursor [65]. As evidenced by the electrochemically active surface area, the resulting NiFe single-bond $NiCoO_2$ composite exhibits a unique hollow nanostructure, which can not only provide abundant mass transfer channels but also increase the contact area of NiFe single-bond NiCoO. For example, the specific capacity of $NiFeCoO_4$ nanoparticles under 1 A g^{-1} is 980 F g^{-1}, the capacity retention rate under 20 A g^{-1} is 72%, and the capacity retention rate after 5000 cycles is 85% (Figure 6.7).

Kvetuse prepared ternary nickel–cobalt–manganese oxide catalysts using a combination of sputtering techniques [66]. A new method was employed to deposit nickel–cobalt–manganese oxide thin films on stainless steel meshes, which combines plasma jet sputtering of a hollow nickel–cobalt cathode with simultaneous magnetron sputtering of manganese (Figure 6.8). The resulting thin-film catalysts are suitable for various oxidation reactions, such as the complete oxidation of volatile organic compounds. Catalysts with a nickel/cobalt molar ratio of 1:1 and different manganese contents were compared with binary nickel–cobalt oxide catalysts in terms of their ability to oxidize ethanol and toluene. Powder XRD showed the presence of spinel-type mixed oxides and nickel oxide-like oxides in the oxide coating. TPR curves indicated that Ni–Co–Mn oxides are more easily reducible than Ni–Co oxides, and the addition of Mn has a synergistic effect. Due to the mixed oxide phase with a disordered structure and the presence of Mn^{4+}, Ni–Co–Mn oxide catalysts exhibit higher catalytic activity than binary catalysts. Nickel-cobalt-manganese-layered double hydroxides (NiCoMn-LDH) regulate the interlayer spacing and electronic structure through the introduction of Mn, expanding the interlayer spacing to 0.85 nm and increasing the ion diffusion rate. For example, the specific capacity of NiCoMnLDH nanosheets under 1 A g^{-1} is 1250 F g^{-1}, the capacity retention rate under 10 A g^{-1} is 78%, and the capacity retention rate after 3000 cycles is 82%.

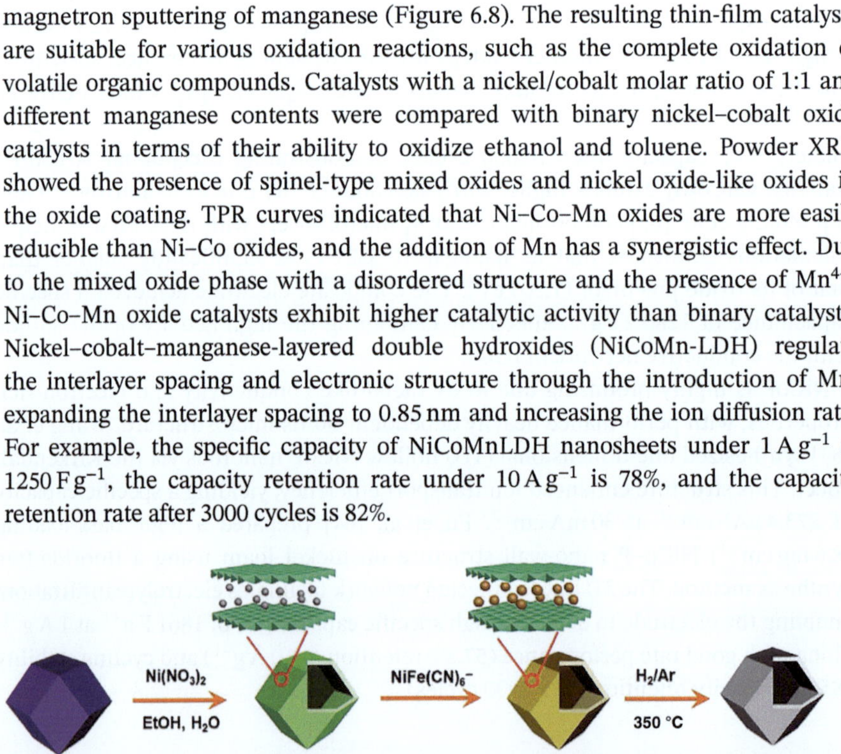

Figure 6.7 Schematic illustration of the synthetic process of the NiFe-$NiCoO_2$ composite. *Source:* Reproduced from Ref. [65] / with permission of Elsevier.

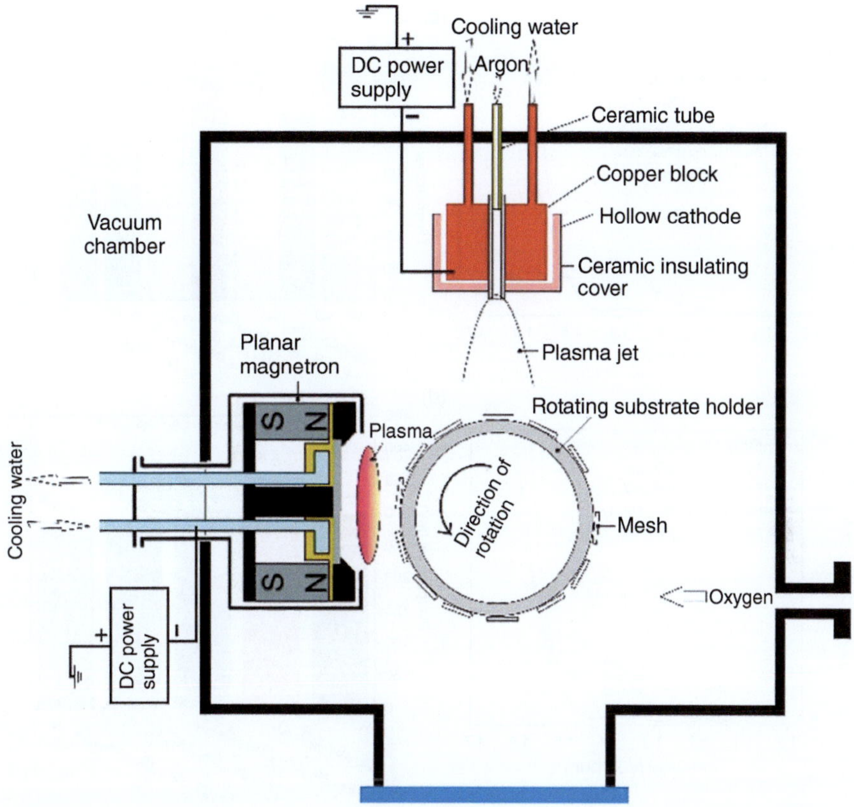

Figure 6.8 Scheme of the experimental arrangement of the setup. *Source:* Reproduced from Ref. [66] / with permission of American Chemical Society.

6.4.3.2 Ternary Sulfides/Phosphides

Soheil synthesized iron-containing nickel sulfide B through a simple two-step hydrothermal reaction [67]. The study shows that iron-containing nickel–cobalt sulfides and selenides exhibit better oxygen evolution reaction (OER) performance than iron-free nickel–cobalt sulfides and selenides, with lower overpotentials. This highlights the important role of iron in enhancing the performance of nickel–cobalt-based OER electrocatalysts. The specific data are as follows: $Fe_{0.1}Ni_{1.4}Co_{2.9}(S_{0.87}O_{0.13})_4$, η_{50} = 318 mV. Nickel–iron–cobalt sulfide (NiFeCo-S) is prepared by the ion exchange method to form heterogeneous ultra-thin nanometers. The rice-flake structure is composed of Ni_2S_3, FeS, $CoNi_2S_4$, $FeCoS_2$, and Ni-Fe-Co hydroxide, with a thickness of only 1.8 nm, providing abundant electrochemical interfaces and active sites. At a current density of 10 mA cm^{-2}, its specific capacity is as high as 973 μA h cm^{-2} (174 mA h g^{-1}) (Figure 6.9). Even when the current density is increased to 100 mA cm^{-2}, it can still retain 68.3% of the initial capacity. The assembled ASC exhibits an energy density of 56.3 Wh kg^{-1}.

A simple hydrothermal strategy was used to prepare NiCoZnS hierarchical nanostructures on a nickel foam substrate [68]. This results in excellent bifunctional

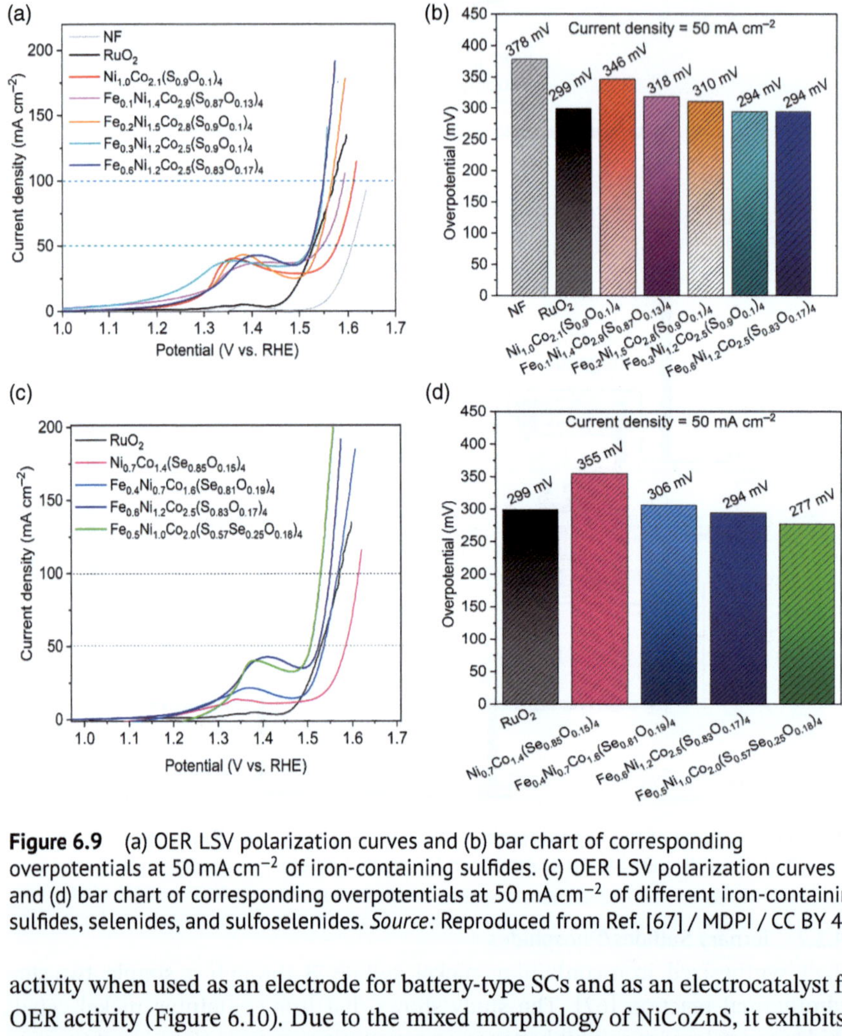

Figure 6.9 (a) OER LSV polarization curves and (b) bar chart of corresponding overpotentials at 50 mA cm^{-2} of iron-containing sulfides. (c) OER LSV polarization curves and (d) bar chart of corresponding overpotentials at 50 mA cm^{-2} of different iron-containing sulfides, selenides, and sulfoselenides. *Source:* Reproduced from Ref. [67] / MDPI / CC BY 4.0.

activity when used as an electrode for battery-type SCs and as an electrocatalyst for OER activity (Figure 6.10). Due to the mixed morphology of NiCoZnS, it exhibits a higher specific capacity of 1171.3 C g^{-1} at a current density of 0.5 A g^{-1}. Even after 5000 cycles at 4 A g^{-1}, the cycle stability is as high as 94.9%. Nickel–cobalt–zinc sulfide (NiCoZn-S) has a 3D porous structure. The introduction of Zn optimizes the conductivity and ion diffusion paths, with a specific capacity of 1100 F g^{-1} under 1 A g^{-1}, and a capacity retention rate of 75% under 20 A g^{-1}.

6.4.4 Fe–Co–Ni-Based Derivatives and Framework Materials

6.4.4.1 Prussian Blue Analogues (PBAs)

Prussian blue analogues (PBAs) are a class of 3D coordination polymers composed of metal ions (e.g., Fe, Co, and Ni) and cyanide ligands, with the general formula $A_xM[M'(CN)_6]_y \cdot zH_2O$ (where A represents an alkali metal ion, and M/M′ denote transition metal ions such as Fe, Co, and Ni). Their open 3D framework structure and high specific surface area (>1000 m^2 g^{-1}) facilitate the rapid

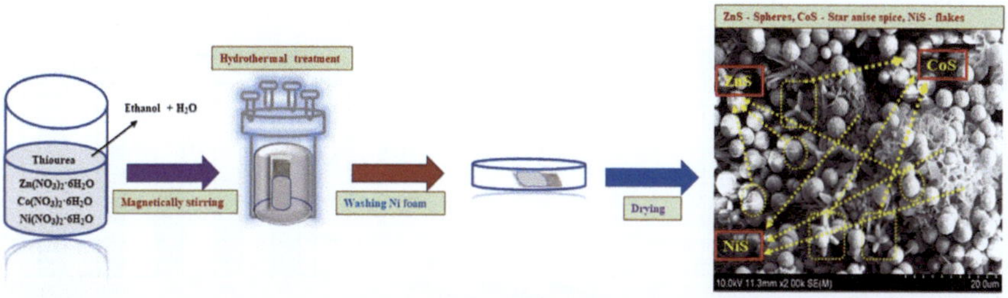

Figure 6.10 Schematic representation of the hydrothermal synthesis process for generating mixed morphologies of NiCoZnS nanostructures. *Source:* Reproduced from Ref. [68] / with permission of Elsevier.

intercalation/deintercalation of electrolyte ions, while the multiple valence states of metal ions provide abundant active sites for redox reactions [69].

For instance, the charge storage of CuFe-PBA@NiFe-PBA is based on the synergistic redox reactions of Ni^{2+}/Ni^{3+}, Cu^{2+}/Cu, and Fe^{2+}/Fe^{3+}. It exhibits a specific capacitance of $576\,F\,g^{-1}$ at a current density of $0.5\,A\,g^{-1}$, with 93.1% capacity retention after 10,000 cycles [70]. NiCo-LDH/PBA (Figure 6.11a) inhibits the agglomeration of NiCo-LDH and enhances its structural stability during charge–discharge processes through the PBA component (Figure 6.11b). The open structure of PBA expands the contact between electrodes and electrolyte ions, promoting ion/charge transfer characteristics and overall electrochemical performance. It delivers specific capacitances of 1973.0 and $1604.4\,F\,g^{-1}$ at 2 and $10\,A\,g^{-1}$, respectively (Figure 6.11c), with 80.7% capacity retention after 9000 cycles, demonstrating excellent cycle stability [71]. The main challenge of PBAs lies in their poor electrical conductivity, which usually requires composite modification with carbon materials or calcination-derived treatment to improve performance [74].

6.4.4.2 Metal-Organic Frameworks (MOFs) and Their Derivatives

MOFs are porous crystalline materials self-assembled from metal ions (e.g., Fe, Co, and Ni) and organic ligands (such as terephthalic acid and imidazole-based ligands). They possess ultrahigh specific surface area (up to $\sim 3000\,m^2\,g^{-1}$), tunable pore structures, and abundant metal active sites [75]. As electrode materials for SCs, MOFs can be used directly or converted into metal oxides, sulfides, phosphides, or carbon composites via treatments such as calcination, sulfidation, and phosphidation.

The directly used Ni/Co-MOF, with a regulated Ni/Co molar ratio of 7:1, achieves a specific capacitance of $755\,F\,g^{-1}$ at $0.5\,A\,g^{-1}$. The combination of diffusion-controlled and capacitive charge storage mechanisms (Figure 6.11d) endows Ni/Co-MOF with excellent cycle performance, showing a high-capacity retention rate of 97.9% after 5000 cycles at $5\,A\,g^{-1}$ (Figure 6.11e) [72]. MOF derivatives integrate the porous structure of MOFs and the high activity of metal compounds. For example, carbon-modified NiS/NiS_2 derived from the sulfidation of Ni-soc-MOF exhibits a specific capacitance of $1082\,C\,g^{-1}$ at $1\,A\,g^{-1}$, with 85.6% capacity retention after 5000 cycles at $5\,A\,g^{-1}$ (Figure 6.11f) [73]. NiO/Co_3O_4 formed by the calcination of NiCo-MOF nanoflowers (Figure 6.11g) delivers a specific capacitance of $1971.4\,F\,g^{-1}$ at $1\,A\,g^{-1}$, maintaining 92.85% capacity after 5000 cycles [76].

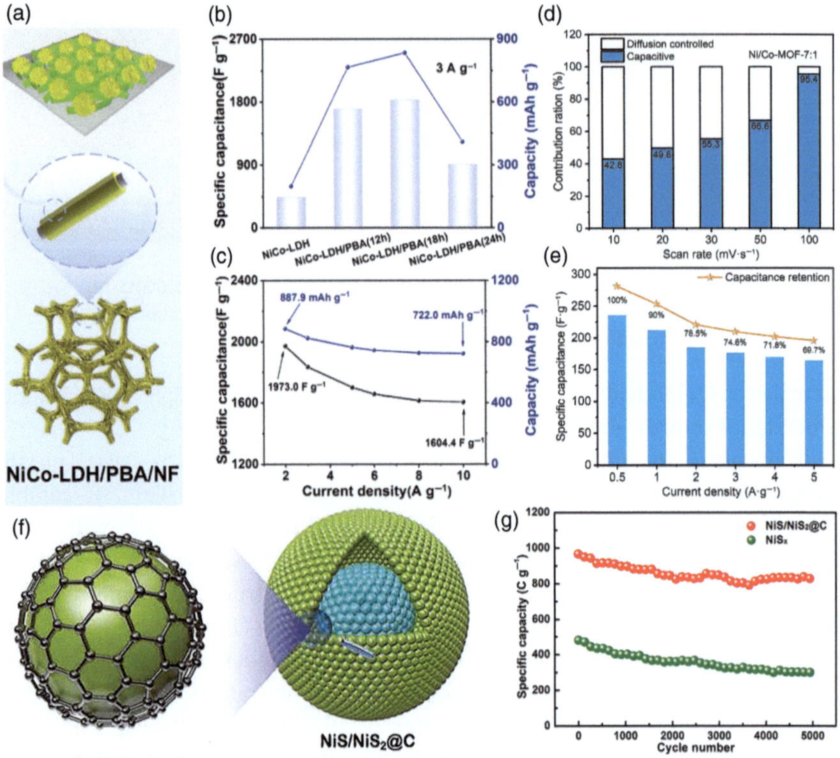

Figure 6.11 (a) Schematic diagram of NiCo-LDH/PBA. (b) Specific capacitances and capacities of NiCo-LDH/PBA and the contrast electrode at $3\,A\,g^{-1}$. (c) Specific capacitance and capacity of NiCo-LDH/PBA. *Source:* Reproduced from Ref. [71] / with permission of Elsevier. (d) Relative contributions of diffusion-controlled and capacitive-controlled charge storage of Ni/Co-MOF at different scan rates; e) capacitive performance and capacity retention capability of Ni/Co-MOF at different current densities. *Source:* Reproduced from Ref. [72] / Tsinghua University Press Ltd / CC BY 4.0. (f) Schematic diagram of $NiS/NiS_2@C$; (g) cycling stability of the $NiS/NiS_2@C$ electrode. *Source:* Reproduced from Ref. [73] / with permission of Springer Nature.

The main drawback of MOFs and their derivatives is poor mechanical stability, which tends to cause structural collapse during charge–discharge processes. This issue needs to be addressed through composite modification or structural optimization (e.g., constructing core–shell structures and introducing carbon matrices).

6.5 Key Modification Strategies

6.5.1 Nanostructure Design (from Zero-Dimensional to Three-Dimensional [0D to 3D])

Nanostructure design is one of the core strategies for improving the electrochemical performance of iron–cobalt–nickel-based compounds. Constructing zero-dimensional (0D), 1D, 2D, and 3D nanostructures can increase specific surface area, shorten ion diffusion paths, expose more active sites, and mitigate volume expansion.

6.5.1.1 Zero-Dimensional Nanostructures

0D nanomaterials such as quantum dots have a high specific surface area and abundant active sites but are susceptible to agglomeration. As an optimized example, carbon-encapsulated distorted iron oxide quantum dots (d-Fe_2O_3 QDs@C) inhibit agglomeration and enhance structural stability while maintaining high activity via the synergistic effect of carbon matrix and PVP molecules. Fabricated through chemical foaming-annealing, the material forms a regular honeycomb-like macrostructure (Figure 6.12a), with ultra-thin carbon walls encapsulating ~5 nm

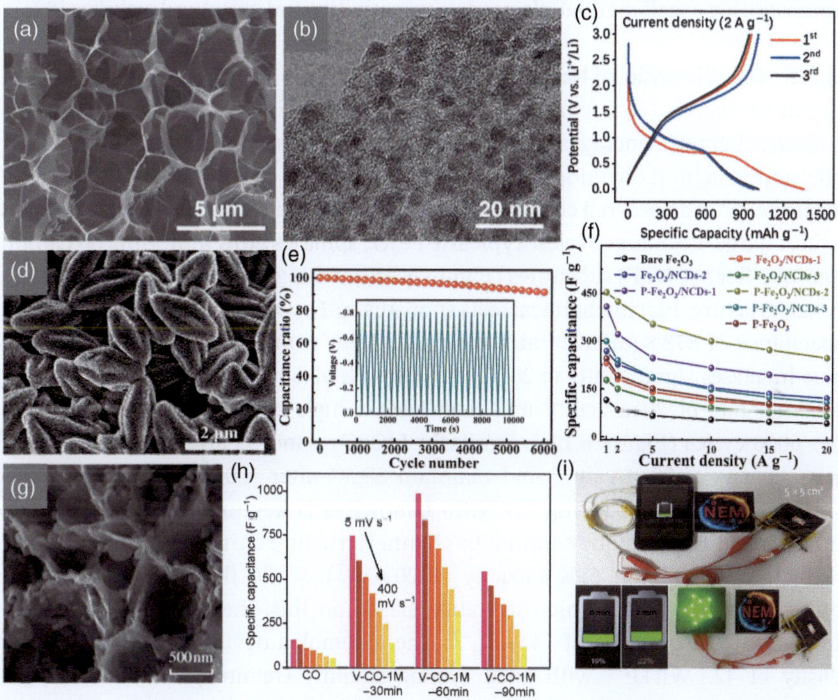

Figure 6.12 (a) SEM image of d-Fe_2O_3 QDs@C. (b) HRTEM images of d-Fe_2O_3 QDs@C. (c) GCD profiles of d-Fe_2O_3 QDs@C at 2 A g^{-1} for the initial three cycles. *Source:* Reproduced from Ref. [77] / with permission of John Wiley & Sons. (d) SEM image of the α-Fe_2O_3/CNTF composite. (e) Cycling performance of the MoS_2@α-Fe_2O_3/CNTF electrode at 4 mA cm^{-2} (the inset shows the corresponding GCD curve). *Source:* Reproduced from Ref. [78] / with permission of John Wiley & Sons. (f) Specific capacitances of bare Fe_2O_3, P-Fe_2O_3, Fe_2O_3/NCDs, and P-Fe_2O_3/NCDs. *Source:* Reproduced from Ref. [79] / with permission of Elsevier. (g) SEM images of Fe@Fe_3O_4 at different magnifications. *Source:* Reproduced from Ref. [80] / Springer Nature / CC BY 4.0. (h) Specific capacitances of CO and V-CO electrodes at various scan rates. (i) A mobile phone being charged by a packaged ASC system consisting of three series-connected 5 × 5 cm^2 single cells. *Source:* Reproduced from Ref. [81] / with permission of Elsevier.

Fe_2O_3 quantum dots (Figure 6.12b). It exhibits outstanding electrochemical performance: a specific capacity of 937 mA h g^{-1} at 2 A g^{-1} (Figure 6.12c), close to the theoretical value; excellent fast-charging capability, delivering 475 mA h g^{-1} in 56 seconds at 30 A g^{-1} and 182 mA h g^{-1} in 7 seconds at 100 A g^{-1}; and superior cycling stability with no capacity decay after 1000 cycles at 30 A g^{-1} [77]. Similarly, 3DNC@Fe_2O_3-700 employs a 3D nitrogen-doped carbon framework to encapsulate Fe_2O_3 nanoparticles, achieving uniform dispersion and structural stability of 0D materials with excellent dual-electrode performance: the anode reaches 946.6 mA h g^{-1} at 0.1 A g^{-1} with a capacity retention of 106.5% after 5000 cycles at 20 A g^{-1}, while the cathode delivers 82.1 mA h g^{-1} at 0.1 A g^{-1} with 100% retention after 5000 cycles at 5 A g^{-1}. The assembled symmetric lithium-ion capacitor achieves an energy density of 206 Wh kg^{-1}, retains 32 Wh kg^{-1} at a high power of 22.6 kW kg^{-1}, and maintains a capacity retention of 118.2% after 5000 cycles, demonstrating excellent dual-electrode compatibility and structural stability [82].

6.5.1.2 One-Dimensional Nanostructures

1D nanostructures (e.g., nanowires and nanorods) exhibit remarkable advantages in electrochemical energy storage due to their unique axial electron transport paths and structural directionality. These structures can effectively reduce interface resistance, provide ordered channels for ion diffusion, and alleviate volume changes during charge–discharge cycles. Typical α-Fe_2O_3 spindles grow uniformly on carbon nanotube fiber (CNTF) substrates (Figure 6.12d), with a size of ~2.5 μm and intact crystal structure. Electrochemical tests show that α-Fe_2O_3/CNTF achieves a specific capacitance of 878.8 mF cm^{-2} at 2 mA cm^{-2}. After compounding with MoS_2, this value increases significantly to 2077.5 mF cm^{-2}. It also retains 85.2% of its capacity at a high rate of 20 mA cm^{-2} and maintains a high-capacity retention of 91.5% after 6000 cycles (Figure 6.12e). Similarly, $LiCoO_2$ nanorod arrays retain 74.2% of their capacity at 20 mA cm^{-2} and maintain 88.5% after 1000 cycles [78]. Studies have proven that introducing 1D NiCo nanowires is key to enhancing electrode performance. Their unique continuous channel structure ensures excellent electron conductivity (retaining 61% capacity at 20 A g^{-1}), while their physical support optimizes the electrode's microstructure. Based on this, the electrode achieves a high specific capacitance of 1449 F g^{-1}. The assembled device delivers an energy density of 42.1 Wh kg^{-1} with good cycling stability (retaining ~79% after 3000 cycles), highlighting the application value of 1D nanomaterials in electrochemical energy storage [83].

6.5.1.3 Two-Dimensional Nanostructures

2D nanostructures (e.g., nanosheets and nanoribbons) possess ultra-high specific surface area and short ion diffusion paths, enabling the exposure of numerous surface atoms as active sites. P–Fe_2O_3/NCDs 2D nanosheets combine the dual-functional advantages of phosphoric acid and nitrogen-doped carbon dots. They feature a suitable specific surface area (25.2 m^2 g^{-1}) and mesoporous structure (3.8 nm). The material exhibits high specific capacitance (453 F g^{-1} @ 1 A g^{-1}), good rate performance (retaining 54.7% at 20 A g^{-1}), and excellent cycling stability

(91.0% capacity retention after 10,000 cycles; Figure 6.12f) [79]. Another iron-based representative is the $Fe@Fe_3O_4$ nanosheet array (Figure 6.12g), which is prepared by direct oxidation of Fe wires. The 2D nanosheets grow vertically on the Fe wire surface, forming an interconnected open structure. This structure ensures tight contact with the substrate and provides abundant active sites. It achieves a specific capacitance of 20.8 $mF\,cm^{-1}$ at 10 $mV\,s^{-1}$, with only 8.3% capacitance loss after 2500 cycles at 0.9 mA. The assembled linear SC delivers an energy density of 9 $\mu Wh\,cm^{-2}$, capable of powering an LED lamp for 2 minutes, demonstrating potential for flexible energy storage [80].

6.5.1.4 Three-Dimensional Hierarchical Porous Structures

3D hierarchical porous structures are self-assembled from low-dimensional nanostructures (e.g., nanowires and nanosheets). They integrate high specific surface area, multilevel pore architectures, and favorable mechanical stability, enabling the simultaneous realization of high capacity, excellent rate performance, and long cycle life. The iron-based material ASV-FO is fabricated via the hydrothermal method combined with $NaBH_4$ treatment. It forms a 3D network composed of 1D nanorod arrays on carbon cloth. Its unique "crystalline core–amorphous shell" heterostructure and surface-enriched oxygen defects work together to increase specific surface area and optimize ion transport pathways. In terms of performance, ASV-FO composite with graphene achieves a specific capacitance of 701 $F\,g^{-1}$ at 1 $A\,g^{-1}$—double that of traditional materials. It also retains 200 $mF\,cm^{-2}$ at a high current density of 20 $mA\,cm^{-2}$, with a capacity retention rate exceeding 90% after 5000 cycles. The cobalt-based material V-CO similarly forms a 3D porous structure by stacking 2D nanosheets on carbon cloth. It exhibits an ultra-high specific capacitance of 990 $F\,g^{-1}$ at 5 $mV\,s^{-1}$ (Figure 6.12h) and remarkable rate performance. The flexible solid-state ASC assembled using this strategy maintains 85.3% capacity retention after 5000 cycles, with a maximum volumetric energy density of 0.95 $mWh\,cm^{-3}$. Series-connected multiple devices can quickly charge mobile phones and power LED lighting (Figure 6.12i), fully demonstrating the significant advantages of 3D hierarchical structures in enhancing comprehensive electrochemical performance and practical application potential [81]. Ma et al. developed a 3D hierarchical porous NCF/NiO composite. This material combines a high specific surface area (412.3 $m^2\,g^{-1}$) with uniformly dispersed NiO nanoparticles. It shows excellent capacitive performance (1074 $F\,g^{-1}$ at 1 $A\,g^{-1}$) as well as superior rate capability and cycle stability. The device assembled based on it delivers an energy density of 40.18 $Wh\,kg^{-1}$ [84].

6.5.2 Construction of Composite Materials

6.5.2.1 Compositing with Carbon Materials

Carbon materials (such as graphene, carbon nanotubes, porous carbon, and carbon cloth) possess high electrical conductivity, large specific surface area, and excellent mechanical stability. When compounded with iron–cobalt–nickel-based compounds, they can effectively improve electrical conductivity, mitigate volume expansion, and provide electric double-layer capacitance contributions.

In graphene composite systems, the 2D sheet-like structure of graphene can inhibit the agglomeration of iron–cobalt–nickel-based compounds, forming a conductive network. For example, NiCo-LDHs@rGO composite materials were prepared via an ammonia diffusion method, where the incorporation of rGO improved the electrical conductivity of NiCo-LDHs. When 30 mg of rGO was added, the specific capacitance of the material reached 1920.5 $F g^{-1}$ at a current density of 1 $A g^{-1}$. After assembling it into an asymmetric two-electrode SC, an energy density of 50.5 $Wh kg^{-1}$ was achieved at a power density of 900 $W kg^{-1}$ [85].

In carbon nanotube composite systems, the 1D structure of carbon nanotubes can construct rapid electron transport channels. For example, the FeNiP/N-CNTs electrocatalyst exhibits a specific capacitance of 1479 $F g^{-1}$ in 3 M KOH electrolyte at a current density of 1 $A g^{-1}$, with a capacity retention rate of 90.8% after 5000 cycles; FeNiCoP/N-CNTs in a two-electrode system achieves a specific capacity of 245 $F g^{-1}$ at 1 $A g^{-1}$, with a power density of 10567 $W kg^{-1}$ and an energy density of 66.7 $Wh kg^{-1}$, and maintains a capacity retention rate as high as 91.5% after 5000 cycles, significantly outperforming traditionally multi-step synthesized metal phosphide materials [86]. The $NiCo_2S_4$/CNT composite material exhibits a specific capacity of 890 $F g^{-1}$ at 1 $A g^{-1}$, with a capacity retention rate of 83.5% at 50 $A g^{-1}$, significantly higher than that of pure $NiCo_2S_4$ [87] (Figure 6.13a).

In porous carbon composite systems, the hierarchical pore structure of porous carbon can facilitate electrolyte infiltration and ion diffusion. For example, the $CoFe_2O_4$-Fe@NC composite material carbonized at 700°C achieves a specific capacitance of 3960.9 $F g^{-1}$ at a current density of 1 $A g^{-1}$ (Figure 6.13b) [88]. When assembled into an ASC with AC, the device delivers a high energy density of 84.9 $Wh kg^{-1}$ at a power density of 291.6 $W kg^{-1}$. Furthermore, the composite material maintains a capacitance retention rate as high as 94.9% after 10,000 cycles [89] (Figure 6.13c).

6.5.2.2 Compositing with Conducting Polymers

Conducting polymers (such as PANI, PPy, and polythiophene [PEDOT]) possess high electrical conductivity and excellent flexibility. When compounded with iron–cobalt–nickel-based compounds, they can simultaneously provide pseudocapacitive contributions, improving both electrical conductivity and mechanical stability.

In PANI composite systems, the conjugated structure of PANI can construct a conductive network while providing additional capacity through redox reactions. For example, the $NiMn_2O_4$/PANI composite material exhibits a specific capacitance (C_2) of 1530 ± 2 $F g^{-1}$ at a current density of 1 $A g^{-1}$. After 10,000 long-term cycles, the material maintains a capacity retention rate of 93.61%, demonstrating exceptional cycling stability. In a two-electrode system, the composite material shows a specific capacitance of 764 $F g^{-1}$ at a scan rate of 5 $mV s^{-1}$ and reaches 826 $F g^{-1}$ at a current density of 1 $A g^{-1}$, with a charge transfer resistance (R_{ct}) of only 0.67 Ω [90]. The POM/α-Fe_2O_3/PANI composite material exhibits a specific capacity of 528 $F g^{-1}$ at 0.2 $A g^{-1}$, while achieving a high energy density of 73.4 $Wh kg^{-1}$ and maintaining a capacity retention rate of 91.62% [91].

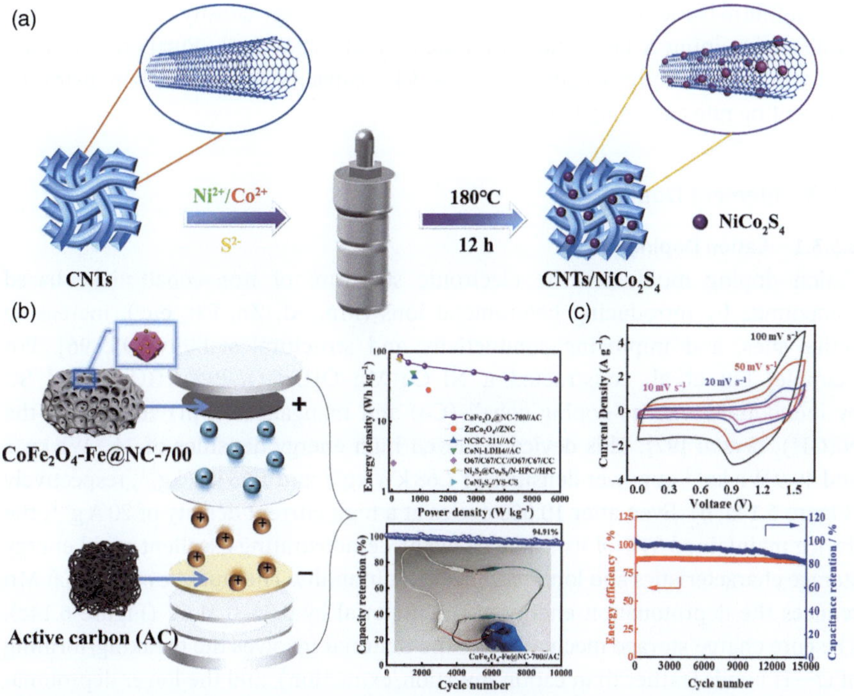

Figure 6.13 (a) Schematic diagram of the preparation process for CNTs/$NiCo_2S_4$ composite material. *Source:* Reproduced from Ref. [87] / The Royal Society of Chemistry / CC BY-NC 3.0. (b) N-doped porous carbon loaded with $CoFe_2O_4$-Fe particles for supercapacitors. *Source:* Reproduced from Ref. [88] / with permission of Elsevier. (c) Electrochemical Performance of Fe_2O_3@Ppy Nanocomposite as an Effective Electrode Material for Supercapacitors. *Source:* Reproduced from Ref. [89] / with permission of Springer Nature.

In PPy composite systems, the high flexibility of PPy can enhance the mechanical properties of the composite material, making it suitable for flexible devices. For example, the rGO/Ni-Co ferrite@PPy composite material exhibits a high specific capacitance of 250 $F\,g^{-1}$, outperforming binary composites such as rGO/$MnFe_2O_4$ [92].

6.5.2.3 Composite with Other Transition Metal Compounds

Compositing with other transition metal compounds can form heterostructures, leveraging interfacial synergistic effects to optimize electronic structure and reaction kinetics. For example, the coral-like Cu_3P/CoP heterostructure enhances the metallic-like conductivity of the composite material, exhibiting a higher number of electron-occupied states near the Fermi level while improving the charge transfer efficiency of the heterostructure. At a current density of 1 $A\,g^{-1}$, the specific capacitance reaches 804.3 $F\,g^{-1}$; when the current density increases to 10 $A\,g^{-1}$, it maintains 92.8% capacitance retention after 10,000 cycles [93]. The HfO_2@Co_3O_4 electrode exhibits an exceptional capacitance of 1553 $F\,g^{-1}$ at a scan rate of 5 $mV\,s^{-1}$ in a 2 M NaOH aqueous electrolyte, which is nearly six times that of pure hafnium oxide. This electrode achieves a high energy density of 49.16 Wh kg^{-1} while maintaining a favorable power density of 522 W kg^{-1}. Furthermore, after 3000 cycling tests, the electrode retains 94% capacity retention, demonstrating outstanding rate capability [94].

6.5.3 Element Doping

6.5.3.1 Cation Doping

Cation doping modulates the electronic structure of iron–cobalt-nickel-based compounds by introducing heterometal ions (Mn, Al, Zn, Cu, etc.), increasing active sites, and improving conductivity and structural stability [95, 96]. For example, Liu et al. constructed a Ni–Co–Mn-OH/rGO//PPD/rGO hybrid SC by individually and co-doping cobalt (Co) and manganese (Mn) atoms into the $Ni(OH)_2$ crystal [97]. This device achieved high energy densities of 74.7 Wh kg^{-1} and 49.9 Wh kg^{-1} at power densities of 1.68 kW kg^{-1} and 18.5 kW kg^{-1}, respectively (Figure 6.14a, b). Even after 10,000 cycles at a high current density of 20 A g^{-1}, the device maintained 91% of its initial capacity, demonstrating excellent rapid energy storage characteristics and long-term cycling durability. The study revealed that Mn reduces the deprotonation energy of the material by 0.26–0.31 eV (Figure 6.14c). The core charge storage mechanism of the material involves the breaking/forming of O—H bonds (rather than cation insertion/extraction), and the lower deprotonation energy directly accelerates reaction kinetics, significantly enhancing capacity. Furthermore, the electronic characteristics of Mn's 3d orbitals fill the region near the top of the valence band close to the Fermi level, narrowing the material's band gap. Simultaneously, strong charge exchange between Mn and O (Figure 6.14d) facilitates easier electron transfer from H to Mn, increasing electron conduction rates and improving capacity retention at high current densities (stable discharge capability even at 20 A g^{-1}). In contrast, Co doping significantly enhances the overall structural stability of the electrode.

Al doping primarily serves a structural supporting role, mitigating volume expansion. For instance, Sun et al. proposed a dual-doping strategy with Al^{3+} and Nb^{5+} for lithium nickel cobalt manganese oxide cathodes [98]. In this approach, Al doping suppresses lattice volume expansion by reinforcing the crystal structure and reducing Li^+–Ni^{2+} cation mixing. In Al-doped lithium nickel cobalt manganese aluminum oxide (Al-NCM92), the lattice volume contraction during charging decreased from −8.20% in pure NCM92 to −7.25% (Figure 6.14e, f), significantly alleviating volume expansion. Meanwhile, in Al and Nb co-doped lithium nickel cobalt manganese aluminum niobium oxide (AlNb-NCM92), not only was the lattice volume contraction further optimized to −7.58%, but Al doping also compensated for the increased cation mixing induced by Nb doping, greatly enhancing structural stability. In terms of performance, the AlNb-NCM92 cathode demonstrated the best overall performance, achieving a capacity retention of 88.3% after 1000 cycles, along with outstanding stability under fast-charging and high-temperature conditions.

Zinc (Zn) is abundant in natural resources and environmentally friendly. Its ionic radius and electronegativity are both similar to those of nickel (Ni),

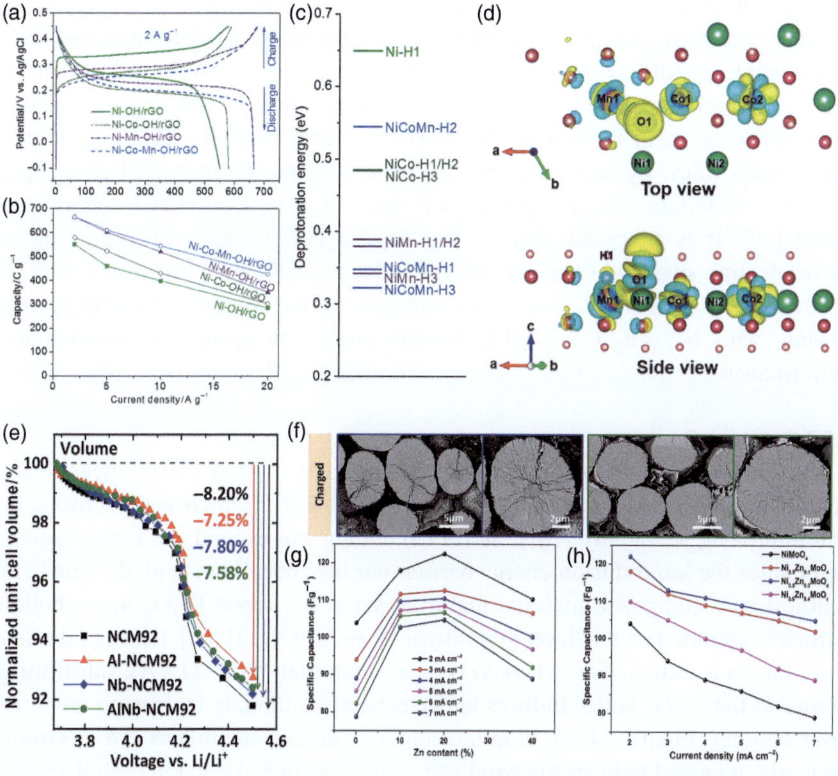

Figure 6.14 (a) Galvanostatic charge–discharge curves of the electrode series at a current density of $2\,A\,g^{-1}$; (b) capacities of the electrode series at different current densities; (c) calculated deprotonation energy per formula unit ($M(OH)_2$) during the deprotonation process of nickel hydroxide-based structural sheets at 298.15 K; (d) density of states diagram of the material and 3D charge density difference isosurface plot during deprotonation. *Source:* Reproduced from Ref. [97] / with permission of John Wiley & Sons. (e) Normalized lattice volume change curves of the corresponding cathode materials measured by in situ X-ray diffraction (in situ XRD); (f) cross-sectional SEM images of particles from various cathode materials in the charged state after long-term cycling; *Source:* Reproduced from Ref. [98] / with permission of American Chemical Society. (g) Effect of zinc doping concentration on specific capacitance; (h) variation trend of specific capacitance at different current densities. *Source:* Reproduced from Ref. [99] / with permission of American Chemical Society.

which facilitates the atomic-level doping of zinc into nickel-based transition metal oxides (BTMOs). For example, zinc-doped $NiMoO_4$ (i.e., $Ni_{1-x}Zn_xMoO_4$) has been successfully prepared via a wet chemical method [100]. Calculations indicate that the surface of $Ni_{1-x}Zn_xMoO_4$ remains stable only when $x \leq 0.25$; excessive zinc doping induces the formation of zinc-based secondary phases, leading to increased resistance and deteriorated electrochemical performance. In contrast, moderate zinc doping supplies additional carriers to the conduction band and improves surface charge transfer, thereby enhancing conductivity and electrochemical activity. As a result, $Ni_{0.8}Zn_{0.2}MoO_4$ achieves a specific capacitance of

460 $F g^{-1}$ at a current density of 2 $mA cm^{-2}$, outperforming both pure $NiMoO_4$ (255 $F g^{-1}$) and $Ni_{0.6}Zn_{0.4}MoO_4$ (256 $F g^{-1}$) (Figure 6.14g, h). Furthermore, ASCs assembled from $NiMoO_4$ samples with different zinc-doping levels and AC exhibit the same trend in capacitive performance. In addition to the doping amount, the doping site also influences the electrochemical behavior of binary transition metal oxides. By adjusting the raw material ratio, two types of zinc-doped $NiCo_2O_4$ materials—$Ni_{0.5}Zn_{0.5}Co_2O_4$/CF and $Ni_{1.0}Zn_{0.5}Co_{1.5}O_4$/CF—were prepared [99]. It is proposed that in $Ni_{0.5}Zn_{0.5}Co_2O_4$/CF, zinc substitutes nickel at octahedral sites, whereas in $Ni_{1.0}Zn_{0.5}Co_{1.5}O_4$/CF, zinc replaces cobalt at octahedral or tetrahedral sites. Both samples exhibit significant bandgap narrowing, with $Ni_{1.0}Zn_{0.5}Co_{1.5}O_4$/CF demonstrating the optimal electrochemical performance.

6.5.3.2 Anion Doping

Anion doping modifies the surface electronic states of iron–cobalt–nickel-based compounds by introducing heteroanions (S, N, P, F, etc.), increasing active sites, and improving wettability [101]. Sulfur (S) doping can expand the lattice spacing and reduce the ion diffusion energy barrier. For instance, Zhao et al. designed and synthesized a composite material featuring surface S-doped Fe_2O_3 micro-hollow particles supported on few-layer graphdiyne (S-Fe_2O_3-GDY) [102]. During S doping, since the ionic radius of S^{2-} (1.84 Å) is larger than that of O^{2-} (1.4 Å), substituting O sites in the Fe_2O_3 lattice induces lattice expansion, directly increasing the interlayer spacing (Figure 6.15a). Simultaneously, S doping modulates the electronic structure of Fe and reduces the band gap, and the lattice expansion optimizes Na^+ transport channels, collectively lowering the diffusion energy barrier for Na^+. In terms of performance, S_1-Fe_2O_3-GDY delivered a specific capacity of 617.5 $mAh g^{-1}$ at 0.1 $A g^{-1}$ and maintained 92.8% of its capacity after 765 cycles at 1.0 $A g^{-1}$ (Figure 6.15b).

Nitrogen (N) doping primarily enhances redox activity by altering the surface charge distribution. For example, N-doped $NiCo_2O_4$ microporous nanograss with rich oxygen vacancies (N-Ov-NCO MiNG) was prepared via a plasma activation strategy [38]. N doping replaces some oxygen atoms in the lattice; due to the higher electronegativity of N, it modifies the surface charge distribution, reduces electron transport energy, and simultaneously introduces oxygen vacancies, synergistically improving electronic conductivity and reaction kinetics. This material exhibited an ultrahigh specific capacitance (2986.25 $F g^{-1}$) and excellent cycling stability. The assembled ASC achieved an energy density of 103.2 $Wh kg^{-1}$ (Figure 6.15c). In related research using N-doped Fe-C@Nb_2CT_x MXene material [103], N doping caused electron delocalization in surrounding carbon atoms, altered the surface charge distribution, and enhanced electron transport rate and ion migration efficiency, while activating more active sites. This material demonstrated a reversible capacity of 471.8 $mAh g^{-1}$ in sodium-ion batteries, along with excellent cycling stability (Figure 6.15d, e).

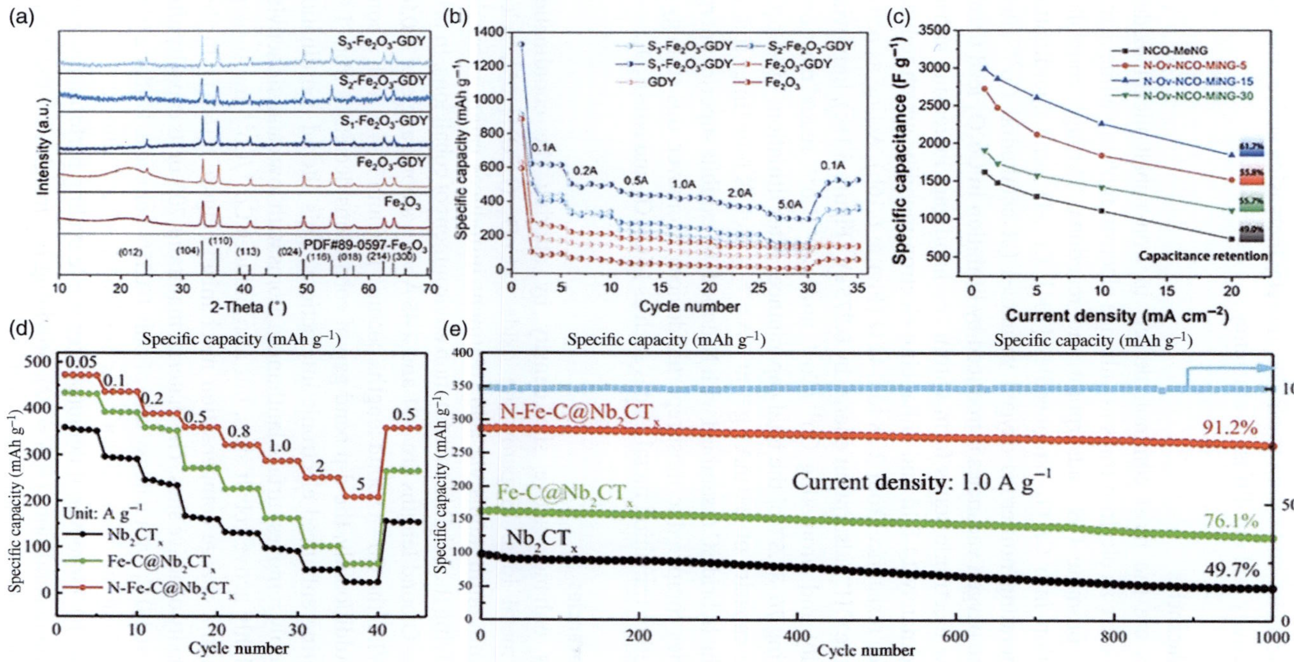

Figure 6.15 (a) X-ray diffraction (XRD) patterns; (b) rate performance at different current densities; *Source:* Reproduced from Ref. [102] / with permission of John Wiley & Sons. (c) Specific capacitance of different electrodes at various current densities; *Source:* Reproduced from Ref. [38] / with permission of Elsevier. (d) Rate capability at different current densities at 0.05 A g^{-1}; (e) cycling stability of Nb_2CT_x, Fe-C@Nb_2CT_x, and N-Fe-C@Nb_2CT_x after 1000 cycles at a current density of 1 A g^{-1}. *Source:* Reproduced from Ref. [103] / with permission of John Wiley & Sons.

6.5.4 Defect Engineering (Oxygen Vacancies, Cation Vacancies, etc.)

Defect engineering modulates the electronic structure and surface chemical properties of iron–cobalt–nickel-based compounds by introducing crystal defects such as oxygen vacancies and cation vacancies, thereby increasing active sites and improving conductivity and ion diffusion efficiency.

6.5.4.1 Oxygen Vacancies

Oxygen vacancies are the most common defects in iron–cobalt–nickel oxides/hydroxides. They serve as electron traps to enhance electron mobility while acting as active sites to promote OH^- adsorption and reactions. Through controlled lithium intercalation into Co_3O_4 nanoneedles [104], Co—O bond coordination is weakened, generating numerous oxygen vacancies (at octahedral Co^{2+} sites) (Figure 6.16a). The oxygen vacancies introduced by lithiation in Co_3O_4 lead to two major effects: new electronic states form within the band gap, increasing carrier density by approximately 14-fold and enhancing electron mobility, which reduces the charge transfer resistance from 2.18 to 0.42 Ω (Figure 6.16b). Acting as active sites, they lower the OH^- adsorption energy by 0.275 eV (Figure 6.16c), improving adsorption stability and promoting Co^{2+}/Co^{3+} and Co^{3+}/Co^{4+} redox reactions, thereby contributing 76–84.5% of the total capacitance. After lithiation, Li_1–Co_3O_4 delivers a specific capacity of 260 mA h g^{-1} at 1 A g^{-1}, which is four times that of pristine Co_3O_4. The hybrid SC assembled with N-doped AC exhibits superior energy and power densities compared to most reported devices. Moreover, this strategy is also effective for other transition metal oxides such as $NiCo_2O_4$ and $NiMn_2O_4$.

6.5.4.2 Cation Vacancies

Cation vacancies enhance active site density by altering the coordination environment of metal ions. In perovskite oxide $LaCoO_3$ (LCO) [105], selective etching of lanthanum atoms via acid treatment introduces cation vacancies (LCO-VLa), modifying the original coordination environment of cobalt ions. In pristine LCO, the Co—O bond lengths are 1.91 and 2.45 Å with a bond angle of 170.0°, whereas in LCO-VLa, the Co—O bond lengths become 1.90 and 1.84 Å with a bond angle of 170.6°. Additionally, the d-p band gap of cobalt ions decreases from 1.94 to 1.57 eV, indicating enhanced electronic interactions. This altered coordination environment not only removes surface lanthanum atoms with low catalytic activity, exposing more highly active cobalt sites, but also induces Co^{3+}/Co^{4+} multivalent cobalt ions through a charge compensation mechanism (Figure 6.16d–f), further increasing the density of active sites. The material maintains 70% of its capacity after nearly 500 cycles at 1.0 C, with a decay rate of only 0.055% per cycle. Under a high sulfur loading of 4.5 mg cm^{-2}, it retains a capacity of 466 mA h g^{-1} after 400 cycles at 4.0 C (Figure 6.16g, h). Moreover, it demonstrates stable cycling performance even under lean electrolyte conditions (E/S = 12), providing key insights for the design of high-performance catalysts for lithium–sulfur batteries.

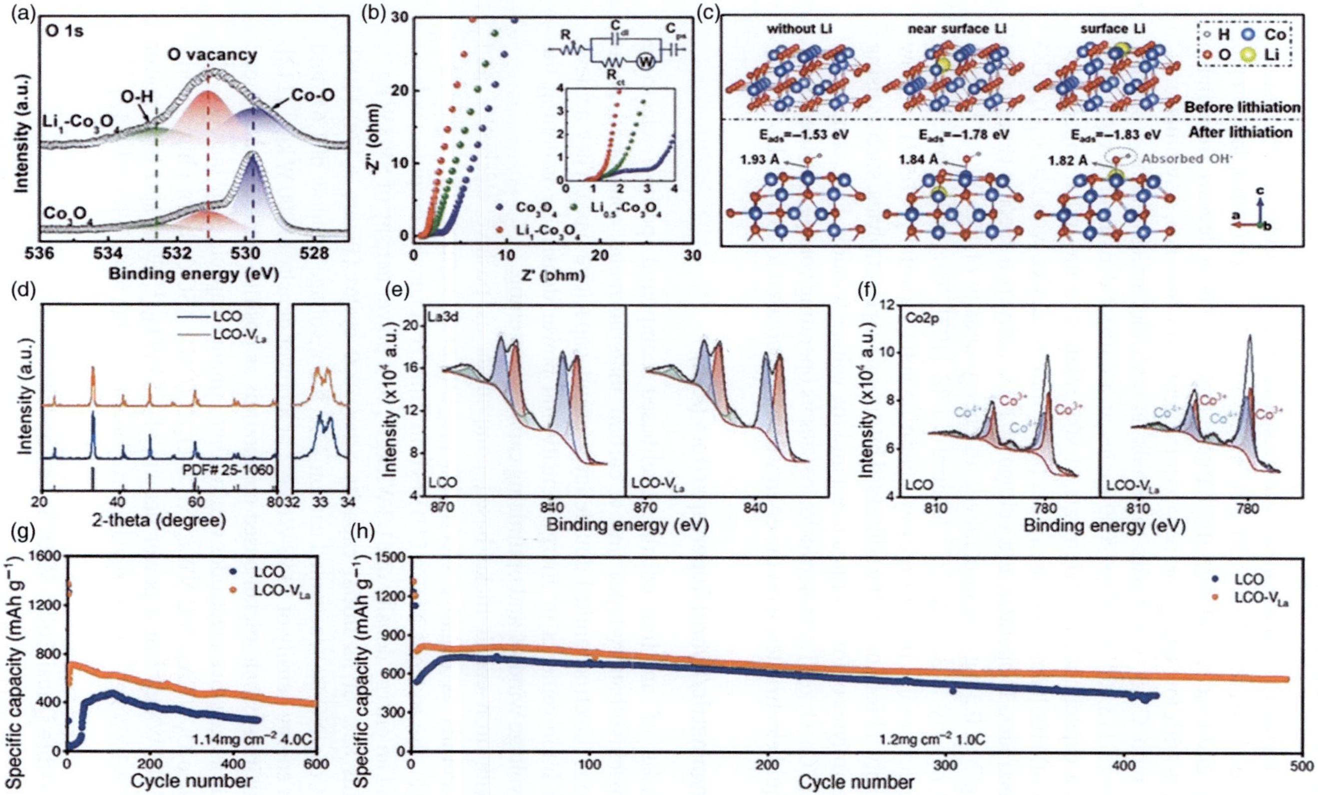

Figure 6.16 (a) High-resolution O 1s X-ray photoelectron spectroscopy (XPS) spectra; (b) Nyquist plots of Co_3O_4, $Li_{0.5}$-Co_3O_4, and Li_1-Co_3O_4 electrodes; (c) optimized structural models of Co_3O_4 at different lithiation states and corresponding hydroxyl (OH^-) adsorption energies on Co sites. *Source:* Reproduced from Ref. [104] / with permission of John Wiley & Sons. (d) X-ray diffraction (XRD) patterns; (e) La 3d X-ray photoelectron spectroscopy (XPS) spectra; (f) Co 2p X-ray photoelectron spectroscopy (XPS) spectra; (g) cycling performance at a current density of 4.0 C; (h) long-term cycling stability of batteries at a sulfur loading of 1.2 mg cm^{-2} and current density of 1.0 C. *Source:* Reproduced from Ref. [105] / with permission of John Wiley & Sons.

6.6 Device Integration and Application Fields

6.6.1 Symmetric Supercapacitors

Symmetric SCs use the same iron–cobalt–nickel-based compounds for both positive and negative electrodes. They offer advantages like simple structure, low preparation cost, and excellent charge–discharge symmetry. These devices are suitable for scenarios with low energy density requirements. For instance, a symmetric SC assembled with Co_9S_8 composite materials as both electrodes retains a specific capacitance of 113.02 $F\,g^{-1}$ after 2000 cycles at a current density of 1 $A\,g^{-1}$. This represents a capacitance retention rate of 90% relative to the initial capacitance (Figure 6.17a) [106]. Another example is the symmetric device with $MnFe_2O_4$/rGO as both electrodes. It operates at a voltage of 1.3 V, delivers a specific capacitance of 271.5 $F\,g^{-1}$ at 0.5 $A\,g^{-1}$, and achieves an energy density of 15.9 $Wh\,kg^{-1}$ when the power density is 324.5 $W\,kg^{-1}$ (Figure 6.17b) [107]. The main limitation of symmetric devices lies in their narrow operating voltage (usually <1.5 V) and low energy density. This issue can be addressed by optimizing electrolytes, such as using organic electrolytes or ionic liquids to expand the voltage window. For example, the Co_3O_4@Ni//Co_3O_4@Ni symmetric device sees its operating voltage extended to 2 V in ionic liquid electrolytes, with the energy density increased to 29.4 $Wh\,kg^{-1}$ [111].

6.6.2 Asymmetric/Hybrid Supercapacitors (ASC)

ASCs consist of an iron–cobalt–nickel-based compound positive electrode (pseudocapacitive/battery-type) and a carbon material/metal compound negative electrode (electrochemical double layer/pseudocapacitive-type). They achieve synergistic improvements in energy density and power density by widening the operating voltage window and optimizing capacity matching.

ASCs with iron–cobalt–nickel-based compounds as positive electrodes and carbon materials as negative electrodes are the most common configuration. For example, the aqueous Ni–Co LDH hybrid films//rGO device outputs a voltage of 1.6 V and an energy density of 61.2 $Wh\,kg^{-1}$. At a power density of 5593 $W\,kg^{-1}$, it maintains 82% of its initial capacity after 5000 cycles (Figure 6.17c) [108]. The NiS_2/CoS_2/NC-500//activated carbon ASC has a voltage window of 1.6 V and reaches an energy density of 53.93 $Wh\,kg^{-1}$ at a power density of 800 $W\,kg^{-1}$ [112]. ASCs using iron–cobalt–nickel-based compounds as positive electrodes and metal compounds as negative electrodes can further enhance capacity. For instance, the SiC NWs@Fe_2O_3 NNAs//SiC NWs@$NiCo_2O_4$/$Ni(OH)_2$ ASC delivers a high energy density of 103 $Wh\,kg^{-1}$ at a power density of 3.5 $kW\,kg^{-1}$. Even at a high power density of 26.1 $kW\,kg^{-1}$, its energy density remains as high as 45 $Wh\,kg^{-1}$ [113].

6.6.3 Flexible Quasi-Solid-State Supercapacitors

Flexible quasi-solid-state SCs adopt gel electrolytes such as PVA/KOH and PVA/H_2SO_4. They combine high safety, good flexibility, and stable electrochemical performance, making them suitable for flexible electronic devices.

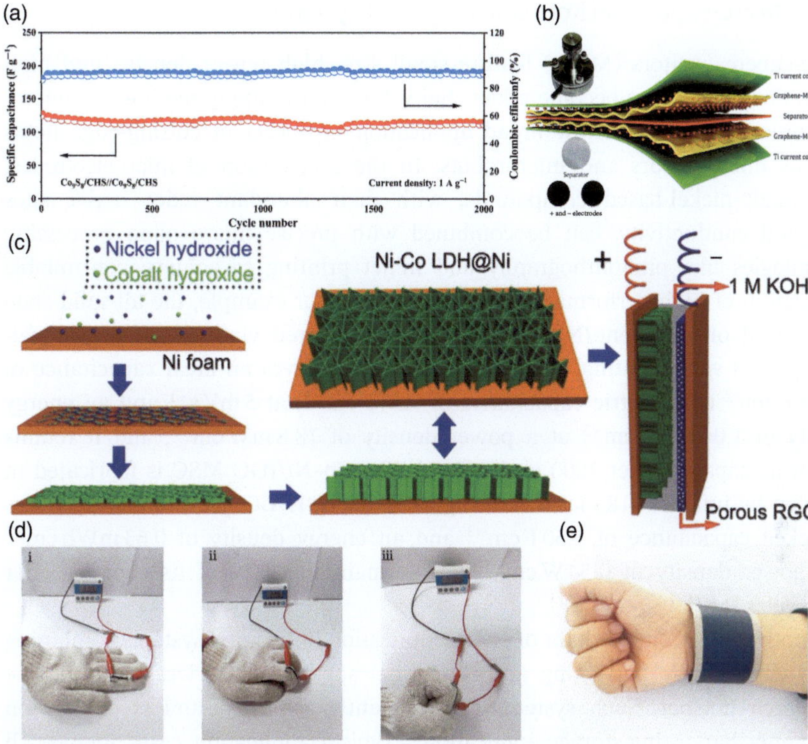

Figure 6.17 (a) Cycling performance of Co_9S_8/CHS//Co_9S_8/CHS symmetric supercapacitors. *Source:* Reproduced from Ref. [106] / with permission of Elsevier. (b) Schematic of the sandwich-type supercapacitor of the $MnFe_2O_4$/rGO nanocomposite. *Source:* Reproduced from Ref. [107] / with permission of The Royal Society of Chemistry. (c) Formation mechanism of Ni–Co LDH hybrid films supported on Ni foams and the illustration scheme of the asymmetrical supercapacitors. *Source:* Reproduced from Ref. [110] / with permission of John Wiley & Sons. (d) The coaxial fiber asymmetric supercapacitor performs different movements in the fabric and lights up a timer. *Source:* Reproduced from Ref. [111] / with permission of John Wiley & Sons. (e) Digital photograph of the assembled solar cell/flexible supercapacitor hybrid power pack with the pulse. *Source:* Reproduced from Ref. [110] / with permission of Elsevier.

Iron–cobalt–nickel-based compounds can be combined with flexible substrates (e.g., carbon cloth and carbon fiber paper) to prepare self-supporting flexible electrodes. For example, the $NiMoO_4$/$CoMoO_4$@carbon cloth flexible electrode exhibits a specific capacitance of 778.1 F g^{-1} at 0.5 A g^{-1} and an energy density of 33.1 Wh kg^{-1} at a power density of 199.6 W kg^{-1} [114]. Connecting multiple flexible devices in series can meet the voltage requirements of practical applications. For example, two solid-state ASCs based on porous graphene sheets/nickel hydroxide (PGNH) films//carbon nanotube (CNT) connected in series can light up a red light-emitting diode [115]. A coaxial fiber ASC ($Ni(OH)_2$@$NiCo_2O_4$/CNT fiber//PVA-KOH//MoS_2@Fe_2O_3/CNT) can power a timer. Even after being woven into fabrics, it maintains excellent mechanical and electrochemical performance (Figure 6.17d) [109].

6.6.4 Micro-supercapacitors and Integrated Systems

Micro-supercapacitors (MSCs) feature small size, high power density, and high integration. They have become a key choice for energy supply modules in microelectronic devices and show broad application prospects in cutting-edge fields such as microsensors and microrobots. In the preparation of microelectrodes, iron–cobalt–nickel-based compounds, with their abundant redox active sites and good conductivity, can be combined with precise micro–nano processing technologies like photolithography and inkjet printing to achieve controllable preparation of high-performance microelectrodes. For example, the all-solid-state MSC based on graphene/NiOOH/$Ni(OH)_2$ is prepared via CVD, maskless patterning, and spray coating. The optimal device achieves an areal capacitance of $0.75\,mF\,cm^{-2}$ (volumetric capacitance of $7.54\,F\,cm^{-3}$) at $5\,mV\,s^{-1}$ and an energy density of $1.04\,mWh\,cm^{-3}$ at a power density of $18.8\,mW\,cm^{-3}$, and it retains 80% of its capacity after 3000 cycles [116]. The Co–Ni/rGO MSC is fabricated in one step by infrared (IR) laser scribing a Co–Ni LDH/GO precursor film. It has a stacked capacitance of $5.50\,F\,cm^{-3}$ and an energy density of $0.63\,mWh\,cm^{-3}$ at a power density of $0.54\,W\,cm^{-3}$, and it maintains 90% of its capacity after 3000 cycles [117].

To optimize the entire chain of energy utilization, integrated systems combining MSCs with energy harvesting and functional application devices have become a research hotspot. Such systems can efficiently achieve seamless connection of "energy harvesting-storage-application". Typical application cases include: Ji et al. prepared a flexible all-solid-state ASC device, which integrates an ASC with a flexible solar cell. The constructed self-powered system has been successfully applied to smart wearable devices, solving the problem of frequent charging of traditional wearable devices (Figure 6.17e) [110]. Hu et al. developed the Ni-Co LDH@CMS composite material. With the high energy density and cycle stability of SCs, as well as the fast response (20 ms) and excellent cycle performance of pressure sensors, this material can be used for energy storage and human health monitoring in flexible wearable electronic devices [118].

References

1 Chuhadiya, S., Himanshu, Suthar, D. et al. (2021). Metal organic frameworks as hybrid porous materials for energy storage and conversion devices: a review. *Coordination Chemistry Reviews* 446: 214115.

2 Ren, X., Du, Y., Song, M. et al. (2019). Facile preparation of mesoporous $NiCo_2S_4$ microaggregates constructed by nanoparticles via puffing $NiCo_2O_4$ cubes for high performance asymmetric supercapacitors. *Journal of Alloys and Compounds* 806: 1481–1490.

3 Zhang, Y., Wang, D., Lü, S. et al. (2020). CoO@CoS/Ni_3S_2 hierarchical nanostructure arrays for high performance asymmetric supercapacitor. *Applied Surface Science* 532: 147438.

4 Liu, W., Liu, H., Sun, Y. et al. (2025). Low-temperature, low-pressure Zn-ion hybrid supercapacitor in extreme near-space application. *Materials Horizons* 12: 3979–3990.

5 Kang, H., Lee, H., Jung, G. et al. (2023). A flexible asymmetric supercapacitor with organohydrogel electrolyte for high voltage operation over wide temperature range. *Applied Surface Science* 638: 158150.

6 Kour, S., Tanwar, S., and Sharma, A.L. (2022). A review on challenges to remedies of MnO_2 based transition-metal oxide, hydroxide, and layered double hydroxide composites for supercapacitor applications. *Materials Today Communications* 32: 104033.

7 Chodankar, N.R., Pham, H.D., Nanjundan, A.K. et al. (2020). True meaning of pseudocapacitors and their performance metrics: asymmetric versus hybrid supercapacitors. *Small* 16: 2002806.

8 Lv, S., Ma, L., Shen, X., and Tong, H. (2021). Recent design and control of carbon materials for supercapacitors. *Journal of Materials Science* 56: 1919–1942.

9 Wang, G., Zhang, L., and Zhang, J. (2012). A review of electrode materials for electrochemical supercapacitors. *Chemical Society Reviews* 41: 797–828.

10 Jiang, Y. and Liu, J. (2019). Definitions of Pseudocapacitive Materials: A Brief Review. *Energy & Environmental Materials* 2: 30–37.

11 Wang, T., Chen, H.C., Yu, F. et al. (2019). Boosting the cycling stability of transition metal compounds-based supercapacitors. *Energy Storage Materials* 16: 545–573.

12 Brousse, T., Bélanger, D., and Long, J.W. (2015). To be or not to be pseudocapacitive? *Journal of the Electrochemical Society* 162: A5185–A5189.

13 Liu, H., Liu, X., Wang, S. et al. (2020). Transition metal based battery-type electrodes in hybrid supercapacitors: a review. *Energy Storage Materials* 28: 122–145.

14 Sahoo, S., Kumar, R., Joanni, E. et al. (2022). Advances in pseudocapacitive and battery-like electrode materials for high performance supercapacitors. *Journal of Materials Chemistry A* 10: 13190–13240.

15 Liu, F., Chen, Y., Liu, Y. et al. (2019). Integrating ultrathin and modified NiCoAl-layered double-hydroxide nanosheets with N-doped reduced graphene oxide for high-performance all-solid-state supercapacitors. *Nanoscale* 11: 9896–9905.

16 Zeng, W.W., Huang, K.L., Yang, Y.P. et al. (2008). Solvothermal synthesis and capacitance performance of Co_3O_4 with different morphologies. *Acta Physico-Chimica Sinica* 24: 263–268.

17 Naveenkumar, P., Maniyazagan, M., Kang, N. et al. (2025). MoF-derived $CuCo_2S_4$@FeS_2 nanohybrids for supercapacitor applications. *Electrochimica Acta* 513: 145546.

18 Hussain, I., Hussain, T., Ahmed, S.B. et al. (2021). Binder-free trimetallic phosphate nanosheets as an electrode: theoretical and experimental investigation. *Journal of Power Sources* 513: 230556.

19 Ren, X., Zhou, Y., Du, Y. et al. (2020). Facile ion exchange to construct Ni-Fe-Co sulfides and hydroxides ultrathin nanosheets with rich interfaces

for advanced all-solid-state asymmetric supercapacitors. *Applied Surface Science* 514: 145951.

20 Sahoo, M.K., Gusain, M., Thangriyal, S. et al. (2020). Energy storage study of trimetallic Cu_2MSnS_4 (M: Fe, Co, Ni) nanomaterials prepared by sequential crystallization method. *Journal of Solid State Chemistry* 282: 121049.

21 Kong, D.Z., Luo, J.S., Wang, Y.L. et al. (2014). Three-dimensional Co_3O_4@MnO_2 hierarchical nanoneedle arrays: morphology control and electrochemical energy storage. *Advanced Functional Materials* 24: 3815–3826.

22 Lv, J.L., Liang, T.X., Meng, Y. et al. (2017). Performance comparison of $NiCo_2O_4$ and $NiCo_2S_4$ formed on Ni foam for supercapacitor. *Composites Part B: Engineering* 123: 28–33.

23 Miao, Y.D., Zhang, X.P., Zhan, J. et al. (2020). Hierarchical NiS@CoS with controllable core-shell structure by two-step strategy for supercapacitor electrodes. *Advanced Materials Interfaces* 7: 1901618.

24 Gogotsi, Y. and Penner, R.M. (2018). Energy storage in nanomaterials—capacitive pseudocapacitive, or battery-like? *ACS Nano* 12: 2081–2083.

25 Seenivasan, S., Adhikari, S., Sivagurunathan, A.T., and Kim, D. (2025). Supercapatteries: unlocking the potential of battery-supercapacitor fusion. *Energy & Environmental Science* 18: 1054–1095.

26 Liu, S.X., Sarwar, S., Zhang, H.P. et al. (2020). One-step microwave-controlled synthesis of $CoV_2O_6{\cdot}2H_2O$ nanosheet for super long cycle-life battery-type supercapacitor. *Electrochimica Acta* 364: 137320.

27 Iqbal, M.Z. and Aziz, U. (2022). Supercapattery: merging of battery-supercapacitor electrodes for hybrid energy storage devices. *Journal of Energy Storage* 46: 103823.

28 Yi, R., Chen, S.R., Song, J.X. et al. (2014). High-performance hybrid supercapacitor enabled by a high-rate Si-based anode. *Advanced Functional Materials* 24: 7433–7439.

29 Ren, X., Du, Y., Song, M. et al. (2019). In-situ transformation of Ni foam into sandwich nanostructured $Co_{1.29}Ni_{1.71}O_4$ nanoparticle@$CoNi_2S_4$ nanosheet networks for high-performance asymmetric supercapacitors. *Chemical Engineering Journal* 375: 122063.

30 Uzun, D., Albas, A., Madakbas, S., and Yetimoglu, E.K. (2025). Electrodeposition of CuNiS as battery type electrode for supercapacitor applications. *Electrochemistry Communications* 179: 108024.

31 Banerjee, S., Mordina, B., Sinha, P., and Kar, K.K. (2025). Recent advancement of supercapacitors: a current era of supercapacitor devices through the development of electrical double layer, pseudo and their hybrid supercapacitor electrodes. *Journal of Energy Storage* 108: 115075.

32 Zhao, W., Xu, X., Wu, N. et al. (2023). Dandelion-like $CuCo_2O_4$@ NiMn LDH core/shell nanoflowers for excellent battery-type supercapacitor. *Nanomaterials* 13: 730.

33 Li, M., Yu, N., Xu, L. et al. (2024). Controllable synthesis of flower-like hierarchical $CuCo_2S_4$ nanostructure arrays for high-performance hybrid supercapacitors. *Metals* 14: 145.

34 Huang, R. and Xie, Y. (2024). Matching Faradaic reaction of multi-transition metal compounds as supercapacitor electrode materials: a review. *Journal of Alloys and Compounds* 1002: 175367.

35 Li, P., Jiao, Y., Yao, S. et al. (2019). Dual role of nickel foam in NiCoAl-LDH ensuring high-performance for asymmetric supercapacitors. *New Journal of Chemistry* 43: 3139–3145.

36 Zhang, W., Zhu, Z., Zhu, L. et al. (2025). Defect-engineered heterostructured NiFe-LDH@CoNi-LDH nanocages via synergistic interface-oxygen vacancy modulation for high-energy-density supercapacitors. *Chemical Engineering Journal* 522: 167235.

37 Goubard-Bretesché, N., Crosnier, O., Douard, C. et al. (2020). Unveiling pseudocapacitive charge storage behavior in $FeWO_4$ electrode material by operando X-ray absorption spectroscopy. *Small* 16: 2002855.

38 Wei, S., Wan, C., Zhang, L. et al. (2022). N-doped and oxygen vacancy-rich $NiCo_2O_4$ nanograss for supercapacitor electrode. *Chemical Engineering Journal* 429: 132242.

39 Guan, B., Li, Y., Yin, B. et al. (2017). Synthesis of hierarchical NiS microflowers for high performance asymmetric supercapacitor. *Chemical Engineering Journal* 308: 1165–1173.

40 Dhandapani, P., Nayak, P.K., and Maruthapillai, A. (2023). Improved electrochemical performance and charge storage mechanism of $NiMnCoO_4$ by XPS study. *Materials Chemistry and Physics* 297: 127287.

41 Wu, N.-L., Wang, S.-Y., Han, C.-Y. et al. (2003). Electrochemical capacitor of magnetite in aqueous electrolytes. *Journal of Power Sources* 113: 173–178.

42 Wang, L.J., Liu, F.H., Pal, A. et al. (2021). Ultra-small Fe_3O_4 nanoparticles encapsulated in hollow porous carbon nanocapsules for high performance supercapacitors. *Carbon* 179: 327–336.

43 Xuan, L., Chen, L., Yang, Q. et al. (2015). Engineering 2D multi-layer graphene-like Co_3O_4 thin sheets with vertically aligned nanosheets as basic building units for advanced pseudocapacitor materials. *Journal of Materials Chemistry A* 3: 17525–17533.

44 Gao, Z., Chen, C., Chang, J. et al. (2018). Balanced energy density and power density: asymmetric supercapacitor based on activated fullerene carbon soot anode and graphene-Co_3O_4 composite cathode. *Electrochimica Acta* 260: 932–943.

45 Kumar, A., Satpathy, B.K., Goyal, P. et al. (2025). Bimetallic MOF-derived CuO-Co_3O_4 heterostructures as high-capacity electrodes for asymmetric supercapacitors. *Chemical Engineering Journal* 520: 165685.

46 Gowdhaman, A., Kumar, S.A., Elumalai, D. et al. (2023). Ni-MOF derived NiO/Ni/r-GO nanocomposite as a novel electrode material for high-performance asymmetric supercapacitor. *Journal of Energy Storage* 61: 106769.

47 Li, N., Yuan, K., Gao, T.J. et al. (2022). Controllable synthesis of hierarchical nanoporous carbon@Ni$(OH)_2$ rambutan-like composite microspheres for high-performance hybrid supercapacitor. *Arabian Journal of Chemistry* 15: 103580.

48 Sudhakaran, M.S.P., Devi, V.S., and Youk, J.H. (2025). β-Co$(OH)_2$@CoNi_2S_4 core-shell heterostructures supported on carbon cloth for high-performance flexible asymmetric supercapacitors. *Journal of Alloys and Compounds* 1022: 180026.

49 Song, H., Wu, M., Tang, X. et al. (2025). Synthesis of Fe_3O_4/FeS_2 composites via MOF-templated sulfurization for high-performance hybrid supercapacitors. *Journal of Alloys and Compounds* 1010: 177658.

50 Owusu, K.A., Qu, L., Li, J. et al. (2017). Low-crystalline iron oxide hydroxide nanoparticle anode for high-performance supercapacitors. *Nature Communications* 8: 14264.

51 Yang, W., Zhang, C., Jiang, B. et al. (2022). Electrochemical activation induced phase and structure reconstruction to reveal cobalt sulfide intrinsic energy storage capacity. *Chemical Engineering Journal* 434: 134473.

52 Jiang, D., Wang, Y., Du, C. et al. (2024). Heterostructured Ni_2P/NiMo-layered double hydroxide nanoarrays with enriched redox active sites for supercapacitors. *Journal of Alloys and Compounds* 970: 172685.

53 Ma, F.-X., Yu, L., Xu, C.-Y., and Lou, X.W. (2016). Self-supported formation of hierarchical $NiCo_2O_4$ tetragonal microtubes with enhanced electrochemical properties. *Energy & Environmental Science* 9: 862–866.

54 Qu, Z., Shi, M., Wu, H. et al. (2019). An efficient binder-free electrode with multiple carbonized channels wrapped by $NiCo_2O_4$ nanosheets for high-performance capacitive energy storage. *Journal of Power Sources* 410–411: 179–187.

55 He, X., Zhao, Y., Chen, R. et al. (2018). Hierarchical $FeCo_2O_4$@polypyrrole core/shell nanowires on carbon cloth for high-performance flexible all-solid-state asymmetric supercapacitors. *ACS Sustainable Chemistry & Engineering* 6: 14945–14954.

56 Li, S., Wang, Y., Sun, J. et al. (2020). Simple preparation of porous $FeCo_2O_4$ microspheres and nanosheets for advanced asymmetric supercapacitors. *ACS Applied Energy Materials* 3: 11307–11317.

57 Wiston, B.R., Prabhakaran, P., and Ashok, M. (2022). Bimetallic NiFe hydroxide coated onto commercial graphite foil as efficient supercapacitor electrode. *Journal of Energy Storage* 50: 104226.

58 Li, Z., Mao, S., Yang, Y. et al. (2021). Controllable synthesis of a hollow core-shell Co-Fe layered double hydroxide derived from Co-MOF and its application in capacitive deionization. *Journal of Colloid and Interface Science* 585: 85–94.

59 Wang, K., Liu, Y., Ding, Z. et al. (2022). Chloride pre-intercalated CoFe-layered double hydroxide as chloride ion capturing electrode for capacitive deionization. *Chemical Engineering Journal* 433: 133578.

60 He, X., Liu, Q., Liu, J. et al. (2017). High-performance all-solid-state asymmetrical supercapacitors based on petal-like $NiCo_2S_4$/Polyaniline nanosheets. *Chemical Engineering Journal* 325: 134–143.

61 Li, W., Zhang, B., Lin, R. et al. (2018). A dendritic nickel cobalt sulfide nanostructure for alkaline battery electrodes. *Advanced Functional Materials* 28: 1705937.

62 Wu, L., Sun, L., Li, X. et al. (2020). $CuCo_2S_4$-rGO microflowers: first-principle calculation and application in energy storage. *Small* 16: e2001468.

63 Wang, Z., Wang, H., Ji, S. et al. (2020). Hollow-structured NiCoP nanorods as high-performance electrodes for asymmetric supercapacitors. *Materials & Design* 193: 108807.

64 Zhang, X., Su, D., Wu, A. et al. (2019). Porous NiCoP nanowalls as promising electrode with high-area and mass capacitance for supercapacitors. *Science China Materials* 62: 1115–1126.

65 Shi, R., Wang, J., Wang, Z. et al. (2019). Unique $NiFeNiCoO_2$ hollow polyhedron as bifunctional electrocatalysts for water splitting. *Journal of Energy Chemistry* 33: 74–80.

66 Jirátová, K., Čada, M., Naiko, I. et al. (2024). Combination of plasma jet and magnetron sputtering-a new method for preparing thin Ni-Co-Mn oxide coatings on stainless steel meshes and their efficiency in the total oxidation of volatile organic compounds. *Industrial & Engineering Chemistry Research* 63: 18759–18772.

67 Abdpour, S., Rademacher, L., Fetzer, M.N.A. et al. (2023). Iron-containing nickel cobalt sulfides, selenides, and sulfoselenides as active and stable electrocatalysts for the oxygen evolution reaction in an alkaline solution. *Solids* 4: 181–200.

68 Kumar, A.S., Naga Sathya Sai, K., Prasad, K. et al. (2024). In-situ constructed NiCoZnS composite on nickel foam with hierarchical structures as bifunctional electrocatalysts for oxygen evolution reaction (OER) and supercapacitors. *Journal of Alloys and Compounds* 1005: 175983.

69 Xiang, J., Jiang, S., Zhang, G. et al. (2026). Prussian blue analogues and their derivatives for electrochemical energy conversion. *Coordination Chemistry Reviews* 548: 217215.

70 Fayaz, M., Liu, X., Chen, B. et al. (2024). Synthesis of CuFe-PBA@NiFe-PBA nanocubes as a battery electrode for Na-ion hybrid supercapacitor. *Journal of Energy Storage* 98: 112999.

71 Sun, A., Chen, W., Yang, J. et al. (2025). Layered double hydroxides/prussian blue analogs toward improved capacitive performances. *Journal of Colloid and Interface Science* 699: 138287.

72 Shan, Y., Shi, Y., Yuan, G. et al. (2025). Construction of MOFs with different dimensions and bimetallic effect for enhanced electrochemical performance. *Nano Research* 18: 94907857.

73 Wang, J., Li, S., Fu, N. et al. (2025). MOF-derived carbon-coated NiS/NiS_2 yolk-shell spheres as a satisfactory positive electrode material for hybrid supercapacitors. *Advanced Composites and Hybrid Materials* 8: 190.

74 Kitchamsetti, N. (2023). A review on recent advances in Prussian blue, its analogues, and their derived materials as electrodes for high performance supercapacitors. *Journal of Energy Storage* 73: 108958.

75 Xu, B., Zhang, H., Mei, H., and Sun, D. (2020). Recent progress in metal-organic framework-based supercapacitor electrode materials. *Coordination Chemistry Reviews* 420: 213438.

76 Chettiannan, B., Arumugam, G., Mathan, S. et al. (2025). Exploring the effect of metal ratio on electrochemical properties of MOF-derived microporous flower-like NiO/Co_3O_4 electrodes for high-performance asymmetric supercapacitor application. *Journal of Energy Storage* 119: 116326.

77 Fan, H., Wang, Y., Luo, Y. et al. (2024). Distorted iron oxide quantum dots with unprecedented fast-charging capability for high-energy and high-power lithium storage. *Advanced Energy Materials* 14: 2400248.

78 Man, P., Zhang, Q., Zhou, Z. et al. (2020). Engineering MoS_2 nanosheets on spindle-like α-Fe_2O_3 as high-performance core–shell pseudocapacitive anodes for fiber-shaped aqueous lithium-ion capacitors. *Advanced Functional Materials* 30: 2003967.

79 Ji, Z., Chen, L., Yu, H. et al. (2022). Dual functionalized Fe_2O_3 nanosheets and Co_9S_8 nanoflowers with phosphate and nitrogen-doped carbon dots for advanced hybrid supercapacitors. *Chemical Engineering Journal* 450: 137942.

80 Li, G., Li, R., and Zhou, W. (2017). A wire-shaped supercapacitor in micrometer size based on Fe_3O_4 nanosheet arrays on Fe wire. *Nano-Micro Letters* 9: 46.

81 Sun, S., Zhai, T., Liang, C. et al. (2018). Boosted crystalline/amorphous $Fe_2O_{3-\delta}$ core/shell heterostructure for flexible solid-state pseudocapacitors in large scale. *Nano Energy* 45: 390–397.

82 Qiao, Y., Jiang, S., Xia, R. et al. (2022). High-performance lithium-ion capacitors based on 3D interconnected nitrogen-doped ultrathin carbon frameworks with encapsulated Fe_2O_3 nanoparticles. *ACS Applied Energy Materials* 5: 4329–4339.

83 Zhang, X., Fan, Q., Liu, S. et al. (2020). A facile fabrication of 1D/2D nanohybrids composed of NiCo-hydroxide nanowires and reduced graphene oxide for high-performance asymmetric supercapacitors. *Inorganic Chemistry Frontiers* 7: 204–211.

84 Ma, L., Sun, G., Ran, J. et al. (2018). One-pot template-free strategy toward 3D hierarchical porous nitrogen-doped carbon framework in situ armored homogeneous NiO nanoparticles for high-performance asymmetric supercapacitors. *ACS Applied Materials & Interfaces* 10: 22278–22290.

85 Wang, X., Zhou, Y., Li, X. et al. (2024). Preparation of NiCo-LDHs@rGO composites with improved electrochemical performance by ammonia diffusion method under closed conditions. *Chemical Physics Letters* 852: 141495.

86 Mayakrishnan, G., Vanaraj, R., Kim, S.C., and Kim, I.S. (2025). Direct solidvapor approach to 2D FeCoNiP@N-graphene-CNTs as robust electrode material for high-performance supercapacitors. *Chemical Engineering Journal* 503: 158229.

87 Chen, J., Du, Z., Cheng, K. et al. (2022). Engineering $NiCo_2S_4$ nanoparticles anchored on carbon nanotubes as superior energy-storage materials for supercapacitors. *RSC Advances* 12: 34904–34909.

88 Zheng, L., Gao, S., Yao, S. et al. (2024). N-doped porous carbon with ZIF-67-derived $CoFe_2O_4$-Fe particles for supercapacitors. *Journal of Colloid and Interface Science* 674: 735–744.

89 Wang, J., Pan, X., Peng, P. et al. (2024). Using metal-organic frameworks to create carbon-encased Ni@$Ni(OH)_2$ for high-performance supercapacitors. *Journal of Nanoparticle Research* 26: 112.

90 Abdullah, M., Shah, S.I.A., Jabbour, K. et al. (2024). Synthesis of $NiMn_2O_4$/PANI nanosized composite with increased specific capacitance for energy storage applications. *Dalton Transactions* 53: 8680–8691.

91 Pujari, S., Desai, N., Y.N, S. et al. (2025). Polyoxometalate/α-Fe_2O_3/polyaniline composite: tailored approaches for high-performance supercapacitors. *Journal of Alloys and Compounds* 1010: 177306.

92 Nabeel, A.N., Jain, A., Kumar, R. et al. (2025). Electrochemical properties of PPy/rGO/$NiCoFe_2O_4$ composites as advanced electrode materials for supercapacitors: a state-of-the-art review. *Ionics* 31: 1233–1253.

93 Zhou, H., Nasser, R., Zhang, L. et al. (2024). Copper/cobalt based metal phosphide composites as positive material for supercapacitors. *Chemical Engineering Journal* 496: 154102.

94 Mumtaz, A., Iqbal, J., Mumtaz, M. et al. (2025). Novel hafnium oxide-cobalt oxide composite electrode with improved performance for supercapacitor applications. *Journal of Alloys and Compounds* 1012: 178459.

95 Lin, Z., Fan, K., Liu, T. et al. (2023). Mitigating lattice distortion of high-voltage $LiCoO_2$ via core-shell structure induced by cationic heterogeneous co-doping for lithium-ion batteries. *Nano-Micro Letters* 16: 48.

96 Patil, A.M., Moon, S., Jadhav, A.A. et al. (2023). Modifying electronic structure of cation-exchanged bimetallic sulfide/metal oxide heterostructure through in situ inclusion of silver (Ag) nanoparticles for extrinsic pseudocapacitor. *Advanced Functional Materials* 33: 2305264.

97 Zhao, B., Zhang, L., Zhang, Q. et al. (2017). Rational design of nickel hydroxide-based nanocrystals on graphene for ultrafast energy storage. *Advanced Energy Materials* 8: 1702247.

98 Lee, S.-B., Park, N.-Y., Park, G.-T. et al. (2024). Doping strategy in developing Ni-rich cathodes for high-performance lithium-ion batteries. *ACS Energy Letters* 9: 740–747.

99 Sharma, P., Minakshi Sundaram, M., Watcharatharapong, T. et al. (2020). Zn metal atom doping on the surface plane of one-dimesional $NiMoO_4$ nanorods with improved redox chemistry. *ACS Applied Materials & Interfaces* 12: 44815–44829.

100 Gao, K. and Li, S.-D. (2020). Hollow fibrous $NiCo_2O_4$ electrodes with controllable Zn substitution sites for supercapacitors. *Journal of Alloys and Compounds* 832: 154927.

101 Ren, K., Liu, Z., Wei, T., and Fan, Z. (2021). Recent developments of transition metal compounds-carbon hybrid electrodes for high energy/power supercapacitors. *Nano-Micro Letters* 13: 129.

102 Fu, Z., Wang, J., Hou, B. et al. (2026). Rational design of surface s-doped Fe_2O_3 micro-hollow structure on graphdiyne for high-performance sodium-ion batteries. *Advanced Functional Materials* 36: e22685.

103 Fu, H., Lian, Y., Hu, Y. et al. (2025). N-doping Fe-C@Nb_2CT_x MXenes with high stability and strong activity for sodium-ion storage and overall water splitting. *Small* 21: e2406131.

104 Zhang, Y., Hu, Y., Wang, Z. et al. (2020). Lithiation-induced vacancy engineering of Co_3O_4 with improved faradic reactivity for high-performance supercapacitor. *Advanced Functional Materials* 30: 2004172.

105 Bai, Z., Wang, Z., Wang, T. et al. (2024). Cation-vacancy engineering modulated perovskite oxide for boosting electrocatalytic conversion of polysulfides. *Advanced Functional Materials* 35: 2419105.

106 Sun, X., Wang, H., Wang, W. et al. (2024). Cobalt-copolymer-derived Co_9S_8 superstructure as electrode materials for advanced symmetric supercapacitor. *Journal of Energy Storage* 99: 113296.

107 Goljanian Tabrizi, A., Arsalani, N., Mohammadi, A. et al. (2017). Facile synthesis of a $MnFe_2O_4$/rGO nanocomposite for an ultra-stable symmetric supercapacitor. *New Journal of Chemistry* 41: 4974–4984.

108 Chen, H., Hu, L., Chen, M. et al. (2014). Nickel–cobalt layered double hydroxide nanosheets for high-performance supercapacitor electrode materials. *Advanced Functional Materials* 24: 934–942.

109 He, H., Yang, X., Wang, L. et al. (2020). Enhanced energy density of coaxial fiber asymmetric supercapacitor based on MoS_2@Fe_2O_3/carbon nanotube paper and $Ni(OH)_2$@$NiCo_2O_4$/carbon nanotube fiber electrodes. *Chemistry – A European Journal* 26: 17212–17221.

110 Liang, X., Long, G., Fu, C. et al. (2018). High performance all-solid-state flexible supercapacitor for wearable storage device application. *Chemical Engineering Journal* 345: 186–195.

111 Marzouki, M., Zarrougui, R., and Ghodbane, O. (2021). Application of aprotic ionic liquids based on bis(trifluoromethylsulfonyl)imide anion as polymer gel electrolytes for cobalt oxide symmetric supercapacitors. *Journal of Energy Storage* 40: 102761.

112 Liu, H., Guo, H., Yue, L. et al. (2019). Metal-organic frameworks-derived NiS_2/CoS_2/N-doped carbon composites as electrode materials for asymmetric supercapacitor. *ChemElectroChem* 6: 3764–3773.

113 Zhao, J., Li, Z., Yuan, X. et al. (2018). A high-energy density asymmetric supercapacitor based on Fe_2O_3 nanoneedle arrays and $NiCo_2O_4$/$Ni(OH)_2$ hybrid nanosheet arrays grown on SiC nanowire networks as free-standing advanced electrodes. *Advanced Energy Materials* 8: 1702787.

114 Yu, B., Jiang, G., Xu, W. et al. (2019). Construction of $NiMoO_4/CoMoO_4$ nanorod arrays wrapped by Ni-Co-S nanosheets on carbon cloth as high performance electrode for supercapacitor. *Journal of Alloys and Compounds* 799: 415–424.

115 Hou, B.Q., Jin, X., Jiang, L.L. et al. (2021). Flexible porous Graphene/Nickel hydroxide composite films with 3D ion transport channels for high volumetric performance asymmetric supercapacitor. *Applied Surface Science* 569: 151036.

116 Feng, X., Ning, J., Wang, D. et al. (2019). All-solid-state planner micro-supercapacitor based on graphene/NiOOH/$Ni(OH)_2$ via mask-free patterning strategy. *Journal of Power Sources* 418: 130–137.

117 Jung, J., Jeong, J.R., Dang Van, C. et al. (2022). Laser-assisted patterning of Co–Ni alloy/reduced graphene oxide composite for enhanced micro-supercapacitor performance. *ACS Applied Electronic Materials* 4: 4840–4848.

118 Hu, H., Li, J., Jiang, Y. et al. (2023). Nickel–cobalt layered double hydroxide nanoflakes combined carbonized melamine sponge for high-performance supercapacitors and pressure sensors. *Ionics* 29: 4285–4293.

7

Applications of Iron (Fe, Co, Ni)-Group Metal Compound Nanomaterials in Electrocatalysis

With the increasing global energy demand, the growing pressure on fossil energy resources and ecosystems has made the development of renewable and clean energy a critical priority. The intermittent nature of wind, tidal, and solar energy leads to spatiotemporal mismatches between energy supply and demand, highlighting the urgent need for efficient, economical, and environmentally friendly energy storage and conversion systems. Electrocatalytic reactions play a vital role in this field; however, the sluggish reaction kinetics of key electrochemical processes significantly hinder the performance improvement and industrial application of these energy conversion and storage devices. Therefore, developing low-cost, highly active, and durable electrocatalysts is essential for advancing renewable energy applications, promoting energy transition, and effectively addressing energy and environmental challenges.

Iron-group elements (iron, cobalt, nickel) are abundant in reserves and easy to extract, attracting widespread attention in the field of electrocatalysis in recent years. With deepening research into catalytic mechanisms, iron-group-based electrocatalysts have demonstrated excellent performance and possess considerable potential for industrial application. To date, iron-group metal compound nanomaterials have been extensively studied and applied in various electrocatalytic reactions. This chapter focuses on the research of iron-group metal compound nanomaterials in crucial electrocatalytic reactions. The content is organized into two main areas: energy conversion reactions, such as the hydrogen evolution reaction (HER), oxygen evolution reaction (OER), overall water splitting (OWS), and oxygen reduction reaction (ORR); and resource utilization reactions, including the nitrogen reduction reaction (NRR), carbon dioxide reduction reaction (CO_2RR), and nitrate reduction reaction (NO_3RR).

7.1 HER

Green, high-efficiency alternative energy plays a crucial role in advancing sustainable development. Among various options, hydrogen obtained through

Iron-Group Metal Compound Nanomaterials: Preparation, Characterization, Structure, and Applications, First Edition. Jinlong Zheng.

electrochemical water splitting driven by renewable energy stands out as one of the most promising replacements for fossil fuels.

The cathodic HER follows a two-electron transfer mechanism. It includes the Volmer step where H_2O and H_3O^+ molecules adsorb onto the catalyst's active sites. After this adsorption, H_2 desorption from the catalyst surface takes place through two possible routes either the electrochemical Heyrovsky step or the chemical Tafel step.

From a thermodynamic standpoint, acidic electrolytes are beneficial for the HER. Nevertheless, catalysts tend to suffer corrosion under acidic conditions, which in turn impairs their catalytic stability [1]. Up to now, many iron-group compound catalysts have been reported to demonstrate outstanding activity when applied in electrolytes with varying pH values.

When the electrolyte is alkaline or neutral, the reaction process involves water dissociation, adsorption of hydroxyl ions (OH_{ads}), and desorption of adsorbed hydrogen (H_{ads}). Similar to other systems, the desorption of H_{ads} follows either the Volmer/Heyrovsky pathway or the Volmer/Tafel pathway, as indicated in relevant studies [2]:

$$H_2O + e^- \rightarrow OH^- + H_{ads} \text{ (Volmer step)} \tag{7.1}$$

$$H_2O + H_{ads} + e^- \rightarrow OH^- + H_2 \text{ (Heyrovsky step)} \tag{7.2}$$

or

$$2H_{ads} \rightarrow H_2 \text{ (Tafel step)} \tag{7.3}$$

In acidic electrolytes, the Volmer and Heyrovsky steps are coupled with proton transfer processes [1]:

$$H_3O^+ + e^- \rightarrow H_2O + H_{ads} \text{ (Volmer step)} \tag{7.4}$$

$$H_3O^+ + H_{ads} + e^- \rightarrow H_2O + H_2 \text{ (Heyrovsky step)} \tag{7.5}$$

In alkaline environments, the dissociation of the robust covalent H—O bond within water molecules serves as a prerequisite for the formation of adsorbed hydrogen (H_{ads}). This characteristic makes the reduction process considerably more challenging than the reduction of H_3O^+ under acidic conditions. Numerous investigations have demonstrated that four critical factors govern the performance of the HER: the adsorption of water, dissociation of water, hydrogen binding energy, and the adsorption of aqueous OH^-. Consequently, these critical factors should be taken into account during the design of high-efficiency HER electrocatalysts.

The Gibbs free energy of adsorbed H* (ΔG_{H*}) serves as a key descriptor for evaluating both the H* adsorption capability and H_2 desorption efficiency during the HER. Weak H* adsorption results in insufficient interaction between protons and the electrode surface, whereas excessively high ΔG_{H*} values indicate strong bonding between H species and active sites—this strong interaction ultimately impedes the desorption of H_2 molecules [5]. Consequently, electrocatalysts achieve optimal HER activity when their ΔG_{H*} values are close to zero [6]. By integrating density

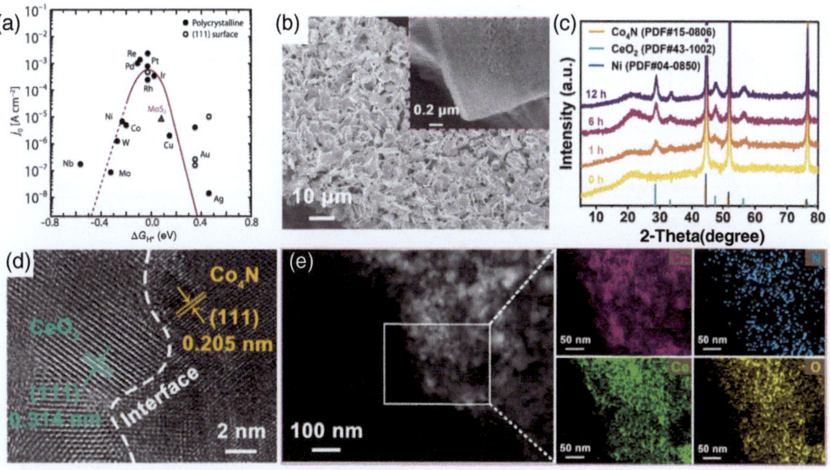

Figure 7.1 (a) Volcano plot of the HER. *Source*: Reproduced from Ref. [3] / with permission of American Chemical Society. (b) SEM image of NF/Co_4N@CeO_2. (c) XRD patterns of NF/Co_4N, NF/Co_4N@CeO_2 (1 h), NF/Co_4N@CeO_2, and NF/Co_4N@CeO_2 (12 h). (d) HRTEM image of Co_4N@CeO_2. (e) STEM-EDX mappings of Co_4N@CeO_2. *Source*: Reproduced from Ref. [4] / with permission of Elsevier.

functional theory (DFT) calculations with experimental data, a classic volcano-type relationship can be established by correlating the experimentally measured current densities of various electrocatalytic materials with their corresponding ΔG_{H^*} values (Figure 7.1a) [3]. This volcano plot not only intuitively reflects the intrinsic HER activity of different catalytic materials but also provides a visual platform for comparing the binding energies of reaction intermediates on various electrode surfaces.

Noble metals (Pt, Rh, Ir) exhibit outstanding HER activity; however, their scarcity and high-cost limit large-scale applications. In recent years, researchers have turned to iron-group materials as alternatives to Pt, driven by three key factors: (i) similar chemical properties, (ii) membership in the same group of the periodic table, and (iii) the low cost and high abundance of nickel [7]. Regulating the electronic structure of catalysts, incorporating atomic configuration, lattice structure, band structure, density of states (DOS), and charge distribution, holds significant importance for balancing the adsorption behavior of reactants, reaction intermediates, and final products [8]. To date, extensive research has focused on designing high-performance HER catalysts by modulating the electronic structure of iron-group compounds.

For instance, Huang et al. put forward taking transition metal (TM) nitride Co_4N as a material design scaffold, and constructed a Co_4N@CeO_2 heterostructure via the strategic introduction of CeO_2 into the Co_4N matrix (Figure 7.1b–e) [4]. In this composite structure, CeO_2 functions as an "electron pump" that draws electrons away from Co_4N, which optimizes the electronic configuration at the heterogeneous interface and thereby endows the catalyst with prominent HER catalytic performance. Electrochemical measurements (Figure 7.2a–c) indicate that the Co_4N@CeO_2 catalyst only needs an overpotential of 49 mV to reach a HER current density of 10 mA cm^{-2}. Moreover, no obvious performance decay was

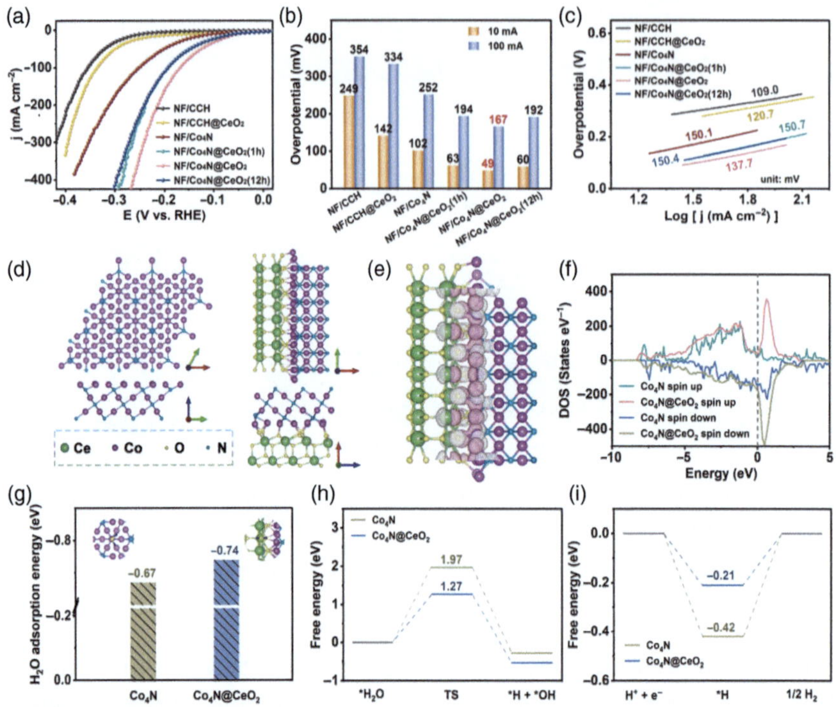

Figure 7.2 (a) Linear sweep voltammetry (LSV) curves. (b) Overpotentials at current densities of 10 and 100 mA cm^{-2}. (c) Tafel slopes. (d) Atomic structure models of Co_4N and $Co_4N@CeO_2$. (e) Interfacial charge density difference profiles of $Co_4N@CeO_2$ (light gray: electron accumulation; light pink: electron depletion). (f) Total DOS of Co_4N and $Co_4N@CeO_2$. The Fermi level is set to zero. (g) Calculated free energies of H_2O adsorption on Co_4N and $Co_4N@CeO_2$. (h) Free energy changes during the H_2O dissociation process. The surface Co sites of Co_4N and the interfacial Ce sites of $Co_4N@CeO_2$ are selected as the adsorption sites for H_2O. (i) Free energy changes during the HER process. The surface Co sites of Co_4N and the interfacial Co sites of $Co_4N@CeO_2$ are selected as the adsorption sites for H intermediates. *Source*: Reproduced from Ref. [4] / with permission of Elsevier.

detected after a 100-hour durability test, manifesting its excellent catalytic stability. DFT calculations (Figure 7.2d–i) comparing $Co_4N@CeO_2$ with pure Co_4N reveal that the tailored interfacial electronic structure of the heterostructure brings two key advantages: on the one hand, it strengthens the adsorption of H_2O; on the other hand, it tunes the Gibbs free energy of H* adsorption to be nearer to the thermoneutral point. Essentially, the regulation effect of CeO_2 on the interfacial electronic structure of $Co_4N@CeO_2$ is the core reason for its enhanced HER activity.

Interface modulation technology serves as a potent strategy for reinforcing the catalytic performance of TM-based materials [9]. Fabricating interfaces on the surface of TM catalysts enables electronic configuration tuning, which not only enriches active sites but also optimizes the chemical adsorption/desorption behaviors of the catalyst surface [10]. Interfaces formed by two phases or compounds with compatible compositions and/or structures possess distinctive atomic and electronic features that differ from their individual components. These unique characteristics induce electronic redistribution and fine-tune the adsorption/desorption intensity of H* on active sites. Rational interface modification of TM catalysts can further create extra active sites, thereby remarkably enhancing the HER catalytic activity [11].

Meanwhile, iron-group phosphides have emerged as promising Pt-free counterparts, attributed to their high earth abundance, facile synthesis routes, as well as excellent reactivity and long-term stability [12]. For clarifying the catalytic mechanisms, it is critical to rationally design interface-modified heterostructures based on iron-group compounds, along with in-depth exploration of interfacial electronic redistribution, electronic structure modulation, and bonding interactions with reaction intermediates.

Lyu et al. successfully fabricated self-supported FeP@CoP core-shell heterogeneous nanoarrays on commercial nickel foam [13]. The layered core–shell architecture not only boosts the specific surface area (SSA) but also facilitates the exposure of abundant surface-active sites. Additionally, the unique self-supported structure ensures close contact between CoP nanowires and the conductive nickel foam substrate, which enhances electron transfer efficiency and mass transport capability—two pivotal factors for achieving outstanding catalytic activity and stability in electrolytic reactions.

HER performance was evaluated in 1 M KOH electrolyte (pH = 14). As shown in Figure 7.3a, the FeP@CoP/NF electrode exhibits superior catalytic activity compared to FeP/NF and CoP/NF electrodes. It requires overpotentials of only 50, 136, and 183 mV to achieve current densities of 10, 50, and 100 mA cm^{-2}, respectively. And significantly lower than those of FeP/NF (97, 164, and 214 mV) and CoP/NF (126, 236, and 298 mV) (Figure 7.3b). This indicates that constructing the FeP@CoP interface remarkably enhances HER activity.

Tafel polarization curves were employed to probe the intrinsic catalytic reaction dynamics of the fabricated electrocatalysts. Consistent with prior discussions, hydrogen evolution in alkaline electrolytes follows either the Volmer-Tafel or Volmer-Heyrovsky pathway, where the Volmer step acts as the rate-determining step (RDS). The theoretical Tafel slope values corresponding to these pathways are 120, 30, and 40 mV dec^{-1}, respectively [14]. As illustrated in Figure 7.3c, the FeP@CoP/NF hybrid catalyst delivered a Tafel slope of 51.1 mV dec^{-1}, which was substantially lower than those of CoP/NF (76.6 mV dec^{-1}) and FeP/NF (61.4 mV dec^{-1}). This result indicates that the FeP@CoP/NF catalyst follows a typical Volmer-Heyrovsky mechanism during the HER process [12], and possesses more favorable reaction kinetics compared to its single-phase CoP/NF and FeP/NF counterparts. The superior catalytic performance of FeP@CoP/NF was primarily ascribed to the robust interfacial interactions between CoP and FeP phases. Such interactions induce interfacial electron redistribution and fine-tune the Gibbs free energy of H* adsorption on the surface-active sites, thereby optimizing the key steps of the HER reaction [11].

In order to validate this conjecture, first-principles simulations were carried out based on the constructed FeP@CoP heterostructure model (Figure 7.3d), aiming to corroborate its outstanding HER catalytic performance. To fully clarify the

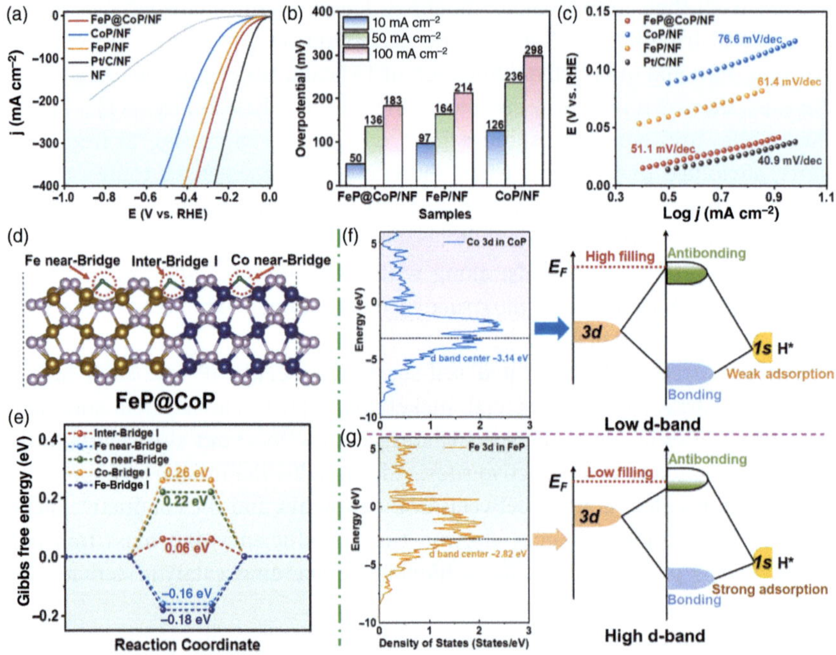

Figure 7.3 HER performance of FeP@CoP/NF in 1 M KOH electrolyte. (a) LSV curves. (b) Overpotential comparison at different current densities. (c) Tafel slopes. (d) Schematic models of H* at Bridge I (interface of FeP@CoP), Co-near-Bridge, and Fe-near-Bridge sites around the interface, where blue, yellow, gray, and green spheres represent Co, Fe, P, and H atoms, respectively. (e) Calculated Gibbs-free energies (ΔG_{H^*}) of adsorbed H* at different adsorption sites. (f) d-PDOS of Co atoms in the Co phase and the corresponding bonding schematic of H atoms at Co sites. (g) d-PDOS of Fe atoms in the FeP phase and the corresponding bonding schematic of H atoms at Fe sites. *Source*: Reproduced from Ref. [13] / with permission of Elsevier.

catalytic roles of CoP and FeP components as well as the interfacial modulation effects in the FeP@CoP hybrid, a comparative analysis of Gibbs free energies was conducted between pure FeP/CoP samples and the interface-engineered sites of the heterostructure.

As presented in Figure 7.3e, the ΔG_{H^*} values of pristine FeP and CoP were calculated to be −0.18 eV and 0.26 eV, respectively. In contrast, the Fe-adjacent and Co-adjacent sites at the FeP@CoP interface exhibited reduced ΔG_{H^*} values of −0.16 and 0.22 eV, indicating more favorable H* adsorption–desorption behaviors than their pure-phase counterparts. Notably, the interfacial site itself achieved the lowest H* adsorption free energy of 0.06 eV, rendering it the optimal catalytic active center for the HER process. Even though the relatively weak adsorption free energy at the interface site is slightly less conducive to H* intermediate adsorption, it still promotes the overall HER kinetics. Therefore, during actual HER operations, H* intermediates may tend to adsorb at regions adjacent to FeP and CoP phases; nonetheless, the ΔG_{H^*} values near the interface are still considerably lower than those of the bridge sites in individual FeP and CoP.

Numerous studies have confirmed that the interaction intensity between reaction intermediates and catalyst surfaces is closely correlated with the electronic configuration of surface catalytically active centers. Modulating the electronic structure within the interfacial region allows precise regulation of the binding energy between H* intermediates and active sites. From the perspective of molecular orbital theory, the interaction between H* species and surface-active centers gives rise to both bonding and antibonding orbitals. According to the d-band center theory, a higher d-band center of active sites leads to the filling of antibonding orbitals at lower energy levels, thereby enhancing the H* adsorption energy. On the contrary, a lower d-band center results in antibonding orbitals being filled at higher energy levels, which weakens the H* adsorption free energy [15].

As shown in Figure 7.3f, g, the simulation results reveal that the d-band center of Co 3d orbitals in pure CoP is −3.14 eV, corresponding to a remarkably weak H* adsorption capacity. For pure FeP, the d-band center of Fe 3d orbitals is −2.82 eV, implying a stronger H* adsorption capability compared to the Co active sites in CoP. This study confirms that the construction of iron-based phosphide interfaces with high crystal plane matching not only optimizes the ΔG_{H^*} values of metal active sites in the vicinity of the interface but also creates novel catalytic active centers, ultimately contributing to the excellent HER catalytic performance.

7.2 OER

Presently, the majority of electrocatalysts employed in industrial OERs are based on Ir and Ru [16]. Yet their scarce earth reserves, cumbersome extraction processes, inferior catalytic stability, and limited operational lifespan led to sustainably high costs. These factors severely impede large-scale commercial application [17]. There is an urgent demand for the development of OER electrocatalysts featuring low cost, long durability, and high performance. This goal still remains a formidable challenge up to now.

The development of high-efficiency OER catalysts to reduce energy consumption relies on clarifying the intrinsic OER mechanism. OER is an inherently complex four-electron transfer reaction with its precise catalytic pathway yet to be fully elucidated. Even so, most proposed OER mechanisms involve oxygen-containing intermediates. These intermediates include hydroxyl ions (OH^-) and oxygen-containing species like O^- and O_2^{2-} [18]. Two main mechanisms are widely accepted for explaining the OER process. They are the adsorbate evolution mechanism (AEM) [19] and the lattice oxygen-mediated mechanism (LOM) [20]. As a traditional OER reaction mechanism, AEM has been extensively applied to explain, simulate, and predict catalytic activity. Its specific reaction routes vary remarkably between acidic and alkaline electrolyte environments [21, 22]:

I. Acidic electrolyte,

$$H_2O + * \rightarrow HO^* + H^+ + e^- \quad (7.6)$$

$$HO^* \rightarrow O^* + H^+ + e^- \quad (7.7)$$

$$O^* + H_2O \rightarrow HOO^* + H^+ + e^- \quad (7.8)$$

$$HOO^* \rightarrow * + O_2 + H^+ + e^- \quad (7.9)$$

II. Alkaline electrolyte,

$$OH^- + * \rightarrow HO^* + e^- \quad (7.10)$$

$$HO^* + OH^- \rightarrow O^* + H_2O + e^- \quad (7.11)$$

$$O^* + OH^- \rightarrow HOO^* + e^- \quad (7.12)$$

$$HOO^* + OH^- \rightarrow * + O_2 + H_2O + e^- \quad (7.13)$$

* Denotes the active sites for catalysis on the electrocatalyst surface, while OH*, O*, and OOH* are three surface-adsorbed reaction intermediates involved in the OER process. In the AEM pathway, the primary factor constraining the overpotential is the adsorption affinity between reaction intermediates and active adsorption sites [23]. Usually, the difference in binding energy between O* and OH* ($\Delta G_{O*} - \Delta G_{HO*}$) acts as a descriptor for evaluating electrocatalytic activity. This value mirrors the variation in the binding intensity of oxygen-containing species with reaction sites. Such variation can be optimized through electronic structure regulation, thereby overcoming the sluggish reaction dynamics [24].

In practical research, a volcano plot is widely adopted to establish the correlation between the OER catalytic activity of different catalysts and the binding energy of O* ($\Delta G_{O*} - \Delta G_{HO*}$). This approach facilitates the efficient evaluation and comparison of catalytic performance across various materials [25]. Although noble metal oxides such as IrO_2 and RuO_2 inherently possess excellent OER activity, recent studies have demonstrated that iron-group oxides also hold substantial application potential in this field [26].

As a fundamental basis for efficient catalyst screening, the AEM closely associates the binding energies of oxygen intermediates with catalytic activity, leading to the attainment of a theoretical overpotential of 370 mV [19], and it cannot account for reports of certain lower overpotentials. As research into the OER mechanism deepens, researchers have put forward an alternative advanced mechanism that incorporates the redox of lattice oxygen to drive the OER cycle [27]. The involvement of lattice oxygen is capable of circumventing the overpotential limitation of AEM, which suggests that metal centers are not the sole source of active sites. This type of OER mechanism is defined as the lattice oxygen-mediated mechanism LOM and its specific steps are as follows:

$$OH^- + * \rightarrow HO^* + H_2O + e^- \quad (7.14)$$

$$HO^* + OH^- \rightarrow O^* + H_2O + e^- \quad (7.15)$$

$$O^* + O_L \rightarrow O_2 + V_O \quad (7.16)$$

$$OH^- + V_O \rightarrow HO^* + e^- \quad (7.17)$$

$$H^* + OH^- \rightarrow * + H_2O + e^- \quad (7.18)$$

Here, OL stands for lattice oxygen, and VO refers to surface oxygen vacancies. The lattice oxygen-mediated mechanism (LOM) features a pivotal process of oxygen vacancy generation. This process triggers metal dissolution, which is the dissociation of metal ions and further results in catalyst deactivation. It is widely regarded as the core cause behind the inferior long-term stability of OER electrocatalysts [18]. Furthermore, metal ions dissolved in the electrolyte might introduce an additional impetus that impairs long-term catalytic performance. This phenomenon may also account for the observed degradation in catalytic activity [28].

As for the fundamental difference between AEM and LOM, it resides in the O—O coupling process. In the LOM pathway, lattice oxygen can directly combine with oxygen intermediates to form O—O bonds, and this characteristic endows LOM with enhanced reaction kinetics [27]. Yet in practical OER systems, these two reaction mechanisms often coexist and engage in mutual competition. Importantly, the covalency of the metal–oxygen bond exerts a significant regulatory effect on the dominant catalytic pathway; as the M—O bond covalency intensifies, the LOM mechanism is more likely to take a dominant position [29]. Although LOM exhibits higher OER activity compared to AEM, it is often accompanied by metal dissolution and structural instability [30]. Therefore, achieving a balance between these two pathways is conducive to the development of electrocatalysts with both high efficiency and excellent stability [31].

Beyond AEM and LOM, an oxidation pathway mechanism (OPM) has also been reported in existing research literature [32]:

$$M + OH^- \rightarrow MOH + e^- \tag{7.19}$$

$$2MOH \rightarrow MO + M + H_2O \tag{7.20}$$

$$2MO \rightarrow 2M + O_2 \tag{7.21}$$

In the oxidation pathway mechanism (OPM) route, O* and OH* act as the sole reaction intermediates. The formation of O—O bonds in this pathway undergoes direct coupling without the participation of lattice oxygen. This characteristic avoids the generation of oxygen vacancies, which in turn suppresses the dissolution of the catalyst. Ideally, the OPM catalytic mechanism is able to break through the activity bottlenecks that restrict the AEM while resolving the stability shortcomings of the LOM throughout the OER process.

Extensive research has shown that OER catalysts experience dynamic structural transformations during catalysis. Multiple reconstruction processes result in the formation of an amorphous layer on the catalyst surface featuring more active sites, structural flexibility, and abundant composition. This layer is also recognized as the catalytically active phase [33]. The self-reconstruction process makes the OER catalytic mechanism more complex. Given the distinct properties of amorphous structures, establishing self-reconstruction and catalytic mechanisms remains a challenge.

As a half-reaction complementary to HER and the reverse process of ORR, OER suffers from sluggish reaction dynamics. This drawback severely limits the efficiency

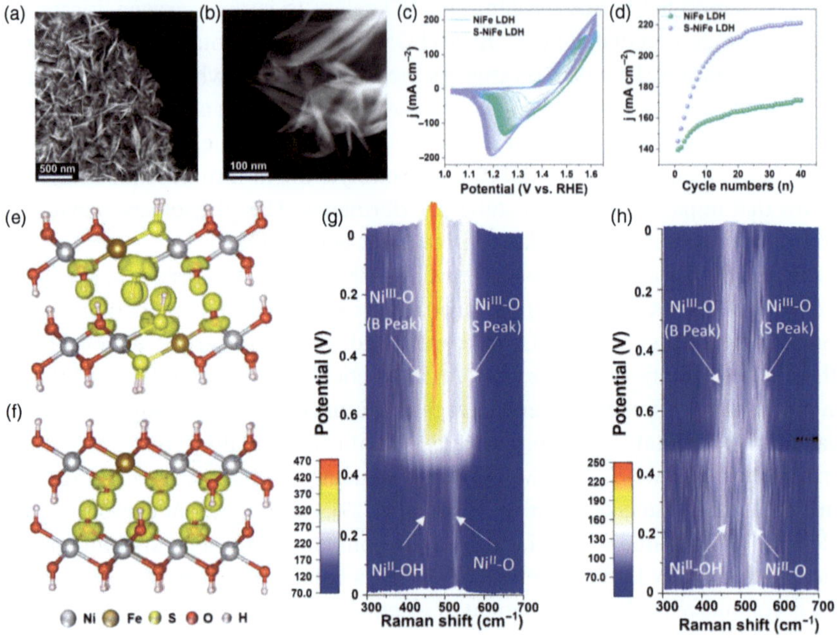

Figure 7.4 (a, b) HAADF-STEM images of S—NiFe LDH. (c) CV curves of 40 cycles at a scan rate of 50 mV s^{-1} in the potential range of 1.025–1.625 V vs. RHE. (d) Current density as a function of CV cycles at 1.625 V vs. RHE. (e) Local electronic structures of S—$Ni_xFe_y(OH)_2$ and (f) $Ni_xFe_y(OH)_2$. (g) In-situ Raman spectra of S—NiFe LDH and (h) NiFe LDH during initial CV cycles. *Source*: Reproduced from Ref. [34] / with permission of John Wiley & Sons.

of hydrogen production via water electrolysis and hinders the practical deployment of metal-air batteries. Consequently, high-efficiency OER catalysts must be characterized by a small Tafel slope and low overpotential. Materials based on Fe, Co, and Ni have demonstrated exceptional OER catalytic performance in alkaline electrolytes, thanks to their abundant earth reserves, low cost, and remarkable activity.

For instance, Mai's research group successfully modulated the intrinsic electronic structure of NiFe LDH through in-situ sulfur doping. Such modifications not only boosted the degree of surface reconstruction of NiFe LDH but also optimized the catalytic activity of the reconstructed active species (Figure 7.4a–f) [34]. The resulting catalyst exhibited superior OER efficiency, requiring merely an overpotential of 219 mV to attain a current density of 10 mA cm^{-2}, alongside a Tafel slope of 24.8 mV dec^{-1}.

Integrating theoretical calculations, in-situ Raman spectroscopy (Figure 7.4g, h), and multiple ex-situ characterization techniques, the team revealed that the introduction of sulfur anions could regulate the electronic structure of NiFe LDH. This regulatory effect effectively reduced the energy barrier for surface reconstruction and facilitated the phase transition of NiFe LDH to highly active S—Ni_xFe_yOOH. The incorporation of sulfur ions significantly enhanced OER activity: the reconstructed S—Ni_xFe_yOOH optimized the adsorption of reaction intermediates, thereby further improving catalytic performance. Specifically, the reconstructed material only needed an overpotential of 259 mV to reach a current density of 100 mA cm^{-2}. Moreover, no obvious performance degradation was detected after a 360-hour stability test, confirming its excellent long-term OER stability.

Fe, Co, and Ni feature unique electronic configurations that endow them with moderate binding affinities toward electrocatalytic intermediates. This inherent characteristic grants them prominent catalytic performance in heterogeneous reactions such as OER. Since the binding strength between active sites and reactants or intermediates directly dictates OER catalytic efficacy, rational optimization of the structure and composition of Fe-, Co-, and Ni-based catalysts proves effective in boosting electrocatalytic performance [36].

For example, porous cobalt oxide nanoplates fabricated via a ligand-aided polyol reduction approach delivered satisfactory OER performance, achieving a current density of 10 mA cm^{-2} at a low overpotential of merely 306 mV (Figure 7.5) [35]. Another case in point is Fe-doped β-$Ni(OH)_2$ nanosheets, which demonstrated outstanding OER catalytic activity—requiring an overpotential of only 219 mV to attain the same current density of 10 mA cm^{-2}. This superior performance stems from their unique wrinkled nanosheet network and special microstructure featuring crystalline-amorphous interfaces. Such structures not only facilitate gas desorption but also provide an unsaturated coordination environment that optimizes the adsorption of OER reactants and intermediates [37]. Despite their inherent suitability for OER, the catalytic activity of Fe Co, Ni-based materials

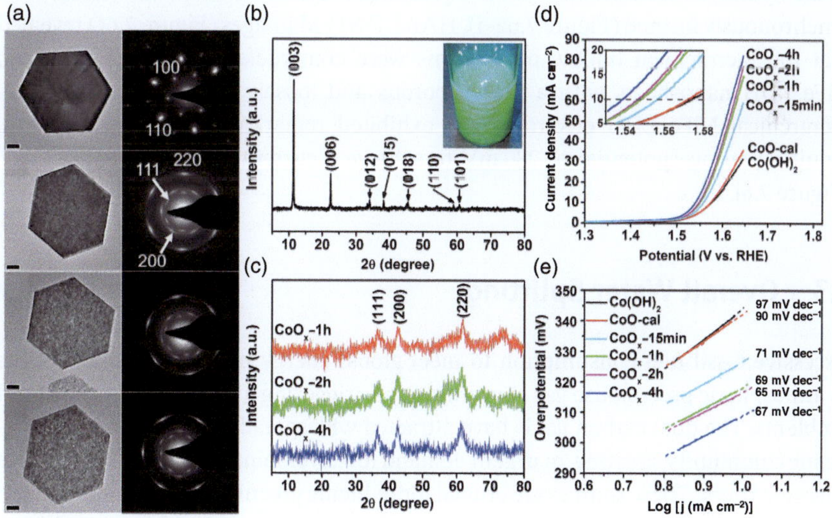

Figure 7.5 (a) TEM images (scale bar: 500 nm) and SAED patterns of α-$Co(OH)_2$ and CoO after reduction for 1, 2, and 4 h (from top to bottom). (b) XRD pattern of α-$Co(OH)_2$ (inset: photograph of α-$Co(OH)_2$). (c) XRD patterns of CoO_x obtained after reduction for 1 h (red), 2 h (green), and 4 h (blue). (d) LSV curves (inset: magnification of the region near η_{10}). (e) Corresponding Tafel plots of the samples. *Source*: Reproduced from Ref. [35] / with permission of Elsevier.

still fails to meet practical demands. In pursuit of high-efficiency TM-based OER catalysts, researchers have employed diverse regulatory tactics, including morphological modulation, defect engineering, and the incorporation of metallic or nonmetallic dopants. These strategies can endow the catalysts with tailored functionalities and remarkably enhance their catalytic capacities.

Co/Ni-based LDHs have emerged as highly promising OER electrocatalyst candidates, owing to their flexibly tunable structural features and chemical compositions, coupled with distinctive physicochemical characteristics and superior catalytic activity. Enhancing the density of active sites is widely recognized as the most direct and effective strategy to promote catalytic performance across various reaction systems, including OER. However, the intrinsic stacked architecture of LDHs restricts the accessibility and exposure of active sites. Exfoliation has been validated as a viable approach to address this challenge.

To boost the OER catalytic activity of CoFe LDH nanosheets, Wang et al. proposed two high-performance liquid-phase and gas-phase exfoliation methods assisted by plasma technology [39]. Specifically, they applied a water-plasma coupled exfoliation strategy to CoFe LDH nanosheets, successfully obtaining ultrathin nanosheet structures. These exfoliated sheets featured highly exposed active sites, as well as abundant defects and vacancies. Such structural merits collectively contributed to excellent OER catalytic performance, delivering an overpotential of merely 232 mV to reach a current density of 10 mA cm^{-2} and a Tafel slope of 36 mV dec^{-1}.

In another study, Lou et al. first synthesized Ni precursor nanoprisms, which served as self-consuming templates [38]. These templates were gradually consumed in the hydrolytic environment of $FeSO_4$, during which NiFe LDH nanosheets were synchronously formed (Figure 7.6a–i). HAADF-STEM images (Figure 7.6f) revealed that the hierarchical hollow nanoprisms were constructed from interconnected NiFe LDH nanosheets with a highly porous and loosely packed structure. This hierarchical NiFe LDH electrocatalyst exhibited remarkable OER performance, requiring an overpotential of 280 mV to achieve a current density of 10 mA cm^{-2} (Figure 7.6j, k).

7.3 Overall Water Splitting

Excessive fossil fuel consumption to meet global energy needs has led to a surge in atmospheric greenhouse gas concentrations, triggering a series of environmental problems. The dual carbon goals have attracted widespread attention from the academic community, creating an urgent demand for developing sustainable alternative energy sources. These sources are crucial for reducing greenhouse gas emissions and replacing nonrenewable energy [40]. As a clean energy carrier, hydrogen H_2 boasts a high energy density of 142 MJ kg^{-1} along with zero-carbon emissions and renewability, making it an ideal and efficient alternative to fossil fuels [41]. Yet hydrogen synthesis still faces bottlenecks. Conventional industrial hydrogen production relies on fossil fuel steam reforming and partial oxidation, which demands high temperatures, 950°C–1100°C, for water splitting, emits carbon dioxide, and depletes fossil

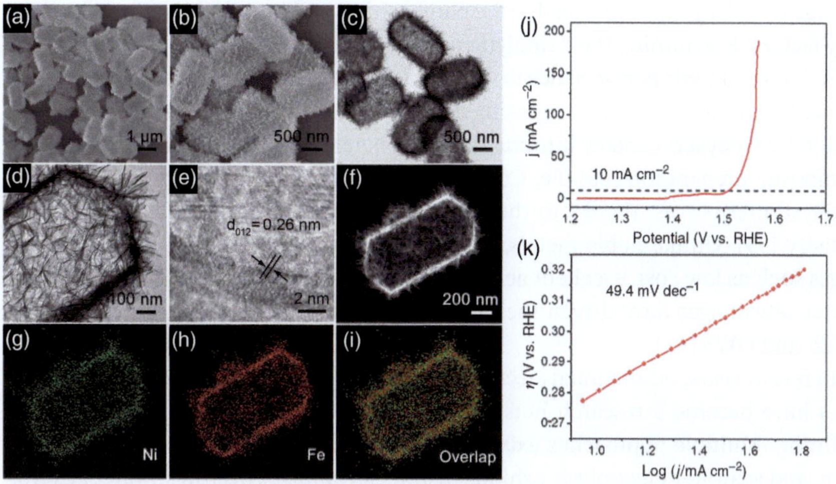

Figure 7.6 (a, b) FESEM images and (c, d) TEM images of NiFe hollow pillars after the hydrolysis reaction. (e) Lattice fringes of the NiFe LDH shell. (f–i) HAADF-STEM image and corresponding elemental mapping images of a single NiFe LDH hollow prism. (j) LSV curves and (k) Tafel plots of NiFe LDH hollow prisms on a glassy carbon (GC) electrode. *Source*: Reproduced from Ref. [38] / with permission of John Wiley & Sons.

fuels. This route yields so-called gray hydrogen. In contrast, green hydrogen refers to high-purity hydrogen produced via electrochemical water splitting using wind, solar, and other renewable energies without generating harmful by-products [42].

In electrochemical water splitting, the cathodic HER and anodic OER are the two core half-cell reactions. Under reversible hydrogen electrode (RHE) conditions, the thermodynamic equilibrium potential TEP for water splitting is 1.23 V [43]. However, the slow reaction kinetics at both electrodes reduce the overall efficiency of water electrolyzers. To overcome kinetic barriers, the operating potential must be significantly higher than the TEP [44]. Termed as overpotential (η), this supplementary potential can be mitigated to a minimum by adopting high-efficiency electrocatalysts, which accelerate the sluggish electrode reaction kinetics [45]. Currently, Pt-based catalysts and Ir/Ru oxides act as benchmark materials for HER and OER, respectively, owing to their outstanding catalytic capabilities. Nevertheless, their scarcity and prohibitive cost have posed a major barrier to large-scale commercial advancement [16].

Thus, research focus has shifted toward developing electrocatalysts with high activity, low cost, and long-term stability. In recent years, numerous electrocatalysts with excellent water electrolysis performance have been reported, encompassing TM oxides [46], hydroxides [47], nitrides [48], sulfides [49], phosphides [50], carbides [51], selenides [52], metal-organic frameworks (MOFs) [53], and heteroatom-doped carbon materials [54], all of which have undergone in-depth research. Generally, porous structures with large SSA, high electrical conductivity, and large pore volume are conducive to excellent electrocatalytic water splitting performance. Such structures provide abundant active sites and enable efficient charge transfer [55]. Additionally, the composition and morphology of catalysts are key factors determining their catalytic performance [56].

To boost catalytic performance to a higher level, researchers have employed a suite of modification tactics, encompassing defect engineering [57], heteroatom incorporation [58], oxygen vacancy introduction [59], heterostructure fabrication, and alloy composite preparation [60]. Fe, Co, and Ni stand out as highly promising electrocatalyst alternatives, thanks to their superior stability and robust electrocatalytic activity [59]. Unlike noble metals, iron-group compounds possess obvious advantages such as low cost, excellent activity, and abundant reserves in the earth's crust. These advantages have driven their extensive application in the research of HER, OER, and OWS [61].

In recent years, bifunctional electrocatalysts integrating both HER and OER activities have become a research hotspot for efficient hydrogen production via water splitting. Multiple approaches exist for water electrolysis-based hydrogen production, and seawater electrolysis exhibits distinct advantages over freshwater electrolysis, thus becoming an intensively studied direction [7]. Significant progress has been made in developing low-cost, high-performance bifunctional HER/OER electrocatalysts based on Fe Co Ni materials. For example, Ghasem et al. prepared a three-dimensional 3D heterostructured MnCo/NiSe/NF catalyst through electrodeposition, which enables high-performance bifunctional water and seawater electrolysis [62].

As illustrated in Figure 7.7a, b, the MnCo/NiSe catalyst achieves ultra-low overpotentials of 22.1 mV at current densities of −10, −100, −500, and −1000 mA cm^{-2} in 1 M KOH with a Tafel slope of 45.0 mV dec^{-1}. In alkaline seawater electrolyte, this catalyst shows an overpotential of 31.4 mV at −10 mA cm^{-2} and a Tafel slope of 58.2 mV dec^{-1}. These findings confirm that MnCo modification remarkably boosts the HER performance of NiSe. Additionally, MnCo/NiSe exhibits excellent OER activity as shown in Figure 7.7c, d. It only needs an overpotential of 225.8 mV to reach 10 mA cm^{-2} in alkaline aqueous solution and 261.3 mV in alkaline seawater. The OER Tafel slopes of the MnCo/NiSe catalyst reach as low as 69.7 mV dec^{-1} in alkaline aqueous solution and 81.2 mV dec^{-1} in alkaline seawater. Notably, the two-electrode electrolyzer assembled with MnCo/NiSe as both cathode and anode (MnCo/NiSe || MnCo/NiSe) exhibits low cell voltages across different electrolyte systems. Specifically, to achieve current densities of 10, 100, 500, and 1000 mA cm^{-2}, the required cell voltages are 1.41, 1.67, 1.88, and 1.92 V in 1.0 M KOH electrolyte, respectively. In the mixed electrolyte of 1.0 M KOH + seawater, the corresponding cell voltages are 1.44, 1.82, 1.99, and 2.11 V (Figure 7.7e). Chronopotentiometry tests were performed to assess the long-term stability of the MnCo/NiSe || MnCo/NiSe electrolyzer (Figure 7.7f). Continuous operation at 500 mA cm^{-2} in 1.0 M KOH + seawater for 200 hours resulted in only a slight increase in cell voltage. The voltage elevation was less than 8%, confirming that the MnCo/NiSe-based electrolyzer possesses excellent long-term operational stability.

DFT was carried out to clarify the intrinsic mechanism through which MnCo enhances the catalytic activity of NiSe. The MnCo/NiSe heterostructure exhibits a higher carrier density near the Fermi level. Specifically, in this Co-dominated

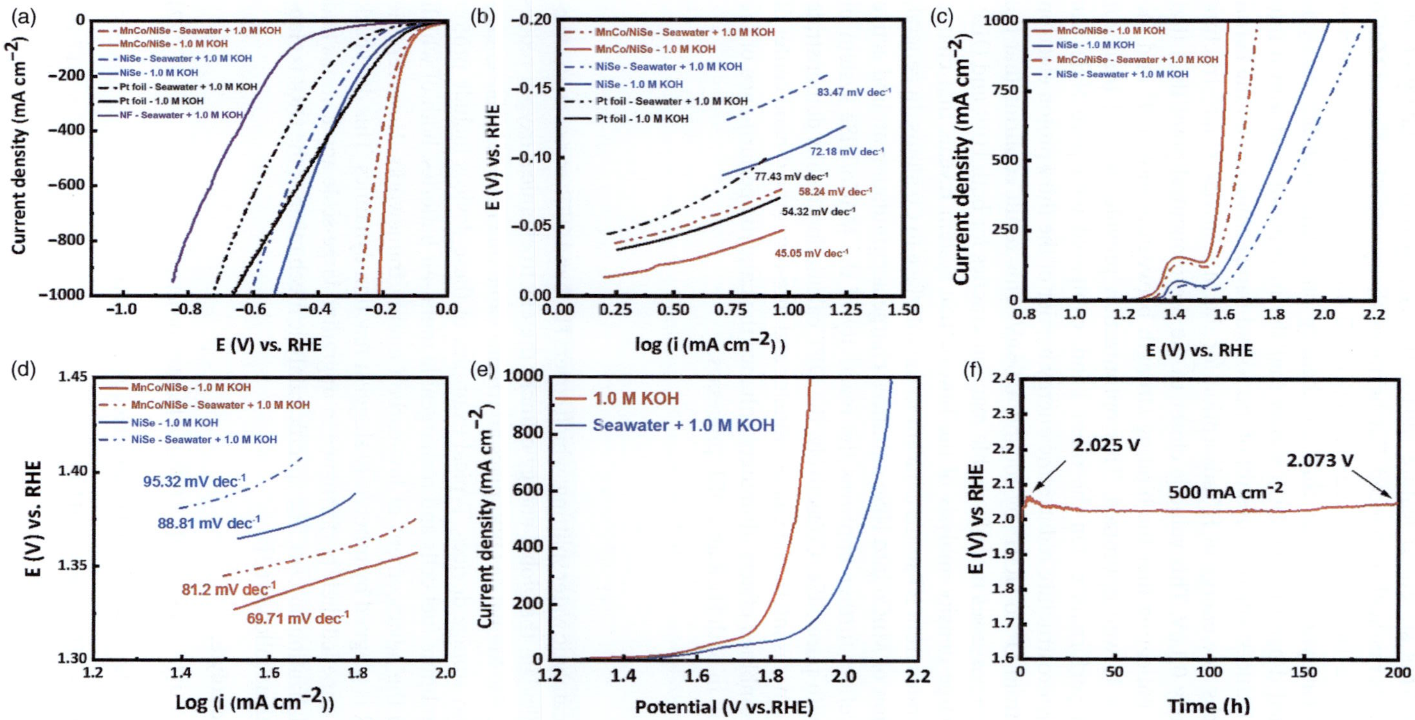

Figure 7.7 (a) LSV curves of nickel foam (NF), MnCo/NiSe, Pt foil, and NiSe electrodes in 1.0 M KOH and 1.0 M KOH + seawater electrolytes at a scan rate of 5 mV s^{-1}. (b) Tafel slopes of the electrocatalysts derived from the LSV curves. (c) Polarization curves for the OER of MnCo/NiSe and NiSe electrocatalysts in 1.0 and 1.0 M KOH + seawater electrolytes at a scan rate of 5.0 mV s^{-1}. (d) Tafel slopes of the electrocatalysts derived from the LSV curves. (e) LSV curves for overall water splitting of the MnCo/NiSe || MnCo/NiSe system in 1.0 and 1.0 M KOH + seawater electrolytes. (f) Chronopotentiometric stability test of the MnCo/NiSe || MnCo/NiSe cell at a constant current density of 500 mA cm^{-2} in 1.0 M KOH + seawater electrolyte (without iR correction). *Source*: Reproduced from Ref. [62] / with permission of Elsevier.

system, the d states are positioned closer to the Fermi level compared with those in pure NiSe. This structural feature means the electrical conductivity of the heterostructure is markedly enhanced. Furthermore, the synergistic effect of MnCo further boosts the DOS of d orbitals related to Ni atoms. This enhancement is beneficial for catalytic processes that rely heavily on charge transfer. Analysis of the projected d-band partial DOS shows that the incorporation of MnCo into pristine NiSe induces a distinct upshift of the NiSe d-band center toward the Fermi level. The Gibbs free energy of H* adsorption for the MnCo/NiSe heterostructure is approximately 0.1 eV. This value is closer to the thermoneutral point than that of pure NiSe, indicating that the binding strength between adsorbed H* and the catalyst surface is well optimized. This optimization provides a solid basis for efficient HER performance. On the other hand, compared with pure NiSe, the MnCo/NiSe heterostructure reduces the energy barrier of the third proton electron transfer step from O* to OOH* by roughly 0.06 eV. This result confirms that the heterostructure possesses more favorable reaction kinetics for both HER and OER. Difference charge density analysis of the MnCo/NiSe system reveals that charge primarily migrates from Ni atoms adjacent to the interface to Co atoms. In general, the introduction of MnCo into NiSe enhances charge accumulation around active species. This effect further improves the reactivity of the MnCo/NiSe system in comparison with pure NiSe. Collectively, the DFT calculation results demonstrate that MnCo promotes interfacial charge transfer and fine-tunes the d-band center of NiSe. These combined effects ultimately enhance the catalytic performance of the heterostructure in both HER and OER processes.

7.4 ORR

While electrocatalytic water splitting, which covers HER and OER, serves as an effective way to alleviate the global energy crunch, the electrochemical oxygen reduction reaction, also known as ORR, plays an irreplaceable role in sustainable energy conversion and storage devices. Typical examples of these devices include proton exchange membrane fuel cells and rechargeable metal-air batteries, both of which are critical to the advancement of low-carbon energy frameworks. However, the cathodic ORR is plagued by inherently sluggish reaction dynamics. This drawback has seriously impeded their performance in upgrading large-scale applications and the commercialization process. The specific catalytic mechanism of the ORR is elaborated in detail as follows [63]:

I. Acidic electrolyte,

$$O_2 + 4H^+ + 4e^- \rightarrow 2H_2O \text{ (four-electron process)} \tag{7.22}$$

$$O_2 + 2H^+ + 2e^- \rightarrow H_2O_2 \text{ (two-electron process)} \tag{7.23}$$

$$H_2O_2 + 2H^+ + 2e^- \rightarrow 2H_2O \tag{7.24}$$

II. Alkaline medium,

$$O_2 + 2H_2O + 4e^- \rightarrow 4OH^- \text{ (four-electron process)} \tag{7.25}$$

$$O_2 + H_2O + 2e^- \rightarrow HO_2^- + OH^- \text{ (two-electron process)} \tag{7.26}$$

$$H_2O + HO_2^- + 2e^- \rightarrow 3OH^- \tag{7.27}$$

Proton exchange membrane fuel cells commonly referred to as PEMFCs have entered the commercialization phase as clean energy devices. Even so, the ORR still ranks as the rate-limiting step in PEMFC reaction mechanisms. This is mainly due to the high overpotential associated with ORR and its inherently sluggish reaction dynamics. Pt-based materials are well recognized for their superior electrochemical activity toward ORR in PEMFCs. They have also been widely employed in the hydrogen oxidation reaction at the anode and ORR at the cathode of such devices [64].

The prohibitive manufacturing cost of Pt-based PEMFCs remains a major bottleneck. This cost burden stems from the scarce reserves and high market price of noble metals. Thus, the core challenge in advancing PEMFC technology lies in developing low-cost electrocatalytic electrodes that can deliver enhanced electrocatalytic performance. Fe-based materials have drawn extensive research attention in recent years owing to their abundant reserves in the Earth's crust.

Recent research findings have highlighted iron-group compounds as promising alternatives to commercial Pt-based ORR catalysts. Among these, TM-nitrogen-carbon catalysts, known as TM—N—C materials, with Fe as the primary TM, have become a research focus for addressing ORR's kinetic sluggishness. In a representative study, Xu and his colleagues employed ZIF-L nanorods as self-sacrificing templates. They obtained ZIF-8 hollow nanorods with an open-pore structure through a ZIF phase transition strategy. Subsequent controlled carbonization and iron atom immobilization processes transformed these ZIF-8 hollow nanorods into atomically dispersed Fe—N—C hollow nanorods, denoted as Fe_1—N—C HNRs, Figure 7.8 [65].

Electrochemical characterizations verify that Fe_1—N—C HNRs exhibit excellent ORR activity across both alkaline and acidic media, outperforming commercial Pt/C and pristine NC HNRs in multiple key indicators. Under alkaline conditions, Fe_1—N—C HNRs achieve an initial potential of 1.13 V vs. RHE, a half-wave potential of 0.91 V vs. RHE, and a limiting current density of 6.0 mA cm^{-2}. At 0.85 V vs. RHE, its kinetic current density reaches 148 mA cm^{-2}, with a Tafel slope of 69 mV dec^{-1} and a turnover frequency of 0.23 e $site^{-1}$ s^{-1}. All these metrics are significantly better than those of commercial Pt/C and NC HNRs. Koutecky-Levich equation analysis and rotating ring-disk electrode tests confirm that Fe_1—N—C HNRs follow a four-electron ORR pathway. The average electron transfer number is close to 4, and the H_2O_2 yield remains below 1% in the 0.2–0.8 V vs. RHE range, showing higher reaction efficiency than commercial Pt/C. In acidic electrolyte, Fe_1—N—C HNRs still maintain prominent catalytic performance with a half-wave potential of 0.8 V vs. RHE, a limiting current density of 6.7 mA cm^{-2}, and a Tafel slope of 100 mV dec^{-1}. These values are superior to commercial Pt/C. Its kinetic current density and turnover frequency at 0.80 V vs. RHE also exceed those of Pt/C and NC HNRs, with the electron transfer number ranging from 3.91 to 3.95 and H_2O_2 yield below 5%. Practical application tests in liquid zinc-air batteries show that Fe_1—N—C HNRs-based air cathodes deliver an open-circuit voltage of about 1.46 V,

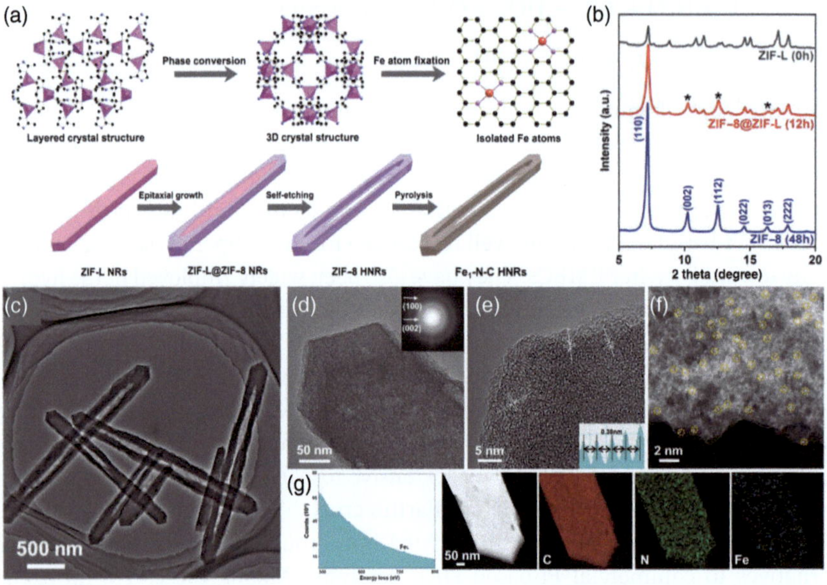

Figure 7.8 (a) Schematic illustration of the synthesis process of Fe_1–N–C HNRs. (b) XRD pattern and SEM image. (c, d) TEM images with SAED patterns as insets. (e) HRTEM image and comparative intensity profile of the (002) lattice fringes. (f) HAADF-STEM image, where yellow dotted circles indicate isolated Fe atoms. (g) EELS spectrum, HAADF-STEM image, and corresponding elemental mapping images. *Source*: Reproduced from Ref. [65] / with permission of John Wiley & Sons.

a maximum power density of 208 mW cm^{-2}, and a specific capacity of 818 mAh $g_{Zn}{}^{-1}$ at 20 mA cm^{-2}. All these performances surpass Pt/C-based batteries. They also exhibit excellent discharge stability, with the discharge voltage nearly recovering to the initial value when the current density is reduced to 2 mA cm^{-2}. The outstanding ORR activity and stability of Fe_1—N—C HNRs stem from four core advantages. First, the unique one-dimensional hollow carbon structure with abundant micro and mesopores and open cavities facilitates electron and electrolyte transfer and promotes active-site exposure. Second, the impregnation and anchoring strategy ensures uniform dispersion of single Fe atoms on the carbon framework, providing sufficient accessible Fe—N_4 active sites. Third, high-density pyridinic and graphitic N doping regulates the microenvironment of isolated Fe atoms, reducing the ORR activation energy barrier. Fourth, the high graphitization degree of the carbon matrix enhances electrical conductivity and the structural stability of Fe—N_4 sites.

Lu et al. designed and prepared a high-performance bifunctional ORR/OER electrocatalyst (Fe—N—C/Fe_3C) via a seed-mediated method, which uses Fe_3C as the substrate and uniformly anchors Fe single atoms (SAs). This work offers important references for the development of high-activity, low-cost non-noble metal electrocatalysts [26]. Electrocatalytic ORR tests were conducted in an O_2-saturated 0.1 M KOH electrolyte. As shown in Figure 7.9a, by regulating the electronic structure of Fe via tuning the Fe_3C-to-Fe SA ratio, the Fe—N—C/Fe_3C catalyst achieves

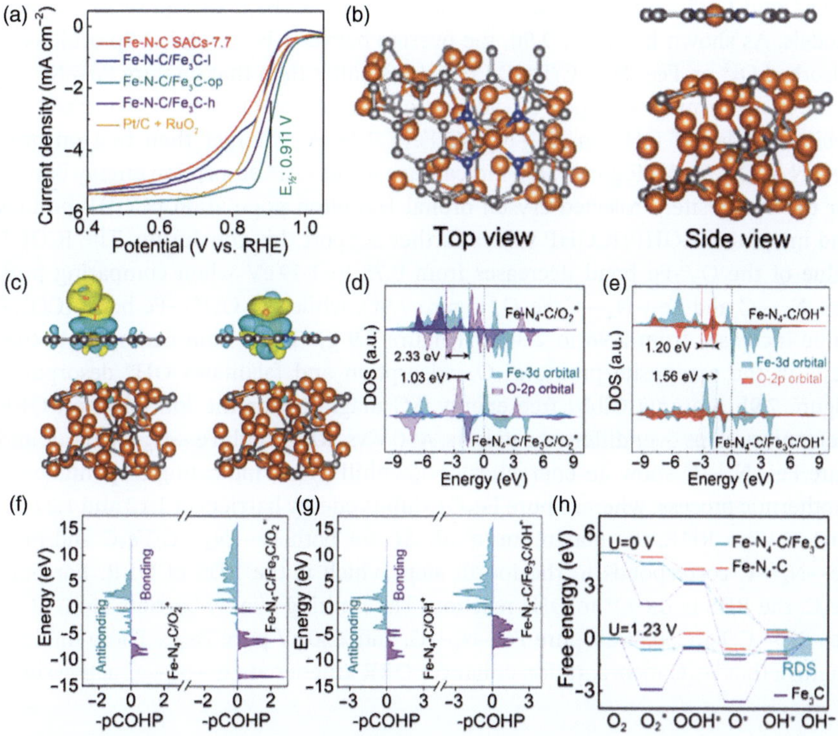

Figure 7.9 (a) LSV curves for ORR, and (b) structural model of the Fe_3C-modified $Fe-N_4-C$ electrocatalyst ($Fe-N_4-C/Fe_3C$). (c) Differential charge density maps of the $Fe-N_4-C/Fe_3C$ model with adsorbed O_2 (O_2*) and (d) adsorbed OH^- (OH*), where cyan and yellow contour lines represent charge depletion and charge accumulation in real space, respectively. (e) Projected crystal orbital Hamilton population (pCOHP) plot of O_2^* and f) OH*. (g) ΔG diagrams for ORR on Fe sites at U = 0 V and 1.23 V vs. RHE. *Source*: Reproduced from Ref. [26] / John Wiley & Sons / CC BY 4.0.

a half-wave potential ($E_{1/2}$) of 0.911 V vs. RHE, outperforming the benchmark Pt/C + RuO_2 ($E_{1/2}$ = 0.874 V vs. RHE). These results indicate that optimizing the Fe_3C-to-Fe SA ratio is beneficial to regulating the electronic structure of $Fe—N_4—C$ active sites, thereby improving catalytic activity.

To clarify the distinct roles of $Fe—N_4—C$ and Fe_3C in ORR activity, DFT calculations were performed on the geometrically optimized $Fe—N_4—C/Fe_3C$ composite model (Figure 7.9b). TM centers in SA catalysts typically exhibit strong O_2 binding energy and high energy barriers for the four-electron ORR process, resulting in suboptimal catalytic performance. Electrochemical tests have confirmed that the interfacial effect from Fe_3C can reduce the ORR energy barrier at $Fe—N_4—C$ active sites. To verify this, different charge densities of key ORR intermediates O_2* and OH* adsorbed on $Fe—N_4—C/Fe_3C$ were calculated (Figure 7.9c). Results show that charge transfer from Fe_3C to $Fe—N_4—C$ enhances the electronic interaction between $Fe—N_4—C$ and O_2*/OH*, thereby promoting OOH* intermediate formation. For qualitative analysis of electronic interactions on bond strength, DOS of O_2* and OH* were computed for both Fe—N_4—C/Fe_3C and pure Fe—N_4—C models. As shown in Figure 7.9d, the overlap between Fe 3d and O 2p orbitals of adsorbed O_2* on Fe—N_4—C/Fe_3C 1.03 eV is smaller than that on pure Fe—N_4—C 2.33 eV, confirming stronger O_2 adsorption on the composite. In contrast, the orbital overlap of OH* on Fe—N_4—C/Fe_3C 1.56 eV is larger than that on pure Fe—N_4—C 1.20 eV (Figure 7.9e), indicating a lower OH* desorption energy barrier for the composite. Projected crystal orbital Hamilton population pCOHP profiles and integrated COHP/ICOHP values further support this conclusion. The ICOHP value of the O_2*-Fe bond decreases from 0.75 to 1.79 eV when comparing pure Fe—N_4—C and Fe—N_4—C/Fe_3C (Figure 7.9f), while the OH*—Fe bond ICOHP value increases from 2.96 to 2.34 eV (Figure 7.9 g). These data demonstrate that Fe_3C in the composite promotes O_2 adsorption and facilitates OH* desorption. Figure 7.9h presents Gibbs free energy ΔG diagrams for the four-electron ORR elementary steps over different catalysts. At 0 V vs. RHE, both Fe—N_4—C/Fe_3C and pure Fe—N_4—C show an energetically downhill trend, indicating a spontaneous exothermic process, whereas pure Fe_3C exhibits energy barriers of 1.12 and 1.27 eV. At 1.23 V vs. RHE, the maximum uphill ΔG for both Fe—N_4—C/Fe_3C and pure Fe—N_4—C corresponds to the fourth step, which is the RDS of ORR. For pure Fe_3C, the RDS is the O* to OH* reaction. The limiting energy barriers are 0.73 eV Fe—N_4—C/Fe_3C, 0.86 eV pure Fe—N_4—C, and 2.50 eV pure Fe_3C. These findings confirm that Fe_3C incorporation enhances ORR kinetics at Fe—N_4—C active sites.

7.5 NRR

Ammonia (NH_3) is a vital chemical raw material extensively applied in industrial and agricultural production, including the manufacture of aqueous ammonia, dyes, plastics, explosives, and fertilizers [66]. The global annual ammonia production surpasses 200 million tons, with 80% of this output devoted to fertilizer synthesis. This application serves a crucial function in the rapid development of the global economy and sustained population growth [67]. Additionally, NH_3 is deemed a potential hydrogen carrier (17.6 wt% H content) on account of its high energy density (4.3 kWh kg^{-1}), zero-carbon attribute, and ease of transportation, which can be liquefied at −33°C [68]. Industrial ammonia production predominantly relies on the Haber-Bosch technology, which utilizes nitrogen and hydrogen as feedstocks. This method demands extreme reaction conditions, including elevated pressure of 200–400 atm, high temperature of 400–550°C, and iron-based catalytic systems [69]. A critical drawback lies in the hydrogen source: most reactant H_2 comes from industrial gray hydrogen, derived from high-temperature reactions between water and carbonaceous materials like coal, CO, methane, and methanol. Such dependency makes the Haber-Bosch process energy-intensive and a major greenhouse gas emitter, contributing 2% of global energy consumption and 1% of total greenhouse gas releases worldwide. Drawing inspiration from biological nitrogenases, ambient temperature and pressure, ammonia synthesis has emerged as a cost-effective and environmentally benign alternative [70]. Among diverse approaches, the electrochemical NRR stands out: it enables NH_3 formation via electrochemical reactions between N_2 and water under mild conditions, with no greenhouse gas emissions. As a promising replacement for the Haber-Bosch process, NRR highlights the pressing demand for high-performance electrocatalysts to boost its reaction efficiency [71].

Nitrogen atoms have five valence electrons (electronic configuration: $2s^2 2p^3$) and three unoccupied 2p orbitals, making isolated N atoms highly reactive. Nevertheless, orbital hybridization between two N atoms forms the extremely stable N_2 molecule. Breaking the N≡N triple bond requires an energy input of 945 kJ mol^{-1}, which explains the harsh conditions of the Haber-Bosch process [67]. Additionally, N_2 features a wide band gap (10.82 eV) between the lowest unoccupied molecular orbital (LUMO) and the highest occupied molecular orbital (HOMO), along with a high ionization energy (15.58 eV), both of which hinder electron transfer [67].

The fundamental mechanisms of N_2 conversion to NH_3 are generally divided into two types: dissociative and associative pathways [72]. Employed primarily in the Haber-Bosch process, the dissociative pathway demands that significant energy be input to first cleave the N≡N triple bond, generating isolated N atoms that subsequently combine with H atoms to form NH_3. In contrast, hydrogenation is involved in the associative pathway prior to the formation of the first NH_3 molecule, and the N—N bond remains intact during the initial steps of the reaction.

The associative pathway can be further categorized into end-on and side-on modes based on the bonding configuration between N_2 and catalyst active sites [72]. In the end-on mode, only one N atom binds to the active site; in the side-on mode, both N atoms interact with the active center. Depending on the hydrogenation sequence, the end-on mode encompasses three sub-pathways for NH_3 generation: distal, alternating, and hybrid pathways, while the side-on mode exclusively follows the enzymatic pathway.

In the distal pathway, proton–electron pairs (PET pairs) first attack the N atom of N_2 that is farther from the active site, reducing it to NH_3. Facilitating electron exchange between N_2 and the catalyst surface is crucial here. After N_2 adsorption, electrons are donated from its bonding orbitals to form one NH_3 molecule, followed by hydrogenation of the remaining N atom. Within the alternating pathway, PET pairs react successively with the two N atoms, and the second NH_3 is liberated after the first has been released. The hydrogenation mechanism of the enzymatic pathway is analogous to that of the alternating pathway. The hybrid pathway, a variant of the end-on mode, switches from the distal to the alternating pathway during the third hydrogenation step, driven by changes in relative free energy [72].

For electrochemical NRR, choosing suitable catalysts to offer favorable active sites is critical for enhancing N_2 adsorption. TMs are capable of adsorbing and activating N_2 by leveraging their abundant d-orbital electrons and unoccupied orbitals to drive the "acceptor-donor" process of N_2 [73]. Moreover, regulating the charge density redistribution of catalysts is another viable approach to strengthen N_2 adsorption, including vacancy engineering [74], edge adsorption [75], heterojunction fabrication [76], and heteroatom doping [77]. Naturally, catalysts with a large SSA also serve as an effective way to boost N_2 adsorption by exposing more active sites [78].

Another key factor is the N_2 concentration in the reaction medium, as dynamic gas–liquid–solid interfacial mass transfer can significantly enhance N_2 adsorption efficiency [79].

N_2 activation mostly adheres to the "acceptor-donor" principle, but most catalysts lack the intrinsic capability to directly activate N_2. Consequently, an applied negative electrode potential is required to drive electron injection into the π^* antibonding orbitals of N_2, achieving complete activation [67]. This critical activation step is generally accompanied by a free energy increase, which is closely associated with the formation of $^*N_2$ or $^*N_2H$ intermediates [72].

At the hydrogenation stage, the catalytic microenvironment exerts a regulatory effect on the hydrogenation sequence and reaction pathway, thereby generating diverse products like NH_3, N_2H_4, and H_2 [67]. For NH_3 desorption, DFT simulations reveal that the instability of the $^*NH_2$ intermediate—particularly on active sites with strong binding ability—plays a pivotal role. It not only reduces the free energy barrier for $^*NH_3$ formation but also lowers the final desorption energy [72]. Practically, the high-water solubility of generated NH_3 effectively releases active sites, enabling the subsequent catalytic cycle to proceed smoothly.

Throughout the NRR process, continuous electron transfer occurs, making a conductive catalyst substrate essential. This ensures adequate electron supply to active sites, sustaining the N_2 reduction reaction [67]. However, NRR efficiency remains low due to sluggish reaction kinetics and the ubiquitous competitive HER, severely hindering its scalable application. This underscores the critical demand for high-performance electrocatalysts. Thus, developing electrocatalysts with high activity, long-term stability, and low cost to accelerate NRR kinetics and inhibit the HER side reaction is of paramount importance.

Noble metal-based catalysts often exhibit outstanding NRR performance, but their high cost and limited reserves restrict large-scale industrial application [80]. This has driven research efforts toward non-noble metal alternatives [81]. For catalyzing the NRR, a key challenge is the activation of the inert N≡N bond. TM-based materials offer a promising solution due to their electronic structure, which includes numerous d-orbital electrons and vacant orbitals capable of facilitating this activation [82]. Electrocatalysts based on iron-group elements (Fe, Co, Ni) are considered ideal substitutes for noble metals [83]. Additionally, iron-group elements are highly reactive and can easily form compounds with excellent catalytic activity and stability through combination with other elements. Notably, iron is a component of biological nitrogenases and is also widely utilized in the industrial Haber-Bosch ammonia synthesis process.

In the search for efficient NRR catalysts, TM-based materials, including oxides, nitrides, carbides, and sulfides, are commonly studied. Metal oxides have garnered significant research interest owing to their merits, which include ease of large-scale synthesis, adjustable catalytic activity, favorable stability, and environmental compatibility. DFT calculations by Nguyen et al. revealed that on the $Fe_2O_3(0001)$ surface, N_2 is adsorbed and activated by Fe ions via an associative mechanism. Notably, the first hydrogenation step, which determines the reaction potential, was calculated to have an energy barrier of 1.14 eV. [84]. Complementing this theoretical work,

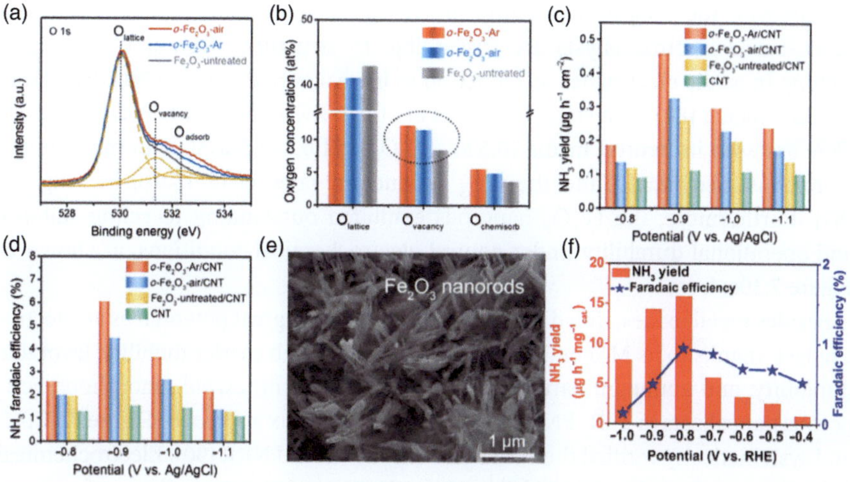

Figure 7.10 (a) O 1 s XPS spectra of Fe_2O_3–Ar, Fe_2O_3-Air, and untreated Fe_2O_3. (b) Concentration comparison of three types of oxygen species derived from the O 1 s XPS spectra. (c) Average NH_3 production rates of different catalysts and (d) their corresponding FEs. *Source*: Reproduced from Ref. [86] / with permission of John Wiley & Sons. e) SEM image of the Fe_2O_3 nanorod catalyst. (f) NH_3 yields and FEs of the Fe_2O_3 nanorod catalyst at different applied potentials in 0.1 M Na_2SO_4 electrolyte. *Source*: Reproduced from Ref. [87] / with permission of John Wiley & Sons.

experimental research by Kong et al. reported that γ-Fe_2O_3 nanoparticle catalysts could produce ammonia from N_2 at temperatures at or below 65°C in 0.1 M KOH. The reported NH_3 yield and Faradaic efficiency for this process were 12.5 nmol h^{-1} mg_{cat}^{-1} and 1.9%, respectively [85].

Modulating the surface structure and electronic properties of catalysts represents a pivotal approach to boosting ammonia (NH_3) production yield and Faradaic efficiency (FE) in the electrocatalytic NRR. Introducing defects to tailor charge density distribution and construct extra active sites has been proven particularly effective. For instance, Cui et al. successfully tailored the surface oxygen vacancy (O_v) concentration of Fe_2O_3 through annealing treatment under an Ar atmosphere [86]. The as-prepared carbon nanotube-supported α-Fe_2O_3_Ar/CNT catalyst exhibited remarkably improved NRR activity in 0.1 M potassium hydroxide (KOH) electrolyte, achieving an NH_3 yield of 0.46 μg h^{-1} cm^{-2} and a FE of 6.04% at a potential of −0.9 V vs. RHE, as depicted in Figure 7.10a–d.

The adoption of highly conductive substrates also plays an indispensable role in optimizing NRR performance. Such substrates can provide abundant electrons, which not only enhance the likelihood of electron injection into N_2 antibonding orbitals but also facilitate the subsequent nitrogen reduction process [88]. Chen et al. conducted systematic investigations on the regulatory effect of catalyst supports, and their results indicated that Fe_2O_3 anchored on carbon nanotubes (Fe_2O_3-CNT) achieved an NH_3 yield of 2.2×10^{-3} g m^{-2} h^{-1}, corresponding to twice the yield of pristine Fe_2O_3 (1.0×10^{-3} g m^{-2} h^{-1}) [89].

Moreover, the long-term stability of NRR catalysts should not be neglected. Fe_2O_3-based electrocatalysts are susceptible to structural degradation, such as etching in acidic electrolytes or reduction to low-valent Fe species (Fe(I)/Fe(II)) in alkaline media [86]. In a study by Sun et al., the NRR activities of Fe_2O_3 nanorods were assessed in neutral media (0.5 M $LiClO_4$ and 0.1 M Na_2SO_4) to mitigate the aforementioned issues, and the NH_3 production rates were 15.9 μg h^{-1} mg_{cat}^{-1} [87]. Furthermore, the Fe_2O_3 nanorods exhibited outstanding recycling stability and operational durability under neutral electrochemical conditions, as shown in Figure 7.10e, f.

Besides metal oxides, iron diboride (FeB_2) has shown great potential as an electrocatalyst, thanks to its MXene-like layered structure, high carrier mobility, favorable durability, and abundant earth reserves. Chu et al. synthesized amorphous FeB_2 porous nanosheets (a-FeB_2 PNSs) through a simple reflux approach (Figure 7.11a–e) and systematically verified their catalytic activity toward NRR [90]. Electrochemical NRR tests conducted in 0.5 M $LiClO_4$ electrolyte revealed outstanding performance of a-FeB_2 PNSs (Figure 7.11f–h): an NH_3 yield of 39.8 μg h^{-1} mg^{-1} cat at −0.3 V vs. RHE and a Faradaic efficiency (FE) of 16.7% at −0.2 V vs. RHE. Notably, no hydrazine (N_2H_4) by-product was detected after electrolysis at all tested potentials, confirming the high selectivity of a-FeB_2 PNSs.

Beyond iron-group compounds, atomically precise metal nanoclusters (NCs) have garnered increasing interest from researchers. Their distinctive characteristics—including ultra-tiny dimensions, ample surface-active sites, and discontinuous electronic energy levels—synergistically confer excellent catalytic performance. Nevertheless, a large proportion of NC-derived catalysts rely on noble metal components, and their exorbitant cost has become a major bottleneck that impedes their translation into practical industrial applications [91]. Therefore, considering cost-effectiveness and practical applicability, it is crucial to incorporate non-noble metals into NC-based catalysts, exploring their roles in regulating catalytic performance and reducing production costs. Han et al. selected Ni as the non-noble metal dopant to synthesize M_4Ni_2 NCs (M = Au, Ag) (Figure 7.12a, b), aiming to modulate the electronic structure of the catalyst via heteroatom doping, thereby suppressing the HER and promoting NRR [92].

As depicted in Figure 7.12b–f, the electrocatalytic NRR capabilities of the target nanoclusters were evaluated in a 0.1 M HCl solution. A more significant elevation in current density was observed for Ag_4Ni_2 NCs in N_2-saturated electrolyte compared to Ar-saturated conditions (Figure 7.12c), which serves as direct evidence for the distinct NRR activity of Ag_4Ni_2 NCs. Chronoamperometric measurements were employed to quantify the ammonia yield and FE of the nanoclusters. Across the entire tested potential range, Ag_4Ni_2 NCs consistently delivered higher FE values than their Au_4Ni_2 counterparts (Figure 7.12d). The two M_4Ni_2 clusters exhibited a similar variation tendency in NH_3 yield, with the sole discrepancy being the inverted yield values at a potential of −0.4 V vs. RHE (Figure 7.12e). Specifically, at −0.2 V vs. RHE, Ag_4Ni_2 NCs achieved a maximum NH_3 yield of 23.32 μg mg^{-1}cat h^{-1} and a peak FE of 78.97%. These performance metrics outperformed those of Au_4Ni_2 NCs,

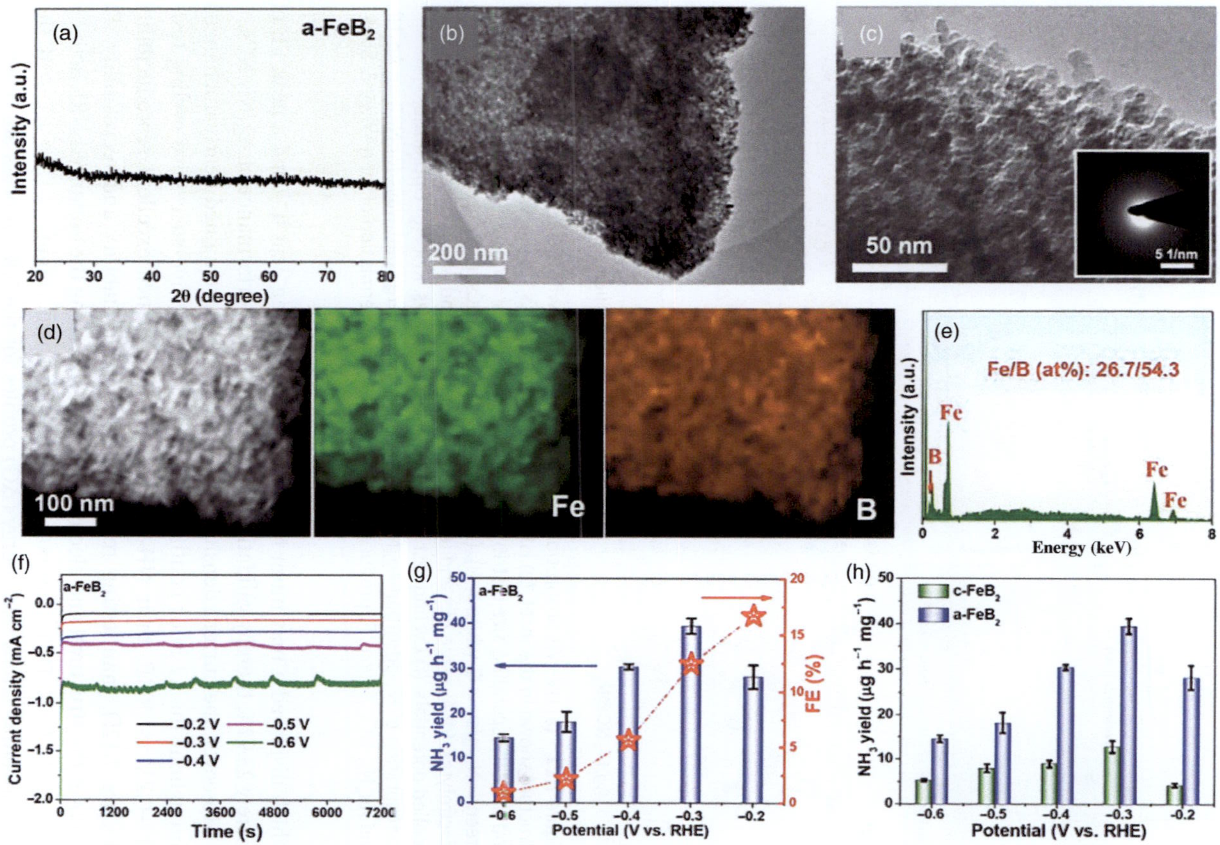

Figure 7.11 (a) XRD pattern of FeB_2. (b, c) TEM images and SAED patterns of FeB_2. (d) STEM and elemental mapping images of FeB_2. (e) EDX analysis of FeB_2. (f) Chronoamperometric tests of a-FeB_2 for NRR electrolysis at different applied potentials for 2 h. (g) Corresponding NH_3 yields and FEs of a-FeB_2. (h) Comparison of NH_3 yields between c-FeB_2 and a-FeB_2 at different applied potentials. *Source*: Reproduced from Ref. [90] / with permission of Elsevier.

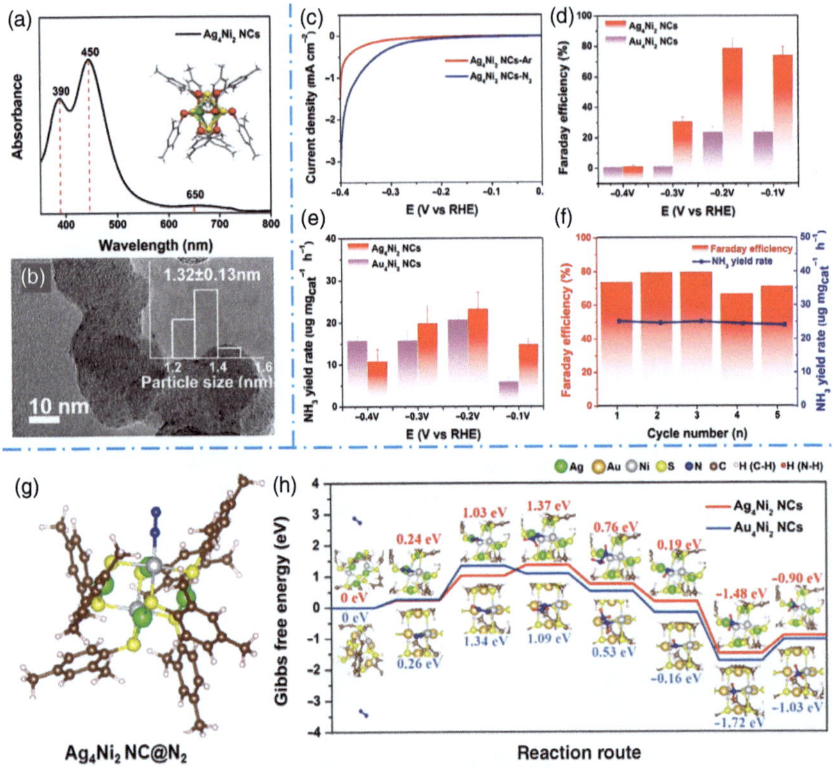

Figure 7.12 (a) UV–Vis absorption spectra of Ag_4Ni_2 NCs, with the corresponding crystal structure shown in the inset. (b) HRTEM image of Ag_4Ni_2 NCs supported on activated carbon and the corresponding size distribution (color labels: yellow = Ag, blue = Au, red = S, green = Ni, gray = C, white = H). (c) LSV curves of Ag_4Ni_2 NCs in Ar- or N_2-saturated electrolytes. (d) Average FEs and (e) NH_3 yields of Ag_4Ni_2 and Au_4Ni_2 NCs at different applied potentials. (f) Cycling test of Ag_4Ni_2 NCs at −0.2 V vs. RHE. (g) Optimized configuration of N_2-adsorbed Ag_4Ni_2 NCs. (h) Gibbs free energy pathways for NRR on Ag_4Ni_2 and Au_4Ni_2 NCs. *Source*: Reproduced from Ref. [92] / with permission of John Wiley & Sons.

which displayed a maximum NH_3 yield of 20.77 μg mg^{-1}cat h^{-1} and a peak FE of merely 23.92%, thereby validating the superior NRR performance of Ag_4Ni_2 NCs.

Moreover, the current densities of both M_4Ni_2 NCs remained nearly unchanged throughout the 12-hour continuous NRR test. After five successive cycling tests at −0.2 V vs. RHE, their NH_3 yields and FEs retained excellent reproducibility (Figure 7.12f), manifesting remarkable long-term stability and recyclability. To uncover the structural and electronic origins governing the enhanced NRR activity of Ag_4Ni_2 NCs, DFT simulations were performed (Figure 7.12 g–h), and the corresponding Gibbs free energy (ΔG) profiles for the NRR process are illustrated in Figure 7.12h. For both nanoclusters, the Gibbs free energy of the NNH_2* intermediate was lower than that of NHNH*, while the $NHNH_2$* species possessed a lower free energy relative to N*. Additionally, the NH* intermediate was found to be more stable than NH_2NH_2*. These findings indicate that among the three possible reaction pathways, the most thermodynamically favorable route is as follows: $N_2 + * \rightarrow N_2* \rightarrow NNH* \rightarrow NNH_2* \rightarrow NHNH_2* \rightarrow N* \rightarrow NH_2* \rightarrow NH_3*$. Notably, in the rate-determining step (RDS) of $N_2* \rightarrow NNH*$, the energy barrier (ΔG uphill) of Ag_4Ni_2 NCs was 0.79 eV, which is substantially lower than that of Au_4Ni_2 NCs (1.08 eV). This result further corroborates the superior NRR catalytic activity of Ag_4Ni_2 NCs from a thermodynamic perspective.

7.6 CO_2RR

The escalating issues of global warming and environmental degradation have made CO_2 emissions a central challenge. In this context, electrochemical CO_2 conversion into storable fuels and valuable chemicals emerges as a promising solution. Over the past several decades, extensive research endeavors have been devoted to the development and advancement of the electrochemical CO_2RR for the synthesis of carbon-containing products, spanning catalyst fabrication, electrolyte optimization, and device configuration [93]. Nevertheless, the widespread industrial application of this technology is still severely hampered by persistently low energy and carbon utilization efficiencies, as well as insufficient selectivity in the formation of multi-carbon (C_{2+}) products [94].

CO_2 conversion is challenging due to the high bond energy of the C=O bond (806 kJ mol^{-1}). The linear structure of the CO_2 molecule contributes to its relative stability and chemical inertness. Concurrently, the multi-electron transfer required for its conversion faces a high free energy barrier [95]. These factors together make the development of efficient pathways to transform CO_2 into valuable chemicals and fuels a significant scientific and technological challenge. Various methods are being explored to achieve this conversion, such as thermocatalytic hydrogenation, along with biological, photochemical, photoelectrochemical, and electrochemical approaches [96]. In recent years, electrochemical CO_2RR has been extensively studied and shows greater potential compared to other processes, with the following advantages: (i) functions efficiently under ambient temperature and pressure conditions, eliminating the need for complex high-temperature or high-pressure equipment and reducing operational costs; (ii) the progression of the reduction reaction is readily regulable via tuning the applied potential, enabling precise control over reaction kinetics and product distribution; (iii) features favorable scalability, allowing for straightforward expansion from laboratory-scale setups to industrial-grade production lines without substantial technical barriers; (iv) possesses a compact structural configuration, coupled with excellent tunability that facilitates flexible adjustments according to diverse application scenarios and performance requirements [97].

Although remarkable headway has been achieved in CO_2 emission abatement, its large-scale industrial deployment is still plagued by a suite of challenges: sluggish reaction dynamics, excessive overpotential demands, considerable energy barriers, inadequate mass transfer efficacy, and intricate multi-stage electron transfer mechanisms. Additionally, CO_2RR competes with the HER in aqueous electrolytes, reducing product efficiency and selectivity [93]. Thus, there is an imperative requirement for highly efficient and durable electrocatalysts that can facilitate CO_2 reduction under low overpotential conditions while delivering high current densities. In contrast to the HER and OER, the electrochemical CO_2RR represents a sophisticated multistep electron transfer process, encompassing 2–12 electron transfer pathways. This inherent complexity gives rise to a broad spectrum of reaction products, with the final distribution of these products being predominantly governed by the specific type of electrocatalyst utilized [98].

The heterogeneous catalysis pathway of the electrochemical CO_2 reduction reaction (CO_2RR) typically adheres to a four-stage mechanism: (i) adsorption step: CO_2 molecules diffuse from the CO_2-saturated electrolyte and undergo chemisorption onto the electrocatalyst surface; (ii) activation and bond rearrangement: via sequential electron and proton transfer, the cleavage of C—O bonds and formation of C—H bonds take place. This stage serves as the RDS of the entire CO_2RR process, primarily because it entails breaking the high-energy C=O double bond and generating the CO_2·− radical intermediate, which are thermodynamically unfavorable processes; (iii) product generation: a series of reduced carbon-containing species are formed through subsequent reaction steps; (iv) desorption step: the formed products detach from the catalyst surface to free up active sites for subsequent reactions [98]. Multiple factors collectively modulate the CO_2RR reaction outcomes, including the intrinsic properties of the electrocatalyst, the physicochemical characteristics of the electrolyte, and the magnitude of the applied electrode potential. The CO_2RR yields a diverse array of carbon-containing products with varying oxidation states, such as carbon monoxide (CO), formic acid (HCOOH), oxalic acid ($C_2H_2O_4$), methanol (CH_3OH), ethanol (C_2H_5OH), propanol (C_3H_7OH), methane (CH_4), and ethylene (C_2H_4) [99]. Notably, the equilibrium potentials of the CO_2RR itself and its respective products are not fixed; instead, they are closely related to the electrolyte pH value and the number of electrons transferred during the reaction process.

Triggering the activation of thermodynamically stable linear CO_2 molecules into bent CO_2^- radical intermediates necessitates considerable energy input, with the thermodynamic equilibrium potential for CO_2^- generation measured at −1.49 V vs. RHE [100]. As a multi-electron transfer reaction, electrochemical CO_2RR inherently proceeds under elevated overpotential conditions [101]. Particularly in aqueous electrolyte systems, the sluggish reaction dynamics of CO_2RR tend to facilitate the competing HER, rendering H_2 the dominant side product. Thermodynamically, therefore, the design of CO_2RR electrocatalysts should enhance CO_2RR activity while suppressing HER, prioritizing $*CO_2$ adsorption and inhibiting *H adsorption [98].

Beyond suppressing the formation of H_2, enhanced selectivity for target high-value-added products holds greater significance in curbing energy dissipation and streamlining subsequent product separation processes. Nevertheless, attaining high selectivity for desired products remains a formidable challenge, largely stemming from the comparable binding energies among different reaction intermediates [102]. Product selectivity is modulated by a multitude of factors, encompassing the category of electrocatalyst, the electronic and geometric configurations of the catalyst, the physicochemical properties of the electrolyte solution, electrolyte pH value, reaction temperature, and operating pressure [103]. For instance, metals like Sn, Bi, Pb, and In have been shown to exhibit high selectivity for formic acid (HCOOH) production. Similarly, Ag, Au, and Pd serve as benchmark catalysts for CO_2-to-CO reduction, as they form weak bonds with *CO but strong bonds with the *COOH intermediate, resulting in high CO selectivity [102]. Nevertheless, their large-scale application is restricted by high cost and scarcity.

In-depth research has revealed that electron transfer and proton attachment convert the *COOH intermediate to CO, while the *OCHO intermediate is transformed into HCOOH. In recent years, both experimental and theoretical studies have indicated that iron-group metal (Ni, Co, Fe)-based catalysts display high activity for CO formation during CO_2RR. Research endeavors have therefore centered on fabricating cost-effective, high-selectivity, low-overpotential, and durable electrocatalysts, with a key strategy being the rational design of iron-group metal (Fe, Co, Ni)-based catalytic materials [104].

Heat-treated iron-group metal (Fe, Co, Ni)-nitrogen co-doped carbon (M—N—C) catalysts have attracted widespread interest owing to their high selectivity in CO conversion, abundant earth reserves, stability, and great conductivity. Ying Li et al. fabricated iron-group metal/nitrogen-doped carbon nanosheets through molecular-level coordination chemistry to construct atomic M—N coordination, followed by solid-phase pyrolysis (Figure 7.13a–c) [105]. Characterization results regarding overpotential, FE, and TOF (Figure 7.13d–g) show that among M—N—C catalysts, Fe and Ni display high activity in CO_2-to-CO reduction.

Fe mainly functions to lower the overpotential: Fe—N—C exhibits a minimum overpotential of merely 0.19 V vs. RHE for CO production. In contrast, Ni effectively improves CO selectivity and reaction kinetics. At 0.65 V vs. RHE, Ni—N—C reaches a maximum FE_{CO} of 96%, along with a partial current density of $-8.2\ mA\ cm^{-2}$ and a TOF of $1060\ h^{-1}$. During long-term CO_2 reduction at −0.75 V vs. RHE, Ni—N—C sustains FE_{CO} above 96% with no decay in CO current density, indicating excellent electrochemical stability.

Iron-group metals show considerable potential in electrochemical CO_2RR. Besides single-atom structures, the reasonable design of metal-based electrocatalysts with customized nanoscale sizes is essential to attain the desired activity in CO_2RR. Chen and coworkers reported the fabrication of nanosized iron carbide clusters (Figure 7.14a–c). These nanomaterials possess multiple structural merits: they not only expose unsaturated carbon sites, elevate the Fermi energy level, and further downshift the d-band center, but also offer a rich repertoire of surface-active sites [106]. The electrocatalytic CO_2RR capabilities of the clusters were assessed in a CO_2-saturated 0.1 M $KHCO_3$ electrolyte. LSV measurements (Figure 7.14d) revealed that the cathodic current density of Fe_2C—Cs@DC in CO_2-saturated $KHCO_3$ solution was markedly higher than that observed in Ar-saturated electrolyte, which attests to its superior CO_2RR catalytic activity.

Stepped potential electrolysis was conducted in the potential range of −0.4 to −1.0 V vs. RHE, with online product monitoring. Notably, the only products of all catalysts are carbon monoxide (CO) and hydrogen (H_2). As illustrated in Figure 7.14e, Fe_2C—Cs@DC maintains a high CO Faradaic efficiency (FE_{CO})

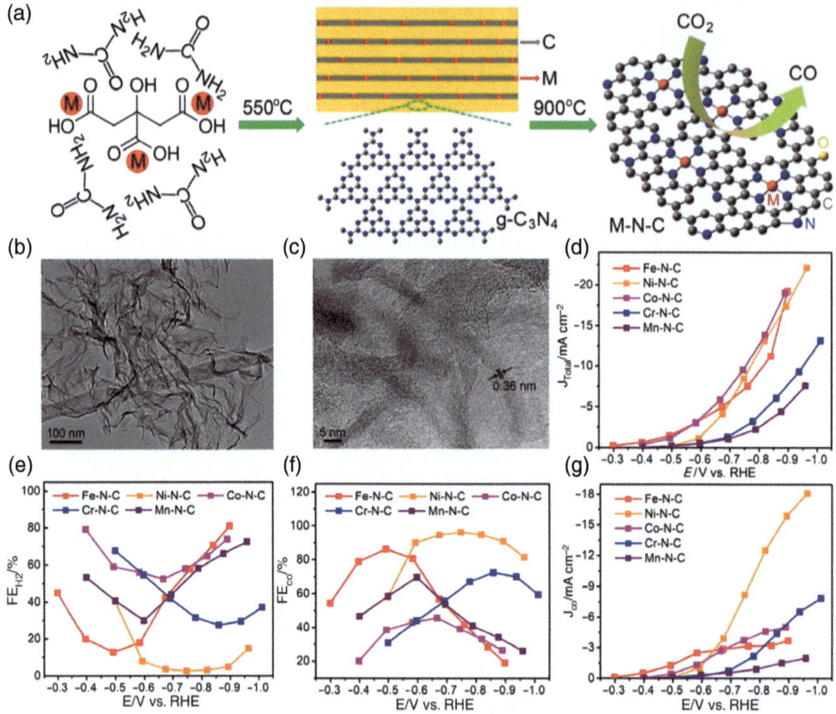

Figure 7.13 (a) Schematic illustration of the fabrication process for TM—N—C catalysts. (b, c) TEM and HRTEM images of Fe—N—C. (d) Total current density, (e) FEs for H_2 production, (f) FEs for CO production, and (g) partial current density for CO production of M—N—C catalysts. *Source*: Reproduced from Ref. [105] / with permission of Elsevier.

plateau exceeding 90% between −0.5 and −0.8 V vs. RHE, reaching the maximum FE_{CO} at −0.7 V vs. RHE. At this potential, Fe_2C—Cs@DC achieves a high current density of 8.29 mA cm^{-2} and a CO conversion selectivity of 97.1%. In stark contrast, FeN—Cs@NC demonstrates inferior catalytic activity, with a CO partial current density (j) of 1.06 mA cm^{-2} and a CO selectivity of 83.4%. Notably, the j value of Fe_2C—Cs@DC is approximately 8 times higher than that of FeN—Cs@NC, which confirms that Fe_2C clusters serve as more robust active centers for CO_2RR in comparison to FeN_x moieties. Among all the synthesized catalytic materials, Fe_2C—Cs@DC stands out with the most prominent CO_2RR activity and product selectivity. As illustrated in Figure 7.14f, g, Fe_2C/Fe_4N@DC reaches a maximum Faradaic efficiency for FE_{CO} of 71.3% and a j of 2.08 mA cm^{-2} at a potential of −0.7 V vs. RHE, corresponding to merely 25.1% of the values achieved by Fe_2C—Cs@DC. This result underscores that the ultrafine nanocluster structure of the Fe_2C core exerts a decisive effect on boosting CO_2RR catalytic performance. Furthermore, Tafel plot analysis (Figure 7.14 h) indicates that Fe_2C—Cs@DC possesses a smaller Tafel slope (89.5 mV dec^{-1}) than Fe—N—C (109 mV dec^{-1}), implying more rapid reaction kinetics during CO_2RR. In addition to its outstanding activity and selectivity, Fe_2C—Cs@DC exhibits remarkable electrocatalytic stability

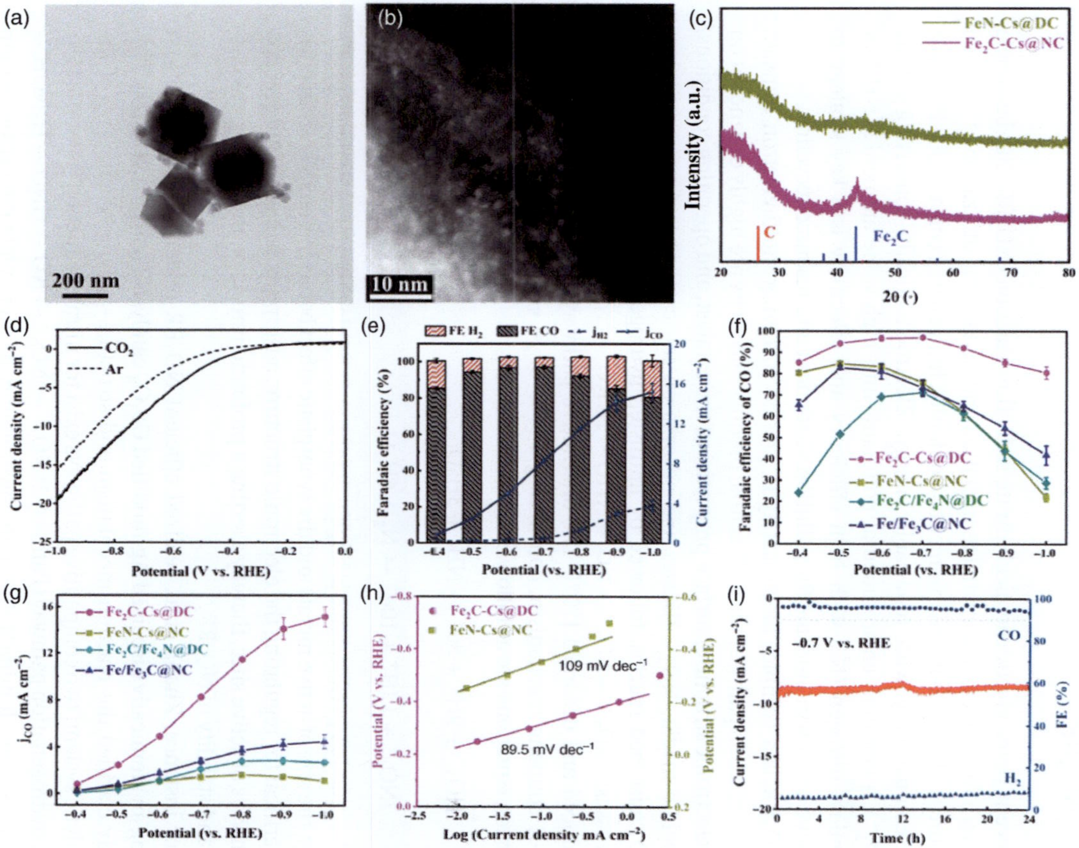

Figure 7.14 (a) TEM image of Fe_2C-Cs@DC. (b) HRTEM image of Fe_2C-Cs@DC. (c) XRD patterns of Fe_2C-Cs@DC and FeN-Cs@NC. (d) LSV curves of Fe_2C-Cs@DC in CO_2-saturated (solid lines) and Ar-saturated (dashed lines) 0.1 M $KHCO_3$ aqueous electrolyte. (e) FEs for CO (gray) and H_2 (red), as well as the partial current density for CO, over Fe_2C-Cs@DC over a range of applied potentials. (f) CO FEs over various catalysts at different applied potentials. (g) CO partial current densities over various catalysts at different applied potentials. (h) Tafel plots of CO partial current densities for Fe_2C-Cs@DC and FeN-Cs@NC in 0.1 M $KHCO_3$ aqueous electrolyte. (i) 24 h CO_2RR stability test of Fe_2C-Cs@DC at −0.7 V vs. RHE. *Source*: Reproduced from Ref. [106] / with permission of John Wiley & Sons.

throughout the CO_2RR process. As presented in Figure 7.14i, during 24 hours of continuous operation at −0.7 V vs. RHE, its FE remains above 95%, and neither FE nor current density shows an obvious downward trend, manifesting excellent long-term operational durability.

7.7 NO_3RR

In recent years, electrochemical nitrate reduction reaction (NO_3RR) has become a research focus in environmental remediation and sustainable utilization of nitrogen resources, thanks to its environmental benignity, high energy conversion efficiency, and compatibility with renewable energy. Electrochemically converting nitrate (NO_3^-), a common environmental pollutant, into high-value ammonia (NH_3) or harmless nitrogen (N_2) with high efficiency and selectivity is a key strategy to achieve a win-win outcome for pollution abatement and resource recycling.

NO_3RR is a complex electrocatalytic process involving multi-electron transfer, and its reaction pathways as well as final product selectivity are highly dependent on the electronic structure, surface properties of the catalyst, and operating conditions. In acidic media, NO_3^- diffuses and adsorbs on the catalyst surface to generate $*NO_3$ adsorbed species, followed by multistep hydrogenation and deoxygenation processes: $NO_3^- \rightarrow NO_2^- \rightarrow NO \rightarrow NHO \rightarrow NH_2O \rightarrow *NH_3$, and finally NH_3 is produced and desorbed [107]. The alternative pathway generates N_2 through an N—N coupling mechanism, consuming a total of 10 electrons [108]. The overall reaction equations are as follows:

$$NO_3^- + 9H^+ + 8e^- \rightarrow NH_3 + 3H_2O \tag{7.28}$$

$$2NO_3^- + 12H^+ + 10e^- \rightarrow N_2 + 6H_2O \tag{7.29}$$

The reaction performance is largely determined by electrocatalyst design. Specifically, the size of active metals and the synergistic effect between metals and supports are critical for regulating the electronic structure, adsorption behavior, and reaction pathways of active sites, thereby exerting a profound impact on the activity, selectivity, and stability of NO_3RR.

For instance, Zhang et al. realized efficient NO_3RR under alkaline conditions using an atomically dispersed Ag-modified Co_3O_4 catalyst (Figure 7.15a) [109]. This study clarified the reaction mechanism and structure–activity relationship of the catalytic system: two competitive pathways exist for nitrate reduction (Figure 7.15b). The hydrogenation pathway (Pathway 1: $NO_3^- \rightarrow NO_2^- \rightarrow NO \rightarrow HNO \rightarrow H_2NOH \rightarrow NH_3$) becomes dominant due to its lower energy barrier, while the deoxygenation pathway (Pathway 2) is suppressed owing to excessively high energy barriers. Experimental findings demonstrate that Ag doping triggers the generation of high-spin octahedral Co^{2+} active sites via interparticle electron transfer, which significantly enhances the adsorption affinity toward the crucial NO_2^- intermediate. Specifically, the NO_2^- adsorption energy of 1Ag—2O_v—Co_3O_4 is −0.998 eV, representing a 0.196 eV increment compared with that of the pristine P-Co_3O_4

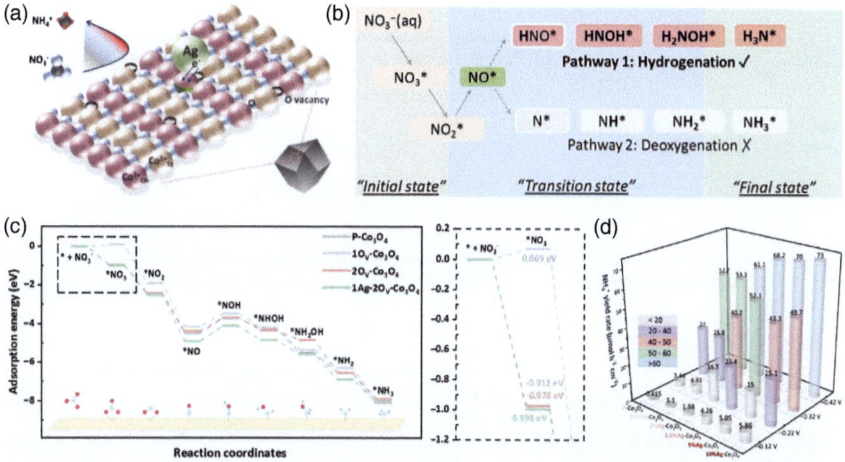

Figure 7.15 (a) Schematic illustration of Ag single-atom decorated Co_3O_4 catalyst for nitrate reduction to ammonia under alkaline conditions. (b) Two competing pathways for nitrate reduction. (c) Adsorption energies of NO_3^-RR intermediates on P-Co_3O_4, 1O_v-Co_3O_4, 2O_v-Co_3O_4, and 1Ag-2O_v-Co_3O_4 (blue spheres: N; red spheres: O; gray spheres: H). (d) NH_4^+ Faradaic efficiencies of Ag—Co_3O_4 at different applied potentials. *Source*: Reproduced from Ref. [109] / with permission of American Chemical Society.

without Ag modification. Meanwhile, Ag modification promotes the enrichment of O_V, which effectively captures the oxygen atoms of $^*NO_2^-$ and facilitate N—O bond cleavage (Figure 7.15c).

Catalytic performance evaluations (Figure 7.15d) reveal that the 2.5%Ag—Co_3O_4 catalyst attains a peak ammonia production rate of 52 μmol h^{-1} cm^{-2} alongside a product selectivity of 88% at an applied potential of −0.32 V vs. RHE. In contrast, an excessive Ag loading content (10%) induces a precipitous drop in ammonia output to merely 5.9 μmol h^{-1} cm^{-2}, which is attributed to the severe aggregation of Ag nanoparticles. These results validate that Ag modification modulates the electronic configuration and surface characteristics of the Co_3O_4 catalyst, selectively promotes the hydrogenation reaction route, and consequently realizes efficient ammonia synthesis.

Building upon the aforementioned research groundwork, Jin and colleagues further expanded the paradigm for catalyst design, devising a mechanochemical approach to accurately construct Ir SAs and clusters anchored on a Co_3O_4 support. Their findings revealed that atomically dispersed Ir SAs markedly promote charge transfer and lower energy barriers [110]. As illustrated in Figure 7.16a, the current density of Ir SAC-Co_3O_4 (Ir single-atom catalyst supported on Co_3O_4) rises sharply upon the addition of 0.1 M NO_3^-. As the potential shifts more negatively, the current density difference between the systems with and without 0.1 M NO_3^- gradually expands, indicating enhanced NO_3^- reduction activity at more negative potentials. The corresponding NH_3 yield and Faradaic efficiency are presented in Figure 7.16b. Notably, Ir SAC-Co_3O_4 exhibits excellent cyclic stability, with negligible fluctuations in NH_3 yield and FE even after 10 consecutive cycles. Furthermore, theoretical

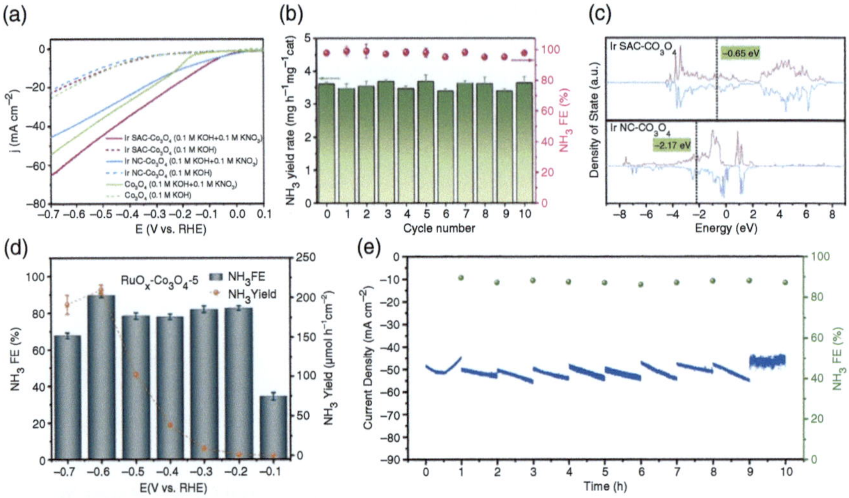

Figure 7.16 (a) LSV curves of Ir SAC-Co_3O_4, Ir NC-Co_3O_4, and Co_3O_4 in Ar-saturated 0.1 M KOH electrolyte with/without 0.1 M KNO_3; (b) Cycling stability and corresponding FEs of Ir SAC-Co_3O_4; (c) PDOS and d-band centers of Ir SAC-Co_3O_4 (top) and Ir NC-Co_3O_4 (bottom). *Source*: Reproduced from Ref. [110] / with permission of The Royal Society of Chemistry. (d) NH_3 yields and FEs of $RuO_x-Co_3O_4-5$ at specific applied potentials; (e) 10-cycle cycling stability test of $RuO_x-Co_3O_4-5$ with corresponding NH_3 FEs (green dots), where the initial NO_3^- concentration is 0.5 M per cycle. *Source*: Reproduced from Ref. [111] / with permission of The Royal Society of Chemistry.

analysis confirmed its advantageous electronic configuration. Specifically, the d-band center of the Ir SAs was calculated to be 1.065 eV, a value closer to the Fermi level than the 1.217 eV observed for the cluster form. This closer proximity correlates with its enhanced catalytic performance, detailed in Figure 7.16c.

In parallel, Liu et al. adopted an alternative approach to construct a RuO_x-Co_3O_4 nanoparticle interface [111]. The central feature of this design is the formation of Ru—O—Co bonds. These bonds play a crucial role in modulating the electronic structure, which in turn greatly boosts the catalyst's nitrate adsorption capability. This synergistic effect culminated in a record-high NH_3 production rate of 210.5 μmol h^{-1} cm^{-2} and a Faradaic efficiency of 89.7% (Figure 7.16d). Stability tests indicated that the catalyst maintains a Faradaic efficiency above 86% after 10 successive catalytic cycles, verifying the outstanding stability of RuO_x-Co_3O_4–5 (Figure 7.16e).

The strategies of single-atom dispersion, interface engineering, and mechanochemical synthesis have collectively advanced the understanding of electronic structure modulation, active-site synergy, and reaction pathways in catalysts. The synergistic modification of active species (Ag, Ir, Ru) with Co_3O_4 supports has proven effective in enhancing NO_3RR activity and selectivity, providing a solid basis for efficient electrocatalytic nitrate reduction. Building on these insights, critical next steps include fine-tuning the support-metal interface, elucidating intrinsic reaction mechanisms, and transitioning lab-scale catalysts into functional devices.

References

1 Wang, J., Xu, F., Jin, H. et al. (2017). Non-noble metal-based carbon composites in hydrogen evolution reaction: fundamentals to applications. *Advanced Materials* 29: 1605838.

2 Jiao, S.L., Fu, X.W., Wang, S.Y., and Zhao, Y. (2021). Perfecting electrocatalysts *via* imperfections: towards the large-scale deployment of water electrolysis technology. *Energy & Environmental Science* 14: 1722–1770.

3 Skúlason, E., Tripkovic, V., Björketun, M.E. et al. (2010). Modeling the electrochemical hydrogen oxidation and evolution reactions on the basis of density functional theory calculations. *Journal of Physical Chemistry C* 114: 18182–18197.

4 Zhou, P.Y., Hai, G.T., Zhao, G.C. et al. (2023). CeO_2 as an "electron pump" to boost the performance of Co_4N in electrocatalytic hydrogen evolution, oxygen evolution and biomass oxidation valorization. *Applied Catalysis B: Environmental* 325: 122364.

5 Yu, M.Q., Budiyanto, E., and Tüysüz, H. (2022). Principles of water electrolysis and recent progress in cobalt-, nickel-, and iron-based oxides for the oxygen evolution reaction. *Angewandte Chemie International Edition* 61: e202103824.

6 Pu, Z.H., Liu, T.T., Amiinu, I.S. et al. (2020). Transition-metal phosphides: activity origin, energy-related electrocatalysis applications, and synthetic strategies. *Advanced Functional Materials* 30: 2004009.

7 Liu, X., Yu, Q., Qu, X. et al. (2024). Manipulating electron redistribution in Ni_2P for enhanced alkaline seawater electrolysis. *Advanced Materials* 36: 2307395.

8 Gao, F., Zhang, Y.P., Wu, Z.Y. et al. (2021). Universal strategies to multi-dimensional noble-metal-based catalysts for electrocatalysis. *Coordination Chemistry Reviews* 436: 213825.

9 Chen, X.H., Li, X.L., Wu, L.L. et al. (2021). Nb_2O_5-Ni_3N heterojunction tuned by interface oxygen vacancy engineering for the enhancement of electrocatalytic hydrogen evolution activity. *Journal of Materials Chemistry A* 9: 11563–11570.

10 Chen, J.Y., Li, X., Ma, B. et al. (2022). Cu_3P@Ni core-shell heterostructure with modulated electronic structure for highly efficient hydrogen evolution. *Nano Research* 15: 2935–2942.

11 Ma, M.J., Xu, J.H., Wang, H.Q. et al. (2021). Multi-interfacial engineering of hierarchical $CoNi_2S_4/WS_2/Co_9S_8$ hybrid frameworks for robust all-pH electrocatalytic hydrogen evolution. *Applied Catalysis B: Environmental* 297: 120455.

12 Qu, M.J., Jiang, Y.M., Yang, M. et al. (2020). Regulating electron density of NiFe-P nanosheets electrocatalysts by a trifle of Ru for high-efficient overall water splitting. *Applied Catalysis B: Environmental* 263: 118324.

13 Lyu, C., Cheng, J., Wu, K. et al. (2022). Interfacial electronic structure modulation of CoP nanowires with FeP nanosheets for enhanced hydrogen evolution under alkaline water/seawater electrolytes. *Applied Catalysis B: Environmental* 317: 121799.

14 Yao, Y.H., Zhang, Z.Y., and Jiao, L.F. (2022). Development strategies in transition metal borides for electrochemical water splitting. *Energy & Environmental Materials* 5: 470–485.

15 Shen, S.J., Wang, Z.P., Lin, Z.P. et al. (2022). Crystalline-amorphous interfaces coupling of $CoSe_2$/CoP with optimized d-band center and boosted electrocatalytic hydrogen evolution. *Advanced Materials* 34: 2110631.

16 Shi, Q.R., Zhu, C.Z., Du, D., and Lin, Y.H. (2019). Robust noble metal-based electrocatalysts for oxygen evolution reaction. *Chemical Society Reviews* 48: 3181–3192.

17 Yao, Y.C., Hu, S.L., Chen, W.X. et al. (2019). Engineering the electronic structure of single atom Ru sites via compressive strain boosts acidic water oxidation electrocatalysis. *Nature Catalysis* 2: 304–313.

18 Zeng, F., Mebrahtu, C., Liao, L.F. et al. (2022). Stability and deactivation of OER electrocatalysts: a review. *Journal of Energy Chemistry* 69: 301–329.

19 Song, J.J., Wei, C., Huang, Z.F. et al. (2020). A review on fundamentals for designing oxygen evolution electrocatalysts. *Chemical Society Reviews* 49: 2196–2214.

20 Xu, X.M., Pan, Y.L., Zhong, J. et al. (2022). New undisputed evidence and strategy for enhanced lattice-oxygen participation of perovskite electrocatalyst through cation deficiency manipulation. *Advanced Science* 9: 2200530.

21 An, L., Wei, C., Lu, M. et al. (2021). Recent development of oxygen evolution electrocatalysts in acidic environment. *Advanced Materials* 33: 2006328.

22 Du, J., Li, F., and Sun, L.C. (2021). Metal-organic frameworks and their derivatives as electrocatalysts for the oxygen evolution reaction. *Chemical Society Reviews* 50: 2663–2695.

23 Yu, J., Yu, F., Yuen, M.F., and Wang, C.D. (2021). Two-dimensional layered double hydroxides as a platform for electrocatalytic oxygen evolution. *Journal of Materials Chemistry A* 9: 9389–9430.

24 Zhang, K. and Zou, R. (2021). Advanced transition metal-based OER electrocatalysts: current status, opportunities, and challenges. *Small* 17: e2100129.

25 Exner, K.S. (2020). Recent progress in the development of screening methods to identify electrode materials for the oxygen evolution reaction. *Advanced Functional Materials* 30: 2005060.

26 Chang, J.W., Zhang, Q., Yu, J.K. et al. (2023). A Fe single atom seed-mediated strategy toward Fe_3C/Fe-N-C catalysts with outstanding bifunctional ORR/OER activities. *Advanced Science* 10: 2301656.

27 Huang, Z.F., Song, J.J., Du, Y.H. et al. (2019). Chemical and structural origin of lattice oxygen oxidation in Co-Zn oxyhydroxide oxygen evolution electrocatalysts. *Nature Energy* 4: 329–338.

28 Exner, K.S. (2021). On the lattice oxygen evolution mechanism: avoiding pitfalls. *ChemCatChem* 13: 4066–4074.

29 Guo, T.Q., Li, L.D., and Wang, Z.C. (2022). Recent development and future perspectives of amorphous transition metal-based electrocatalysts for oxygen evolution reaction. *Advanced Energy Materials* 12: 2200827.

30 Yang, C.H., Yang, Z.D., Dong, H. et al. (2019). Theory-driven design and targeting synthesis of a highly-conjugated basal-plane 2D covalent organic framework for metal-free electrocatalytic OER. *ACS Energy Letters* 4: 2251–2258.

31 Alia, S.M., Shulda, S., Ngo, C. et al. (2018). Iridium-based nanowires as highly active, oxygen evolution reaction electrocatalysts. *ACS Catalysis* 8: 2111–2120.

32 Bockris, J.O.M. (2004). Kinetics of activation controlled consecutive electrochemical reactions: anodic evolution of oxygen. *The Journal of Chemical Physics* 24: 817–827.

33 Chen, K., Cao, Y.H., Yadav, S. et al. (2023). Electronic structure reconfiguration of nickel-cobalt layered double hydroxide nanoflakes *via* engineered heteroatom and oxygen-vacancies defect for efficient electrochemical water splitting. *Chemical Engineering Journal* 463: 142396.

34 Lei, H., Ma, L., Wan, Q. et al. (2022). Promoting surface reconstruction of NiFe layered double hydroxide for enhanced oxygen evolution. *Advanced Energy Materials* 12: 2202522.

35 Xu, W.J., Lyu, F.L., Bai, Y.C. et al. (2018). Porous cobalt oxide nanoplates enriched with oxygen vacancies for oxygen evolution reaction. *Nano Energy* 43: 110–116.

36 Kim, J.S., Kim, B., Kim, H., and Kang, K. (2018). Recent progress on multimetal oxide catalysts for the oxygen evolution reaction. *Advanced Energy Materials* 8: 1702774.

37 Wang, A.L., Xu, H., and Li, G.R. (2016). NiCoFe layered triple hydroxides with porous structures as high-performance electrocatalysts for overall water splitting. *ACS Energy Letters* 1: 445–453.

38 Yu, L., Yang, J.F., Guan, B.Y. et al. (2018). Hierarchical hollow nanoprisms based on ultrathin Ni-Fe layered double hydroxide nanosheets with enhanced electrocatalytic activity towards oxygen evolution. *Angewandte Chemie, International Edition* 57: 172–176.

39 Liu, R., Wang, Y.Y., Liu, D.D. et al. (2017). Water-plasma-enabled exfoliation of ultrathin layered double hydroxide nanosheets with multivacancies for water oxidation. *Advanced Materials* 29: 1701546.

40 Guo, B.R., Ding, Y.N., Huo, H.H. et al. (2023). Recent advances of transition metal basic salts for electrocatalytic oxygen evolution reaction and overall water electrolysis. *Nano-Micro Letters* 15: 57.

41 Zhu, J., Hu, L.S., Zhao, P.X. et al. (2020). Recent advances in electrocatalytic hydrogen evolution using nanoparticles. *Chemical Reviews* 120: 851–918.

42 Zhang, R.D., Xu, X.L., Zhang, Y.H., and Dong, Y. (2023). Analysis and forecast of the substitution potential of China's wind power-hydrogen production for fossil fuel hydrogen production. *Journal of Cleaner Production* 422: 138410.

43 Kang, Z., Guo, H.J., Wu, J. et al. (2019). Engineering an Earth-abundant element-based bifunctional electrocatalyst for highly efficient and durable overall water splitting. *Advanced Functional Materials* 29: 1807031.

44 Lyu, F.L., Wang, Q.F., Choi, S.M., and Yin, Y.D. (2019). Noble-metal-free electrocatalysts for oxygen evolution. *Small* 15: 1804201.

45 Bhunia, K., Chandra, M., Sharma, S.K. et al. (2023). A critical review on transition metal phosphide based catalyst for electrochemical hydrogen evolution reaction: Gibbs free energy, composition, stability, and true identity of active site. *Coordination Chemistry Reviews* 478: 214956.

46 Zhang, H.J., Maijenburg, A.W., Li, X.P. et al. (2020). Bifunctional heterostructured transition metal phosphides for efficient electrochemical water splitting. *Advanced Functional Materials* 30: 2003261.

47 Sahoo, D.P., Das, K.K., Mansingh, S. et al. (2022). Recent progress in first row transition metal layered double hydroxide (LDH) based electrocatalysts towards water splitting: a review with insights on synthesis. *Coordination Chemistry Reviews* 469: 214666.

48 Batool, M., Hameed, A., and Nadeem, M.A. (2023). Recent developments on iron and nickel-based transition metal nitrides for overall water splitting: a critical review. *Coordination Chemistry Reviews* 480: 215029.

49 Zhang, Q., Chen, W., Chen, G.L. et al. (2020). Bi-metallic nitroxide nanodot-decorated tri-metallic sulphide nanosheets by on-electrode plasma-hydrothermal sprouting for overall water splitting. *Applied Catalysis B: Environmental* 261: 118254.

50 Song, H.Q., Yu, J.K., Tang, Z.Y. et al. (2022). Halogen-doped carbon dots on amorphous cobalt phosphide as robust electrocatalysts for overall water splitting. *Advanced Energy Materials* 12: 2102573.

51 Chen, J.P., Ren, B.W., Cui, H., and Wang, C.X. (2020). Constructing pure phase tungsten-based bimetallic carbide nanosheet as an efficient bifunctional electrocatalyst for overall water splitting. *Small* 16: 1907556.

52 Wang, J.J., Huang, J., Chen, G.L. et al. (2022). In-situ engineered heterostructured nickel tellur-selenide nanosheets for robust overall water splitting. *Chemical Engineering Journal* 446: 137297.

53 Wang, Y., Wang, S., Ma, Z.L. et al. (2022). Competitive coordination-oriented monodispersed ruthenium sites in conductive MOF/LDH Hetero-nanotree catalysts for efficient overall water splitting in alkaline media. *Advanced Materials* 34: e2107488.

54 Lv, X.S., Wei, W., Wang, H. et al. (2020). Holey graphitic carbon nitride (g-CN) supported bifunctional single atom electrocatalysts for highly efficient overall water splitting. *Applied Catalysis B: Environmental* 264: 118521.

55 Duan, J.J., Chen, S., and Zhao, C. (2017). Ultrathin metal-organic framework array for efficient electrocatalytic water splitting. *Nature Communications* 8: 15341.

56 Joo, J., Kim, T., Lee, J. et al. (2019). Morphology-controlled metal sulfides and phosphides for electrochemical water splitting. *Advanced Materials* 31: e1806682.

57 Zhu, C., Zhu, M.M., Sun, Y. et al. (2018). Defects induced efficient overall water splitting on a carbon-based metal free photocatalyst. *Applied Catalysis B: Environmental* 237: 166–174.

58 Zhang, B., Yang, F., Liu, X.D. et al. (2021). Phosphorus doped nickel-molybdenum aerogel for efficient overall water splitting. *Applied Catalysis B: Environmental* 298: 120494.

59 Zhang, B., Shan, J.W., Wang, W.L. et al. (2022). Oxygen vacancy and core-shell heterojunction engineering of anemone-like CoP@CoOOH bifunctional electrocatalyst for efficient overall water splitting. *Small* 18: 2106012.

60 Lin, X.L., Liu, J.L., Qiu, X.Q. et al. (2023). Ru-FeNi alloy heterojunctions on lignin-derived carbon as bifunctional electrocatalysts for efficient overall water splitting. *Angewandte Chemie International Edition* 62: e202306333.

61 Suryanto, B.H.R., Wang, Y., Hocking, R.K. et al. (2019). Overall electrochemical splitting of water at the heterogeneous interface of nickel and iron oxide. *Nature Communications* 10: 5599.

62 Andaveh, R., Rouhaghdam, A.S., Ai, J.P. et al. (2023). Boosting the electrocatalytic activity of NiSe by introducing MnCo as an efficient heterostructured electrocatalyst for large-current-density alkaline seawater splitting. *Applied Catalysis B: Environment and Energy* 325: 122355.

63 Zhang, J.T., Xia, Z.H., and Dai, L.M. (2015). Carbon-based electrocatalysts for advanced energy conversion and storage. *Science Advances* 1: 1500564.

64 Zeng, Y.C., Liang, J.S., Li, C.Z. et al. (2023). Regulating catalytic properties and thermal stability of Pt and PtCo intermetallic fuel-cell catalysts via strong coupling effects between single-metal site-rich carbon and Pt. *Journal of the American Chemical Society* 145: 17643–17655.

65 Ma, F.-X., Liu, Z.-Q., Zhang, G. et al. (2022). Isolating Fe atoms in N-doped carbon hollow nanorods through a ZIF-phase-transition strategy for efficient oxygen reduction. *Small* 18: 2205033.

66 Schlögl, R. (2003). Catalytic synthesis of ammonia: a "never-ending story"? *Angewandte Chemie International Edition* 42: 2004–2008.

67 Wang, S.Y., Ichihara, F., Pang, H. et al. (2018). Nitrogen fixation reaction derived from nanostructured catalytic materials. *Advanced Functional Materials* 28: 1803309.

68 Klerke, A., Christensen, C.H., Norskov, J.K., and Vegge, T. (2008). Ammonia for hydrogen storage: challenges and opportunities. *Journal of Materials Chemistry* 18: 2304–2310.

69 Qi, P., Gao, X.X., Wang, J. et al. (2022). A minireview on catalysts for photocatalytic N_2 fixation to synthesize ammonia. *RSC Advances* 12: 1244–1257.

70 Spatzal, T., Aksoyoglu, M., Zhang, L.M. et al. (2011). Evidence for interstitial carbon in nitrogenase FeMo cofactor. *Science* 334: 940–940.

71 Wen, L.L., Sun, K., Liu, X.S. et al. (2023). Electronic state and microenvironment modulation of metal nanoparticles stabilized by MOFs for boosting electrocatalytic nitrogen reduction. *Advanced Materials* 35: 2210669.

72 Wan, Y.C., Xu, J.C., and Lv, R.T. (2019). Heterogeneous electrocatalysts design for nitrogen reduction reaction under ambient conditions. *Materials Today* 27: 69–90.

73 Ling, C.Y., Ouyang, Y.X., Li, Q. et al. (2019). A general two-step strategy-based high-throughput screening of single atom catalysts for nitrogen fixation. *Small Methods* 3: 1800376.

74 Lai, F.L., Feng, J.R., Ye, X.B. et al. (2020). Oxygen vacancy engineering in spinel-structured nanosheet wrapped hollow polyhedra for electrochemical nitrogen fixation under ambient conditions. *Journal of Materials Chemistry A* 8: 1652–1659.

75 Zhang, L., Ji, X.Q., Ren, X. et al. (2018). Electrochemical ammonia synthesis via nitrogen reduction reaction on a MoS_2 catalyst: theoretical and experimental studies. *Advanced Materials* 30: 1800191.

76 Liu, Y.T., Chen, X.X., Yu, J.Y., and Ding, B. (2019). Carbon-nanoplated CoS@TiO_2 nanofibrous membrane: an interface-engineered heterojunction for high-efficiency electrocatalytic nitrogen reduction. *Angewandte Chemie International Edition* 58: 18903–18907.

77 Wu, T.W., Kong, W.H., Zhang, Y. et al. (2019). Greatly enhanced electrocatalytic N_2 reduction on TiO_2 via V doping. *Small Methods* 3: 1900356.

78 Tong, W., Huang, B.L., Wang, P.T. et al. (2021). Exposed facet-controlled N_2 electroreduction on distinct Pt_3Fe nanostructures of nanocubes, nanorods and nanowires. *National Science Review* 8: nwaa088.

79 Ali, M., Zhou, F.L., Chen, K. et al. (2016). Nanostructured photoelectrochemical solar cell for nitrogen reduction using plasmon-enhanced black silicon. *Nature Communications* 7: 11335.

80 Zhao, R.B., Liu, C.W., Zhang, X.X. et al. (2020). An ultrasmall Ru_2P nanoparticles-reduced graphene oxide hybrid: an efficient electrocatalyst for NH_3 synthesis under ambient conditions. *Journal of Materials Chemistry A* 8: 77–81.

81 Wu, T.W., Zhao, H.T., Zhu, X.J. et al. (2020). Identifying the origin of Ti^{3+} activity toward enhanced electrocatalytic N_2 reduction over TiO_2 nanoparticles modulated by mixed-valent copper. *Advanced Materials* 32: 2000299.

82 Jia, H.P. and Quadrelli, E.A. (2014). Mechanistic aspects of dinitrogen cleavage and hydrogenation to produce ammonia in catalysis and organometallic chemistry: relevance of metal hydride bonds and dihydrogen. *Chemical Society Reviews* 43: 547–564.

83 Jaccarino, V., Shulman, R.G., and Stout, J.W. (1957). Nuclear magnetic resonance in paramagnetic iron group fluorides. *Physical Review* 106: 602–603.

84 Nguyen, M.T., Seriani, N., and Gebauer, R. (2015). Nitrogen electrochemically reduced to ammonia with hematite: density-functional insights. *Physical Chemistry Chemical Physics* 17: 14317–14322.

85 Kong, J., Lim, A., Yoon, C. et al. (2017). Electrochemical synthesis of NH_3 at low temperature and atmospheric pressure using a γ-Fe_2O_3 catalyst. *ACS Sustainable Chemistry & Engineering* 5: 10986–10995.

86 Cui, X.Y., Tang, C., Liu, X.M. et al. (2018). Highly selective electrochemical reduction of dinitrogen to ammonia at ambient temperature and pressure over iron oxide catalysts. *Chemistry—A European Journal* 24: 18494–18501.

87 Xiang, X.J., Wang, Z., Shi, X.F. et al. (2018). Ammonia synthesis from electrocatalytic N_2 reduction under ambient conditions by Fe_2O_3 nanorods. *ChemCatChem* 10: 4530–4535.

88 Liu, C.W., Li, Q.Y., Wu, C.Z. et al. (2019). Single-boron catalysts for nitrogen reduction reaction. *Journal of the American Chemical Society* 141: 2884–2888.

89 Chen, S.M., Perathoner, S., Ampelli, C. et al. (2017). Electrocatalytic synthesis of ammonia at room temperature and atmospheric pressure from water and nitrogen on a carbon-nanotube-based electrocatalyst. *Angewandte Chemie, International Edition* 56: 2699–2703.

90 Chu, K., Gu, W., Li, Q. et al. (2021). Amorphization activated FeB_2 porous nanosheets enable efficient electrocatalytic N_2 fixation. *Journal of Energy Chemistry* 53: 82–89.

91 Yao, C.H., Guo, N., Xi, S.B. et al. (2020). Atomically-precise dopant-controlled single cluster catalysis for electrochemical nitrogen reduction. *Nature Communications* 11: 4389.

92 Han, M., Guo, M.H., Yun, Y.P. et al. (2022). Effect of heteroatom and charge reconstruction in atomically precise metal nanoclusters on electrochemical synthesis of ammonia. *Advanced Functional Materials* 32: 2202820.

93 Woldu, A.R., Huang, Z.L., Zhao, P.X. et al. (2022). Electrochemical CO_2 reduction (CO_2RR) to multi-carbon products over copper-based catalysts. *Coordination Chemistry Reviews* 454: 214340.

94 Xu, L., Trogadas, P., and Coppens, M.-O. (2023). Nature-inspired electrocatalysts for CO_2 reduction to C_{2+} products. *Advanced Energy Materials* 13: 2302974.

95 Han, N., Ding, P., He, L. et al. (2020). Promises of main group metal-based nanostructured materials for electrochemical CO_2 reduction to formate. *Advanced Energy Materials* 10: 1902338.

96 Nielsen, D.U., Hu, X.M., Daasbjerg, K., and Skrydstrup, T. (2018). Chemically and electrochemically catalysed conversion of CO_2 to CO with follow-up utilization to value-added chemicals. *Nature Catalysis* 1: 244–254.

97 Wu, J.J., Ma, S.C., Sun, J. et al. (2016). A metal-free electrocatalyst for carbon dioxide reduction to multi-carbon hydrocarbons and oxygenates. *Nature Communications* 7: 13869.

98 Zhu, D.D., Liu, J.L., and Qiao, S.Z. (2016). Recent advances in inorganic heterogeneous electrocatalysts for reduction of carbon dioxide. *Advanced Materials* 28: 3423–3452.

99 Mohanty, B., Basu, S., and Jena, B.K. (2022). Transition metal-based single-atom catalysts (TM-SACs); rising materials for electrochemical CO_2 reduction. *Journal of Energy Chemistry* 70: 444–471.

100 Lü, F.L., Bao, H.H., Mi, Y.Y. et al. (2020). Electrochemical CO_2 reduction: from nanoclusters to single atom catalysts. *Sustainable Energy & Fuels* 4: 1012–1028.

101 Gong, L.L., Zhang, D.T., Lin, C.Y. et al. (2019). Catalytic mechanisms and design principles for single-atom catalysts in highly efficient CO_2 conversion. *Advanced Energy Materials* 9: 1902625.

102 Zhang, X.L., Guo, S.X., Gandionco, K.A. et al. (2020). Electrocatalytic carbon dioxide reduction: from fundamental principles to catalyst design. *Materials Today Advances* 7: 100074.

103 Wang, G.X., Chen, J.X., Ding, Y.C. et al. (2021). Electrocatalysis for CO_2 conversion: from fundamentals to value-added products. *Chemical Society Reviews* 50: 4993–5061.

104 Nitopi, S., Bertheussen, E., Scott, S.B. et al. (2019). Progress and perspectives of electrochemical CO_2 reduction on copper in aqueous electrolyte. *Chemical Reviews* 119: 7610–7672.

105 Pan, F.P., Deng, W., Justiniano, C., and Li, Y. (2018). Identification of champion transition metals centers in metal and nitrogen-codoped carbon catalysts for CO_2 reduction. *Applied Catalysis B: Environmental* 226: 463–472.

106 Su, J.W., Pan, D.H., Dong, Y. et al. (2023). Ultrafine Fe_2C iron carbide nanoclusters trapped in topological carbon defects for efficient electroreduction of carbon dioxide. *Advanced Energy Materials* 13: 2204391.

107 Wei, H., Tang, C., Wang, Y. et al. (2025). Intermetallic compounds for nitrogen electrochemistry. *Green Energy & Environment* 10: 268–291.

108 Yin, S., Guan, Z., Zhu, Y. et al. (2024). Highly efficient electrocatalytic nitrate reduction to ammonia: group VIII-based catalysts. *ACS Nano* 18: 27833–27852.

109 Zhang, M., Ma, Z., Zhou, S. et al. (2024). Surface engineering on Ag-decorated Co_3O_4 electrocatalysts for boosting nitrate reduction to ammonia. *ACS Catalysis* 14: 11231–11242.

110 Jin, T., Wang, J., Gong, Y. et al. (2023). Mechanochemical-tuning size dependence of iridium single atom and nanocluster toward highly selective ammonium production. *Chem Catalysis* 3: 100477.

111 Liu, Y., Jiang, X., Zhang, Y. et al. (2024). The interface-mediated electron structure tuning of RuO_x-Co_3O_4 nano-particles for efficient electrocatalytic nitrate reduction. *Dalton Transactions* 53: 162–170.

8

Applications of Iron (Fe, Co, and Ni)-Group Metal Compound Nanomaterials in Photocatalysis

The efficient conversion of solar energy into chemical fuels represents a pivotal strategy for addressing global energy and environmental challenges. Photocatalytic water splitting, in particular, offers a direct pathway for producing hydrogen, a clean and storable energy carrier. The widespread implementation of this technology, however, is hindered by the low efficiency and high cost associated with conventional precious metal-based catalysts. Iron-group metal (Fe, Co, Ni) compound nanomaterials have thus emerged as a highly promising class of materials, leveraging their Earth abundance, tunable electronic structures, and multifunctional catalytic properties. In photocatalytic systems, these compounds serve as critical components, acting as photosensitizers to broaden light absorption, facilitating the separation and migration of photogenerated charge carriers, and providing optimized active sites for surface redox reactions. Concurrently, in dye-sensitized solar cells (DSSCs), they function as efficient and stable counter electrodes (CEs), demonstrating electrocatalytic performance for triiodide reduction that rivals or even surpasses that of platinum. The rational design of these materials, encompassing dimensional control, heterojunction engineering, and interface modulation, is fundamental to enhancing their performance. Despite significant progress, challenges remain regarding long-term stability under operational conditions, the precise identification of active sites, and the scalability of synthesis methods. Future advancements in this field are anticipated to rely on the integration of advanced in situ characterization and theoretical simulations to unravel mechanistic details, guiding the design of next-generation, high-performance, and economically viable solar energy conversion systems based on iron-group metal compounds.

8.1 Photocatalytic Fuel Production

Environmental pollution and energy shortage represent core challenges to the sustainable development of human society today. Against this backdrop, the development of clean and renewable energy conversion technologies has become a global consensus. Solar energy, as a widely distributed and abundant clean energy source, stands as a key pathway to addressing energy and environmental issues.

Iron-Group Metal Compound Nanomaterials: Preparation, Characterization, Structure, and Applications, First Edition. Jinlong Zheng.

It is estimated that the total amount of solar energy reaching the Earth's surface far exceeds current global energy consumption by several orders of magnitude [1]. Therefore, the efficient conversion of solar energy into chemical fuels that are easy to store and transport is regarded as an ideal solution for achieving sustainable development. Among these, photocatalytic water-splitting technology can directly utilize sunlight to convert water into hydrogen. The produced hydrogen, as a clean fuel with high energy density, can serve as an energy carrier or a basic raw material for the chemical industry, offering new perspectives for carbon cycle regulation [2–5].

Photocatalysis represents the chemical behavior of photogenerated charge carriers, manifesting as the conversion of solar energy into chemical energy. Driven by the electric field force, these carriers migrate to the catalyst surface and participate in water oxidation reactions. The principle relies on the strong reducing and oxidizing capabilities of photogenerated electrons and holes, respectively, which can drive the redox reactions of water to produce hydrogen and oxygen. Figure 8.1 illustrates the general mechanism of semiconductor-based photocatalysts.

The milestone for photocatalytic water-splitting technology was set in 1972 when Fujishima and Honda first reported that a TiO_2 electrode could catalyze water splitting to produce hydrogen under ultraviolet light [7]. Since then, semiconductor photocatalysis has attracted extensive research interest. A typical photocatalytic process involves three key steps: (i) the semiconductor photocatalyst absorbs photons with energy equal to or greater than its bandgap, generating photogenerated electron–hole pairs; (ii) the photogenerated charge carriers separate and migrate under the influence of the built-in electric field; (iii) the separated electrons and holes migrate to the catalyst surface, driving the hydrogen evolution reaction (HER) and oxygen evolution reaction (OER), respectively [8].

However, the solar energy conversion efficiency remains very low due to several challenges: most semiconductor materials have wide bandgaps, allowing only ultraviolet light (which constitutes about 5% of the solar spectrum) to be absorbed, resulting in very low utilization of visible light; the rapid recombination of photogenerated electron–hole pairs; and the presence of reversible reactions between the formed hydrogen and oxygen [9–13].

To overcome these bottlenecks, researchers have developed various strategies. Doping and sensitization strategies can significantly enhance light absorption

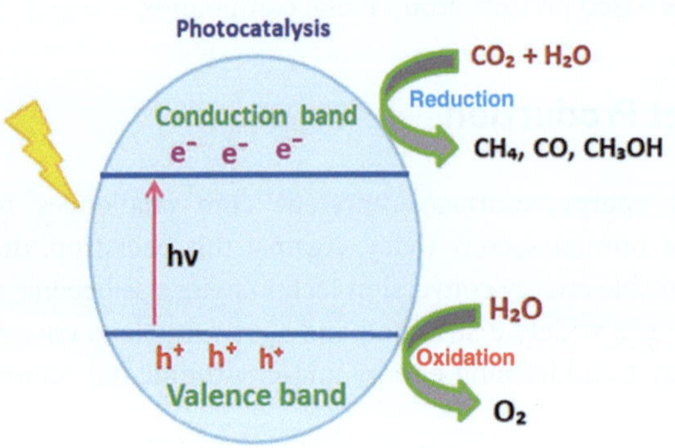

Figure 8.1 General mechanism of photocatalysis in semiconductor-based catalysts. *Source:* Reproduced from Ref. [6] / with permission of Elsevier.

efficiency, thereby improving photocatalytic performance [14, 15]; alternatively, constructing heterostructures using semiconductors with suitable bandgaps or metals can improve the separation and migration efficiency of photoexcited charge carriers [16, 17]. Furthermore, the development of low-cost, high-efficiency cocatalysts is particularly important. Selecting appropriate metals or semiconductor photocatalysts as cocatalysts can not only enhance the separation and transfer of photoexcited charge carriers but, most importantly, improve the utilization efficiency of photogenerated electrons and holes on the surface of composite semiconductor photocatalysts [18–20]. Although precious metals (e.g., Pt and Au) have been extensively studied due to their exceptional catalytic activity, their high cost and scarcity severely limit large-scale applications [21, 22]. Therefore, exploring efficient alternatives based on non-precious metals has become a research focus.

Various photocatalysts, such as metal-free catalysts [23], metal oxides [24], metal sulfides [25], metal phosphides [26], and metal selenides [27], have been extensively studied. However, most of them still fail to achieve the desired hydrogen evolution rate and yield. In this context, iron-group metal (Fe, Co, Ni) compound nanomaterials have emerged as a highly promising class of photocatalytic materials due to their abundant reserves, low cost, and unique electronic structures and catalytic properties [19, 28–30]. Analysis of the photocatalytic mechanism reveals that to improve solar energy conversion efficiency, a photocatalyst must meet the following conditions: (i) appropriate conduction and valence band structures (i.e., bandgap width); (ii) high separation and migration rates of photogenerated electron–hole pairs; and (iii) a high number of active reaction sites. Therefore, the search for efficient water-splitting catalysts should focus on these three aspects. Iron-group metals can serve as photosensitizing components to broaden the spectral absorption range. Doping iron-group metal ions (e.g., Fe^{3+} and Co^{2+}) into wide-bandgap semiconductors (e.g., TiO_2) introduces impurity energy levels within the bandgap, effectively extending the light response into the visible region. Furthermore, some iron-group metal oxides themselves are high-performance visible-light-responsive semiconductors and can act as the primary light-absorbing materials. Iron-group metal compounds (e.g., NiO, CoP, and NiS) can serve as trapping centers for electrons or holes, effectively suppressing charge carrier recombination. Due to their variable valence states and partially filled d-orbitals, iron-group metal ions often exhibit optimized adsorption/desorption energies for reaction intermediates, thereby significantly reducing the overpotential for HER or OER. Certain Fe- and Co-based compounds (e.g., Fe_3O_4 and $CoFe_2O_4$) possess superparamagnetic or ferromagnetic properties, enabling rapid and efficient catalyst recovery and recycling post-reaction via an external magnetic field. This addresses the engineering challenge of separating nano-powder catalysts.

Despite the promising prospects of iron-group metal compounds, their application still faces numerous challenges. These include the potential biotoxicity of certain metal ions (e.g., Co^{2+}), chemical instability under specific reaction conditions (e.g., photocorrosion), and the need for further in-depth analysis of the fine structure of active sites and reaction mechanisms using advanced in situ techniques. In summary, iron-group metal compound nanomaterials offer rich possibilities for constructing efficient, stable, and low-cost photocatalytic systems, owing to their tunable band structures, efficient charge separation capabilities, superior surface reaction activity, and unique magnetic recovery properties. Currently, various iron-group metal-based compounds, such as oxides, sulfides, phosphides, and selenides, have been extensively explored for photocatalytic hydrogen production. This chapter will systematically review the latest research progress in Fe, Co, and Ni-based metal compound nanomaterials in the field of photocatalysis, particularly in photocatalytic water splitting for hydrogen production. It will focus on material design, mechanistic insights, and performance optimization strategies, while also providing an outlook on future directions and challenges in this field.

8.1.1 Fe-Based Photocatalyst

Iron-based nanomaterials are characterized by abundant reserves, low cost, excellent redox overpotential, and narrow band gaps. These include various forms such as oxides, sulfides, nanocomposites, and Fe-doped materials (Figure 8.2). Iron-based oxide semiconductors have proven to be one of the most promising materials for photocatalytic and photoelectrocatalytic applications in the fields of environmental remediation and solar fuel production [32].

Hematite (α-Fe_2O_3) has become a research focus due to its advantages such as narrow bandgap, nontoxicity, abundant reserves, and low cost. However, its inherent limitations, including low charge separation efficiency and short carrier lifetime, restrict the activity of the pristine material [33, 34]. To address these issues, strategies such as morphology control, doping, and surface engineering have been employed to modify hematite, with further extensions to structurally tunable novel material systems like Fe-based perovskites and spinels. Hematite possesses a corundum-like (α-Al_2O_3) structure, composed of FeO_6 octahedral layers. These

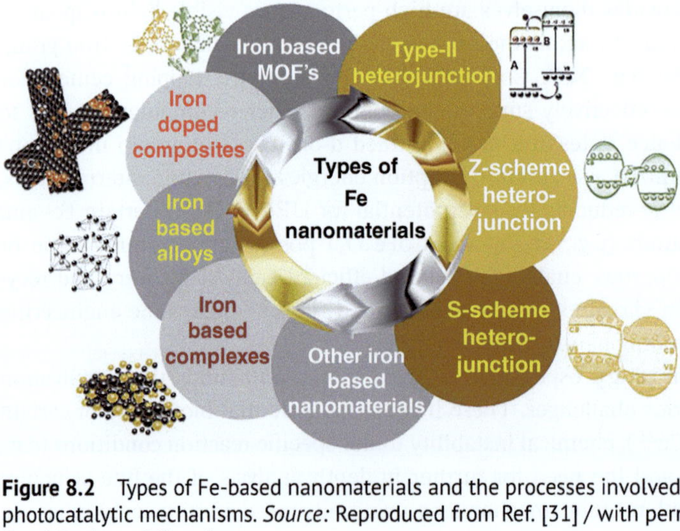

Figure 8.2 Types of Fe-based nanomaterials and the processes involved in their photocatalytic mechanisms. *Source:* Reproduced from Ref. [31] / with permission of The Royal Society of Chemistry.

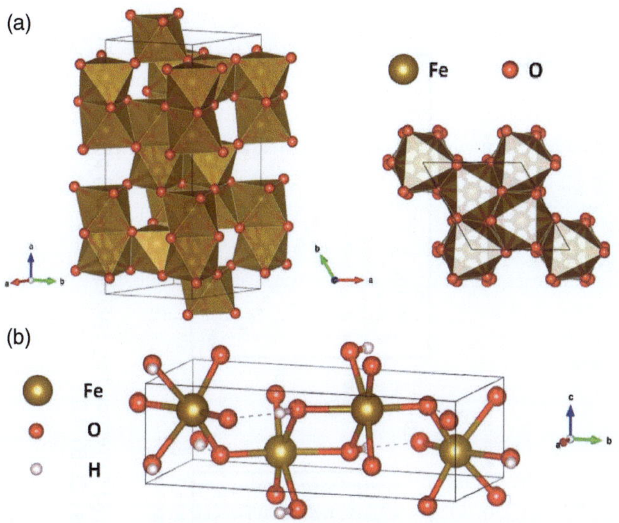

Figure 8.3 (a) Hematite unit cell (left). Its structure consists of layered FeO_6 octahedra connected via edge- and face-sharing; the octahedra are slightly distorted to optimize face-sharing. The FeO_6 units stack perpendicular to the c-axis, as visible in the view along the c-axis (right). (b) Goethite unit cell, showing the distinctive $FeO_3(HO)_3$ octahedra present in all iron oxyhydroxides (FeOOH). *Source:* Reproduced from Ref. [35] / with permission of Elsevier.

octahedra are connected via edge- and face-sharing and stack perpendicular to the c-axis (Figure 8.3a). This results in a hexagonal close-packed crystal system where two-thirds of the octahedral interstices are occupied by Fe^{3+} ions, while all tetrahedral sites remain unoccupied [36]. The high density (5260 kg m^{-3}) of hematite arises from the distortion of FeO_6 octahedra caused by the regular displacement of Fe^{3+} ions. This distortion enables optimal face-sharing of FeO_6 units.

Various crystalline phases of FeOOH (α, β, γ, δ, and high-pressure phases) are mostly semiconductors with photocatalytic potential [37]. The common structural unit of these isomers is the $FeO_3(HO)_3$ octahedron, and their different spatial arrangements determine the final crystal configuration, such as the orthorhombic unit cell of goethite (α-FeOOH) shown in Figure 8.3b [38]. However, the electrochemical stability of these materials, particularly goethite, under alkaline conditions, is a major bottleneck for their application. Under cathodic conditions, goethite is reduced to Fe(II), while under anodic potentials, it is rapidly oxidized to form soluble FeO_4^{2-} [39]. This severely limits its application in photoelectrocatalytic water splitting [40].

Although hematite (α-Fe_2O_3) faces inherent limitations such as low charge separation efficiency in photocatalytic applications, its performance can be significantly enhanced through sophisticated interface engineering design. Wu et al. [41] successfully constructed an M-Fe_2O_3@MXene heterojunction (Figure 8.4a) via an electrostatic self-assembly strategy, combining KH-550-modified hollow Fe_2O_3 nanocubes (M-Fe_2O_3) with monolayer Ti_3C_2 MXene. As shown in Figure 8.4b,

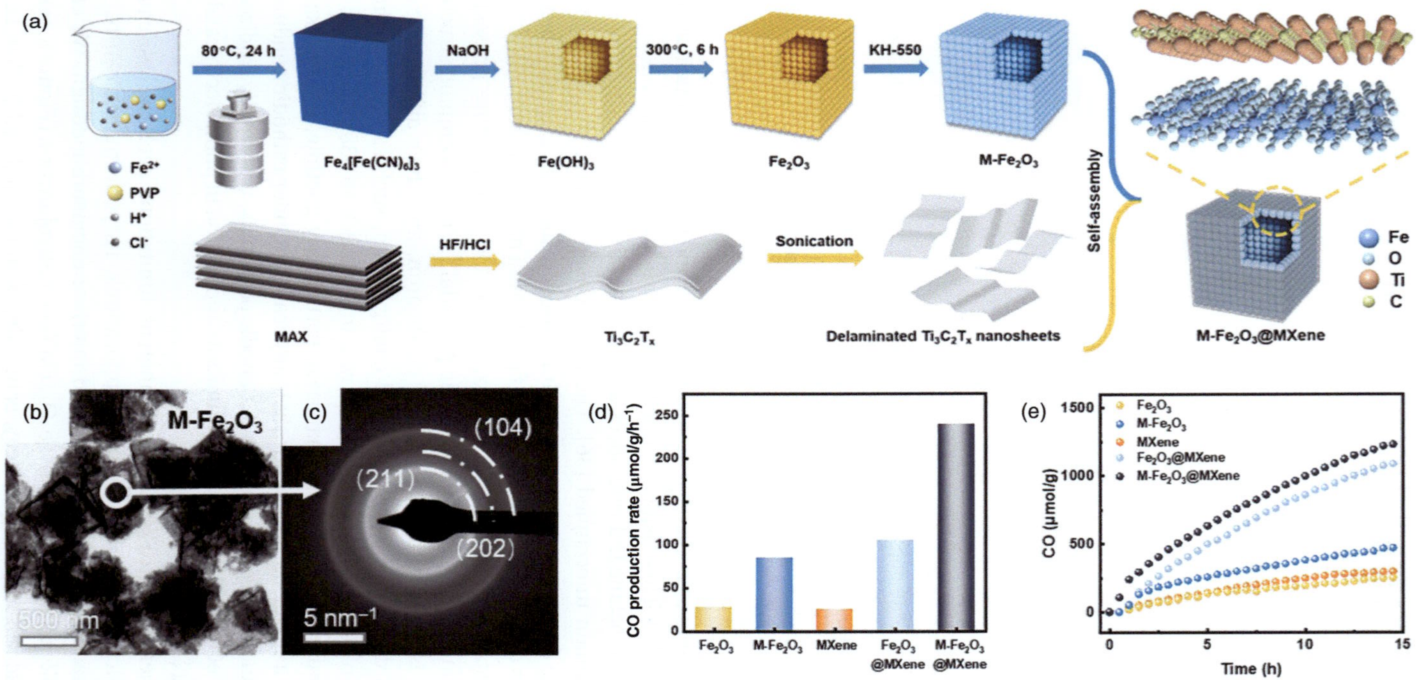

Figure 8.4 (a) Schematic diagram of the synthesis of M-Fe_2O_3@MXene; (b) TEM image of M-Fe_2O_3; (c) corresponding SAED pattern; (d) photocatalytic CO_2 conversion yield of the catalysts; (e) CO production amount. *Source:* Ref. [41] / John Wiley & Sons / CC BY 4.0.

the synthesized Fe-based material exhibits uniform cubic morphology with a side length of approximately 600 nm. The selected area electron diffraction (SAED) pattern of M-Fe_2O_3 confirms its polycrystalline α-Fe_2O_3 structure composed of small particles (Figure 8.4c). This hollow nanostructure provides a high specific surface area and short diffusion pathways, exposing more Fe–O active sites, while the surface modification effectively improves dispersion and promotes charge transport. The innovation of this design lies in the formation of critical Fe—O—Ti chemical bonds at the interface. This bonding not only serves as an efficient charge transfer channel, significantly promoting the separation and migration of photogenerated carriers, but more importantly, it modulates the electronic structure of the Fe^{3+} active sites, rendering them more electron-deficient. This electronic modulation substantially reduces the energy barrier for the formation of the key intermediate *COOH during CO_2 reduction (DFT calculations confirm a 22% reduction in the energy barrier). Ultimately, this catalyst demonstrates exceptional activity in photocatalytic CO_2 reduction, achieving a CO production rate of 240 $\mu mol\,g^{-1}\,h^{-1}$, which is 8.6 times higher than that of pristine Fe_2O_3, ranking among the highest performances reported for non-precious metal Fe-based photocatalysts (Figure 8.4d–e).

8.1.2 Co-Based Photocatalyst

Co-based materials exhibit exceptional photocatalytic performance in various applications, including photocatalytic water splitting, CO_2 reduction, environmental pollutant degradation, and hydrogen peroxide production, owing to their unique structural characteristics, strong light absorption capability, rapid charge carrier transfer, and efficient spatial separation. Wang et al. [42] constructed a NiCoOP-NPs@MHCFs catalyst by confining highly dispersed Ni–Co oxyphosphide nanoparticles (NiCoOP NPs) within multichannel hollow carbon fibers (MHCFs) for efficient CO_2 photoreduction (Figure 8.5a). In this catalyst, the ultrasmall and highly dispersed bimetallic oxyphosphide nanoparticles provide abundant catalytic active sites for redox reactions due to their unique properties. Meanwhile, the multichannel hollow carbon matrix with high electrical conductivity and open ends effectively promotes mass/charge transfer and enhances CO_2 adsorption (Figure 8.5b). Furthermore, this design helps prevent the aggregation of metal oxyphosphide nanoparticles, ensuring long-term stability and high efficiency of the catalyst. The optimized bimetallic oxyphosphide catalyst exhibits considerable activity for photosensitized CO_2 reduction, achieving a high CO evolution rate of 16.6 $\mu mol\,h^{-1}$ (Figure 8.5c).

In photocatalytic hydrogen production systems, the introduction of cocatalysts is an effective strategy to enhance the performance and selectivity of semiconductor photocatalysts. By establishing synergistic effects with the host semiconductor, cocatalysts significantly improve the separation efficiency of photogenerated charge carriers and the kinetics of surface reactions [43]. Co-based materials have shown broad application prospects in the HER due to their diverse compositions, tunable structures, and relatively low cost. Studies have demonstrated that various Co-based

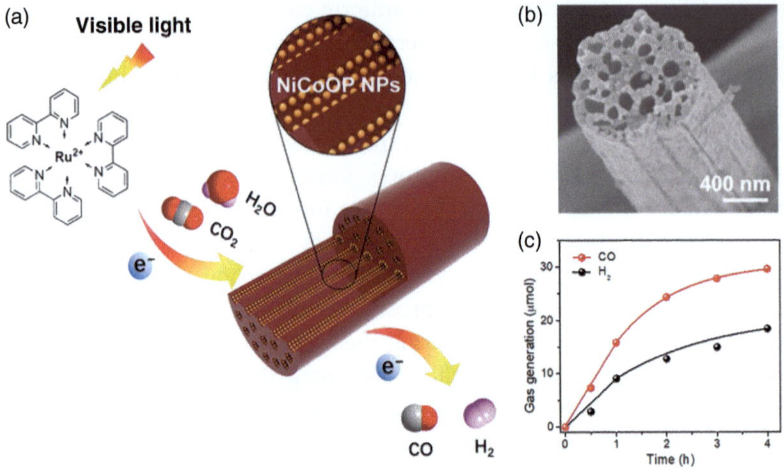

Figure 8.5 (a) Schematic diagram of NiCoOP NPs for CO_2 photoreduction. (b) FESEM image of NiCoOP-NPs@C_x fibers. (c) Relationship between reaction time and the generation of CO and H_2 catalyzed by NiCoOP-NPs@MHCFs. *Source:* Reproduced from Ref. [42] / with permission of John Wiley & Sons.

materials—including metallic Co [44, 45], oxides [46–51], hydroxides [52, 53], phosphides [47, 54, 55], and sulfides—can serve as efficient cocatalysts, exhibiting excellent properties in terms of electrical conductivity, band structure, active site density, and chemical stability.

Co-based cocatalysts primarily fulfill several key functions in photocatalytic hydrogen evolution: they act as electron trapping centers, facilitating the migration of photogenerated electrons from the semiconductor to the reaction interface, and they provide abundant surface active sites, promoting the reduction of protons to hydrogen atoms and ultimately the formation of hydrogen molecules (Figure 8.6).

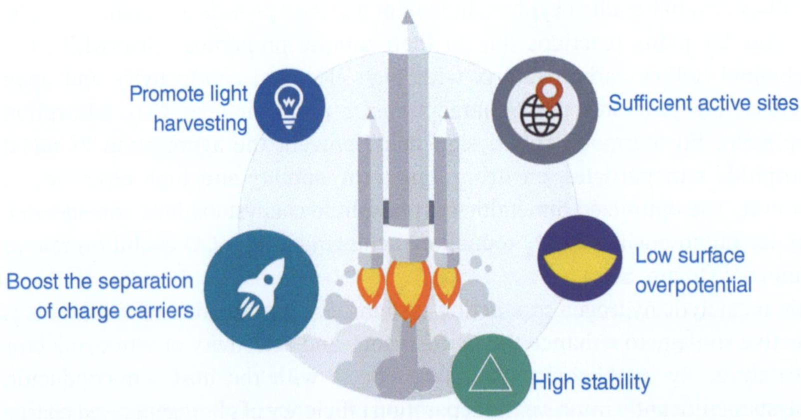

Figure 8.6 Schematic diagram of photocatalytic hydrogen production using a semiconductor photocatalyst: the role of Co-based cocatalysts. *Source:* Reproduced from Ref. [56] / with permission of Elsevier.

The reaction pathway mainly involves the adsorption of protons on active sites, the combination of electrons and protons to form hydrogen atoms, and the desorption of hydrogen molecules. To achieve efficient and sustained hydrogen evolution performance, an ideal Co-based cocatalyst should possess a hydrogen adsorption free energy close to zero, thereby optimizing both the thermodynamics and kinetics of the reaction [22, 57].

Understanding the mechanism of cocatalysts is crucial for constructing photocatalytic composite systems with well-matched band structures. Their electronic structure, band alignment, and surface active site characteristics collectively determine the separation and transport behavior of electron–hole pairs [58–60]. Therefore, the systematic regulation of intrinsic parameters of cocatalysts—such as chemical composition, microstructure, crystal phase, exposed facets, and particle size—is key to optimizing their functionality. Furthermore, Co-based cocatalysts can extend the light absorption range of the host semiconductor, enhance photon utilization efficiency, and significantly reduce the overpotential and activation energy barrier of the HER, thereby comprehensively improving photocatalytic performance [61]. It is noteworthy that even with identical elemental composition, Co-based compounds with different stoichiometries can exhibit distinctly different catalytic behaviors due to variations in their electronic structures and metal oxidation states. Thus, precise control over the composition of cocatalysts is essential for optimizing their surface properties and enhancing photocatalytic performance.

The intrinsic microstructure of cocatalysts fundamentally governs the transfer direction, pathways, and distance of charge carriers, as well as the specific surface area, which originates from their dimensional characteristics such as zero-dimensional (0D) quantum dots, one-dimensional (1D), two-dimensional (2D), and three-dimensional (3D) structures [62]. As shown in Figure 8.7a, one study utilized ZIF-67-derived CoP quantum dots anchored on a graphitized carbon skeleton as a cocatalyst supported on CdS nanorods, significantly enhancing photocatalytic hydrogen production performance [63]. This enhancement is primarily attributed to the formation of a Schottky junction between CoP and CdS, which facilitates the efficient separation and migration of photogenerated charge carriers. In this composite structure, the photocatalytic hydrogen production rate of CoP/CdS reached 104,947 $\mu mol\ h^{-1}\ g^{-1}$, far exceeding that of pristine CdS (1916 $\mu mol\ h^{-1}\ g^{-1}$). However, despite the excellent electron modulation capability of quantum dots, their uneven distribution on the semiconductor surface may still limit further performance improvement. In comparison, 1D nanostructures such as nanorods and nanowires, due to their high specific surface area and well-defined geometry, are more conducive to promoting rapid transport of photogenerated charges and exposure of reactive sites. As shown in Figure 8.7b, one study significantly improved photocatalytic hydrogen production performance by combining $Zn_{0.50.5}Cd_{0.5}S$ with 1D CoP nanowires, mainly benefiting from the high metallic conductivity of CoP nanowires and the synergistic effects conferred by the 1D nanostructure [68]. The hydrogen production rate of $Zn_{0.5}Cd_{0.5}S$/CoP NWs reached 12,175.8 $\mu mol\ h^{-1}\ g^{-1}$, approximately twice that of $Zn_{0.5}Cd_{0.5}S$/CoP NPs, demonstrating performance superior to nanoparticle cocatalyst systems. Nevertheless,

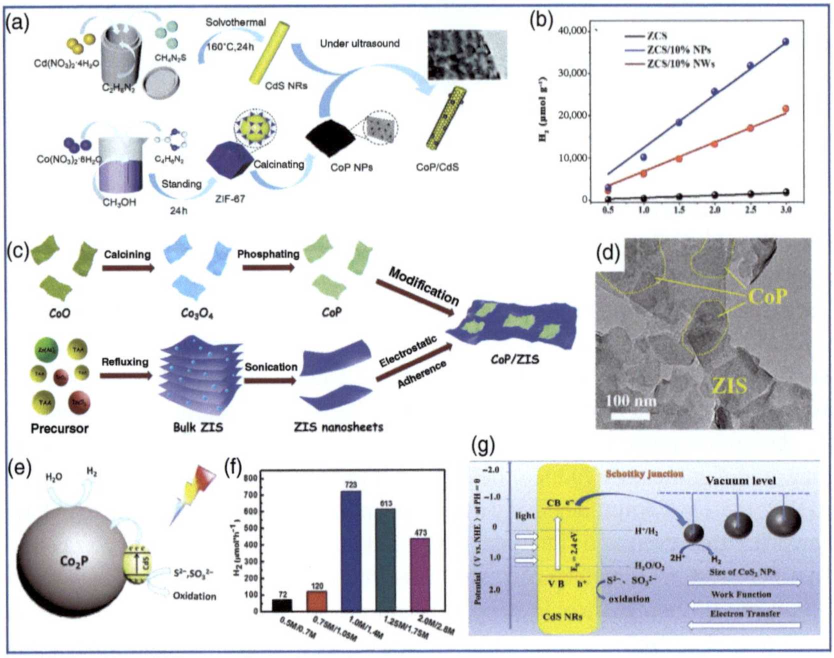

Figure 8.7 (a) 0D CoP/CdS composite photocatalyst. *Source:* Reproduced from Ref. [63] / with permission of The Royal Society of Chemistry. (b) Photocatalytic hydrogen production curves of ZCS, ZCS/10% CoP NWs, and ZCS/10% CoP NPs. *Source:* Reproduced from Ref. [64] / with permission of Elsevier. (c) Schematic diagram of the synthesis of 2D-2D CoP/$ZnIn_2S_4$ photocatalyst; (d) TEM image. *Source:* Reproduced from Ref. [65] / with permission of John Wiley & Sons. (e) Schematic diagram of photocatalytic hydrogen evolution over CdS/Co_2P; (f) H_2 evolution of CdS-15/Co_2P photocatalyst under different concentrations of Na_2S/Na_2SO_3. *Source:* Reproduced from Ref. [66] / with permission of Elsevier. (g) Schematic diagram of the H_2 production mechanism for CoS_2/CdS composites with different CoS_2 particle sizes. *Source:* Reproduced from Ref. [67] / with permission of Elsevier.

limitations in the interfacial contact area of zero-dimensional cocatalysts and the charge transport path length in 1D structures still constrain photocatalytic reaction efficiency to some extent. In contrast, 2D nanostructures show greater potential in promoting photocatalytic HERs due to their larger interfacial contact area and multidimensional electron transport channels [20].

As shown in Figure 8.7c, d, Xiang et al. constructed a 2D CoP/$ZnIn_2S_4$ heterojunction via electrostatic self-assembly, where the intimate interfacial contact effectively promoted charge separation and transport, increasing the hydrogen evolution rate to 8.775 mmol h^{-1} g^{-1}—significantly higher than that of pure $ZnIn_2S_4$ (1.447 mmol h^{-1} g^{-1}) [65]. This performance enhancement stems from improved light absorption, optimized charge transfer dynamics, and the active sites provided by the Schottky junction. It should be noted that the interfacial interactions in 2D structures require precise regulation to prevent agglomeration. In contrast, 3D structures exhibit unique advantages in photocatalytic hydrogen evolution due to their porous nature, multiple light scattering, and high light-harvesting capability. As illustrated by the mechanism in Figure 8.7e, Co_2P microspheres act as efficient electron acceptors in the composite, significantly accelerating the separation and migration of photogenerated charges. The 3D structure also provides abundant reactive interfaces and optimized proton diffusion pathways. Under visible light, the optimal hydrogen evolution rate reached 723 μmol h^{-1} g^{-1}, with a quantum efficiency of 23.6% at 470 nm (Figure 8.7f). This improvement is attributed to the synergistic effect between the electronic structure of Co_2P and the composite architecture [69]. As summarized in Figure 8.7g, Qiu et al. systematically investigated the size effect of CoS_2 cocatalysts in the CoS_2/CdS system. By tuning the CoS_2 particle size between 9 and 45 nm, they found that the sample with 9 nm CoS_2 exhibited the best performance, achieving a hydrogen production rate 13 times that of pure CdS. This phenomenon is attributed to the larger work function difference between the smaller CoS_2 nanoparticles and CdS, which effectively promotes interfacial electron transfer. However, smaller cocatalyst size is not always better; a balance must be struck between the electron capture capability at the semiconductor/metal interface and the electron release rate at the metal/solution interface [70]. Therefore, optimizing the size of cocatalysts is essential for achieving efficient photocatalysis.

In photocatalytic processes, variations in the particle size of Pt cocatalysts exert differential effects on distinct kinetic steps of charge transfer. As the particle size increases, the electron capture efficiency at the semiconductor/metal interface is enhanced; however, the electron injection rate at the metal/solution interface concurrently decreases [71]. Therefore, precise regulation of cocatalyst particle size is necessary to achieve an optimal balance between electron capture capability and interfacial electron injection rate, thereby meeting the practical application requirements of different semiconductor materials. It is noteworthy that the specific location of the cocatalyst on the host semiconductor surface directly determines the direction of charge transfer, due to the precise orientation of adsorption and activation sites for reactant molecules [22, 55, 72, 73].

The structure of cocatalysts across different dimensions significantly influences interfacial properties. Zero-dimensional quantum dots (e.g., CoP QDs) enable atomic-level dispersion and efficient electron modulation; 1D nanowires (e.g., CoP NWs) provide directional electron transport pathways through their linear structures; 2D nanosheets (e.g., $CoP/ZnIn_2S_4$) form large-area interfaces via face-to-face contact; while 3D structures (e.g., Co_2P microspheres) facilitate proton and reactant transport owing to their porous characteristics. These structural designs across various dimensions collectively demonstrate the central role of interface engineering in optimizing charge separation and surface reactions.

Co phosphides (e.g., CoP) enhance hydrogen proton adsorption through their phosphorus components, while Co oxides (e.g., CoO_x) optimize interfacial contact via core-shell structural design. Both achieve improved electron transfer pathways and enhanced surface reaction kinetics through precise interface regulation [43]. These findings collectively demonstrate that rational interfacial structure design serves as a crucial factor in achieving efficient photocatalytic hydrogen evolution.

In summary, by precisely regulating the dimensionality, size, and interfacial structure of cocatalysts, the separation and transport of photogenerated charges can be effectively optimized, thereby enhancing photocatalytic hydrogen evolution performance [51, 74–76]. Future research should focus on developing novel heterojunction interfaces and delving into the charge transfer mechanisms at these interfaces to provide theoretical guidance for designing efficient photocatalytic systems. Despite significant progress, several key directions require in-depth exploration to achieve large-scale practical application.

1) **Expanding Material Dimensions:** from composition innovation to multi-parameter synergistic control. Current research on Co-based cocatalysts has expanded from initial oxides like CoO and Co_3O_4 to various forms such as hydroxides, sulfides, phosphides, and molecular complexes, fully demonstrating the diversity of this material family. However, studies on emerging systems like Co nitrides, carbides, and borides are still in their infancy. These materials may possess higher conductivity and unique surface properties, representing an important source of future material innovation. Furthermore, single-metal components often face performance bottlenecks, whereas multi-metal alloys or composite compounds (e.g., CoMo-, CoW-, FeCo-, or NiCo-based materials) can optimize the d-band center and adsorption energy of reaction intermediates through electronic synergy between components, thereby surpassing the activity limits of single Co-based materials. Future research should not only focus on synthesizing new materials but also emphasize the integrated design and precise control of multiple parameters such as material size, microstructure (e.g., hollow and porous), exposed crystal facets, and crystallinity. Maximizing intrinsic catalytic activity through multifactor synergy is essential.
2) **Coupling with Novel Semiconductors and Interface Engineering:** Currently, the successful coupling of Co-based cocatalysts is mainly limited to a few classic semiconductors such as CdS, g-C_3N_4, and TiO_2. Future development should follow two parallel paths: first, actively exploring coupling with novel semiconductors (e.g., halide perovskites, covalent organic frameworks [COFs], and MXene) to construct unprecedented efficient composite photocatalytic systems. Second, deepening the understanding and optimization of existing composite material systems. Moving beyond the simple concept of "loading", systematic research should focus on deeper aspects such as multidimensional microstructure design, rational construction of heterojunction types (Type II, Z-scheme, or S-scheme), interfacial chemical bonding, and built-in electric fields. Clarifying the structure-activity relationship between reaction parameters and final structural characteristics is key to maximizing the matching of light absorption, carrier lifetime, and the number of active sites.
3) Unveiling microscopic mechanisms with advanced characterization and theoretical calculations. The exceptional co-catalytic performance of Co-based materials has been confirmed by numerous experiments, but the underlying microscopic mechanisms—particularly the catalytic kinetics on different semiconductor surfaces, the precise active centers, and the dynamic separation and

injection processes of photogenerated charges—remain challenging. Addressing this requires the deep integration of in situ/operando advanced characterization techniques and theoretical calculations. For example, in situ X-ray absorption spectroscopy (XAS) can track the evolution of cobalt's valence state and coordination environment; femtosecond transient absorption spectroscopy (fs-TAS) can directly observe ultrafast interfacial charge transfer kinetics; and synchrotron radiation techniques can resolve atomic-level local structures. Meanwhile, density functional theory (DFT) calculations can accurately determine key parameters such as hydrogen adsorption free energy (ΔG_{H*}) and simulate reaction pathways, providing a solid scientific basis for understanding surface reactions and interfacial charge transfer processes. These approaches are essential for elucidating structure-activity relationships at the molecular and electronic levels and guiding the rational design of high-performance catalysts.

4) **Material and Process Optimization for Large-scale Production:** Scaling up Co-based photocatalytic materials from the laboratory to mass production and application is the ultimate goal and currently the weakest link. Achieving large-scale production faces dual challenges: first, the materials themselves must maintain high activity and long-term stability in complex aqueous environments, requiring integrated optimization of the atomic configuration, band alignment, and macroscopic structure of Co-based materials to resist photocorrosion and ion poisoning. Second, the synthesis processes must be simple, controllable, low-cost, and easily scalable. The scalability potential and cost-effectiveness of laboratory techniques such as hydrothermal/solvothermal methods, electrodeposition, and photodeposition need reassessment. Future research should place greater emphasis on performance evaluation under conditions close to practical applications and synergistically consider raw material costs, equipment requirements, and production cycles to pave the way for the industrialization of photocatalytic hydrogen production technology.

In conclusion, with their flexible band structures, diverse microstructures, and stable chemical properties, Co-based nanomaterials play a crucial role in the family of iron-group photocatalytic materials. Future research, driven by composition innovation, system integration, mechanistic investigation, and process optimization, holds promise for achieving major breakthroughs in the near future, advancing Co-based photocatalytic materials from fundamental research to practical applications in the field of solar energy conversion.

8.1.3 Ni-Based Photocatalyst

The introduction of cocatalysts in photocatalytic hydrogen production reactions enables the development of efficient composite photocatalysts, as suitable cocatalysts can reduce the activation energy of the reaction, provide active sites, and suppress reverse reactions. Research on photocatalytic cocatalysts is undergoing a paradigm shift from noble metals to earth-abundant elements. Among various transition metal materials, Ni-based cocatalysts (including elemental Ni, phosphides, and sulfides) have emerged as an important class of hydrogen evolution cocatalysts due to their low cost, high stability, and excellent hydrogen evolution performance [77–79]. Studies have deepened the understanding of their mechanism: metallic Ni, with its high work function of approximately 5.35 eV, acts as an efficient electron trapping center. It forms a Schottky junction at the interface with semiconductors, thereby significantly promoting the spatial separation of photogenerated charge carriers [80, 81]. Further research indicates that reducing the size of Ni to the nanoscale effectively utilizes its surface effects and quantum confinement effects, markedly enhancing charge transfer efficiency and surface reaction activity. For example, Chai et al. [82] constructed a high-performance Ni/CdS photocatalyst by loading ultrafine Ni nanocrystals (~5 nm) onto CdS, which exhibited high activity in hydrogen production via alcohol cleavage, fully demonstrating the effectiveness of nanostructure design and electronic structure modulation.

Nickel hydroxide ($Ni(OH)_2$) has been widely studied due to its strong CO_2 adsorption capacity, high catalytic activity, and low cost. Tao et al. [83] constructed an efficient photocatalytic system capable of synergistically promoting CO_2 adsorption and activation by precisely anchoring ruthenium (Ru) single atoms on vacancy-rich $Ni(OH)_x$ nanoplates (Figure 8.8a, b). A two-step method was employed: first, preparing the vacancy-rich $Ni(OH)_x$ support via a solvothermal method, followed by selectively anchoring Ru single atoms at the vacancy sites through photoreduction. Characterization results showed that the catalyst maintained its nanoplate morphology (Figure 8.8c), with uniform distribution of Ru, Ni, and O elements (Figure 8.8d). The key structural feature is the precise anchoring of single atoms around the vacancies, forming synergistic active centers. Performance tests indicated (Figure 8.8e) that the catalyst activity exhibited a "volcano-type" dependence, with optimal performance at 0.60 wt% Ru loading, achieving a CO production rate of 27.6 mmol g^{-1} h^{-1} and effectively suppressing the competing HER. Insufficient loading led to a lack of active sites, while excessive loading hindered charge transport due to over-occupation of vacancies. DFT calculations revealed the synergistic mechanism: the vacancies are responsible for the physical adsorption of CO_2, while the adjacent Ru single atoms significantly reduce the energy barrier for forming the key intermediate *COOH from 1.06 to 0.19 eV, thereby accelerating the reaction (Figure 8.8f). This mechanism reasonably explains the "volcano-type" trend in performance, indicating that a balanced quantity of vacancies and single-atom sites is crucial for achieving highly efficient catalysis.

Nickel oxides have long been investigated as cocatalysts in the field of photocatalysis and continue to demonstrate strong vitality. Early studies (such as the exploration of $NiO/SrTiO_3$ in 1986) revealed their potential for application in photocatalytic water splitting [84]. Subsequent research has deepened this understanding while simultaneously expanding the scope of investigation. Both NiO and $Ni(OH)_2$ have been successfully applied to modify various semiconductors such as TiO_2 and $NaTaO_3$, confirming their broad applicability. Wang et al. [85] successfully prepared TiO_2/NiO p-n junction hollow photocatalysts using a template-assisted method. The study demonstrated that the thickness of the TiO_2 shell could be precisely controlled by adjusting the amount of the titanium source TBOT, thereby

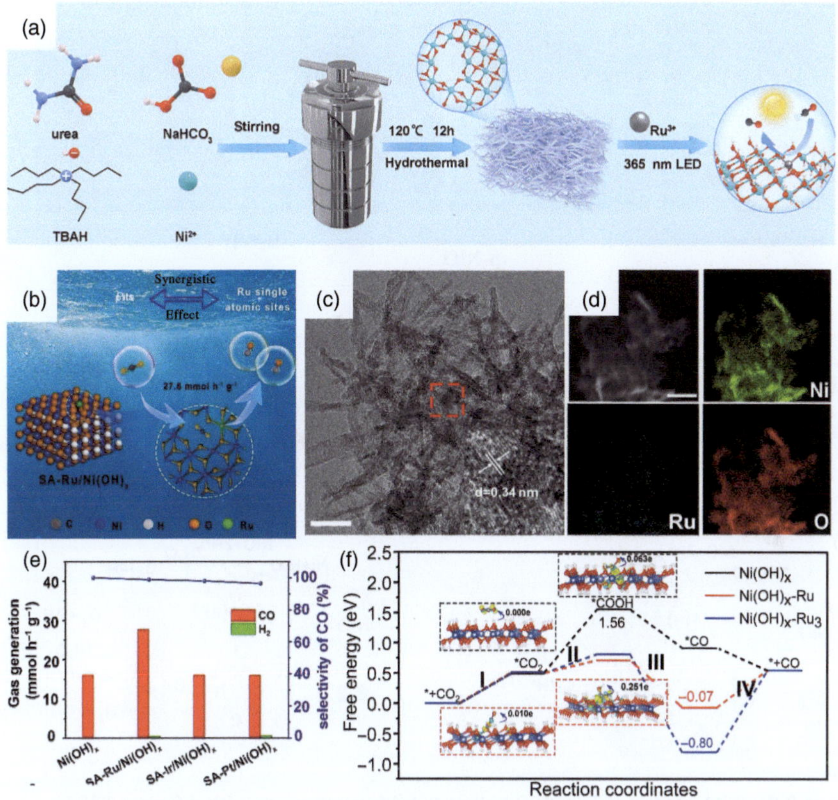

Figure 8.8 (a) Schematic diagram of the synthesis process of w-SA-Ru/Ni$(OH)_x$ catalyst; (b) schematic diagram of CO_2RR on SA-Ru/Ni$(OH)_x$; (c) HR-TEM image of 0.60SA-Ru/Ni$(OH)_x$; (d) STEM image and corresponding EDS elemental mapping of 0.60SA-Ru/Ni$(OH)_x$; (e) CO_2RR performance on Ni$(OH)_x$, SA-Ru/Ni$(OH)_x$, SA-Ir/Ni$(OH)_x$, and SA-Pt/Ni$(OH)_x$; (f) relative free energy of each elementary reaction of CO_2RR on Ni$(OH)_x$, Ni$(OH)_x$-Ru1, and Ni$(OH)_x$-Ru3. Electron density difference for $^*CO_2$ and *COOH adsorption on Ni$(OH)_x$ (black rectangles) and Ni$(OH)_{xx}$-Ru1 (red rectangles). Yellow and cyan represent electron accumulation and depletion, respectively. *Source:* Reproduced from Ref. [83] / with permission of John Wiley & Sons.

optimizing the material structure (Figure 8.9a–d). Performance tests showed that the optimized sample achieved a hydrogen production rate of 393 μmol h^{-1} g^{-1} under simulated sunlight, which is 26 times that of pure TiO_2, while exhibiting excellent cycling stability. The key to the performance enhancement lies in the successful construction of the p-n heterojunction, as shown in Figure 8.9e, the built-in electric field at the interface drives the migration of photogenerated holes to NiO and the accumulation of electrons in TiO_2, achieving efficient charge separation. This work provides an important reference for the design of efficient heterojunction photocatalysts. Furthermore, as shown in Figure 8.9f, cubic $NaTaO_3$ modified with NiO exhibits excellent performance in photocatalytic overall water splitting [86].

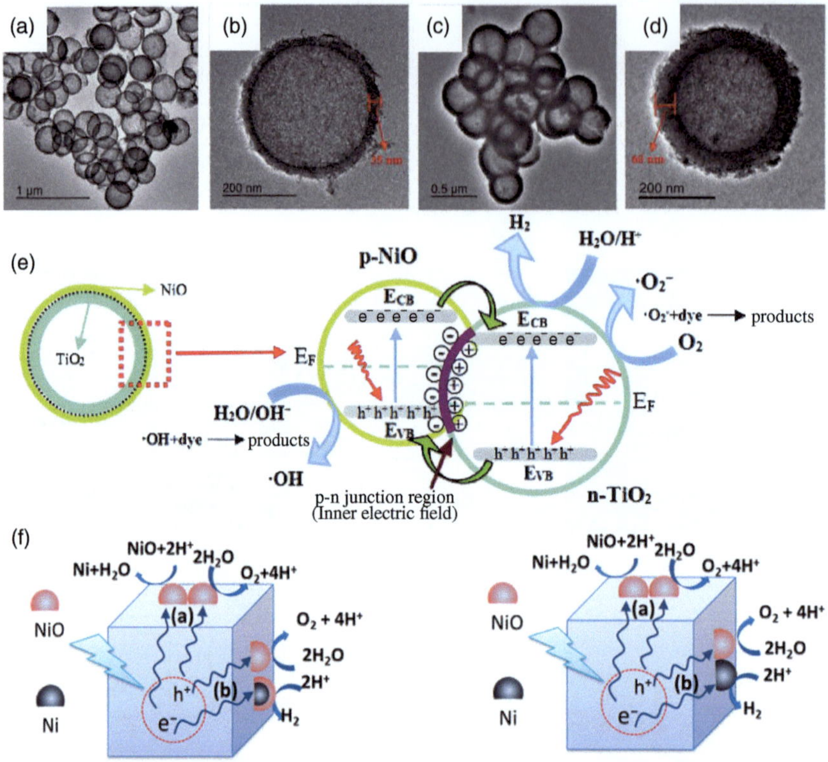

Figure 8.9 TEM images of TiO_2/NiO hollow hybrids synthesized with different TBOT additions: (a and b) 0.2 mL and (c, d) 0.5 mL, with a concentration of 0.015 mol L^{-1} $NiCl_2 \cdot 6H_2O$; (e) schematic diagram of the charge transfer and separation mechanism during photocatalysis in TiO_2/NiO hollow hybrids. *Source:* Reproduced from Ref. [85] / with permission of American Chemical Society. (f) Schematic diagram of overall water splitting on cubic $NaTaO_3$ loaded with equivalent-faceted Ni/NiO dual cocatalysts. *Source:* Reproduced from Ref. [86] / with permission of American Chemical Society.

Nickel sulfides (including NiS, NiS_2, and Ni_3S_2) have emerged as highly promising photocatalytic cocatalysts due to their narrow bandgap, suitable band positions, high electrical conductivity, and unique $(S_2)^{2-}$ chemical characteristics (Figure 8.10a) [88–90]. Studies have shown that NiS/CdS composites constructed via a simple hydrothermal method exhibit excellent hydrogen production performance, achieving a hydrogen evolution rate of 230.6 μmol h^{-1}, which is 8.42 and 1.72 times higher than that of unmodified CdS and CdS-DETA hybrid materials, respectively [91]. This enhancement is primarily attributed to the role of NiS as a cocatalyst, which effectively promotes the separation and migration of photogenerated charge carriers and reduces the overpotential for hydrogen evolution. Similarly, significant improvements in hydrogen production performance have been observed when NiS is introduced into systems such as g-C_3N_4, TiO_2, WO_3, and CdS/ZnS. Recent research has further developed core-shell structured CdS@Ni_3S_2 nanorod arrays grown in situ on Ni foam (Figure 8.10b, c). In this structure, the Ni_3S_2 shell not only

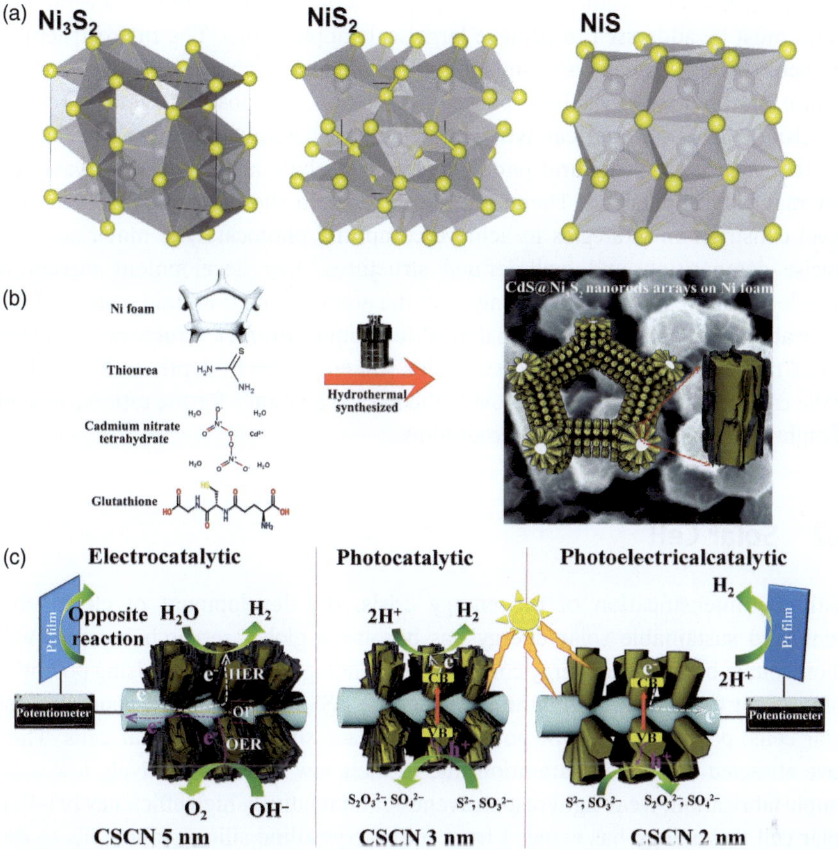

Figure 8.10 (a) Schematic diagram of the crystal structure. (b) Schematic diagram of the hydrothermal synthesis of CdS@Ni_3S_2 heterostructured nanorod composites; (c) proposed charge transfer mechanisms and H_2 generation mechanisms for EC (left), PC (middle), and PEC (right) on samples CSNS-5, CSNS-3, and CSNS-2 (CdS@Ni_3S_2 nanorod arrays with different Ni_3S_2 contents), respectively. *Source:* Reproduced from Ref. [87] / with permission of The Royal Society of Chemistry.

acts as an efficient cocatalyst to enhance hydrogen production performance but also effectively suppresses the photocorrosion of the CdS core, thereby improving the stability of the system. The optimized composite material exhibits an apparent quantum efficiency of 1.36% and a hydrogen evolution rate of 8.2 μmol cm^{-2} h^{-1} under monochromatic light irradiation at 475 nm.

In the field of photocatalytic hydrogen production, significant breakthroughs have been made in recent years in the study of transition metal-based semiconductor materials. Constructing composite photocatalytic systems has been proven to be an effective strategy for enhancing performance, among which the selection and design of cocatalysts are particularly critical. Nickel phosphide-based materials demonstrate remarkable advantages in this field due to their unique electronic structure and catalytic activity. Although existing studies have confirmed their potential in enhancing photocatalytic activity and stability, several key issues must be addressed to achieve large-scale applications. The microstructural characteristics of cocatalysts—including parameters such as size distribution, surface morphology, crystalline phase composition, and loading methods—have a decisive impact on their catalytic performance. Currently, the intrinsic relationship between these structural parameters and catalytic activity has not yet been systematically understood. Therefore, future research should focus on developing novel construction strategies to achieve composite photocatalytic materials with precise composition and well-defined structures. Key development directions include precise regulation of cocatalysts at the atomic or near-atomic scale, and the application of advanced in situ analytical techniques to track structural evolution in real time during catalytic processes. This will enable an in-depth understanding of the catalytic mechanism and provide theoretical guidance for the rational design of high-performance photocatalytic systems.

8.2 Solar Cell

With the intensification of the energy crisis, the development of clean, efficient, and sustainable solar energy has become a global research hotspot [92]. Photovoltaic-based solar energy conversion technologies offer a promising potential solution for energy harvesting and conversion. DSSCs, as the third generation of solar cells, possess significant potential to replace silicon-based solar cells. They have attracted widespread attention due to their low toxicity, relatively low cost, simple fabrication, weak light responsiveness, and relatively high efficiency [93–95]. Solar cell technology has evolved from initial crystalline silicon solar cells to the present day and can be categorized into three generations. The first generation is based on crystalline solar cells, with silicon being the most suitable material for their production. The second generation consists of thin-film solar cells. The third generation represents emerging thin-film solar cells. Among them, DSSCs have consistently garnered extensive attention from the research community due to their low fabrication cost, diverse material options, and good mechanical flexibility. Over the past two decades, DSSCs have made significant progress in terms of photoelectric conversion efficiency, long-term stability, and fabrication processes. A key milestone in this field was set in 1991 when O'Regan and Grätzel used colloidal TiO_2 films coated with a monolayer of charge transfer dyes for light harvesting, achieving a power conversion efficiency (PCE) of 7.12% under simulated sunlight, which laid the foundation for DSSC research [96]. Subsequently, in 2011, the Grätzel team combined a Co(II/III) tris(bipyridine)-based redox electrolyte with a custom-synthesized donor-π-bridge-acceptor zinc porphyrin dye as a sensitizer, boosting the PCE to 12.3% and demonstrating the great potential of synergistically optimizing dyes and electrolytes to enhance device performance [97]. Later, in 2017, they achieved a further breakthrough by increasing the PCE to 28.9% under indoor conditions at 1000 lux, marking a new height in DSSC performance [98]. Recently reported high-performance DSSC devices generally exhibit PCE values exceeding

10%. This achievement is primarily attributed to continuous in-depth research in sensitizer dye molecular engineering, innovation in redox electrolyte systems, and optimization of CE materials.

A standard DSSC mainly consists of three components: a dye-sensitized photoanode, an electrolyte solution containing a redox couple, and a CE. The semiconductor material in the photoanode is generally TiO_2, dye molecules are mostly carboxylated bipyridine ruthenium or organic dyes, the electrolyte is typically an acetonitrile solution containing I_3^-/I^-, and the CE is usually the noble metal platinum [99–102]. The main functions of each component are as follows: dye molecules act as sensitizers, absorbing solar radiation and injecting photoexcited electrons into the conduction band of the semiconductor oxide, leaving the dye in an oxidized state; the injected electrons transfer from the semiconductor to the FTO conducting glass and external circuit, which requires an appropriate band structure of the semiconductor oxide; the electroactive material in the CE collects electrons from the external circuit and transfers them to the electrolyte to achieve catalytic reduction of the redox species (I_3^- or Co^{3+}) present in the electrolyte; and the electrolyte is responsible for providing electrons and reducing the oxidized state of the dye, ensuring the dye returns to its ground state.

As shown in Figure 8.11, when the photoanode is irradiated by sunlight, the dye molecules absorb photons and transition to an excited state. The excited

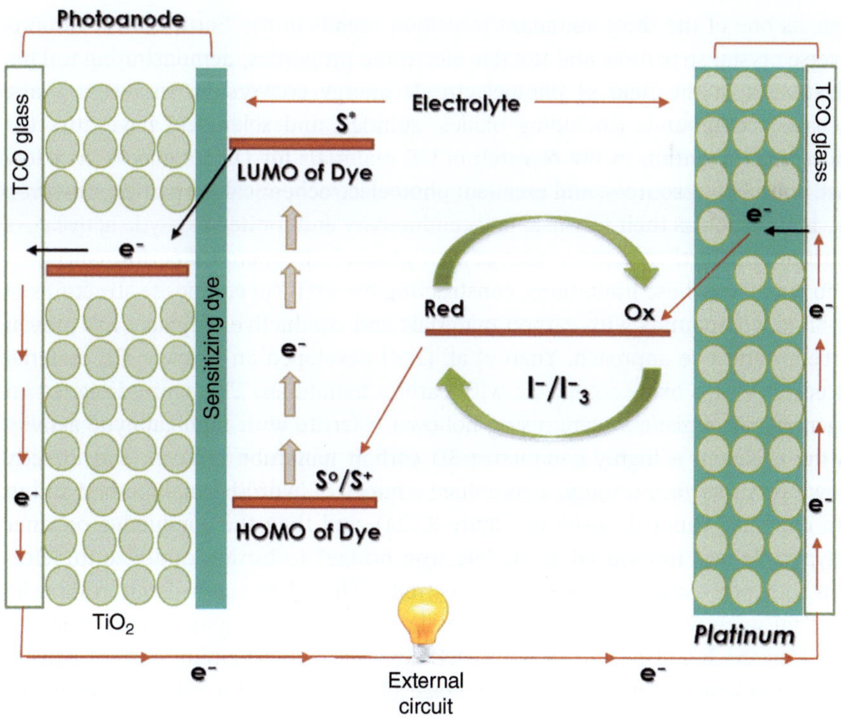

Figure 8.11 Working principle diagram of dye-sensitized solar cells. *Source:* Reproduced from Ref. [103] / with permission of Elsevier.

dye molecules inject electrons into the semiconductor conduction band, which are transmitted through the external circuit to the CE. The oxidized species in the electrolyte receive electrons at the CE and are reduced to the reduced state, while the oxidized dye is reduced back to the ground-state dye by the reduced species of the redox couple, thus completing an electrical circuit cycle.

For DSSCs, the role of the CE is crucial and critically important. The CE collects electrons from the external circuit, and these electrons are used to catalyze the reduction of I^-/I_3^- at the CE/electrolyte interface. Based on this principle, an efficient catalytic CE should possess high conductivity and sufficient catalytic capability, thereby reducing energy loss in the internal circuit and providing high energy conversion efficiency. Pt is by far the most effective CE material in DSSCs, exhibiting excellent electrocatalytic activity for I_3^- reduction, while the high cost of Pt limits the widespread application of DSSCs. Therefore, exploring inexpensive and efficient alternative CE materials to replace Pt is a pathway to reduce costs and achieve commercialization of DSSCs. CE materials must have good electrical conductivity, high catalytic activity, and chemical stability. In the past few years, researchers have investigated various alternative materials for use as CE catalysts, such as transition metal compounds and conductive polymers.

8.2.1 Fe-Based Compound CE

Iron, as one of the most abundant transition metals in the Earth's crust, exhibits diverse crystal structures and tunable electronic properties, demonstrating unique advantages in the field of photoelectronic energy conversion. In recent years, Fe-based compounds (including oxides, sulfides, and selenides) have attracted widespread attention in the research of CE materials for DSSCs due to their low cost, abundant resources, and excellent photoelectrochemical properties. However, key issues such as their intrinsic low conductivity and limited catalytic active sites constrain the further improvement of their photoelectrochemical performance.

To overcome these limitations, constructing hierarchical composite structures of Fe-based compounds with carbon materials and conductive polymers has proven to be an effective approach. Yuan et al. [104] developed an efficient CE material based on Fe–Co oxide composite with carbon nanotubes. This work designed an ingenious synergistic system: using hollow Co ferrite with high catalytic activity as the skeleton, a highly conductive 3D carbon nanotube network was directly grown on its surface through a combined strategy of hydrothermal method and in situ chemical vapor deposition (Figure 8.12a), and then the conductive polymer polypyrrole was introduced as an "electron bridge" to further optimize the electron transport pathways within the electrode. The advantages of this composite material were reflected in its photovoltaic performance (Figure 8.12b). The cell based on the h-$CoFe_2O_4$@CNTs/PPy CE achieved a short-circuit current density of 12.63 mA cm^{-2}, an open-circuit voltage of 0.691 V, and a fill factor of 0.709, ultimately achieving a photoelectric conversion efficiency of 6.19%, significantly superior to other comparative samples. Mechanism analysis (Figure 8.12c, d) revealed that the superiority of this design stems from three synergistic effects.

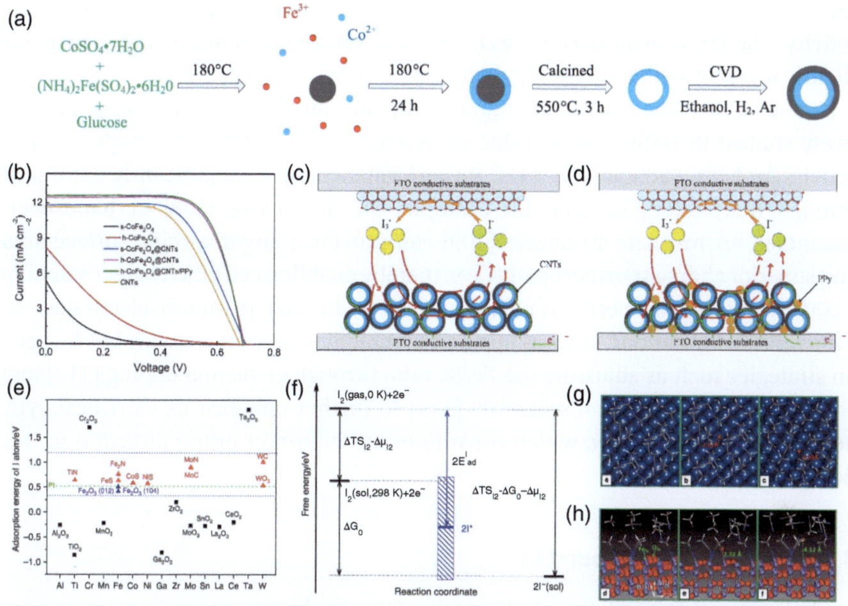

Figure 8.12 (a) Preparation procedure of h-$CoFe_2O_4$@CNTs; (b) photocurrent density–voltage curves of DSSCs based on s-$CoFe_2O_4$, s-$CoFe_2O_4$@CNTs, h-$CoFe_2O_4$, h-$CoFe_2O_4$@CNTs, h-$CoFe_2O_4$@CNTs/PPy, and CNTs CEs; (c) schematic diagram of the catalytic reduction of triiodide by h-$CoFe_2O_4$@CNTs; and (d) h-$CoFe_2O_4$@CNTs/PPy CEs. *Source:* Reproduced from Ref. [104] / with permission of Elsevier. (e) Adsorption energies of I atoms on various compounds and skeletons in gas phase and acetonitrile solvent calculated using density functional method, with red triangles corresponding to reported active materials; blank squares indicate unreported materials that are predicted to have lower catalytic activity; (f) demonstration of the estimation model for applicable electrode range based on I atom adsorption energy; (g) surface structure of Pt (111); (h) surface structure of α-Fe_2O_3 (012) in the presence of acetonitrile solvent, I adsorption structure and transition state structure. *Source:* Reproduced from Ref. [105] / with permission of Springer Nature.

First, hollow $CoFe_2O_4$ provides a highly active surface and catalytic sites. Second, the CNTs network imparts efficient electron conduction capability; when carbon nanotubes are grown in situ on the $CoFe_2O_4$ surface, the conductivity of the h-$CoFe_2O_4$ CE is significantly enhanced, allowing electrons to easily transfer to the catalytic sites through the conductive carbon nanotubes. Third, PPy nanoparticles bridge the composite material, further improving conductivity and shortening electron transport distances. These advantages collectively promote the catalytic performance for I_3^- reduction reaction and the overall cell efficiency.

In addition to enhancing material performance through composite structure design, gaining an in-depth understanding of the intrinsic catalytic properties of materials from the perspective of theoretical calculations provides new research ideas for screening efficient CE materials. Based on first-principles calculations, Hou et al. [105] investigated the electrocatalytic activity of α-Fe_2O_3 nanocrystals through theoretical prediction and experimental verification. These nanocrystals can be used as CEs for DSSCs (Figure 8.12e–g). Electrochemical and photovoltaic test results indicated that α-Fe_2O_3 possesses high conductivity and good electrocatalytic activity. The DSSC using α-Fe_2O_3 as the CE showed an efficiency of 6.96%, which is slightly lower than that of the cell using a platinum CE (7.32%).

Iron selenide ($FeSe_2$), as an emerging non-platinum CE material, has been extensively studied in DSSC research due to its exceptional stability and high catalytic activity for I_3^- reduction [106–111]. Its performance can be significantly optimized through morphology control; for example, 3D flower-like [108] or nanosheet-assembled microsphere structures [109] can provide a larger specific surface area and superior charge transport pathways, thereby enabling cell efficiencies (reaching 8.00% and 8.39%, respectively) that surpass traditional platinum electrodes. To overcome its limitations, such as intrinsic conductivity, recent research has focused on strategies such as adjusting the Fe/Se ratio through elemental doping [112] and compounding with carbon materials [113] to further enhance its electrocatalytic performance and stability, which constitutes an important future direction in this field.

8.2.2 Co-Based Compound CE

In the exploration of platinum-alternative CEs, Co-based materials have demonstrated unique advantages due to their distinctive electronic structure and tunable physicochemical properties. An early breakthrough came from the Zakeeruddin team [114], who constructed CoS nanostructures on flexible substrates, achieving a conversion efficiency of 6.5%, comparable to that of platinum electrodes, while exhibiting superior stability, laying an important foundation for subsequent research.

As research deepened, the integration of material synthesis processes and device manufacturing technologies became a key driver of progress in the field. The work by Chang et al. [115] in 2013 on the gram-scale synthesis of Co_9S_8 nanocrystals holds certain pioneering significance. A standout contribution of this study was the establishment of a complete technical pathway from material to device. As shown in Figure 8.13, they formulated the synthesized nanocrystals into stable ink and, for the first time, successfully prepared uniform, crack-free catalytic films on large-area 10 cm × 10 cm FTO glass using spray coating technology. This work not only confirmed the excellent catalytic performance of Co_9S_8 itself (achieving a cell efficiency of 7.02%) but, more importantly, it demonstrated a scalable manufacturing paradigm of "nanocrystal ink + spray deposition." This provided crucial methodological insights and feasibility verification for subsequent studies utilizing solution-based processing of various metal chalcogenide CEs and advancing DSSCs toward low-cost, large-area, and flexible development. Simultaneously, the development of various novel Co-based materials continued to advance. The Shrestha team [116] prepared porous 1D $CoS_{1.0365}$ nanorods via an ion exchange strategy. Although the efficiency (7.4%) was slightly lower than the platinum benchmark (7.7%), the unique 1D porous structure provided ideal channels for charge transport and electrolyte diffusion, highlighting the potential of structural design in enhancing catalytic performance. These studies collectively outline

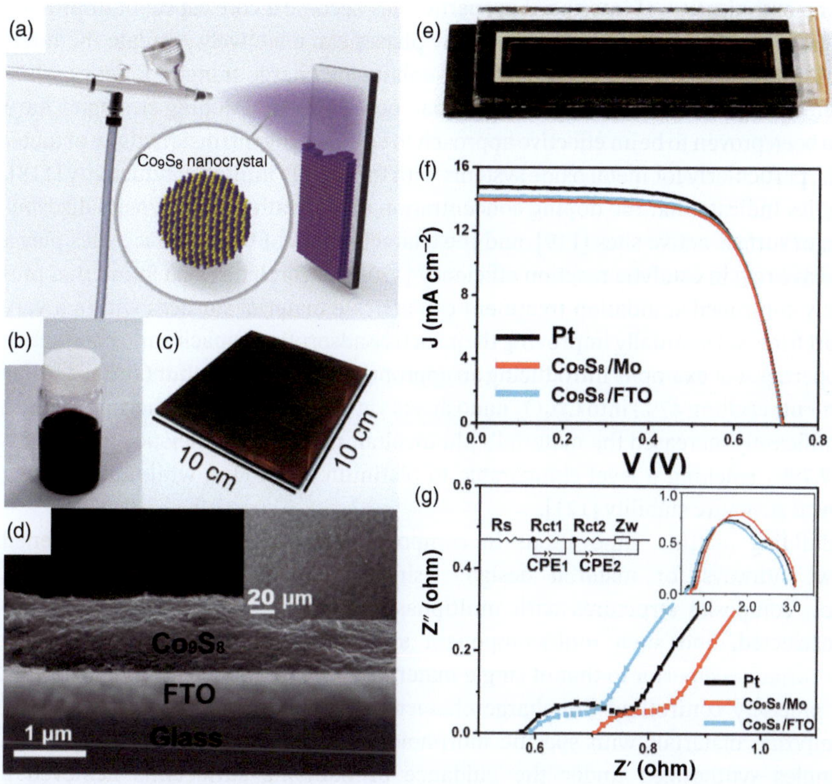

Figure 8.13 (a) Schematic diagram of Co_9S_8 nanocrystals synthesized on FTO by spray deposition; (b) photograph of Co_9S_8 nano-ink for thin-film deposition; (c) photograph of Co_9S_8 nanocrystal film deposited on a 10 cm × 10 cm FTO substrate; (d) SEM images of the cross-sectional and planar (inset) views of the Co_9S_8 film/FTO; (e) photograph of the simulated DSSC cell with a working area of 2 cm^2; (f) photocurrent density-voltage diagrams of DSSCs based on platinum, Co_9S_8 on FTO, and Co_9S_8 on Mo CE obtained under AM1.5 illumination; (g) Nyquist curves of DSSCs based on platinum, Co_9S_8 on FTO, and Co_9S_8 on Mo CE. *Source:* Reproduced from Ref. [115] / with permission of American Chemical Society.

the development trajectory of Co-based CE materials: from initial performance validation, to innovations in preparation processes, and further to the fine-tuning of microstructures. Each breakthrough has propelled DSSC technology toward lower cost and higher performance.

In the development process of Co-based chalcogenide CEs, the research focus is gradually shifting from single-component optimization to composite structure design. Compared to sulfides, selenides exhibit unique advantages in charge transport efficiency and intrinsic catalytic activity due to their distinct electronic configuration and stronger metallic character [117]. This characteristic enables Co-selenium hybrid materials with precisely controlled microstructures to achieve photovoltaic conversion performance surpassing that of platinum electrodes, fully demonstrating the critical role of electronic structure modulation in enhancing catalytic activity.

The optimization of interfacial properties has become a core aspect of improving material performance. Constructing alloy phases can effectively regulate the interfacial electron distribution, significantly enhancing charge transfer kinetics while maintaining the material's optical transparency. Elemental doping strategies have also been proven to be an effective approach to enhance the intrinsic activity of materials, particularly for metal oxide systems with inherently limited conductivity [118]. Studies indicate that the doping concentration directly affects the density distribution of surface active sites [119], and the characteristics of these surface sites play a decisive role in catalytic reaction efficiency [120]. Research has also found that precisely controlled sulfidation treatment can activate material surfaces within a very short time, substantially improving their surface adsorption capacity and conductive properties. For example, introducing an appropriate amount of sulfur (surface sulfur content reaching 47%) into Co_3O_4 nanosheets via a gas-phase hydrothermal method significantly increased the material's photovoltaic conversion efficiency from 0.33% to 7.79%, reaching a level comparable to platinum electrodes, while maintaining crystal structure stability [121].

Building on this, innovations in composite phase engineering have opened new pathways for material design. Using surfactants for directional guidance, composite structures with multiphase synergistic catalytic effects can be constructed, and such multicomponent systems often exhibit comprehensive performance superior to that of single materials. Research practice has shown that by precisely controlling the charge characteristics of surfactants, $CoSe_2/CoSeO_3$ composite materials with specific morphologies can be prepared. Among them, samples synthesized under the guidance of nonionic surfactants achieved a conversion efficiency of 9.27%, significantly exceeding the 7.91% of platinum electrodes [122]. These systematic studies not only validate the application potential of Co-based selenides as platinum-alternative materials but, more importantly, deepen the understanding of the intrinsic relationships among material composition-structure-performance, laying a theoretical foundation for the rational design of next-generation high-performance CE materials.

8.2.3 Ni-Based Compound CE

Given that nickel oxide (NiO) faces limitations in its application as a CE in DSSCs due to insufficient conductivity and electrocatalytic activity [123], research focus has gradually shifted toward Ni-based chalcogenides with superior intrinsic properties, such as sulfides and selenides (NiSe, Ni_3Se_4, $NiSe_2$, and $Ni_{0.85}Se$) [124–127]. These materials, leveraging their unique d-electron configurations and tunable anion coordination environments, demonstrate catalytic potential surpassing that of single oxides. By regulating the types of anions and their stoichiometric ratios, the electronic conductivity of the materials and their adsorption–desorption behavior toward the I_3^- reduction reaction can be effectively optimized. Based on this understanding, research has evolved from early single-phase nickel sulfides to the construction of multiphase heterojunctions and synergistic systems compounded with carbon materials, aiming to drive breakthroughs in CE performance [128–137].

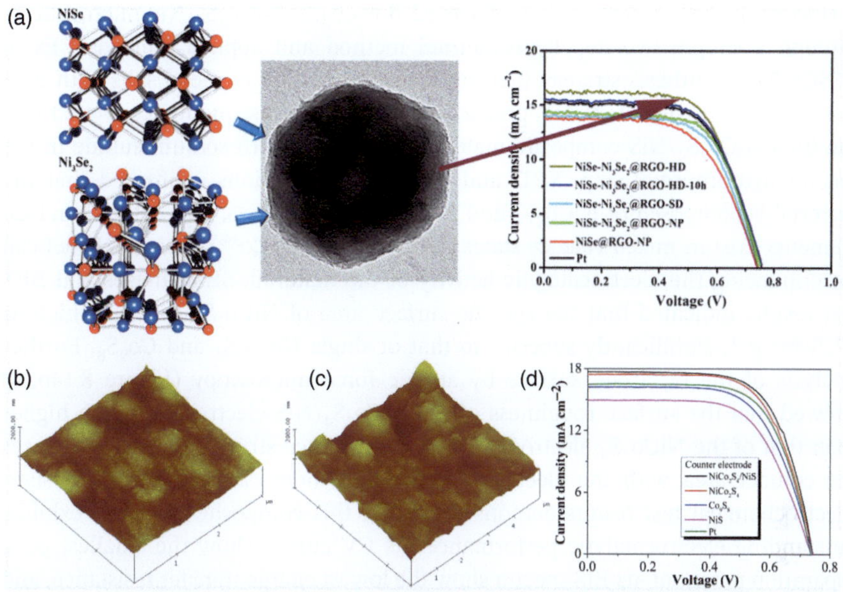

Figure 8.14 (a) TEM image of NiSe-Ni_3Se_2/RGO nanodiscs, and photocurrent density–voltage curves of DSSCs with the NiSe-Ni_3Se_2/RGO series and platinum CEs. *Source:* Reproduced from Ref. [138] / with permission of American Chemical Society. (b) Surface roughness images of $NiCo_2S_4$ and (c) $NiCo_2S_4$/NiS CEs; (d) J-V curves of DSSCs based on $NiCo_2S_4$/NiS, $NiCo_2S_4$, Co_9S_8, NiS, and Pt CEs under full illumination (100 mW cm^{-2}, AM 1.5 G). *Source:* Reproduced from Ref. [139] / with permission of Elsevier.

Along this research path, composite materials with hollow and hybrid structures have shown significant advantages. The hollow hybrid NiSe-Ni_3Se_2/RGO nanostructure (Figure 8.14a) serves as a representative example. This design integrates multiple enhancement mechanisms: the hollow structure significantly increases the specific surface area and provides efficient mass transfer channels; the heterointerface between the NiSe and Ni_3Se_2 phases greatly optimizes charge transfer kinetics; and the reduced graphene oxide (RGO) substrate constructs a high-speed electron conduction network. The synergistic effects of this hierarchical structure endow it with exceptional electrocatalytic performance and very low charge transfer resistance when used as a CE. Ultimately, the DSSC based on this material achieved a photoelectric conversion efficiency of 7.87%, significantly surpassing that of the traditional platinum CE (7.28%). This achievement highlights that through ingenious micro–nano structural design and composite strategies, the performance limitations of single components can be effectively overcome, providing key theoretical foundations and practical solutions for developing efficient and stable non-precious metal CEs. Furthermore, Dong et al. [140] fabricated $Ni_{0.85}Se$ nanosheets grown on RGO with varying concentrations ($Ni_{0.85}Se$/RGO) via a simple hydrothermal technique. Based on the optimal RGO concentration, the cell with the $Ni_{0.85}Se$/RGO CE achieved an efficiency of 9.75%, representing a 19.6% improvement compared to the efficiency of the cell using a platinum CE (8.15%).

Huo et al. [139] successfully synthesized flower-like $NiCo_2S_4/NiS$ microspheres through a simple two-step hydrothermal method and applied them as CEs in DSSCs. This synthesis strategy first involved the reaction of metal salts with urea to form a $(Ni, Co)_2(CO_3)(OH)_2$ precursor, which was subsequently converted into the final $NiCo_2S_4/NiS$ composite material by reacting with sodium sulfide in the second hydrothermal step. XRD and XPS characterizations confirmed that the material is composed of $NiCo_2S_4$ and NiS composite phases, and both Ni and Co elements exist in mixed valence states (Ni^{2+}/Ni^{3+}, Co^{2+}/Co^{3+}), which is beneficial for enhancing the electrocatalytic activity of the material. SEM images and BET test results indicated that the specific surface area of $NiCo_2S_4/NiS$ is as high as $27.79\,m^2\,g^{-1}$, significantly superior to that of single $NiCo_2S_4$ and Co_9S_8. Further analysis of the electrode surface by atomic force microscopy (Figure 8.14b, c) showed that the surface roughness of the $NiCo_2S_4/NiS$ electrode is much higher than that of the $NiCo_2S_4$ electrode. This rough surface structure greatly increases the contact area with the electrolyte, providing more catalytically active sites. Electrochemical test results demonstrated that this composite electrode exhibits outstanding electrocatalytic performance: its CV curves show the smallest peak separation potential, its EIS spectra show the lowest charge transfer resistance, and its Tafel curves show the highest exchange current density and limiting diffusion current density. These data collectively prove that it has superior catalytic activity and conductivity for the I_3^- reduction reaction compared to the platinum electrode. The final photovoltaic performance test (Figure 8.14d) perfectly confirmed the above advantages. The cell based on the $NiCo_2S_4/NiS$ CE achieved a photoelectric conversion efficiency of 8.8%. This value not only significantly exceeds that of the reference cell using a platinum CE but also highlights the great potential of this low-cost, high-efficiency composite material in replacing platinum CEs.

In the optimization process of Ni-based CE materials, morphology control and composition design have been proven to be key pathways to enhance electrocatalytic performance. By controlling the amounts of various source materials and reaction time, NiSe nanomaterials with different morphologies were synthesized [141]. Electrochemical test results showed that due to their small and uniform size, NiSe nanoparticles (NiSe-1) produced better electron conduction and electrocatalytic activity than micro- and submicron particles (NiSe-2). Using the NiSe-1 CE, the DSSC showed a higher PCE (7.82%) than using the NiSe-2 CE (6.90%) and Pt CE (7.62%). Research indicates that constructing hollow structures (such as urchin-like microspheres [142]) or nanostructures with specific crystal facet exposure (such as octahedra [143]) can significantly increase the specific surface area and number of active sites of the material, while promoting electrolyte penetration and charge transport, thereby enabling cell efficiency to exceed 8% and surpass the performance of traditional platinum electrodes.

In terms of multicomponent composite materials, the combination of Ni-based chalcogenides with conductive substrates (such as graphene) or metal phosphides (such as Ni-P) effectively addresses the issues of insufficient conductivity and limited catalytic sites in single materials. Such composite structures significantly reduce the charge transfer impedance for the I_3^- reduction reaction through interfacial electron coupling and synergistic catalytic mechanisms, enabling the devices to maintain excellent performance even on flexible substrates or fluorine-free transparent conductive oxides. Furthermore, elemental doping (such as sulfur doping) [144] and heteroanion regulation (such as constructing NiS/Ni_3S_2 hybrid phases) further expand the optimization space for material performance. Doping not only improves the intrinsic conductivity of the material but also optimizes the adsorption behavior of reaction intermediates by modulating its surface electronic structure; while heterophase compounding achieves synergistic enhancement of catalytic activity through band matching and interfacial charge rearrangement, enabling Ni-based CEs to exhibit comprehensive performance comparable to platinum in various device configurations. By constructing a $Ni_xS_y/NiSe_2$ nano-hybrid CE on F-doped SnO_2 conductive glass, research by Chen et al. [145] showed that this material, due to the synergistic effect between components, exhibits electrocatalytic activity for I_3^- reduction superior to that of single Ni_xS_y or $NiSe_2$. The DSSC based on this CE achieved a photoelectric conversion efficiency of 8.64%, significantly higher than that of cells using Ni_xS_y (7.95%) and $NiSe_2$ (7.47%) CEs.

In summary, iron-group metal compounds, as a new generation of energy materials, hold strategic significance that extends beyond simply replacing precious metals, embodying the core concept of addressing energy challenges through electronic structure modulation. These materials, leveraging their abundant reserves and unique 3d electronic configurations, exhibit remarkable flexibility in composition and structural design. In applications such as photocatalytic hydrogen production and DSSCs, although their specific roles differ, their excellent performance stems from the same core mechanism: effectively promoting the separation of photogenerated charges and interfacial reactions through their unique d-electron structures, systematically overcoming the efficiency bottlenecks in solar energy conversion.

Breakthroughs in performance fundamentally originate from the precise construction of multilevel material structures. This includes nanoscale morphology control (e.g., constructing low-dimensional nanostructures to increase active site exposure and shorten charge diffusion paths), microlevel interface engineering (e.g., building built-in electric fields to drive directional charge separation), and electronic-level composition design (e.g., modulating the Fermi level and adsorption strength of reaction intermediates through doping or alloying). This multi-scale, cross-dimensional synergistic design transforms materials from carriers of single functions into integrated systems capable of efficiently responding to and managing light, electricity, and chemical signals simultaneously.

Looking ahead, the research frontier in this field is transitioning from "performance-oriented" exploration to a new phase of "mechanism deepening" and "technology integration." The primary task is to utilize advanced in situ/operando characterization techniques and theoretical simulations to track dynamic structural evolution during catalytic processes in real time at the atomic scale, clarifying the true active centers and reaction pathways, thereby establishing precise structure–property relationships and providing a solid blueprint for the rational design of materials. Concurrently, the long-term chemical stability and mechanical durability of materials have become key barriers restricting their transition from the laboratory to practical applications. Developing effective surface/interface passivation strategies, exploring novel stable phases, and designing self-healing mechanisms are inevitable directions to address this challenge. Ultimately, to realize their industrial value, the excellent properties of these materials must be integrated with green, low-cost, large-scale preparation processes and device engineering designs that match actual application scenarios, achieving a leap in product performance. Through such multidisciplinary, multi-objective collaborative innovation, iron-group metal compounds are destined to contribute irreplaceable strength to building a sustainable energy system for the future.

References

1 Hisatomi, T., Yamada, T., Nishiyama, H. et al. (2025). Materials and systems for large-scale photocatalytic water splitting. *Nature Reviews Materials* 10: 769–782.

2 Bhatt, M.D. and Lee, J.S. (2017). Nanomaterials for photocatalytic hydrogen production: from theoretical perspectives. *RSC Advances* 7: 34875–34885.

3 Yu, J.G., Qi, L.F., and Jaroniec, M. (2010). Hydrogen production by photocatalytic water splitting over Pt/TiO_2 nanosheets with exposed (001) facets. *Journal of Physical Chemistry C* 114: 13118–13125.

4 Ganguly, P., Harb, M., Cao, Z. et al. (2019). 2D nanomaterials for photocatalytic hydrogen production. *ACS Energy Letters* 4: 1687–1709.

5 Corredor, J., Rivero, M.J., Rangel, C.M. et al. (2019). Comprehensive review and future perspectives on the photocatalytic hydrogen production. *Journal of Chemical Technology & Biotechnology* 94: 3049–3063.

6 Pan, Y., Abazari, R., Tahir, B. et al. (2024). Iron-based metal–organic frameworks and their derived materials for photocatalytic and photoelectrocatalytic reactions. *Coordination Chemistry Reviews* 499: 215538.

7 Fujishima, A. and Honda, K. (1972). Electrochemical photolysis of water at a semiconductor electrode. *Nature* 238: 37–38.

8 Liu, X. and Zhuang, H. (2020). Recent progresses in photocatalytic hydrogen production: design and construction of Ni-based cocatalysts. *International Journal of Energy Research* 45: 1480–1495.

9 Du, C., Yan, B., Lin, Z., and Yang, G. (2020). Enhanced carrier separation and increased electron density in 2D heavily N-doped $ZnIn_2S_4$ for photocatalytic hydrogen production. *Journal of Materials Chemistry A* 8: 207–217.

10 Gao, W., Lu, J., Zhang, S. et al. (2019). Suppressing photoinduced charge recombination via the Lorentz force in a photocatalytic system. *Advanced Science* 6: 1901244.

11 Goto, Y., Hisatomi, T., Wang, Q. et al. (2018). A particulate photocatalyst water-splitting panel for large-scale solar hydrogen generation. *Joule* 2: 509–520.

12 Qi, S., Fan, Y., Wang, J. et al. (2020). Metal-free highly efficient photocatalysts for overall water splitting: C_3N_5 multilayers. *Nanoscale* 12: 306–315.

13 Wang, Z., Liu, Y., Huang, B. et al. (2014). Progress on extending the light absorption spectra of photocatalysts. *Physical Chemistry Chemical Physics* 16: 2758–2774.

14 Zhuang, H., Liu, X., Li, F. et al. (2019). Construction of CdSe@TiO_2 core-shell nanorod arrays by electrochemical deposition for efficient visible light photoelectrochemical performance. *International Journal of Energy Research* 43: 7197–7205.

15 Zhuang, H., Zhang, Y., Chu, Z. et al. (2016). Synergy of metal and nonmetal dopants for visible-light photocatalysis: a case-study of Sn and N co-doped TiO_2. *Physical Chemistry Chemical Physics* 18: 9636–9644.

16 Zhuang, H., Cai, Z., Xu, W. et al. (2019). In situ construction of WO_3/g C_3N_4 composite photocatalyst with 2D–2D heterostructure for enhanced visible light photocatalytic performance. *New Journal of Chemistry* 43: 17416–17422.

17 Zhuang, H., Chen, W., Xu, W., and Liu, X. (2020). Facile synthesis of MoS_2 QDs/TiO_2 nanosheets via a self-assembly strategy for enhanced photocatalytic hydrogen production. *International Journal of Energy Research* 44: 3224–3230.

18 Chen, Y., Feng, X., Guo, X., and Zheng, W. (2019). Toward a fundamental understanding of factors affecting the function of cocatalysts in photocatalytic water splitting. *Current Opinion in Green and Sustainable Chemistry* 17: 21–28.

19 Xiao, N., Li, S., Li, X. et al. (2020). The roles and mechanism of cocatalysts in photocatalytic water splitting to produce hydrogen. *Chinese Journal of Catalysis* 41: 642–671.

20 Chen, R., Ao, Y., Wang, C., and Wang, P. (2019). 2D ultrathin CoP modified $Mn_xCd_{1-x}S$ with controllable band structure and robust photocatalytic performance for hydrogen generation. *Dalton Transactions* 48: 14783–14791.

21 Lu, X., Xie, J., Chen, X., and Li, X. (2019). Engineering MP_x (M = Fe, Co or Ni) interface electron transfer channels for boosting photocatalytic H_2 evolution over g-C_3N_4/MoS_2 layered heterojunctions. *Applied Catalysis B: Environmental* 252: 250–259.

22 Chen, W., Wang, Y., Liu, M. et al. (2018). In situ photodeposition of cobalt on CdS nanorod for promoting photocatalytic hydrogen production under visible light irradiation. *Applied Surface Science* 444: 485–490.

23 Yang, Y., Zhang, C., Huang, D. et al. (2019). Boron nitride quantum dots decorated ultrathin porous g-C_3N_4: intensified exciton dissociation and charge transfer for promoting visible-light-driven molecular oxygen activation. *Applied Catalysis B: Environmental* 245: 87–99.

24 Sharma, A., Thuan, D.V., Pham, T.-D. et al. (2020). Advanced surface of fibrous activated carbon immobilized with FeO/TiO_2 for photocatalytic evolution of hydrogen under visible light. *Chemical Engineering & Technology* 43: 752–761.

25 Sun, X. and Du, H. (2019). Acid-exfoliated metallic $Co_{0.85}Se$ nanosheets as cocatalyst on $Cd_{0.5}Zn_{0.5}S$ for photocatalytic hydrogen evolution under visible-light irradiation. *ACS Sustainable Chemistry & Engineering* 7: 16320–16328.

26 Zhang, L., Wang, G., Hao, X. et al. (2020). MOFs-derived Cu_3P@CoP p-n heterojunction for enhanced photocatalytic hydrogen evolution. *Chemical Engineering Journal* 395: 125113.

27 Du, S., Lin, X., Li, C. et al. (2020). $CoSe_2$ modified Se-decorated CdS nanowire Schottky heterojunctions for highly efficient photocatalytic hydrogen evolution. *Chemical Engineering Journal* 389: 124431.

28 Chen, W., Wang, Y., Liu, S. et al. (2018). Non-noble metal Cu as a cocatalyst on TiO_2 nanorod for highly efficient photocatalytic hydrogen production. *Applied Surface Science* 445: 527–534.

29 Ji, C., Yin, S.-N., Sun, S., and Yang, S. (2018). An in situ mediator-free route to fabricate $Cu_2O/g\text{-}C_3N_4$ type-II heterojunctions for enhanced visible-light photocatalytic H_2 generation. *Applied Surface Science* 434: 1224–1231.

30 Meng, A., Zhang, J., Xu, D. et al. (2016). Enhanced photocatalytic H_2-production activity of anatase TiO_2 nanosheet by selectively depositing dual-cocatalysts on {101} and {001} facets. *Applied Catalysis B: Environmental* 198: 286–294.

31 Arif, N., Zafar, M.N., Batool, M. et al. (2024). Recent advances and perspectives on iron-based photocatalysts. *Journal of Materials Chemistry C* 12: 12653–12691.

32 Mishra, M. and Chun, D.-M. (2015). $\alpha\text{-}Fe_2O_3$ as a photocatalytic material: a review. *Applied Catalysis A: General* 498: 126–141.

33 Tilley, S.D., Cornuz, M., Sivula, K., and Grätzel, M. (2010). Light-induced water splitting with hematite: improved nanostructure and iridium oxide catalysis. *Angewandte Chemie* 122: 6549–6552.

34 Wang, G., Ling, Y., Wheeler, D.A. et al. (2011). Facile synthesis of highly photoactive alpha-Fe_2O_3-based films for water oxidation. *Nano Letters* 11: 3503–3509.

35 AlSalka, Y., Granone, L.I., Ramadan, W. et al. (2019). Iron-based photocatalytic and photoelectrocatalytic nano-structures: facts, perspectives, and expectations. *Applied Catalysis B: Environmental* 244: 1065–1095.

36 Petit, S., Melissen, S.T.A.G., Duclaux, L. et al. (2016). How should iron and titanium be combined in oxides to improve photoelectrochemical properties? *The Journal of Physical Chemistry C* 120: 24521–24532.

37 Ali, F.M., Hmadeh, M., O'Brien, P.G. et al. (2016). Photocatalytic properties of all four polymorphs of nanostructured iron oxyhydroxides. *ChemNanoMat* 2: 1047–1054.

38 Yang, H., Lu, R., Downs, R.T., and Costin, G. (2006). Goethite, [alpha]-FeO(OH), from single-crystal data. *Acta Crystallographica Section E* 62: i250–i252.

39 Fracchia, M., Visibile, A., Ahlberg, E. et al. (2018). α- and γ-FeOOH: stability, reversibility, and nature of the active phase under hydrogen evolution. *ACS Applied Energy Materials* 1: 1716–1725.

40 Luo, W., Jiang, C., Li, Y. et al. (2017). Highly crystallized α-FeOOH for a stable and efficient oxygen evolution reaction. *Journal of Materials Chemistry A* 5: 2021–2028.

41 Wu, J., Wang, W., Chen, X. et al. (2025). Boosted efficiency of Fe_2O_3 for photocatalytic CO_2 reduction via engineering Fe–O–Ti bonding. *Advanced Science* 12: e2409002.

42 Wang, Y., Wang, S., and Lou, X.W. (2019). Dispersed nickel cobalt oxyphosphide nanoparticles confined in multichannel hollow carbon fibers for photocatalytic CO_2 reduction. *Angewandte Chemie International Edition* 58: 17236–17240.

43 Weng, C.C., Ren, J.T., and Yuan, Z.Y. (2020). Transition metal phosphide-based materials for efficient electrochemical hydrogen evolution: a critical review. *ChemSusChem* 13: 3357–3375.

44 Shi, R., Tian, C., Zhu, X. et al. (2019). Achieving an exceptionally high loading of isolated cobalt single atoms on a porous carbon matrix for efficient visible-light-driven photocatalytic hydrogen production. *Chemical Science* 10: 2585–2591.

45 Zhang, D., Peng, L., Liu, K. et al. (2020). Cobalt nanoparticle with tunable size supported on nitrogen-deficient graphitic carbon nitride for efficient visible light driven H_2 evolution reaction. *Chemical Engineering Journal* 381: 122576.

46 Chu, J., Sun, G., Han, X. et al. (2019). Ultrafine CoO nanoparticles as an efficient cocatalyst for enhanced photocatalytic hydrogen evolution. *Nanoscale* 11: 15633–15640.

47 Yan, X., Wang, G., Zhang, Y. et al. (2020). 3D layered nano-flower MoS_x anchored with CoP nanoparticles form double proton adsorption site for enhanced photocatalytic hydrogen evolution under visible light driven. *International Journal of Hydrogen Energy* 45: 2578–2592.

48 Zhu, Y., Wan, T., Wen, X. et al. (2019). Tunable Type I and II heterojunction of CoO_x nanoparticles confined in g C_3N_4 nanotubes for photocatalytic hydrogen production. *Applied Catalysis B: Environmental* 244: 814–822.

49 Yan, Z., Wu, H., Han, A. et al. (2014). Noble metal-free cobalt oxide (CoO_x) nanoparticles loaded on titanium dioxide/cadmium sulfide composite for enhanced photocatalytic hydrogen production from water. *International Journal of Hydrogen Energy* 39: 13353–13360.

50 Liu, X., Zhang, Q., Liang, L. et al. (2019). In situ growing of CoO nanoparticles on g-C_3N_4 composites with highly improved photocatalytic activity for hydrogen evolution. *Royal Society Open Science* 6: 190433.

51 Han, Y., Liang, Z., Dang, H., and Dong, X. (2018). Extremely high photocatalytic H_2 evolution of novel $Co_3O_4/Cd_{0.9}Zn_{0.1}S$ p–n heterojunction photocatalyst under visible light irradiation. *Journal of the Taiwan Institute of Chemical Engineers* 87: 196–203.

52 Zhou, X., Jin, J., Zhu, X. et al. (2016). New $Co(OH)_2$/CdS nanowires for efficient visible light photocatalytic hydrogen production. *Journal of Materials Chemistry A* 4: 5282–5287.

53 Zhang, L.J., Zheng, R., Li, S. et al. (2014). Enhanced photocatalytic H_2 generation on cadmium sulfide nanorods with cobalt hydroxide as cocatalyst and insights into their photogenerated charge transfer properties. *ACS Applied Materials and Interfaces* 6: 13406–13412.

54 Qi, K., Lv, W., Khan, I., and Liu, S.Y. (2020). Photocatalytic H_2 generation via CoP quantum-dot-modified g-C_3N_4 synthesized by electroless plating. *Chinese Journal of Catalysis* 41: 114–121.

55 Luo, B., Song, R., Geng, J. et al. (2019). Towards the prominent cocatalytic effect of ultra-small CoP particles anchored on g-C_3N_4 nanosheets for visible light driven photocatalytic H_2 production. *Applied Catalysis B: Environmental* 256: 117819.

56 Wang, W., Li, T., Komarneni, S. et al. (2022). Recent advances in Co-based co-catalysts for efficient photocatalytic hydrogen generation. *Journal of Colloid and Interface Science* 608: 1553–1575.

57 Zhao, W., Chen, Z., Yang, X. et al. (2020). Recent advances in photocatalytic hydrogen evolution with high-performance catalysts without precious metals. *Renewable and Sustainable Energy Reviews* 132: 110040.

58 Mao, J., Ge, M., Huang, J. et al. (2017). Constructing multifunctional MOF@rGO hydro-/aerogels by the self-assembly process for customized water remediation. *Journal of Materials Chemistry A* 5: 11873–11881.

59 Yi, J., Zhu, X., Zhou, M. et al. (2020). Crystal phase dependent solar driven hydrogen evolution catalysis over cobalt diselenide. *Chemical Engineering Journal* 396: 125244.

60 Dong, Y., Kong, L., Wang, G. et al. (2017). Photochemical synthesis of Co_xP as cocatalyst for boosting photocatalytic H_2 production via spatial charge separation. *Applied Catalysis B: Environmental* 211: 245–251.

61 Shi, Y. and Zhang, B. (2016). Recent advances in transition metal phosphide nanomaterials: synthesis and applications in hydrogen evolution reaction. *Chemical Society Reviews* 45: 1529–1541.

62 Wang, W., Yang, R., Li, T. et al. (2021). Advances in recyclable and superior photocatalytic fibers: material, construction, application and future perspective. *Composites Part B: Engineering* 205: 108512.

63 Sun, Q., Yu, Z., Jiang, R. et al. (2020). CoP QD anchored carbon skeleton modified CdS nanorods as a co-catalyst for photocatalytic hydrogen production. *Nanoscale* 12: 19203–19212.

64 Wang, P., Zhan, S., Wang, H. et al. (2018). Cobalt phosphide nanowires as efficient co-catalyst for photocatalytic hydrogen evolution over $Zn_{0.5}Cd_{0.5}S$. *Applied Catalysis B: Environmental* 230: 210–219.

65 Xiang, Z., Guan, H., Zhang, B., and Zhao, Y. (2020). Electrostatic self-assembly of 2D-2D CoP/$ZnIn_2S_4$ nanosheets for efficient photocatalytic hydrogen evolution. *Journal of the American Ceramic Society* 104: 504–513.

66 He, Y., Cui, R., Gao, C. et al. (2019). Cobalt phosphide microspheres integrated with cadmium sulfide nanowires as an efficient photocatalyst for hydrogen evolution reaction. *Molecular Catalysis* 469: 161–166.

67 Qiu, B.Q., Li, C.X., Shen, X.Q. et al. (2020). Revealing the size effect of metallic CoS_2 on CdS nanorods for photocatalytic hydrogen evolution based on Schottky junction. *Applied Catalysis A: General* 592: 117377.

68 Wang, L., Gao, W.P., Liu, Z.Y. et al. (2018). Core-shell nanostructured cobalt-platinum electrocatalysts with enhanced durability. *ACS Catalysis* 8: 35–42.

69 Ma, B., Zhao, J., Ge, Z. et al. (2020). 5 nm NiCoP nanoparticles coupled with g-C_3N_4 as high-performance photocatalyst for hydrogen evolution. *Science China Materials* 63: 258–266.

70 Tanaka, A., Teramura, K., Hosokawa, S. et al. (2017). Visible light-induced water splitting in an aqueous suspension of a plasmonic Au/TiO_2 photocatalyst with metal co-catalysts. *Chemical Science* 8: 2574–2580.

71 Karakus, M., Sung, Y., Wang, H.I. et al. (2017). Correlating carrier dynamics and photocatalytic hydrogen generation in Pt decorated CdSe tetrapods as a function of cocatalyst size. *Journal of Physical Chemistry C* 121: 13070–13077.

72 Bai, S., Yin, W., Wang, L. et al. (2016). Surface and interface design in cocatalysts for photocatalytic water splitting and CO_2 reduction. *RSC Advances* 6: 57446–57463.

73 Su, T., Shao, Q., Qin, Z. et al. (2018). Role of interfaces in two-dimensional photocatalyst for water splitting. *ACS Catalysis* 8: 2253–2276.

74 Liu, H., Yan, T., Jin, Z., and Ma, Q. (2020). CoP nanoparticles as cocatalyst modified the CdS/$NiWO_4$ p–n heterojunction to produce hydrogen efficiently. *New Journal of Chemistry* 44: 1426–1438.

75 He, R., Liang, H., Li, C., and Bai, J. (2020). Enhanced photocatalytic hydrogen production over Co_3O_4@g-C_3N_4 p-n junction adhering on one-dimensional carbon fiber. *Colloids and Surfaces A: Physicochemical and Engineering Aspects* 586: 124200.

76 Huang, Q.-Z., Wang, J.-C., Ye, L.-Q. et al. (2017). A novel p–n heterojunction $Mn_{0.25}$ $Cd_{0.75}S/Co_3O_4$ for highly efficient photocatalytic H_2 evolution under visible light irradiation. *Journal of the Taiwan Institute of Chemical Engineers* 80: 570–577.

77 Zhang, L., Si, R., Liu, H. et al. (2019). Atomic layer deposited Pt-Ru dual-metal dimers and identifying their active sites for hydrogen evolution reaction. *Nature Communications* 10: 4936.

78 Yu, J. and Ran, J. (2011). Facile preparation and enhanced photocatalytic H_2-production activity of $Cu(OH)_2$ cluster modified TiO_2. *Energy & Environmental Science* 4: 1364.

79 Zhang, G., Zang, S., and Wang, X. (2015). Layered $Co(OH)_2$ deposited polymeric carbon nitrides for photocatalytic water oxidation. *ACS Catalysis* 5: 941–947.

80 Digdaya, I.A., Adhyaksa, G.W.P., Trzesniewski, B.J. et al. (2017). Interfacial engineering of metal-insulator-semiconductor junctions for efficient and stable photoelectrochemical water oxidation. *Nature Communications* 8: 15968.

81 Mattsson, S. and Paulus, B. (2020). A theoretical study on the binding and electrolytic splitting of hydrogen fluoride on Ni(111) and Ni(211). *Physical Chemistry Chemical Physics* 22: 4407–4415.

82 Chai, Z., Zeng, T.-T., Li, Q. et al. (2016). Efficient visible light-driven splitting of alcohols into hydrogen and corresponding carbonyl compounds over a

Ni-modified CdS photocatalyst. *Journal of the American Chemical Society* 138: 10128–10131.

83 Tao, Y., Guan, J., Zhang, J. et al. (2024). Ruthenium single atomic sites surrounding the support pit with exceptional photocatalytic activity. *Angewandte Chemie International Edition* 63: e202400625.

84 Kazunari Domen, A.K., Onishi, T., Kosugi, N., and Kuroda, H. (1986). *Photocatalytic Decomposition of Water into H_2 and O_2 over NiO-$SrTiO_3$ Powder. 1. Structure of the Catalyst*, 292–295. American Chemical Society.

85 Wang, M., Hu, Y., Han, J. et al. (2015). TiO_2/NiO hybrid shells: p–n junction photocatalysts with enhanced activity under visible light. *Journal of Materials Chemistry A* 3: 20727–20735.

86 Zhang, Q., Li, Z., Wang, S. et al. (2016). Effect of redox cocatalysts location on photocatalytic overall water splitting over cubic $NaTaO_3$ semiconductor crystals exposed with equivalent facets. *ACS Catalysis* 6: 2182–2191.

87 Guan, H., Zhang, S., Cai, X. et al. (2019). CdS@Ni_3S_2 core–shell nanorod arrays on nickel foam: a multifunctional catalyst for efficient electrochemical catalytic, photoelectrochemical and photocatalytic H_2 production reaction. *Journal of Materials Chemistry A* 7: 2560–2574.

88 Guo, S., Yang, L., Zhang, Y. et al. (2018). Enhanced hydrogen evolution via interlaced Ni_3S_2/MoS_2 heterojunction photocatalysts with efficient interfacial contact and broadband absorption. *Journal of Alloys and Compounds* 749: 473–480.

89 Li, H., Wang, M., Wei, Y., and Long, F. (2019). Noble metal-free NiS_2 with rich active sites loaded g-C_3N_4 for highly efficient photocatalytic H_2 evolution under visible light irradiation. *Journal of Colloid and Interface Science* 534: 343–349.

90 Pan, J., Wang, B., Dong, Z. et al. (2019). The 2D RGO-NiS_2 dual co-catalyst synergistic modified g-C_3N_4 aerogel towards enhanced photocatalytic hydrogen production. *International Journal of Hydrogen Energy* 44: 19942–19952.

91 Zhang, W., Wang, Y., Wang, Z. et al. (2010). Highly efficient and noble metal-free NiS/CdS photocatalysts for H_2 evolution from lactic acid sacrificial solution under visible light. *Chemical Communications* 46: 7631–7633.

92 Chen, M., Shao, L.L., Yuan, Z.Y. et al. (2017). General strategy for controlled synthesis of Ni_xP_y/carbon and its evaluation as a counter electrode material in dye-sensitized solar cells. *ACS Applied Materials & Interfaces* 9: 17949–17960.

93 Azadmanjiri, J., Srivastava, V.K., Kumar, P. et al. (2018). Two- and three-dimensional graphene-based hybrid composites for advanced energy storage and conversion devices. *Journal of Materials Chemistry A* 6: 702–734.

94 Chen, M., Shao, L.-L., Gao, Z.-M. et al. (2015). Cobalt oxide and nitride particles supported on mesoporous carbons as composite electrocatalysts for dye-sensitized solar cells. *Journal of Power Sources* 286: 82–90.

95 Liu, I.P., Teng, H., and Lee, Y.-L. (2017). Highly electrocatalytic carbon black/copper sulfide composite counter electrodes fabricated by a facile method for quantum-dot-sensitized solar cells. *Journal of Materials Chemistry A* 5: 23146–23157.

96 O'Regan, B. and Grätzel, M. (1991). A low-cost, high-efficiency solar cell based on dye-sensitized colloidal TiO_2 films. *Nature* 353: 737–740.

97 Yella, A., Lee, H.W., Tsao, H.N. et al. (2011). Porphyrin-sensitized solar cells with cobalt (II/III)-based redox electrolyte exceed 12 percent efficiency. *Science* 334: 629–634.

98 Freitag, M., Teuscher, J., Saygili, Y. et al. (2017). Dye-sensitized solar cells for efficient power generation under ambient lighting. *Nature Photonics* 11: 372–378.

99 Zhang, X., Bai, J., Zhen, M., and Liu, L. (2016). Ultrathin Ni-Ni_3Se_2 nanosheets on graphene as a high-performance counter electrode for dye-sensitized solar cells. *RSC Advances* 6: 89614–89620.

100 Wu, Y., Zhou, B., Yang, C. et al. (2016). $CuFeS_2$ colloidal nanocrystals as an efficient electrocatalyst for dye sensitized solar cells. *Chemical Communications* 52: 11488–11491.

101 Chen, M. and Shao, L.L. (2016). Review on the recent progress of carbon counter electrodes for dye-sensitized solar cells. *Chemical Engineering Journal* 304: 629–645.

102 Chen, M., Shao, L.L., Qian, X. et al. (2014). Direct synthesis of cobalt nanoparticle-imbedded mesoporous carbons for high-performance dye-sensitized solar cell counter electrodes. *Journal of Materials Chemistry C* 2: 10312–10321.

103 Ahmed, U., Alizadeh, M., Rahim, N.A. et al. (2018). A comprehensive review on counter electrodes for dye sensitized solar cells: a special focus on Pt-TCO free counter electrodes. *Solar Energy* 174: 1097–1125.

104 Yuan, H., Jiao, Q., Zhang, S. et al. (2016). In situ chemical vapor deposition growth of carbon nanotubes on hollow $CuFe_2O_4$ as an efficient and low cost counter electrode for dye-sensitized solar cells. *Journal of Power Sources* 325: 417–426.

105 Hou, Y., Wang, D., Yang, X.H. et al. (2013). Rational screening low-cost counter electrodes for dye-sensitized solar cells. *Nature Communications* 4: 1583.

106 Duan, Y., Tang, Q., Liu, J. et al. (2014). Transparent metal selenide alloy counter electrodes for high- efficiency bifacial dye-sensitized solar cells. *Angewandte Chemie International Edition* 53: 14569–14574.

107 Huang, S., He, Q., Chen, W. et al. (2014). Ultrathin $FeSe_2$ nanosheets: controlled synthesis and application as a heterogeneous catalyst in dye-sensitized solar cells. *Chemistry—A European Journal* 21: 4085–4091.

108 Wang, W., Pan, X., Liu, W. et al. (2014). $FeSe_2$ films with controllable morphologies as efficient counter electrodes for dye-sensitized solar cells. *Chemical Communications* 50: 2618–2620.

109 Huang, S., He, Q., Chen, W. et al. (2015). 3D hierarchical $FeSe_2$ microspheres: controlled synthesis and applications in dye-sensitized solar cells. *Nano Energy* 15: 205–215.

110 Liu, F., Zhu, J., Hu, L. et al. (2015). Low-temperature, solution-deposited metal chalcogenide films as highly efficient counter electrodes for sensitized solar cells. *Journal of Materials Chemistry A* 3: 6315–6323.

111 Liu, J., Wang, L., Ouyang, W. et al. (2015). Fabrication of PMMA nanofluidic electrochemical chips with integrated microelectrodes. *Biosensors and Bioelectronics* 72: 288–293.

112 Rickard, D. and Luther Iii, G.W. (2007). Chemistry of iron sulfides. *Chemical Reviews* 107: 514–562.

113 Xu, H., Zhang, C., Wang, Z. et al. (2014). Nitrogen-doped carbon and iron carbide nanocomposites as cost-effective counter electrodes of dye-sensitized solar cells. *Journal of Materials Chemistry A* 2: 4676–4681.

114 Wang, M., Anghel, A.M., Marsan, B. et al. (2009). CoS supersedes Pt as efficient electrocatalyst for triiodide reduction in dye-sensitized solar cells. *Journal of the American Chemical Society* 131: 15976–15977.

115 Chang, S.H., Lu, M.D., Tung, Y.L., and Tuan, H.Y. (2013). Gram-scale synthesis of catalytic Co_9S_8 nanocrystal ink as a cathode material for spray-deposited, large-area dye-sensitized solar cells. *ACS Nano* 7: 9443–9451.

116 Patil, S.A., Shinde, D.V., Lim, I. et al. (2015). An ion exchange mediated shape-preserving strategy for constructing 1-D arrays of porous $CoS_{1.0365}$ nanorods for electrocatalytic reduction of triiodide. *Journal of Materials Chemistry A* 3: 7900–7909.

117 Zhang, Z., Pang, S., Xu, H. et al. (2013). Electrodeposition of nanostructured cobalt selenide films towards high performance counter electrodes in dye-sensitized solar cells. *RSC Advances* 3: 16528–16533.

118 Li, Y., Zhang, H., Wang, Y. et al. (2014). A self-sponsored doping approach for controllable synthesis of S and N co-doped trimodal-porous structured graphitic carbon electrocatalysts. *Energy and Environmental Science* 7: 3720–3726.

119 Datta, M.K., Kadakia, K., Velikokhatnyi, O.I. et al. (2013). High performance robust F-doped tin oxide based oxygen evolution electro-catalysts for PEM based water electrolysis. *Journal of Materials Chemistry A* 1: 4026–4037.

120 Cui, C.H. and Yu, S.H. (2013). Engineering interface and surface of noble metal nanoparticle nanotubes toward enhanced catalytic activity for fuel cell applications. *Accounts of Chemical Research* 46: 1427–1437.

121 Tan, Z., Liu, P., Zhang, H. et al. (2015). An in situ vapour phase hydrothermal surface doping approach for fabrication of high performance Co_3O_4 electrocatalysts with an exceptionally high S-doped active surface. *Chemical Communications* 51: 5695–5697.

122 Huang, Y.J., Lee, C.P., Pang, H.W. et al. (2017). Microemulsion-controlled synthesis of $CoSe_2/CoSeO_3$ composite crystals for electrocatalysis in dye-sensitized solar cells. *Materials Today Energy* 6: 189–197.

123 Awais, M., Gibson, E., Vos, J.G. et al. (2014). Fabrication of efficient NiO photocathodes prepared via RDS with novel routes of substrate processing for p-type dye-sensitized solar cells. *ChemElectroChem* 1: 384–391.

124 Zhang, X., Zhang, H., Wang, X., and Zhou, X. (2018). Enhanced electrocatalytic performance of nickel diselenide grown on graphene toward the reduction of triiodide redox couples. *RSC Advances* 8: 28131–28138.

125 Feng, C., Jin, Z., Zhang, M., and Wang, Z.S. (2018). Nanocomposites of nickel selenide supported on cube-shaped lidless graphitic boxes as efficient counter electrodes for quasi-solid-state dye-sensitized solar cells. *Electrochimica Acta* 281: 237–245.

126 Duan, Y., Tang, Q., He, B. et al. (2014). Transparent nickel selenide alloy counter electrodes for bifacial dye-sensitized solar cells exceeding 10% efficiency. *Nanoscale* 6: 12601–12608.

127 Jia, J., Wu, J., Tu, Y. et al. (2015). Transparent nickel selenide used as counter electrode in high efficient dye-sensitized solar cells. *Journal of Alloys and Compounds* 640: 29–33.

128 Sun, X., Dou, J., Xie, F. et al. (2014). One-step preparation of mirror-like NiS nanosheets on ITO for the efficient counter electrode of dye-sensitized solar cells. *Chemical Communications* 50: 9869–9871.

129 Wang, X., Batter, B., Xie, Y. et al. (2015). Highly crystalline, small sized, monodisperse α-NiS nanocrystal ink as an efficient counter electrode for dye-sensitized solar cells. *Journal of Materials Chemistry A* 3: 15905–15912.

130 Zheng, J., Zhou, W., Ma, Y. et al. (2015). Facet-dependent NiS_2 polyhedrons on counter electrodes for dye-sensitized solar cells. *Chemical Communications* 51: 12863–12866.

131 Xiao, Y., Han, G., Zhou, H. et al. (2015). Nickel sulfide counter electrodes enhanced by hydrosulphuric acid hydrothermal treatments for use in Pt-free dye-sensitized solar cells. *Electrochimica Acta* 155: 103–109.

132 Wan, Z., Jia, C., and Wang, Y. (2015). In situ growth of hierarchical NiS_2 hollow microspheres as efficient counter electrode for dye-sensitized solar cell. *Nanoscale* 7: 12737–12742.

133 Sun, H., Qin, D., Huang, S. et al. (2011). Dye-sensitized solar cells with NiS counter electrodes electrodeposited by a potential reversal technique. *Energy and Environmental Science* 4: 2630–2637.

134 Chi, W.S., Han, J.W., Yang, S. et al. (2012). Employing electrostatic self-assembly of tailored nickel sulfide nanoparticles for quasi-solid-state dye-sensitized solar cells with Pt-free counter electrodes. *Chemical Communications* 48: 9501–9503.

135 Zhao, W., Lin, T., Sun, S. et al. (2013). Oriented single-crystalline nickel sulfide nanorod arrays: "two-in-one" counter electrodes for dye-sensitized solar cells. *Journal of Materials Chemistry A* 1: 194–198.

136 Ku, Z., Li, X., Liu, G. et al. (2013). Transparent NiS counter electrodes for thiolate/disulfide mediated dye-sensitized solar cells. *Journal of Materials Chemistry A* 1: 237–240.

137 Li, Y., Chang, Y., Zhao, Y. et al. (2016). In situ synthesis of oriented NiS nanotube arrays on FTO as high-performance counter electrode for dye-sensitized solar cells. *Journal of Alloys and Compounds* 679: 384–390.

138 Zhang, X., Zhen, M., Bai, J. et al. (2016). Efficient NiSe-Ni_3Se_2/graphene electrocatalyst in dye-sensitized solar cells: the role of hollow hybrid nanostructure. *ACS Applied Materials and Interfaces* 8: 17187–17193.

139 Huo, J., Wu, J., Zheng, M. et al. (2016). Flower-like nickel cobalt sulfide microspheres modified with nickel sulfide as Pt-free counter electrode for dye-sensitized solar cells. *Journal of Power Sources* 304: 266–272.

140 Dong, J., Wu, J., Jia, J. et al. (2017). Nickel selenide/reduced graphene oxide nanocomposite as counter electrode for high efficient dye-sensitized solar cells. *Journal of Colloid and Interface Science* 498: 217–222.

141 Wu, H., Wang, Y., Zhang, L. et al. (2018). Comparison of two nickel selenides materials with different morphologies as counter electrodes in dye-sensitized solar cells. *Journal of Alloys and Compounds* 745: 222–227.

142 Lee, C.T., Peng, J.D., Li, C.T. et al. (2014). Ni_3Se_4 hollow architectures as catalytic materials for the counter electrodes of dye-sensitized solar cells. *Nano Energy* 10: 201–211.

143 Gong, F., Xu, X., Li, Z. et al. (2013). $NiSe_2$ as an efficient electrocatalyst for a Pt-free counter electrode of dye-sensitized solar cells. *Chemical Communications* 49: 1437–1439.

144 Guai, G.H., Leiw, M.Y., Ng, C.M., and Li, C.M. (2012). Sulfur-doped nickel oxide thin film as an alternative to Pt for dye-sensitized solar cell counter electrodes. *Advanced Energy Materials* 2: 334–338.

145 Liu, Q.-L., Dong, Y.-J., Cao, Y. et al. (2017). $Ni_xS_y/NiSe_2$ hybrid catalyst grown in situ on conductive glass substrate as efficient counter electrode for dye-sensitized solar cells. *Electrochimica Acta* 250: 244–250.

9

Applications of Iron (Fe, Co, and Ni)-Group Metal Compound Nanomaterials in Electrocatalytic Upgrading of Biomass and Degradation of Pollutants

The conversion of biomass into valuable fuels and chemicals has advanced rapidly in recent years, with electrocatalytic methods demonstrating significant advantages in this field. Biomass is generally categorized into inorganic and organic matter, and its sourcing can often be aligned with environmental protection efforts [1, 2]. With ongoing industrialization and urbanization, pollutants are frequently released into the environment, disrupting ecosystems and posing risks to human health [3, 4]. The accumulation of persistent and toxic pollutants increases health hazards, especially under prolonged exposure to high concentrations. Degrading biomass not only drives scientific progress but also helps alleviate environmental pressures. Researchers are therefore focusing on breaking down pollutants while upgrading biomass into useful products for human activities [5–7]. In this context, Fe-group elements (Fe, Co, and Ni) have gained considerable attention due to their abundance, excellent performance, and stability [8, 9].

Traditional biomass utilization often relies on direct combustion for heat or power generation. However, this process suffers from low energy efficiency and excessive greenhouse gas emissions (CO_2), which conflict with current green and low-carbon development goals [10, 11]. Consequently, developing efficient and clean oxidation pathways for biomass conversion has become a key research focus. Electrocatalytic technology enables the targeted conversion of biomass (such as cellulose, lignin, and glucose) into high-value chemicals (like 2,5-Furandicarboxylic Acid [FDCA], ethanol, and acetic acid) by precisely controlling reaction pathways. This offers a sustainable and green alternative to conventional combustion processes [12, 13]. When the biomass oxidation reaction at the anode is coupled with the hydrogen evolution reaction (HER) at the cathode in an electrolysis system, it not only achieves the selective oxidation of biomass efficiently but also reduces the energy consumption for hydrogen production through electron transfer. This coupling also enhances reaction kinetics and system stability [14, 15]. Notably, this electrocatalytic strategy for biomass offers two core advantages: first, it can degrade pollutants (both inorganic and organic) generated during biomass conversion, contributing to environmental remediation [16, 17]. Second, it upgrades biomass into valuable organic molecules, providing a new paradigm for the green and efficient utilization of biomass resources [18, 19].

Iron-Group Metal Compound Nanomaterials: Preparation, Characterization, Structure, and Applications, First Edition. Jinlong Zheng.

9.1 Electrocatalytic Upgrading of Biomass

Shifting the current energy production and consumption structure away from over-reliance on fossil fuels toward the efficient use of renewable resources is an effective pathway to address the energy crisis and achieve carbon neutrality. Biomass stands as one of the most promising renewable alternatives to fossil fuels for producing valuable organic compounds. The electrocatalytic conversion of biomass-derived organic molecules under relatively mild conditions offers a green and efficient strategy for producing high-value chemicals [20–22]. Unlike traditional combustion for heat, which is essentially a "downcycling" process, electrocatalytic conversion aims to selectively cleave specific chemical bonds and introduce functional groups into biomass molecules. This is achieved by controlling electron transfer pathways under mild conditions, enabling the targeted synthesis of high-value products [23, 24].

In heterogeneous electrocatalysis, catalysts play a crucial role in directly modulating the adsorption and desorption behavior of reactants. This regulation influences energy barriers and reaction pathways, ultimately enabling the synthesis of target chemicals [25]. Transition metal electrocatalysts, with their abundant 3d electrons and unique e_g orbitals, enhance the covalency of metal–oxygen bonds, making them prominent candidates for catalyzing C—C bond cleavage. Research indicates that the coordination environment and electronic structure of these catalysts significantly affect product selectivity. By tuning these properties, electrochemical C—C bond cleavage in biomass-derived molecules can be achieved [26–28]. Ni readily forms high-valence species with strong oxidizing power. It is particularly effective in cleaving C—C bonds and deeply oxidizing C—OH bonds, making it highly suitable for converting substrates like glucose into small carboxylic acids or valuable platform compounds [29, 30]. Co leverages its flexible valence states to modulate the local electron density on the catalyst surface. This optimizes the adsorption of reaction intermediates, lowers energy barriers, and stabilizes key species. These properties give Co an advantage in C—H bond activation and selective oxidation, often suppressing excessive C–C cleavage to selectively produce target aldehydes, ketones, or carboxylic acids [31, 32]. Fe, as an eco-friendly and low-cost element, often exhibits good activity at low potentials. Its catalytic mechanism is associated with the Fe^{2+}/Fe^{3+} redox cycle, generating reactive oxygen species [33, 34]. Given the diversity of biomass feedstocks, this chapter categorizes them based on functional groups: monoalcohols (e.g., methanol, ethanol, and benzyl alcohol), polyols (e.g., ethylene glycol [EG], glycerol, and glucose), amines, aldehydes, and the biopolymer lignin [35, 36].

9.1.1 Monohydric Alcohol

Monoalcohols, characterized by their relatively simple molecular structures, include representative compounds such as methanol (CH_3OH), ethanol (CH_3CH_2OH), and benzyl alcohol ($C_6H_5CH_2OH$). The key to their electrocatalytic oxidation lies in the stepwise dehydrogenation process, which involves the activation and cleavage of C—H bonds. The objective is to oxidize the carbon atom bearing the hydroxyl group, ultimately yielding carboxylic acids or aldehydes. A significant feature of monoalcohol electrocatalysis is that their thermodynamic oxidation potential is considerably lower than that of the oxygen evolution reaction (OER). Therefore, substituting the anodic OER with monoalcohol oxidation can significantly reduce the cell voltage in an electrolyzer. The primary challenge lies in controlling the oxidation depth to prevent over-oxidation to CO_2 while achieving high selectivity toward value-added chemicals such as formic acid, acetic acid, or benzoic acid.

Methanol, as the simplest alcohol, is one of the most favored organic feedstocks due to its abundant availability and low cost. The electrocatalytic oxidation of methanol is used to produce formic acid and formate, which are valuable in industries such as pharmaceuticals, textiles, rubber, and agrochemicals. The overall reaction for methanol-assisted water electrolysis, yielding formic acid and hydrogen, can be expressed as: $CH_3OH + H_2O \rightarrow HCOOH + 2H_2$. Based on thermodynamic data under standard conditions, the Gibbs free energy change (ΔG^0) for this reaction is approximately 42.3 kJ mol^{-1}. Using the formula Ucell = $\Delta G^0/nF$ (where n is the number of electrons transferred, and F is the Faraday constant), the corresponding theoretical cell voltage is about 0.110 V, which is significantly lower than the theoretical decomposition voltage of water. Xiang et al. [37] investigated the catalytic performance of a core–shell heterostructure $Co(OH)_2$@HOS/CP for the methanol oxidation reaction (MOR). Figure 9.1a illustrates the material synthesis process, while Figure 9.1c, d show the structural characterization of the heterojunction core–shell architecture. Compared to α-$Co(OH)_2$/CP, fully sulfurized $CoS_x(OH)_y$/CP, and bare CP, the $Co(OH)_2$@HOS/CP catalyst exhibited the lowest onset potential for methanol oxidation at about 1.29 V (vs. RHE). Furthermore, a potential of only 1.385 V was required to achieve a current density of 10 mA cm^{-2}, which is much lower than that needed for the OER (1.57 V) on the same catalyst. In a two-electrode system for methanol-assisted overall water splitting, $Co(OH)_2$@HOS/CP served as the catalyst to enhance the reaction rate. The experiment showed that a cell voltage of only 1.497 V was required to achieve a current density of 10 mA cm^{-2}, lower than the 1.631 V needed for conventional water electrolysis. This indicates that methanol-assisted overall water splitting is more favorable for hydrogen production compared to traditional water electrolysis. Quantitative analysis of gas and liquid products by gas chromatography and ion chromatography (Figure 9.1b, e) confirmed Faradaic efficiencies of nearly 100% for both the HER and the MOR (with formate as the oxidation product).

To further investigate the synergistic effects of reaction intermediates and surface anion elements on the MOR, Fu et al. [38] studied the catalytic performance of hollow NiSe nanocrystals distributed on carbon nanotubes (h-NiSe/CNTs) using X-ray photoelectron spectroscopy (XPS), X-ray absorption fine structure (XAFS) spectroscopy, in situ Raman, and density functional theory (DFT) calculations. XPS and XAFS results indicated a gradual increase in the signals for $Ni(OH)_2$ and SeO_x on the NiSe surface due to hydroxylation/oxidation. In situ Raman spectra revealed that the Ni=O species of NiOOH intermediates appeared only during the OER, as NiOOH can be rapidly reduced by methanol during MOR to form the key intermediate $Ni{=}OCH_3$ (Figure 9.2a–c). According to the Sabatier principle, a d-band

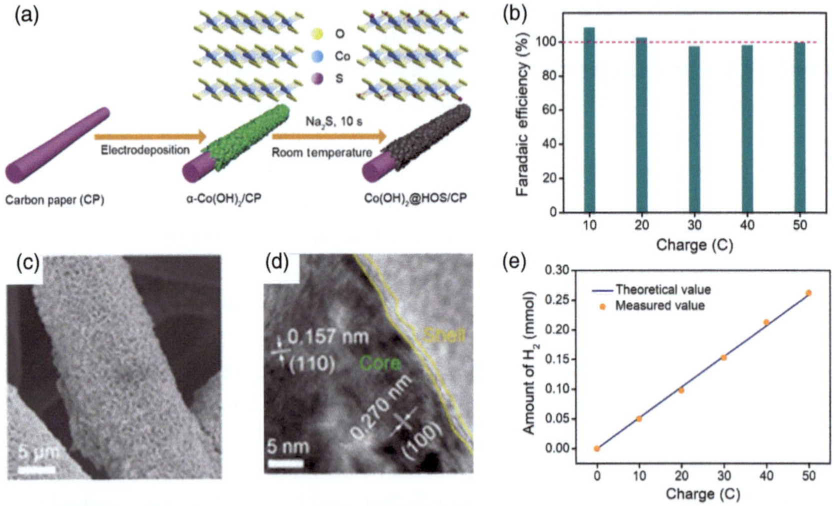

Figure 9.1 (a) Schematic illustration of the synthesis of $Co(OH)_2$@HOS/CP; (b) Faraday efficiency of $Co(OH)_2$@HOS/CP for methanol oxidation at 1.50 V; (c) scanning electron microscope (SEM) image; and (d) high-resolution transmission electron microscope (HRTEM) image of $Co(OH)_2$@HOS/CP; (e) measured and theoretical values of H_2 content. *Source*: Reproduced from Ref. [37] / with permission of John Wiley & Sons.

center closer to the Fermi level indicates stronger adsorption of intermediates. Based on this, five models simulating the evolution of NiSe over reaction time were constructed (Figure 9.2d). Depending on the hydroxylation/oxidation state of the NiSe surface, Model IV was observed to have a moderate εd among all models, suggesting it can effectively adsorb organic molecules with an optimal adsorption energy. DFT results (Figure 9.2e) showed that Model IV-$SeO_x/NiOOH_{opt}$ exhibited the lowest adsorption energy for intermediates, which is also the rate-determining step for most models. The Gibbs free energy diagram indicated that the ΔG_{H*} of Model IV is neither too positive nor too negative, favoring the adsorption and desorption of intermediate products, thereby leading to high activity and selectivity for producing value-added formate. The coexistence of in situ formed NiOOH and SeO_x species synergistically modulates the position of the d-band center in NiSe within h-NiSe/CNTs during MOR. This optimization of intermediate adsorption/desorption energetics facilitates the selective conversion of methanol to the value-added product, formate.

The ethanol oxidation reaction (EOR) is often coupled with the HER. Besides the complete oxidation of ethanol to CO_2, this process can simultaneously produce green hydrogen and high-value products such as acetic acid or even ethyl acetate. Bioethanol, produced from the fermentation of biomass (e.g., sugars and starches), serves not only as a transportation fuel or fuel additive but also as a sustainable feedstock for valuable chemicals like ethylene, acetaldehyde, acetic acid, and ethyl acetate. Ni-based materials are widely used for electrocatalytic ethanol oxidation due to their redox properties. In alkaline media, the EOR on Ni-based catalysts

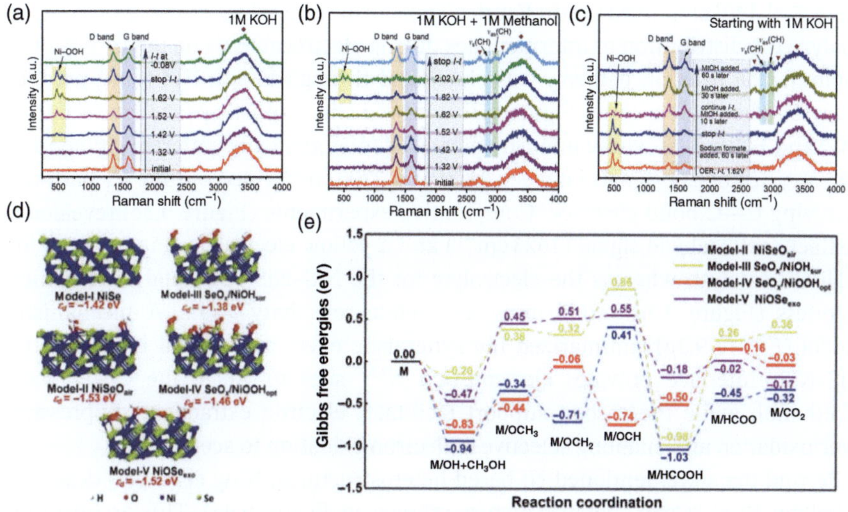

Figure 9.2 (a) In situ Raman spectra of anode potential in 1.0 M KOH; (b) in situ Raman spectra of anode potential in a mixed solution of 1.0 M KOH and 1.0 M methanol; (c) in situ Raman spectra collected after adding formate and methanol to 1.0 M KOH solution (initial) under chronoamperometry (I-t) at 1.62 V; (d) models of NiSe surfaces with different hydroxylation/oxidation states and corresponding Ni 3d orbitals: Model I-NiSe (pristine NiSe surface), Model II-$NiSeO_{air}$ (surface oxidation in air), Model III-$SeO_x/NiOH_{sur}$ (initial hydroxylation/oxidation on the catalyst surface), Model IV-$SeO_x/NiOOH_{opt}$ (equilibrium hydroxylation/oxidation on the catalyst surface), and Model V-$NiOSe_{exo}$ (excessive surface oxidation); (e) Gibbs free energy diagrams during methanol activation under different conditions. *Source*: Reproduced from Ref. [38] / with permission of John Wiley & Sons.

proceeds via the oxidation of Ni^{2+} to Ni^{3+}, followed by the spontaneous reaction of Ni^{3+} with ethanol, which acts as a nucleophile, oxidizing it to acetaldehyde or acetic acid while Ni^{3+} is reduced back to Ni^{2+}. Thus, the interconversion between Ni^{2+} and Ni^{3+} is central to the EOR mechanism. However, Ni oxides and hydroxides typically have wide band gaps, leading to poor electrical conductivity and limited reaction kinetics. Introducing metallic Ni^0 into NiO-based materials can narrow the band gap and accelerate electrochemical charge transfer. Du et al. [39] enhanced the electronic and oxophilic properties of a catalyst by designing a Ni^0/NiO heterostructure. The interfacial channels within this structure support efficient charge transfer, thereby improving both conductivity and catalytic activity. As shown in Figure 9.3a, the electrode's surface hydrophilicity/hydrophobicity was tuned via an electrochemical deposition process to achieve efficient synthesis. The incorporation of Ni^0 promoted the conversion of Ni^{2+} to Ni^{3+}, providing abundant active sites for EOR. Furthermore, it lowered the kinetic barrier for the further oxidation of acetaldehyde, thereby enhancing the overall oxidation of ethanol to acetate. This catalyst demonstrated outstanding performance in the EOR, achieving a current density as high as 573.7 mA cm^{-2} with high selectivity (98.31%) for acetate. To further investigate the conversion of intermediates on heterostructured Ni-based catalysts,

Zhao et al. [40] employed in situ Fourier Transform Infrared Spectroscopy (FTIR) to analyze the distribution of intermediates during electrocatalysis on a NiS–NiS_2/CC catalyst. The reaction pathway was elucidated through in situ FTIR and product analysis. The FTIR spectra (Figure 9.3b) showed characteristic peaks for acetate at 1712 and 1410 cm^{-1}, whose intensity increased monotonically with rising potential. The absence of CO signals between 1800 and 2200 cm^{-1} ruled out the C_1 pathway involving C—C bond cleavage. Comparative experiments (Figure 9.3c) revealed a distinct acetaldehyde signal (1623 cm^{-1}) and a yellow electrolyte (Figure 9.3e) for a Ni/CC catalyst, whereas the electrolyte for the NiS–NiS_2/CC catalyst remained colorless (Figure 9.3d), confirming an acetate selectivity >99%. A mechanistic model (Figure 9.3g) summarized the synergistic effect at the dual interface: the NiS–NiS_2 interface provides electron-rich Ni^{3+} sites to attack the C—H bond of ethanol, while the carbon support facilitates electron extraction, suppressing over-oxidation and enabling selective 4-electron oxidation to acetic acid.

Beyond the aforementioned Ni-based heterostructures, Ning et al. [41] designed a hollow Fe_2O_3/$CoSe_2$ heterostructure (shown in Figure 9.4a). This architecture not only increases the exposure of active sites (Figure 9.4b–e) but also significantly enhances charge and mass transport, leading to markedly improved electrocatalytic performance. The study revealed a "dynamic electron spring effect" of Fe_2O_3 on the Co sites during the electrocatalytic EOR, as illustrated in Figure 9.4f. Specifically, Fe_2O_3 dynamically modulates and stabilizes the valence state of Co: it promotes the oxidation of Co^{2+} to Co^{3+} at lower potentials, while at higher potentials, it delays the over-oxidation of Co^{3+} to Co^{4+}. This dynamic electronic regulation effectively widens the potential window for EOR on the $CoSe_2$ catalyst.

Benzyl alcohol, as the simplest aromatic alcohol, can be oxidized to benzoic acid, a high-value product used as an intermediate in fragrances and cosmetic preservatives. The electrocatalytic oxidation of benzyl alcohol to benzoic acid has a lower thermodynamic potential than the OER. The Qiu group [42] reported nanosheets composed of Ni/Co layered double hydroxide (LDH) supported on Ni foam (A–Ni–Co–H/NF) for benzyl alcohol oxidation. At room temperature, this catalyst achieved a current density exceeding 400 mA cm^{-2} with a Faradaic efficiency close to 100% (Figure 9.5a). In situ Raman spectroscopy revealed that during the reaction, the catalyst undergoes a reversible structural transformation from Ni/Co LDH to Co-doped Ni oxyhydroxide (Co–NiOOH), indicating that Co–NiOOH is the true active species (Figure 9.5b). Experimental data and DFT calculations further showed that abundant vacancies can modulate the electronic structure of the Co_3O_4 nanosurface, increase active sites, promote reactant adsorption and electron transfer, and ultimately enhance electrocatalytic performance. Besides benzoic acid, benzaldehyde is another important oxidation product of benzyl alcohol, which is widely used in pharmaceutical synthesis and the resin industry. Jian et al. [43] designed a 3D electrode consisting of porous $CuCoN_x$ nanosheet arrays with a nitrogen-doped carbon shell grown on carbon fiber (NC@$CuCo_2N_x$/CF). This electrode selectively oxidizes benzyl alcohol to benzaldehyde in an alkaline medium (Figure 9.5c). The NC@$CuCo_2N_x$/CF electrode exhibited excellent performance for selective benzyl alcohol oxidation, requiring only 1.25 V to reach a current

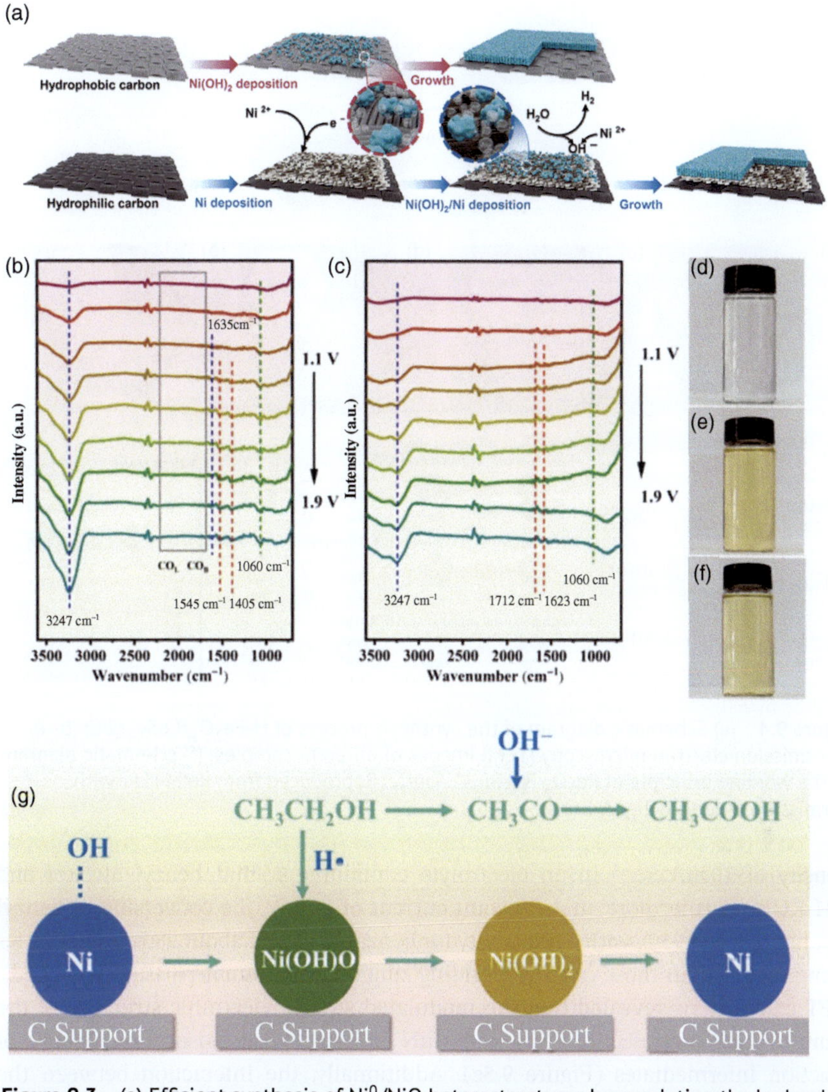

Figure 9.3 (a) Efficient synthesis of Ni^0/NiO heterostructures by regulating the hydrophilic and hydrophobic properties of the electrode surface through the electrochemical deposition process; *Source*: Reproduced from Ref. [39] / with permission of John Wiley & Sons. (b) In situ FTIR of EOR on NiS–NiS_2/CC catalyst during chronoamperometry tests from 1.1 to 1.9 V (vs. reversible hydrogen electrode [RHE]); (c) in situ FTIR of EOR on Ni/CC catalyst; (d) photographs of liquid products from EOR on NiS–NiS_2/CC catalyst; (e) photographs of liquid products from EOR on Ni/CC catalyst; (f) schematic diagram of infrared characteristic peaks of acetaldehyde; (g) mechanistic model of EOR catalyzed by NiS–NiS_2/CC. *Source*: Reproduced from Ref. [40] / with permission of Elsevier.

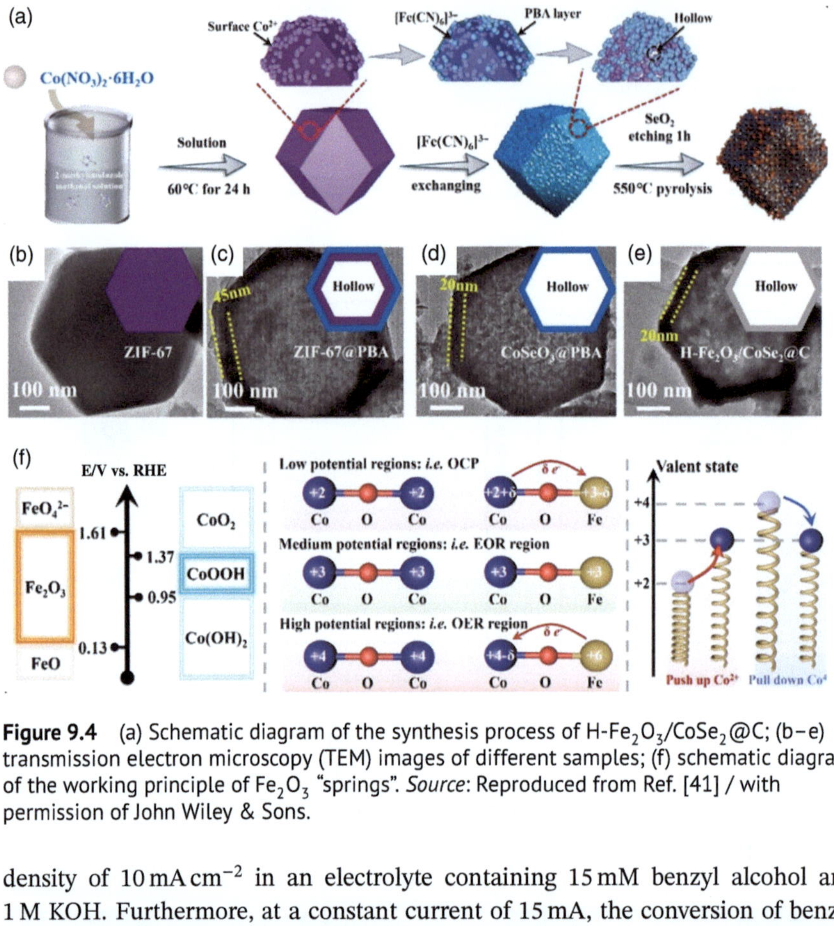

Figure 9.4 (a) Schematic diagram of the synthesis process of H-Fe_2O_3/$CoSe_2$@C; (b–e) transmission electron microscopy (TEM) images of different samples; (f) schematic diagram of the working principle of Fe_2O_3 "springs". *Source*: Reproduced from Ref. [41] / with permission of John Wiley & Sons.

density of 10 mA cm^{-2} in an electrolyte containing 15 mM benzyl alcohol and 1 M KOH. Furthermore, at a constant current of 15 mA, the conversion of benzyl alcohol reached 95%, with a selectivity for benzaldehyde of about 98%, both metrics being higher than those of corresponding single-metal counterparts (Figure 9.5d). DFT calculations revealed that the modulated surface electronic structure of the nano-hybrid composed of CoN and CuN influences the adsorption energy of reaction intermediates (Figure 9.5e). Additionally, the interaction between the metal nitrides and the nitrogen-doped carbon shell, combined with a large specific surface area and abundant porosity, facilitates mass and charge transfer, collectively enhancing the catalytic activity for benzyl alcohol oxidation.

9.1.2 Dihydric Alcohol

Polyols, represented by compounds such as EG ($HOCH_2CH_2OH$), glycerol ($HOCH_2CH(OH)CH_2OH$), and glucose ($CH_2OH(CHOH)_4CHO$), feature complex molecular structures containing multiple hydroxyl groups and longer carbon chains. The primary challenge in the electrocatalytic upgrading of these biomass-derived polyols lies in the selective cleavage of specific C—H and C—OH bonds while preserving the main carbon skeleton (C—C bonds). This selectivity is essential for

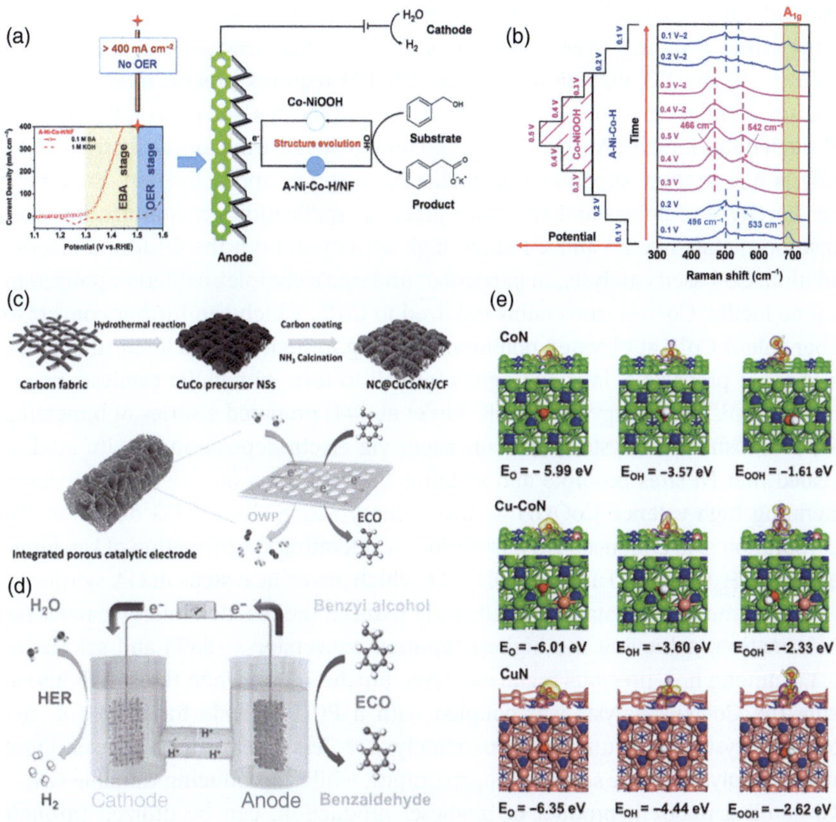

Figure 9.5 (a) LSV curves of A-Ni-Co-H/NF in 1 mM KOH solution and mixed solution of 0.1 mM BA and 1 mM KOH at a scan rate of 10 mV s^{-1}; (b) in situ Raman spectra of A-Ni-Co-H/NF electrode and corresponding structural transformations at varying potentials vs. Ag/AgCl; *Source*: Reproduced from Ref. [42] / with permission of The Royal Society of Chemistry. (c) schematic illustration of NC@CuCo$_2$N$_x$/CF synthesis; (d) schematic diagram of NC@CuCo$_2$N$_x$/CF as a bifunctional electrocatalyst for HER and selective electrocatalytic oxidation of benzyl alcohol (ECO); (e) adsorption energies and charge density differences of O, OH, and OOH intermediates on CoN, Cu-CoN, and CuN; green, brown, red, blue, and white represent Co, Cu, O, N, and H atoms, respectively; yellow and purple represent negative and positive charges, respectively. *Source*: Reproduced from Ref. [43] / with permission of John Wiley & Sons. KGaA.

converting them into valuable oxygenated chemicals, as opposed to their complete oxidation to CO_2. Specifically, an effective catalyst must be able to recognize and preferentially oxidize primary hydroxyl groups (–CH_2OH) over secondary hydroxyl groups (–CHOH), while simultaneously avoiding random cleavage of C—C bonds. This precise control is crucial for achieving a highly selective and targeted conversion process.

The anodic oxidation of biomass-derived EG into high-value glycolic acid (GA), a key raw material for biodegradable plastics, faces the challenge of developing low-cost electrocatalysts for economically viable and selective GA production. A major difficulty lies in selectively oxidizing one of EG's two identical hydroxymethyl groups while avoiding over-oxidation or C—C bond cleavage. Achieving high selectivity in the EG oxidation reaction (EGOR) requires precise modulation of EG adsorption and enhancing the concentration of the $*OC–CH_2OH$ intermediate. Although noble metals (Pd, Pt, and Au) and their oxides are the most effective EGOR catalysts due to their superior intermediate adsorption and activation, their high cost, scarcity, and poor stability hinder practical application. In contrast, low-cost transition metals like Ni and Co show high activity in both the OER and alcohol oxidation. Co-based catalysts, in particular, undergo a complex oxidation process in alkaline media: Co^{2+} is irreversibly oxidized to Co^{3+}, which can further convert to higher-valent Co^{4+} at elevated potentials, leading to high electrochemical activity. Given these properties, introducing Co into Ni to form bimetallic catalysts represents a promising strategy for EGOR. Lin et al. [44] prepared a series of bimetallic Co–Ni nanosheet catalysts on carbon paper via electrodeposition. In situ studies revealed that Ni sites promote and stabilize the formation of active Co^{3+} species, generating high-valence CoOOH as the primary active site for EG oxidation. Ni incorporation also enhanced EG adsorption, accelerating the formation of key intermediates, $CH_2OH–CHO$ and $CH_2OH–CO$, which are critical steps in GA synthesis. As a result, the Co–Ni catalyst significantly reduced the reaction energy barrier for GA formation, achieving the highest reported conversion (>85%) and selectivity (96.3%) among non-precious metal catalysts. Furthermore, when the EGOR anode (using the Co–Ni catalyst) was coupled with a Pt/C cathode for the HER, the integrated system consumed approximately 27% less energy than conventional water electrolysis for the same hydrogen output, while coproducing valuable GA.

Glycerol, a major byproduct of biodiesel production, can be utilized through electrocatalytic oxidation to achieve both waste upcycling and the coproduction of high-value organic chemicals alongside green hydrogen. This process demonstrates significant potential for coupling green chemistry with renewable energy applications. However, existing catalysts often suffer from low current density, poor selectivity, and insufficient stability, primarily due to inadequate control over reactant adsorption behavior. Maintaining an adsorption balance between OH^- and glycerol on the electrode surface is crucial for achieving efficient, selective, and stable electrocatalytic conversion. A balanced adsorption ensures effective collisions between glycerol molecules and active OH^- species on the catalyst surface, thereby sustaining the reaction. This balance helps prevent excessive OH^-, which can lead to competing OER or glycerol over-adsorption that produces impure products, thereby enhancing the yield and Faradaic efficiency of target products like formate. Moreover, it avoids active site blockage from single-species over-adsorption and prevents electrode oxidation/reconstruction, ensuring long-term operational stability. Long et al. [45] constructed a Co_4N–CoO heterostructure catalyst (Figure 9.6a–c). Glycerol adsorbs strongly on Co_4N but weakly on CoO, while CoO preferentially adsorbs OH^- species. This unilateral adsorption characteristic results in slow oxidation kinetics for organics on each individual phase. Creating a Co_4N–CoO heterointerface effectively balanced this adsorption. Due to the moderate co-adsorption of both OH^- and glycerol on the interface, OH^- is primarily utilized for activating

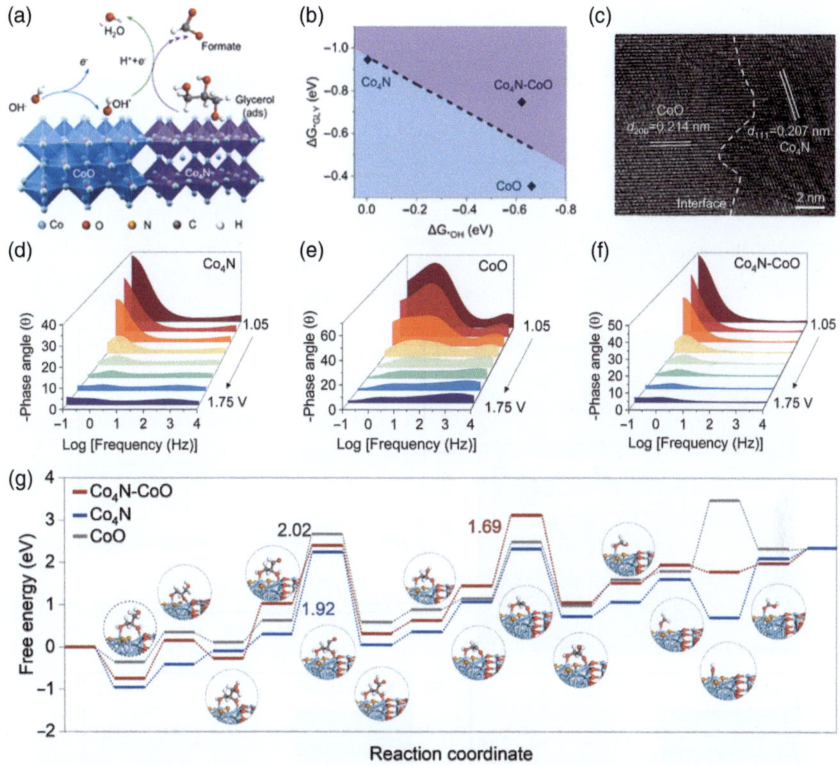

Figure 9.6 (a) Schematic diagram illustrating the balance between OH^- and glycerol adsorption at the Co_4N–CoO interface; (b) free energies of OH and glycerol adsorption on Co_4N, Co_4N–CoO, and CoO, calculated by DFT; (c) high-resolution transmission electron microscopy (HRTEM) image of the CoO/Co_4N interface; (d) Bode plots of Co_4N, (e) CoO, and (f) Co_4N–CoO collected in 1.0 M KOH solution containing 0.3 M glycerol; (g) Gibbs free energy profiles of glycerol oxidation reaction (GOR) on Co_4N–CoO, Co_4N, and CoO, with the inset showing the corresponding adsorption configurations of intermediates.
Source: Ref. [45] / with permission of John Wiley & Sons.

glycerol rather than triggering material reconstruction into high-valence OER active sites (Figure 9.6d–f). The Co_4N–CoO catalyst exhibited a low Tafel slope of 178 mV dec^{-1} and achieved a high formate production rate of 29.40 mmol cm^{-2} h^{-1}. Furthermore, the Faradaic efficiency for formate remained above 90% over 120 hours of continuous electrolysis. DFT simulations indicated that the enhanced glycerol oxidation reaction (GOR) performance is mainly attributed to the balanced co-adsorption of OH^- and glycerol at the Co_4N–CoO interface. In contrast, the GOR activity of traditional Co_3O_4 catalysts in neutral media is limited by their electronic structure and conductivity, highlighting the need for developing novel catalysts to improve both activity and selectivity (Figure 9.6g).

Luo et al. [46] enhanced the GOR activity and glyceraldehyde (GAD) selectivity by incorporating single Ru atoms into the Co_3O_4 lattice, thereby activating the intrinsic catalytic properties of Co_3O_4 (Figure 9.7a). The Ru–Co_3O_4/NF catalyst was synthesized via a one-step hydrothermal method followed by annealing. XRD patterns

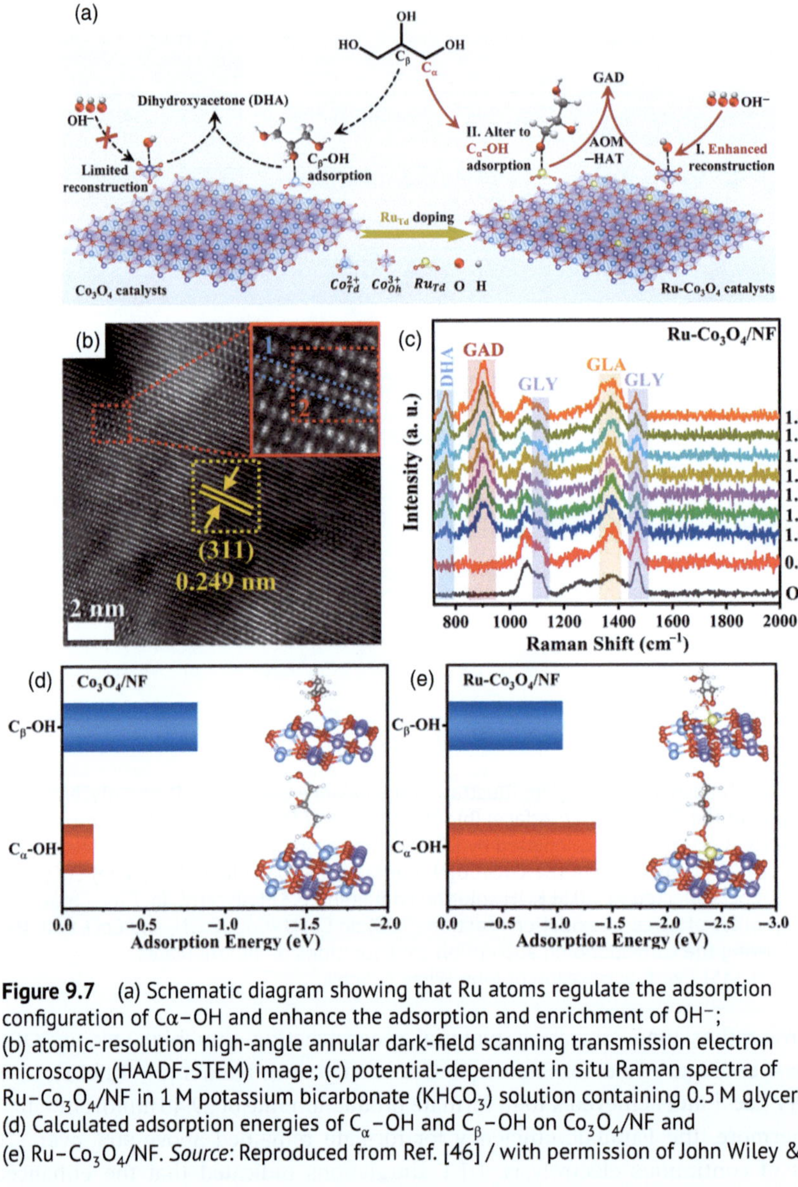

Figure 9.7 (a) Schematic diagram showing that Ru atoms regulate the adsorption configuration of Cα–OH and enhance the adsorption and enrichment of OH^-; (b) atomic-resolution high-angle annular dark-field scanning transmission electron microscopy (HAADF-STEM) image; (c) potential-dependent in situ Raman spectra of Ru–Co_3O_4/NF in 1 M potassium bicarbonate ($KHCO_3$) solution containing 0.5 M glycerol. (d) Calculated adsorption energies of C_α–OH and C_β–OH on Co_3O_4/NF and (e) Ru–Co_3O_4/NF. *Source*: Reproduced from Ref. [46] / with permission of John Wiley & Sons.

indicated high dispersion of Ru without detectable Ru-characteristic diffraction peaks. SEM and TEM images showed uniform growth of Co_3O_4 nanosheets on Ni foam (NF). HAADF-STEM imaging confirmed the homogeneous distribution of Ru single atoms within the Co_3O_4 matrix (Figure 9.7b). Bader charge analysis revealed that Ru doping increased the charge on Co atoms, generating more electron-deficient Co sites. Density of states (DOS) analysis indicated a narrowed band gap in Ru–Co_3O_4/NF, which facilitates improved electronic conductivity (Figure 9.7d, e). DFT calculations further demonstrated that Ru–Co_3O_4/NF exhibits higher adsorption energy toward the C_α–OH group of glycerol compared to Co_3O_4/NF adsorbing the C_β–OH group, confirming that Ru doping promotes preferential adsorption via the C_α–OH moiety (Figure 9.7c).

Gluconic acid (GNA) and its derivatives (e.g., gluconates, glucaric acid (GRA), and glucono-delta-lactone) are value-added chemicals derived from biomass. They can be produced via glucose oxidation and quantified by high-performance liquid chromatography (HPLC). The true catalytically active phases for glucose oxidation are metal oxides. Coupling glucose oxidation with the HER using Fe_2P/SSM as the anode and Pt/C as the cathode can significantly reduce energy consumption. Specifically, in a two-electrode system with Fe_2P/SSM and Pt/C, a cell voltage reduction of 300 mV was achieved at 10 mA cm^{-2} compared to conventional water electrolysis (Figure 9.8a) [47]. To simplify catalyst preparation, Liu et al. applied bimetallic Ni–Fe nitride ($NiFeN_x$) and oxide ($NiFeO_x$), synthesized via the same hydrothermal process, for anodic glucose oxidation and cathodic HER, respectively. The $NiFeN_x$/NF electrode required an overpotential of only 40.6 mV to reach 10 mA cm^{-2} for HER in alkaline electrolyte, with a Tafel slope of 39 mV dec^{-1}, indicating fast electron transfer. Meanwhile, $NiFeO_x$/NF exhibited excellent performance for glucose oxidation, achieving 87.6 mA cm^{-2} at 1.30 V with a Tafel slope of 19 mV dec^{-1}. The glucose oxidation pathway on $NiFeO_x$/NF was proposed to proceed in two steps: initial two-electron oxidation to GNA, followed by a four-electron process forming GRA. In a coupled two-electrode system, a current density of 200 mA cm^{-2} was achieved at just 1.48 V, with high Faradaic efficiency for GNA and GRA production. The use of non-precious metal catalysts for both glucose oxidation and HER enables cost-effective coproduction of GRA and H_2. Bifunctional electrocatalysts can further reduce overall costs. For instance, in a Co–Ni alloy catalyst (Figure 9.8b, c) [48], metallic Ni or Co sites promote HER, while oxides such as NiO, CoO, or Co–NiO are crucial for glucose oxidation.

9.1.3 Amines

Amines, characterized by the amino functional group, can be upgraded electrocatalytically via oxidative dehydrogenation to form high-value nitriles, a process that can be coupled with the HER. Similar to alcohol oxidation, amine oxidation has a lower theoretical potential than the OER, making it an ideal alternative anodic reaction. The target nitrile products are important chemical intermediates. Using primary and secondary amines as examples, Ni-based electrocatalysts demonstrate excellent performance in amine oxidation. Replacing OER with amine oxidation has attracted significant research interest due to its coproduction of valuable nitriles. Electrocatalytic amine oxidation can generate valuable compounds such as imines, nitriles, amides, amine oxides, and azo compounds, which are widely used in the synthesis of pharmaceuticals and agrochemicals.

Zhang and colleagues [49] coupled the oxidation of primary amines ($-CH_2-NH_2$) with the HER in 1.0 M KOH electrolyte (Figure 9.9a). Using Ni-based catalysts as an example for amine oxidation, NiSe nanorod arrays achieved high yield (>93%) and selectivity (>94%) in electrooxidizing various aromatic and aliphatic primary

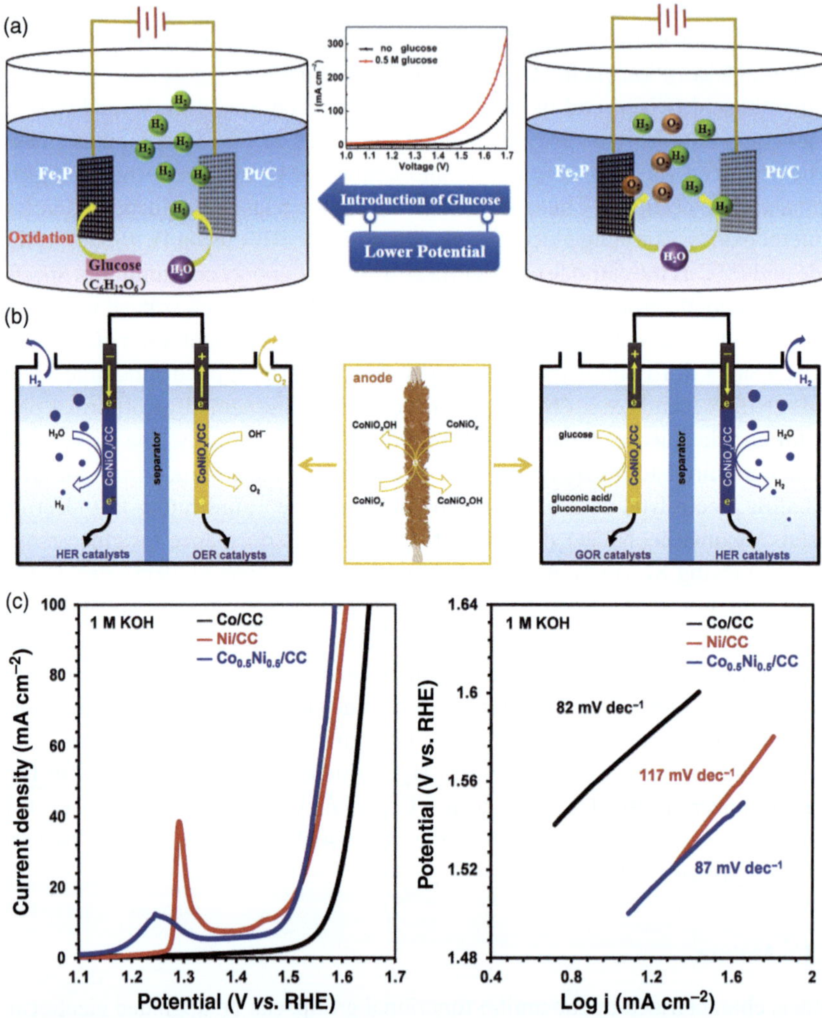

Figure 9.8 Device for coupling HER with glucose oxidation. (a) Two-electrode system for coupling glucose oxidation with HER. *Source*: Reproduced from Ref. [47] / with permission of John Wiley & Sons. (b) schematic diagram of the device for coupling HER with glucose oxidation; (c) LSV curves and Tafel slopes of Co/CC, Ni/CC, and $Co_{0.5}Ni_{0.5}$/CC in 1 M KOH. *Source*: Reproduced from Ref. [48] / with permission of John Wiley & Sons.

amines to their corresponding nitriles. Taking the conversion of benzylamine to benzonitrile as an example, the electrooxidation occurred at approximately 1.34 V, significantly lower than the potential required for the OER. In a two-electrode electrolyzer assembled with NiSe as the anode and CoP as the cathode for HER, a cell voltage of only 1.59 V was required to reach a current density of 20 mA cm^{-2} (Figure 9.9b, c). At 1.5 V, the Faradaic efficiencies for benzonitrile and hydrogen reached 98% and nearly 100%, respectively. In situ Raman spectroscopy at different

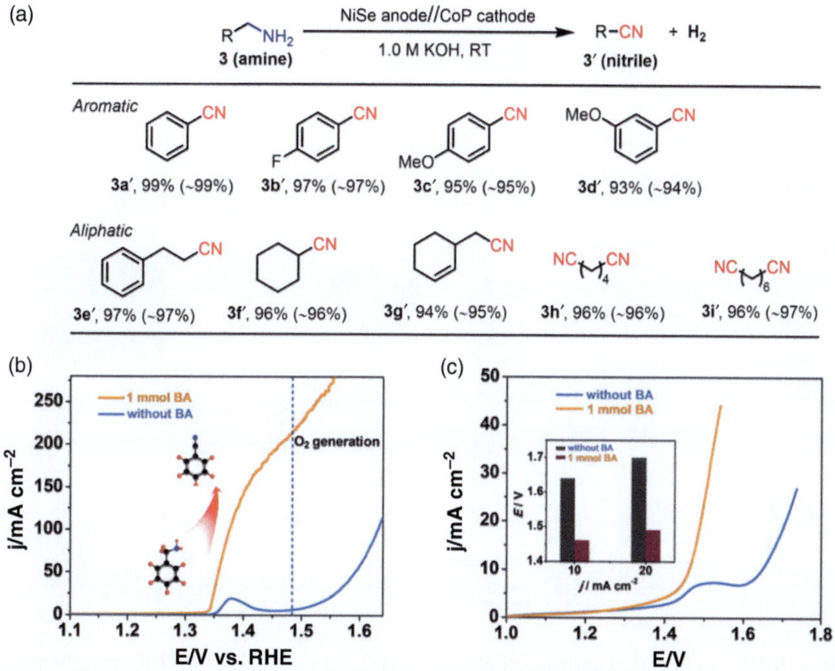

Figure 9.9 (a) Coupling of electrooxidation of primary amines on NiSe anode with HER; (b) LSV curves of NiSe in 1.0 M KOH solution and mixed solution of 1 mmol benzyl alcohol and 1.0 M KOH, respectively; (c) LSV curves of benzyl alcohol and HER conversion rates in NiSe//CoP hybrid electrolytic cell, with the inset showing a comparison of potentials to reach current densities of 10 and 20 mA cm^{-2}. *Source*: Reproduced from Ref. [49] / with permission of John Wiley & Sons.

potentials revealed that in situ formed Ni oxyselenide species act as redox-active centers, facilitating the conversion of primary amines to nitriles.

Nitrogen-centered radicals (NCRs) are valuable intermediates in organic synthesis, particularly for N–N coupling reactions. While direct N—H bond cleavage to generate NCRs offers advantages in atom and step economy compared to traditional two-step methods, the high bond energy of N–H makes this challenging. The electrochemical oxidation dehydrogenation of primary amines on Ni-based catalysts provides a promising route for NCR formation. This process involves the cleavage of both α—C—H and N—H bonds. Structurally, stable NCRs form only when the number of N—H bonds exceeds that of α—C—H bonds (Figure 9.10a). Han et al. [50] used primary amines containing only one α—C—H bond (R_1R_2-CH-NH_2) as substrates. Their oxidative dehydrogenation generated stable NCRs, which subsequently underwent N–N coupling to form ketazines. Theoretical analysis revealed that oxygen vacancies in NiO (V_0-NiO) influence the reaction pathway. The vacancies cause excess electrons to redistribute into the 3d orbitals of adjacent Ni cations, lowering their valence state (Figure 9.10b). This electron-rich character and unsaturated coordination enhance the adsorption of electronegative elements like nitrogen. Differential charge density showed strong adsorption of

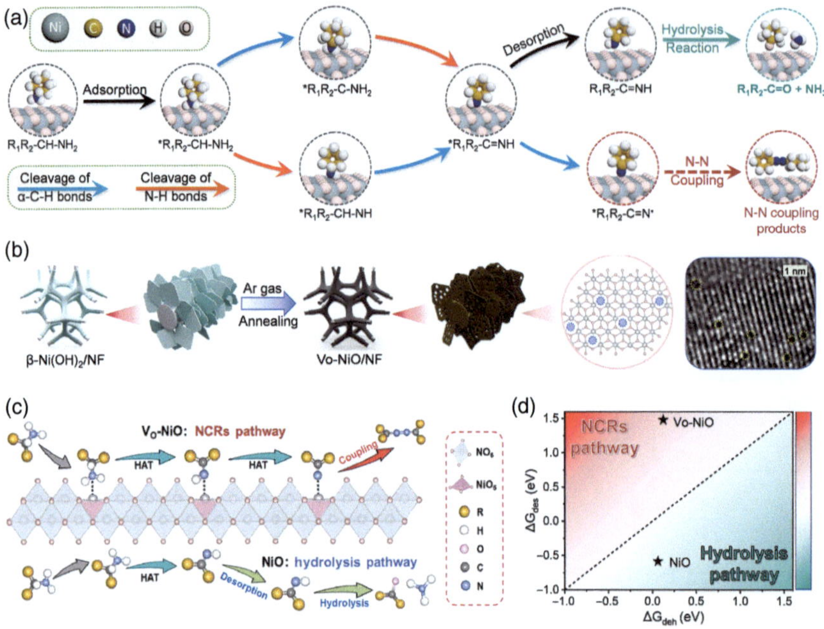

Figure 9.10 (a) Reaction pathway of ePAOR on NiO; (b) synthesis of V_0–NiO; (c) schematic diagram illustrating that oxygen vacancies promote the electrocatalytic primary amine oxidative coupling reaction (ePAOR); (d) comparison of the branching reaction pathways for the dehydrogenation of R_1R_2C=NH to R_1R_2C=N intermediate on NiO (111) and Vo–NiO (111). *Source*: Reproduced from Ref. [50] / with permission of American Chemical Society.

the intermediate R_1R_2—C=NH on V_0–Ni sites via significant charge transfer to the nitrogen atom, with a high desorption energy of 1.48 eV. Since the energy barrier for dehydrogenation is much lower than for desorption, dehydrogenation is favored over desorption on V_0–NiO, leading to a different pathway compared to regular NiO (111) (Figure 9.10c). Free energy diagrams indicate that on NiO (111), the reaction primarily follows a hydrolysis pathway to ketones and ammonia. In contrast, V_0–NiO (111) favors an N–N coupling route, producing ketazines. Furthermore, the energy barrier of the rate-determining step is significantly lower on V_0–NiO (111). DFT calculations confirm that oxygen vacancies alter the reaction pathway, enhancing both activity and selectivity for N–N coupling. Pan et al. [51] synthesized a Ni–ZIF precursor from Ni and dimethylimidazole via hydrothermal reaction. Low-temperature annealing in argon followed by room-temperature oxidation yielded Ni oxide supported on nitrogen-doped carbon (NiO/NC). The d-π interaction between NiO and NC creates positively charged sites on the catalyst surface. DOS analysis showed that the NC support modifies the electronic structure of Ni, shifting its d-band center closer to the Fermi level. This enhances the adsorption and activation of primary amines by Ni sites. In situ experiments and theoretical calculations elucidated the active species and pathway for the stepwise dehydrogenation of amines to nitriles. In situ FTIR spectroscopy detected a sharp peak at 2200 cm^{-1} at 1.40 V vs. RHE, assigned to the C≡N stretch of the nitrile product. In situ Raman spectroscopy revealed that the NC support modulates Ni's electronic structure, increasing its valence state and promoting the formation of the active species γ-NiOOH. Combined in situ infrared (IR) and DFT studies verified the stepwise dehydrogenation mechanism and identified N—H bond cleavage as the rate-determining step. The introduction of the NC support significantly reduces the energy barrier for N–H cleavage, enabling the efficient synthesis of valuable nitriles.

9.1.4 Aldehydes

Furanic compounds, exemplified by 5-hydroxymethylfurfural (HMF), are unique due to the simultaneous presence of an aldehyde group (–CHO), a hydroxymethyl group (–CH_2OH), and a furan ring. This structure enables diverse conversion pathways: oxidation can convert both functional groups to carboxyl groups, yielding the valuable polymer precursor FDCA, while reduction produces high-energy-density fuel precursors. As a key platform molecule, the valorization of HMF proceeds primarily through two core electrocatalytic routes: oxidation and reduction. In the oxidation pathway (HMFOR), the key challenge lies in controlling the oxidation sequence of the aldehyde and hydroxymethyl groups, a process highly dependent on the catalyst's nature and the reaction environment [52]. Under alkaline conditions, Ni-based catalysts typically favor initial oxidation of the hydroxymethyl group, efficiently converting HMF to FDCA via the HMFCA intermediate [53]. Complementarily, the reduction pathway focuses on hydrogenation/hydrogenolysis, involving hydrogenation of the furan ring and the C=O bond, along with deoxygenation. HMF can be sequentially reduced: the aldehyde group is first hydrogenated to form DHMF, which is then converted via deoxygenation to the biofuel DMF, or via furan ring hydrogenation to other valuable products like DHMTHF [54]. These oxidative and reductive routes provide versatile strategies for the value-added conversion of HMF.

The electrocatalytic oxidation of HMF proceeds via two competing pathways, both ultimately yielding the key polymer monomer FDCA. FDCA is a crucial precursor for producing the bio-based plastic polyethylene furanoate (PEF), a sustainable alternative to petroleum-derived PET. Pathway I (dominant at $pH < 13$): Oxidation begins at the hydroxymethyl group, first forming the intermediate 2,5-diformylfuran (DFF). Pathway II (dominant at $pH \geq 13$): Oxidation starts at the formyl group, first generating the intermediate 5-hydroxymethyl-2-furancarboxylic acid (HMFCA). Both intermediates (DFF and HMFCA) are further oxidized to 5-formyl-2-furancarboxylic acid (FFCA) and finally to FDCA. The mechanism can be direct or indirect. Direct Oxidation: The applied potential directly drives substrate oxidation without changing the catalyst's oxidation state. The key step involves adsorbed OH^- abstracting protons from HMF. In strong alkaline conditions, the hydrated formyl group is oxidized first; under milder conditions, dehydrogenation of the hydroxymethyl group is favored [55, 56]. Indirect Oxidation: The catalyst (e.g., Ni-based) acts as a redox mediator, cycling its oxidation state (e.g., Ni^{2+}/Ni^{3+}). The applied potential regenerates the high-valent active species (e.g., NiOOH), which then chemically oxidizes HMF [57].

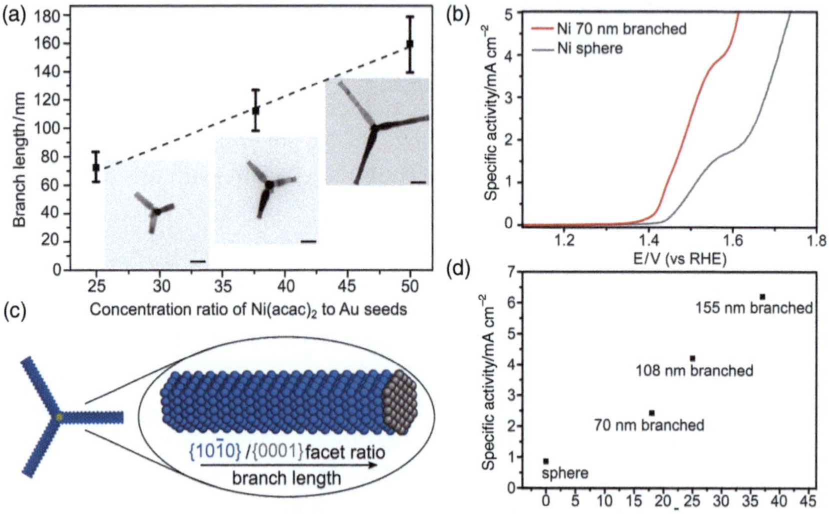

Figure 9.11 (a) Dependence of branch length (nm) on the concentration ratio of $Ni(acac)_2$ to Au seeds, with representative TEM images of each sample (scale bar: 50 nm); (b) LSV curves of branched Ni nanoparticles (red line) and amorphous Ni spheres (gray line) in a mixed solution of 10 mM HMF and 0.1 M KOH; (c) model showing curved mosaic facets along the branches; (d) relationship between specific activity and the ratio of crystal facets. *Source*: Ref. [59] / John Wiley & Sons / CC BY 4.0.

Ni-based catalysts are recognized as high-performance catalysts for the HMF oxidation reaction (HMFOR) due to their abundant 3d electrons and unique eg orbital characteristics. The eg electrons effectively enhance the covalent bonding between the metal and adsorbed intermediates, optimizing the reaction pathway. Sun et al. [58] synthesized a ligand-modified catalyst, $Ni(OH)_2$-TPA (TPA = terephthalic acid), which enables efficient HMF oxidation. Its uncoordinated carboxyl groups act as proton relay centers, boosting the current density by 16-fold compared to pure $Ni(OH)_2$.

Tilley et al [59] synthesized branched Ni nanoparticles (Figure 9.11a) with tunable branch lengths using a combined seed-mediated and high-hydrogen-pressure growth mechanism. These nanoparticles catalyze the conversion of HMF into valuable oxidation products. Longer branches resulted in higher coverage of active facets, leading to significantly improved performance compared to amorphous spherical Ni nanoparticles (Figure 9.11b–d). Various Ni-based compounds, including borides, nitrides, oxides, and hydroxides, have been extensively studied for the electrochemical oxidation of HMF. However, achieving high current density at low potentials remains challenging. Luo et al. [60] synthesized a "nanoarray-on-nanosheet" Ni–Co hydroxide electrocatalyst (Figure 9.12a, b) from a bimetallic NiCo-MOF nanoarray via a simple cyclic voltammetry-assisted method. In 1.0 M KOH with 50 mM HMF, this catalyst achieved a current density of 600 mA cm^{-2} at 1.4 V (Figure 9.12c).

Gan's research group [61] utilized in situ electrochemical pre-oxidation of NiSe to prepare NiSe@NiO_x core–shell nanowires (Figure 9.13a). In this structure, NiSe serves as a highly conductive framework, while the surface NiO_x layer acts as the

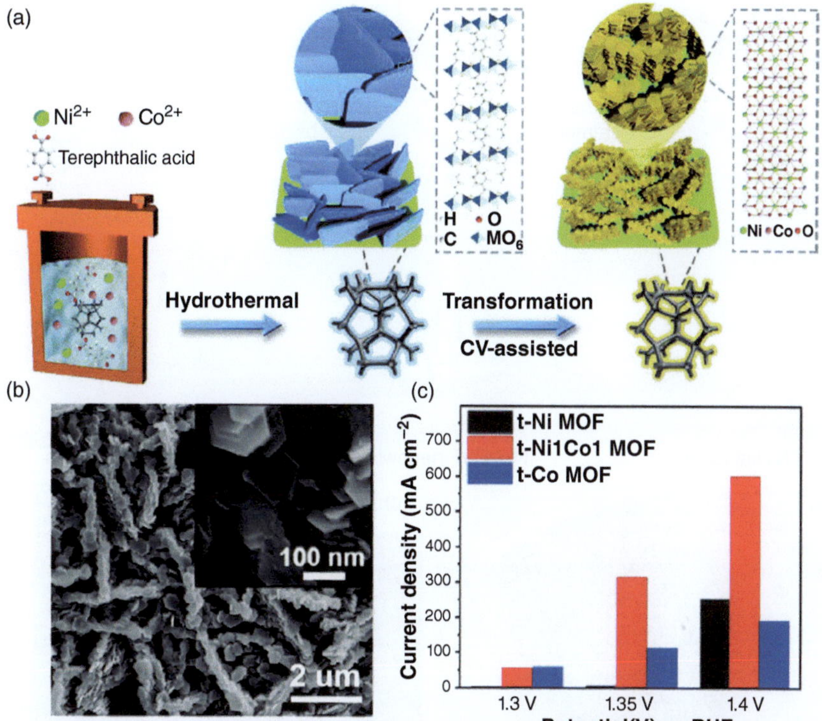

Figure 9.12 (a) Schematic diagram for the preparation of "nanosheet-nanoarray" electrocatalysts via the transformation of hydrothermally synthesized Ni–Co bimetallic MOF nanoarrays on Ni foam; (b) SEM image of transformed NiCo-MOF (t-NiCo-MOF); (c) comparison of current densities at 1.30, 1.35, and 1.40 V. *Source*: (a) & (c): [60] / with permission of Elsevier; (b): Reproduced from Ref. [59] / with permission of John Wiley & Sons.

active site (Figure 9.13b, c). This catalyst achieved 100% HMF conversion with 99% yield and 99% Faradaic efficiency for FDCA. The high activity is attributed to the in situ formed Ni^{3+} species during the reaction, which governs the HMF electrooxidation process. Recent studies using model systems like NiO nanoparticles and Au–NiO heterojunctions, analyzed via in situ Raman spectroscopy, have elucidated the Ni^{2+} to Ni^{3+} transformation. Sun et al. prepared amorphous NiB_x as an HMF oxidation catalyst and confirmed the generation of Ni^{3+} during operation using in situ Raman, highlighting the key role of water as an oxygen source providing OH^- to drive this oxidation. Beyond activity and efficiency, understanding the catalytic mechanism is a major research focus. The electrochemical oxidation of HMF on Ni-based materials predominantly follows an indirect mechanism, where the applied potential generates high-valence Ni^{3+} species that chemically oxidize HMF. In summary, Ni-based catalysts, which facilitate faster kinetics and electron transfer, are among the most active non-precious metal materials for HMF electrooxidation, achieving high conversion, yield, and Faradaic efficiency. However, their high activity often requires a strongly alkaline environment, which

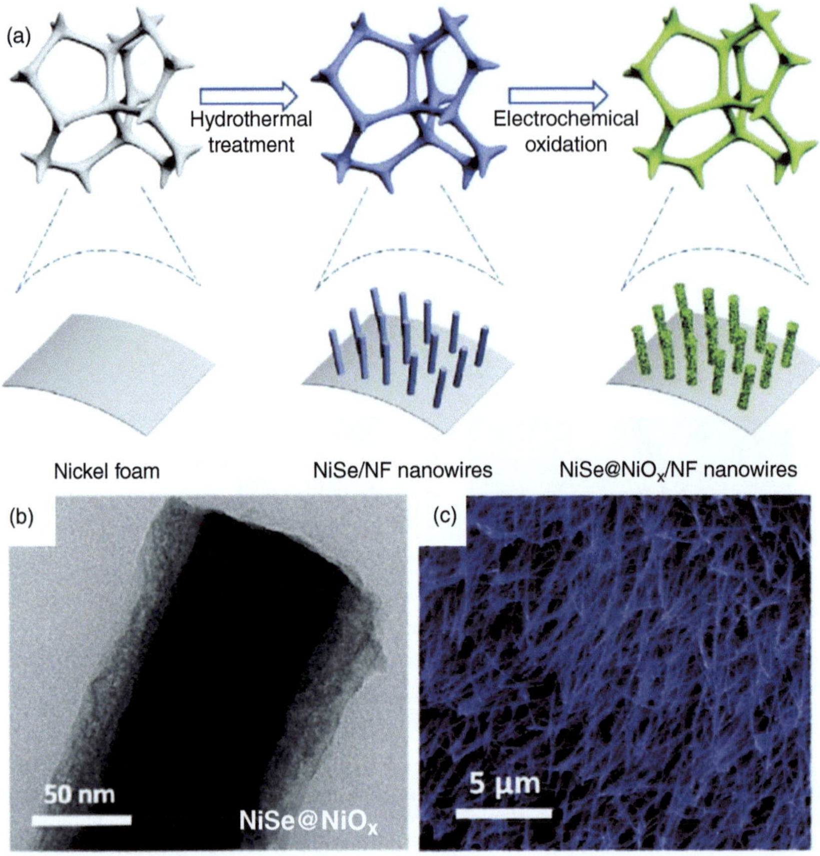

Figure 9.13 (a) Schematic diagram for the preparation of NiSe@NiO$_x$/NF core–shell nanowires; (b) TEM image of NiSe@NiO$_x$ nanowires; (c) SEM image of NiSe@NiO$_x$/NF. *Source*: Reproduced from Ref. [61] / with permission of Elsevier.

can lead to HMF degradation during prolonged operation. Future research should focus on developing practical strategies to enhance HMF oxidation kinetics, shorten reaction time, and maintain high efficiency under milder conditions.

Co-based catalysts are increasingly recognized as active bifunctional catalysts for both the HER and OER, particularly with the growing number of reports on alloys of Co with boron, sulfur, and phosphorus for water splitting. Deng et al. [62] synthesized Co–P on copper foam using the same electrodeposition method. This approach avoided the complexity of multiple alkylation steps and achieved 99% HMF conversion with 85.3% FDCA yield in a 0.5 M $NaHCO_3$ solution (pH = 12). Analysis of the catalysts revealed that the surfaces of Co–P and Co–B readily transform into their corresponding oxides/hydroxides during the reaction, which are likely the true active species for HMF electrooxidation. However, metal oxides generally exhibit lower electrocatalytic efficiency compared to their sulfide, phosphide, boride, and nitride counterparts due to poor conductivity and low active surface area. To address this, researchers have explored strategies to enhance

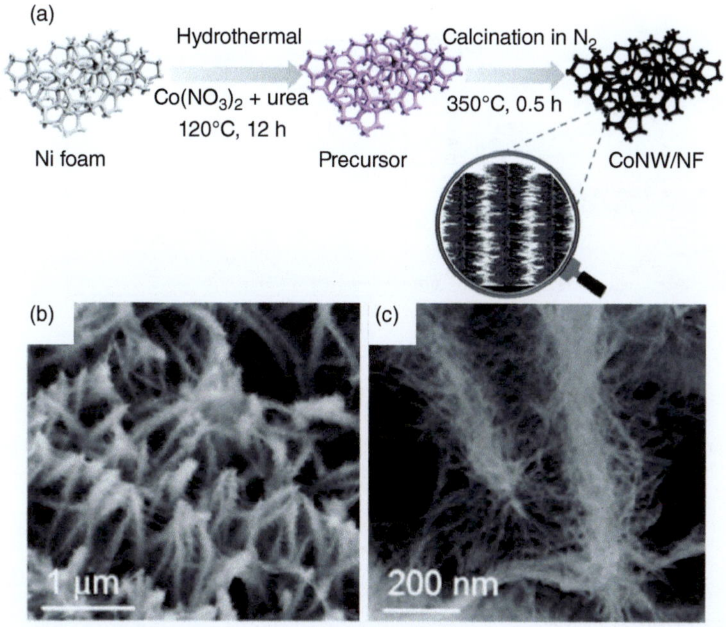

Figure 9.14 (a) Schematic diagram of in situ fabrication of Co_3O_4 nanowires on Ni foam; (b) SEM image of Co_3O_4NW/NF at low magnification; (c) SEM image at high magnification. *Source*: Reproduced from Ref. [63] / with permission of The Royal Society of Chemistry.

the performance of Co-based oxides for HMF oxidation. Morphology engineering is a viable approach. For instance, Zhang et al. [63] designed pine-needle-like Co_3O_4 nanowires on Ni foam to form a three-dimensional monolithic electrode (Co_3O_4NW/NF), as shown in Figure 9.14a–c. This catalyst achieved 100% HMF conversion, 96.8% FDCA yield, and 96.6% Faradaic efficiency.

Wang et al. [64] identified the distinct roles of Co^{2+} and Co^{3+} in the typical spinel Co_3O_4, finding that both ions contribute to the overall catalytic activity. During HMF oxidation, Co^{2+} sites adsorb HMF, while Co^{3+} sites play a decisive role in the oxidation process (Figure 9.15a, b). Co-based materials generally exhibit low onset potentials for HMF electrooxidation, making them ideal candidates for reducing overpotential. However, their slow reaction kinetics often result in unsatisfactory current densities and lower efficiencies. Therefore, enhancing the electrochemical oxidation kinetics of Co-based catalysts is a key direction for future research. Given that Ni-containing materials typically demonstrate faster kinetics (or electron transfer coefficients), developing more efficient NiCo alloys or heterojunctions with open nanostructures to achieve synergistic catalytic effects is a promising strategy. Additionally, more feasible approaches, such as constructing Co-based oxide/hydroxide catalysts with optimized support effects and interface engineering, should be pursued.

Fe is the second most abundant transition metal in the Earth's crust (approximately 5.0%) and is tens to hundreds of times less expensive than Ni or Co. Its multiple accessible redox states and diverse coordination environments make it

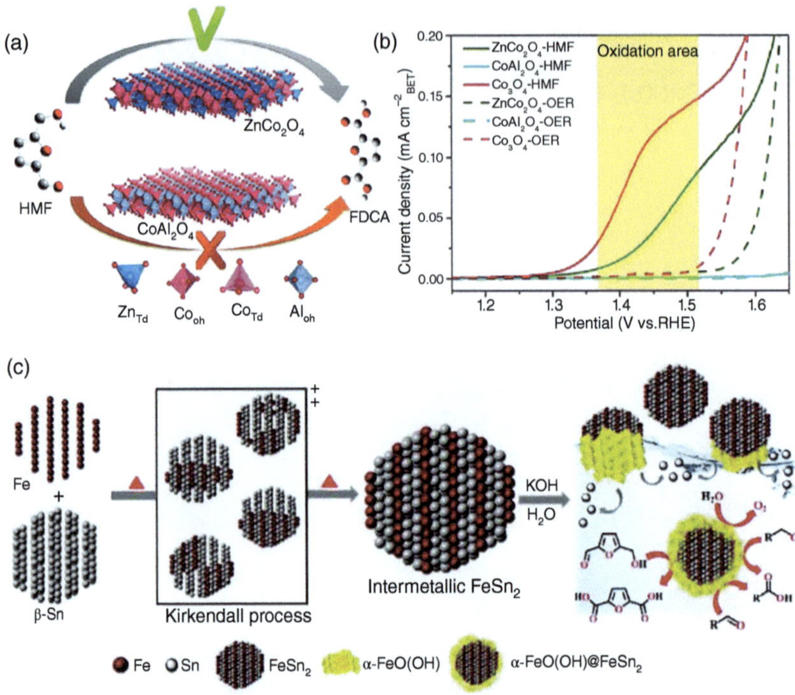

Figure 9.15 (a) Schematic diagram illustrating the construction of tetrahedral (Zn^{2+}) or octahedral (Al^{3+}) geometric sites; (b) LSV curves of $ZnCo_2O_4$, Co_3O_4, and $CoAl_2O_4$ normalized by BET surface area; *Source*: Reproduced from Ref. [64]; © 2020, Wiley-VCH Verlag GmbH & Co. KGaA. (c) Schematic diagram showing the formation of structurally ordered intermetallic $FeSn_2$ and the close packing of separate atomic layers of Fe and Sn; the ordered structure of $FeSn_2$ is established through the intercalation of metallic Fe layers (red spheres) and β-Sn layers (gray spheres) at high temperature via the Kirkendall process. Under alkaline electrochemical reaction conditions, substantial reconstruction occurs at the surface due to the slow depletion of Sn from $FeSn_2$ nanostructures, followed by the in situ formation of highly active crystalline α-FeO(OH)@$FeSn_2$ to catalyze OER and various organic oxidations. *Source*: Reproduced from Ref. [65] / with permission of John Wiley & Sons.

suitable for electrocatalysis, including the OER, HER, and nitrogen reduction reaction. Several studies have reported the promising catalytic performance of Fe-based materials in HMF oxidation. However, compared to Ni- or Co-based catalysts, Fe-based catalysts generally suffer from poor conductivity, suboptimal oxophilicity, and a tendency for ion leaching into solution. These factors contribute to their relatively lower catalytic activity in the electrochemical oxidation of HMF. A common strategy to enhance the intrinsic activity of Fe-based catalysts is to incorporate other metals or metal (oxy)hydroxides. For example, $FeNi(OH)_x$ catalysts can achieve nearly 100% FDCA yield with a Faradaic efficiency of up to 99.4%. Menezes et al. [65] reported an alternative approach using an ordered intermetallic $FeSn_2$ pre-catalyst synthesized via the Kirkendall process (Figure 9.15c). Under anodic bias, a highly crystalline α-FeOOH (goethite) layer forms on the surface, serving as the primary active species. This in situ surface-modified $FeSn_2$ catalyst achieved a Faradaic efficiency close to 90% for HMF oxidation. Fe species can also function as cocatalysts to enhance HMF oxidation. For instance, Fe_3O_4-modified, reduced graphene oxide-supported platinum nanoparticles ($Pt/Fe_3O_4/rGO$) significantly increased the HMF oxidation rate in 0.05 M H_2SO_4 due to strong synergy between Pt and Fe_3O_4. Given the abundance and low cost of Fe, developing effective strategies, such as optimizing Fe-rich OER electrocatalysts to improve their activity for HMF oxidation, remains an important research direction.

9.1.5 Lignin

Lignin, the most abundant aromatic polymer in lignocellulosic biomass, poses a key challenge for biorefining due to the difficulty of achieving efficient depolymerization. Conventional thermocatalytic methods typically require high temperature and pressure while suffering from low product selectivity. Electrocatalytic reduction offers an alternative pathway by utilizing renewable electricity to selectively cleave C—O bonds under mild conditions, generating valuable aromatic monomers and enabling the valorization of lignin.

Commercial Ni foam is widely used for the electrocatalytic valorization of lignin. Sun et al. [22] demonstrated its direct application as an electrocatalyst for the oxidative depolymerization of three types of native ethanol organosolv lignin in a 1.0 M KOH alkaline system. The Ni foam surface undergoes electrochemical oxidation to form NiOOH, which serves as the true active species. This species selectively oxidizes the secondary benzylic alcohol groups at the C_α position in lignin to ketones, while preventing over-oxidation of the resulting aromatic monomers, thereby ensuring high product selectivity. The process efficiently cleaves C—O bonds in lignin, yielding two high-value aromatic aldehydes—vanillin and syringaldehyde—with a combined yield of 17.5%. Despite the mild operating conditions, current electrocatalytic systems face limitations, such as reliance on organic solvents that complicate separation, predominant formation of acidic products, and the generally weak adsorption of OH^- and substrates on $Ni(OH)_2$ catalysts, leading to suboptimal performance. To overcome these challenges, Yang et al. [66] designed a Ni hydroxide electrocatalyst rich in dual vacancies ($Ni(OH)_2$-V), enabling highly efficient and selective oxidative cleavage of the C_α—C_β bond in lignin β-O-4 model compounds under mild conditions. To elucidate the role of the dual vacancies, the researchers combined in situ experiments with DFT calculations, revealing that the vacancies regulate the adsorption/desorption energetics of reactants and products. The synergistic effect of the dual vacancies manifests in three aspects: (i) optimizing the electronic structure of the catalyst and promoting the formation of highly active $NiO(OH)_{ads}$ species; (ii) enhancing interfacial adsorption of both the substrate and hydroxyl groups; and (iii) significantly lowering the energy barrier of the rate-determining step while facilitating timely desorption of phenolic products. The catalyst exhibited excellent performance not only in model compound conversion (85.7% phenol yield) but also in the depolymerization of real lignin sources (poplar and corn stover), achieving a high total aromatic monomer yield of 19.54 wt%.

Lignin contains a C(OH)–C motif, which can be cleaved to produce carboxylic acids via selective C—C bond scission. However, the high dissociation energy of the C(OH)—C bond in lignin derivatives limits its practical application. Zhou et al. [67] developed an electrocatalytic system using Mn-doped Co oxyhydroxide (MnCoOOH) to selectively cleave C(OH)—C and C(O)—C bonds in various lignin derivatives, yielding corresponding carboxylic acids with high efficiency (80–99% yield) and stable catalytic activity over 200 hours. The reaction proceeds via a cascade oxidation process: the alkaline electrolyte converts secondary alcohols and ketone derivatives into reactive nucleophiles, which are then oxidized by electrophilic oxygen species on the MnCoOOH catalyst in the aqueous phase. This electrochemical oxidation of lignin can be coupled with the reduction of furfural to 2-methylfuran, offering a promising route for coproducing high-value products. Nevertheless, designing efficient bifunctional electrocatalysts with noble-metal-like activity remains challenging. Qi et al. [68] addressed this by modulating the electronic structure of $CoMoO_4$ spinel via phosphorus doping (Figure 9.16a). The optimized catalyst enabled selective oxidative cleavage of the C_α—C_β bond in lignin, yielding aromatic monomers with 99% conversion and up to 56% selectivity. Simultaneously, furfural was reduced to 2-methylfuran with 87% conversion and 73% selectivity. In situ FTIR and DFT analysis (Figure 9.16b–d) revealed that phosphorus doping upshifts the d-band center, raising the antibonding orbital energy and enhancing adsorption of the C_α—C_β bond in lignin model compounds. This strengthens the bifunctional electrocatalytic activity. Time-dependent in situ IR spectra showed the gradual disappearance of β-O-4 ether vibrations and the emergence of aromatic ring signals, confirming bond cleavage. Spin-trapping ESR experiments detected O^{2-} and OH radical adducts, indicating the involvement of radical species in C_α—C_β bond scission.

9.2 Pollutant Degradation

The rapid advancement of industrialization and urbanization has led to the continuous release of large quantities of pollutants into the environment, severely disrupting ecological balance, contaminating water and soil, and directly threatening human health and well-being. Consequently, the development of efficient, green, and sustainable pollutant treatment technologies has become an urgent global priority in the environmental field [69, 70]. Various pollutant removal methods have been reported, including redox reactions, chemical precipitation, photocatalytic degradation, ion exchange, reverse osmosis, adsorption, filtration, and electrochemical processes. Among these, adsorption stands out for its low cost, high efficiency, and environmental friendliness, and has been widely applied for the removal of organic and inorganic pollutants from aqueous solutions. The core of this technology lies in the adsorbent materials, and researchers have developed various types such as carbon-based materials, biochar, nano-metal oxides, clay minerals, and inorganic metal sulfides [71, 72]. Some materials (e.g., nano-metal oxides and inorganic metal sulfides) also possess dual capabilities of adsorption and oxidation/reduction, enabling the targeted removal of redox-sensitive pollutants. However, these methods generally face limitations such as the need for regeneration after adsorption

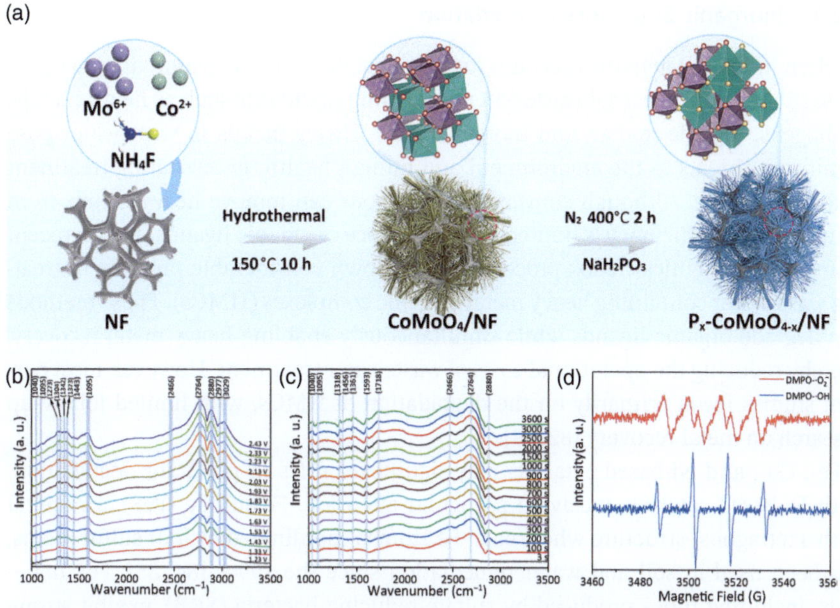

Figure 9.16 (a) Schematic diagram of the synthesis of P_x-$CoMoO_{4-x}$/NF; (b) in situ infrared spectra of the P_x-$CoMoO_{4-x}$/NF electrode surface at different potentials; (c) in situ infrared spectra of the P_x-$CoMoO_{4-x}$/NF electrode surface detected at different time points; (d) electron paramagnetic resonance detection of free radicals in P_x-$CoMoO_{4-x}$/NF using DMPO spin trap characterization of the catalyst after five cycles. *Source*: Reproduced from Ref. [68] / with permission of John Wiley & Sons.

saturation, difficulty in achieving complete mineralization of pollutants, and limited potential for resource recovery [73–75].

Electrocatalysis, as an efficient and environmentally friendly advanced oxidation process, is increasingly being applied in pollutant treatment. This technology uses renewable electricity to drive precise electron transfer at the catalyst interface, enabling targeted conversion of various pollutants [76, 77]. For inorganic pollutants like heavy metal ions, the core mechanism involves cathodic reduction to transform and stabilize their chemical states. For example, highly toxic and mobile hexavalent chromium (Cr(VI)) can be reduced to less toxic and readily precipitable trivalent chromium (Cr(III)). The efficiency of this process depends strongly on the hydrogen evolution overpotential of the cathode catalyst and electron conductivity. For organic pollutants—such as phenols, aldehydes, and platform compounds generated during biomass conversion—electrocatalysis offers flexible anodic oxidation pathways. These include direct electron transfer of pollutants at the anode, as well as indirect oxidation via anode-generated reactive oxygen species or high-valence metal centers, leading to efficient decomposition or even deep mineralization. Notably, by rationally designing transition metal catalysts, such as tuning the d-band center or constructing single-atom sites, the reaction can be guided from simple mineralization toward selective value-added transformation [78–80].

9.2.1 Inorganic Substance Degradation

Modern industrial activities account for 22% of global water consumption and generate substantial industrial wastewater containing pollutants such as heavy metals, pathogens, organic matter, and inorganic salts. Heavy metals in wastewater pose significant threats to the environment and human health, necessitating treatment before discharge. Although conventional methods can remove heavy metals from solutions, their efficiency is limited in the presence of organic ligands [81]. In recent years, electrochemical redox processes have shown considerable progress in treating wastewater containing heavy metal–organic complexes (HMCs). These methods can degrade organic ligands while simultaneously enabling heavy metal recovery, thereby reducing the operational costs of wastewater treatment. However, most current studies focus primarily on the degradation of HMCs, with limited follow-up research on metal recovery [82, 83].

Fe-, Co-, and Ni-based catalysts play significant roles in pollutant degradation, with Fe-based catalysts receiving particular attention. Fe sulfide (FeS), a mineral with a tetragonal structure where each Fe atom is coordinated by four sulfur atoms, has been used in soil and water remediation since the 1990s [84, 85]. FeS materials, including those produced by sulfate-reducing bacteria (SRB), exhibit strong adsorption capabilities for heavy metals such as mercury, cadmium, and copper. FeS contains both high-valence Fe sites and reduced sulfide sites, making chemisorption dominant in the adsorption process. Additionally, FeS acts as an effective reductant, degrading halogenated hydrocarbon pollutants and removing redox-sensitive metals from contaminated soil and water, as Fe^{2+} and S^{2-} species serve as electron donors. It is particularly effective in degrading chlorinated hydrocarbons (CHCs) like vinyl chloride (VC), hexachloroethane (HCA), trichloroethylene (TCE), and carbon tetrachloride (CT). FeS also shows excellent performance in heavy metal removal, especially in reducing toxic Cr(VI) to less harmful and readily precipitable Cr(III) [86, 87]. However, bare FeS nanoparticles face limitations in large-scale applications due to agglomeration and oxidizability. To address these issues, modified nano FeS materials have been developed to improve dispersion and enhance remediation efficiency through adsorption and redox reactions. Over the past decade, nanotechnology has been increasingly applied to remediate contaminated surface water, groundwater, soil, and sediments, leveraging nanomaterials' high reactivity and large surface-area-to-volume ratio. Nano-FeS shows notable potential not only in degrading redox-sensitive pollutants via reduction but also in removing contaminants through chemisorption [88, 89].

Heavy metals are difficult to remove from the environment due to their non-biodegradability, toxicity, and persistence. They tend to accumulate in the food chain, causing harm to organisms and posing risks to human health. Effective control of heavy metal ion concentrations is therefore essential for public safety. Common heavy metal pollutants include chromium (Cr), lead (Pb), mercury (Hg), copper (Cu), cadmium (Cd), zinc (Zn), and nickel (Ni). FeS nanoparticles have been recognized as effective reductants and adsorbents for removing these ions [90]. Chromium, a redox-active metal widely used in electroplating and metallurgy, is a common environmental pollutant. Cr(VI) and Cr(III) are its stable forms, with Cr(VI) being more toxic, carcinogenic, and soluble. Converting Cr(VI) to Cr(III) is critical for remediation. Studies show that FeS nanoparticles act as both adsorbents and reductants, enhancing Cr(VI) removal from soil and water [91]. Modified nanoparticles exhibit even higher removal efficiency [92]. For example, sodium alginate-stabilized FeS (FeS-SA) increased Cr(VI) removal from 65% (7.50 mmol g^{-1}) to 100% (11.54 mmol g^{-1}). Fang et al. [93] demonstrated that CMC-FeS@biochar nanoparticles immobilized Cr(VI) in soil more effectively than biochar or FeS alone, reducing Cr(VI) concentration from 11.9 to 0.63 mg L^{-1}. Mercury, primarily present as methylmercury (MeHg) and Hg(II), is another challenging pollutant. Hg(II) readily converts to highly toxic MeHg, which accumulates in the food chain. Removing Hg(II) is key to controlling mercury pollution. Tetragonal FeS can adsorb Hg(II), inhibiting its methylation. The removal mechanism depends on the Hg(II)/FeS molar ratio: adsorption dominates at low ratios (~0.05), while precipitation prevails at higher ratios. Modified nano FeS not only reduces agglomeration but also improves mercury removal efficiency. For instance, Zhou et al. [94] synthesized chitosan-stabilized magnetic FeS/Fe_3O_4 (CTO-MFeS), achieving an adsorption capacity of 97.86 mg g^{-1} for Hg(II) within 120 minutes, with easy separation due to its magnetism. Divalent metals like Cd, Pb, Cu, and Ni can also be treated via adsorption and coprecipitation with FeS nanoparticles. Ni removal reached 6.4 mmol g^{-1} using 2.25 g L^{-1} nanocrystalline FeS, with rapid initial uptake (<1 hour). Pb and Cd are retained through ion exchange with Fe, forming surface complexes like (Pb and Fe) S or distinct sulfide phases (e.g., PbS). This supports the use of nano FeS for immobilizing divalent metals in contaminated soil and water [95, 96].

9.2.2 Organic Matter Degradation

Halogenated hydrocarbons are widely present in soil and water due to industrial wastewater discharge, posing threats to human health and ecological security. FeS nanoparticles have been extensively applied in degrading halogenated hydrocarbons, primarily chlorinated and brominated compounds [97, 98]. CHCs, such as polychlorinated biphenyls (PCBs), tetrachloroethylene (PCE), TCE, dichloroethane (DCE), VC, and HCA, have been widely used as raw materials in polymers, chemicals, pharmaceuticals, and pesticides. Their presence in soil and water poses risks due to their persistence, bioaccumulation, carcinogenicity, and toxicity, which disrupt ecological balance and harm living organisms. Various technologies focus on the dechlorination of CHCs, including bioremediation, redox processes, permeable reactive barriers (PRBs), and photocatalytic degradation [99-101]. Among these, abiotic dechlorination using Fe(II)-bearing minerals as reducing agents, particularly FeS nanomaterials, is a well-established approach. Notably, nano-sized FeS exhibits higher efficiency in reductive dechlorination of CHCs compared to bulk FeS, making it a promising technology for abiotic degradation. Huo et al. [90] utilized biogenic FeS nanoparticles produced by strain CN32 to enhance CT dechlorination. Pasteurized biogenic FeS nanoparticles showed high reactivity, with a CT dechlorination rate five times higher than that of abiotic FeS (Figure 9.17a).

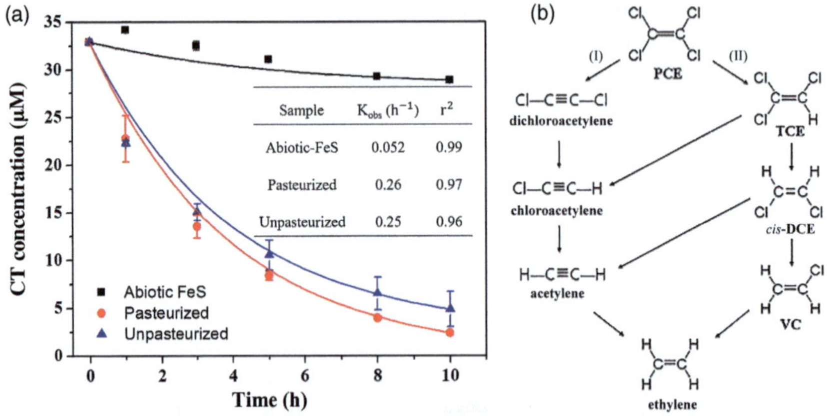

Figure 9.17 (a) CT dechlorination rates of pasteurized, unpasteurized biogenic FeS, and abiogenic FeS; *Source*: Adapted from Ref. [101]. (b) PCE degradation pathways: (I) β-elimination; (II) hydrodechlorination. *Source*: Reproduced from Ref. [102] / with permission of Elsevier.

However, not all FeS nanoparticles exhibit superior dehalogenation performance. To address aggregation and deactivation, recent studies have focused on modified FeS nanoparticles. For example, fungal polymer-stabilized FeS nanoparticles achieved 94% degradation of γ-hexachlorohexane ($5\,mg\,L^{-1}$) within 8 hours, compared to only 25% degradation by unstabilized FeS. These results confirm that stabilized FeS nanoparticles enable more effective CHC degradation. Beyond a standalone application, combining FeS nanoparticles with cyanocobalamin (Cbl(III)) further enhances reductive dechlorination efficiency. The presence of Cbl(III) significantly improves PCE dechlorination, primarily via hydrogenolysis and β-elimination pathways. Higher pH values further improved dechlorination performance. Kim et al. [102] reported that at pH 12, 95% of PCE was degraded within 20 hours in the presence of Cbl(III), compared to only 25% removal without Cbl(III) (Figure 9.17b). Product analysis indicated acetylene and TCE as the primary dechlorination products.

Urea is a common organic pollutant, with approximately 80% of urea-containing industrial and domestic wastewater discharged directly into the environment annually. If not treated properly, urea decomposes naturally, releasing ammonia and nitrate, leading to severe environmental pollution and water eutrophication [103-105]. The urea oxidation reaction (UOR) offers a promising approach by utilizing urea from wastewater, and its low theoretical cell voltage enables coupling with the HER for simultaneous wastewater treatment and efficient hydrogen production [106, 107]. However, UOR involves a six-electron transfer process and suffers from intrinsically slow kinetics, making the development of highly active catalysts challenging. Ni-based catalysts are widely studied due to their favorable UOR performance. Various design strategies have been employed to prepare efficient Ni-based catalysts for urea degradation. Chang et al. [108] prepared $NiFe[Fe(CN)_6]$ catalysts, which exhibited high carrier concentration and fast electron transfer, leading to high activity in 1 M KOH with 0.33 M urea. In urea electrolysis, this catalyst achieved a

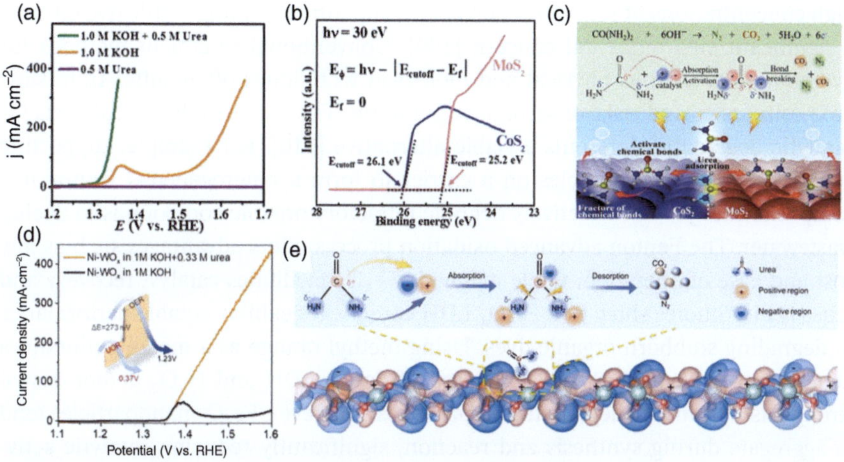

Figure 9.18 (a) Polarization curves of CoS_2–MoS_2 in different media; (b) synchrotron radiation ultraviolet photoelectron spectroscopy (SR-UPS) of CoS_2 and MoS_2; (c) schematic diagram of the UOR catalytic mechanism of CoS_2–MoS_2; *Source*: Reproduced from Ref. [103] / with permission of John Wiley & Sons. (d) Polarization curves of Ni–WO_x in 1 M KOH and a mixed solution of 0.33 M urea and 1 M KOH; (e) charge density difference of Ni–WO_x. *Source*: Reproduced from Ref. [104] / with permission of John Wiley & Sons.

high hydrogen production rate at 1.8 V, significantly outperforming conventional water splitting. Urea is a complex molecule containing both electron-donating amino groups and a carbonyl group. Understanding the relationship between adsorption strength and functional groups aids in designing better UOR catalysts. For example, a CoS_2/MoS_2 heterojunction was used as a model to tune adsorption strength and activate urea bonds. Experimental and theoretical results indicated that electron transfer from CoS_2 to MoS_2 creates nucleophilic regions at the interface, promoting the adsorption of electron-donating/withdrawing groups and facilitating urea decomposition (Figure 9.18a–c). Wang et al. [104] developed a W-doped NiO ($NiWO_x$) catalyst with unprecedented reaction kinetics and high urea degradation efficiency, far surpassing undoped NiO. XPS and in situ XAS revealed that W doping modulates the charge distribution around Ni atoms, promoting the formation of Ni^{3+} sites and creating localized electron-deficient and electron-rich regions. These regions favor the adsorption of —NH_2 and C=O groups, respectively, leading to superior catalytic performance and faster kinetics (Figure 9.18d, e). The Qiao group reported that $Ni_2Fe(CN)_6$ follows a two-stage UOR pathway, distinct from conventional Ni-based catalysts that rely on NiOOH formation. Urea is oxidized at Ni sites to form CO_2 and NH_3 intermediates, with NH_3 subsequently decomposed at Fe sites to N_2. Using an ion-selective ammonia electrode, they detected increasing NH_3 concentration during electrolysis at 1.34 V, confirming the synergistic role of Ni and Fe sites in enhancing UOR rates.

Dyes are widely used in various industries, with the global textile sector discharging approximately 100 tons of dye-containing wastewater into rivers annually. These dyes are characterized by high toxicity, resistance to decolorization, and high concentrations of recalcitrant organic compounds, making textile dye effluent a significant environmental concern [109]. Conventional treatment methods for dye pollutants such as physical and biological techniques often suffer from slow processing rates and risks of secondary pollution. The use of Fe-based catalysts for catalytic degradation presents a viable alternative [110]. For example, supporting magnetic Fe_3O_4 nanoparticles on a carrier to form a heterogeneous Fenton-like composite catalyst can effectively degrade refractory organic compounds in dyeing wastewater. The Fenton advanced oxidation process offers advantages such as low cost and ease of operation, while magnetic Fe_3O_4 facilitates catalyst recovery and reuse [111]. Studies show that Fe_3O_4 LDH composites exhibit notable performance in degrading stubborn organic dyes. Using methyl orange as a model pollutant, a heterogeneous Fenton system composed of Fe_3O_4 LDH and H_2O_2 demonstrated strong adsorption and degradation capabilities. However, Fe_3O_4 nanoparticles tend to aggregate during synthesis and reaction, significantly reducing catalytic activity [112]. To address this, Fe_3O_4 can be supported on substrates with high affinity for dye molecules, such as LDH. This support improves the dispersion of Fe_3O_4 nanoparticles, thereby enhancing their performance in Fenton-like reactions [113].

References

1 Pei, A., Wang, P., Zhang, S. et al. (2024). Enhanced electrocatalytic biomass oxidation at low voltage by Ni^{2+}-O-Pd interfaces. *Nature Communications* 15: 5899.

2 Yan, Y., Zhou, H., Xu, S.M. et al. (2023). Electrocatalytic upcycling of biomass and plastic wastes to biodegradable polymer monomers and hydrogen fuel at high current densities. *Journal of the American Chemical Society* 145: 6144–6155.

3 Gao, Y., Ge, L., Xu, H. et al. (2023). Electrocatalytic refinery of biomass-based 5-hydroxymethylfurfural to fine chemicals. *ACS Catalysis* 13: 11204–11231.

4 Akhade, S.A., Singh, N., Gutiérrez, O.Y. et al. (2020). Electrocatalytic hydrogenation of biomass-derived organics: a review. *Chemical Reviews* 120: 11370–11419.

5 Li, X., Guan, B.Y., Gao, S., and Lou, X.W.(D.) (2019). A general dual-templating approach to biomass-derived hierarchically porous heteroatom-doped carbon materials for enhanced electrocatalytic oxygen reduction. *Energy & Environmental Science* 12: 648–655.

6 Lin, G., Ma, R., Zhou, Y. et al. (2018). KOH activation of biomass-derived nitrogen-doped carbons for supercapacitor and electrocatalytic oxygen reduction. *Electrochimica Acta* 261: 49–57.

7 Ji, K., Liu, Y., Wang, Y. et al. (2024). Steering selectivity in electrocatalytic furfural reduction via electrode-electrolyte interface modification. *Journal of the American Chemical Society* 146: 11876–11886.

8 Liu, Y., Zhou, L., Liu, S. et al. (2024). Fe, N-inducing interfacial electron redistribution in NiCo spinel on biomass-derived carbon for bi-functional oxygen conversion. *Angewandte Chemie International Edition* 63 (e202319983).

9 Yang, Z., Chen, H., Bei, S. et al. (2024). Ultralow RuO_2 doped NiS_2 heterojunction as a multifunctional electrocatalyst for hydrogen evolution linking to biomass organics oxidation. *Small* 20: 2310286.

10 Wu, Y., Ma, L., Wu, J. et al. (2024). High-surface area mesoporous Sc_2O_3 with abundant oxygen vacancies as new and advanced electrocatalyst for electrochemical biomass valorization. *Advanced Materials* 36: 2311698.

11 Wu, Q., Chen, K., Shadike, Z., and Li, C. (2024). Relay-type catalysis by a dual-metal single-atom system in a waste biomass derivative host for high-rate and durable Li-S batteries. *ACS Nano* 18: 13468–13483.

12 Wang, Y., Zhang, M., Liu, Y. et al. (2023). Recent advances on transition-metal-based layered double hydroxides nanosheets for electrocatalytic energy conversion. *Advanced Science* 10 (2207519).

13 Jiang, X., Li, W., Liu, Y. et al. (2023). Electrocatalytic oxidation of 5-hydroxymethylfurfural for sustainable 2,5-furandicarboxylic acid production-from mechanism to catalysts design. *SusMat* 3: 21–43.

14 Fan, L., Ji, Y., Wang, G. et al. (2022). High entropy alloy electrocatalytic electrode toward alkaline glycerol valorization coupling with acidic hydrogen production. *Journal of the American Chemical Society* 144: 7224–7235.

15 Wang, Y., Zhu, Y.Q., Xie, Z. et al. (2022). Efficient electrocatalytic oxidation of glycerol via promoted OH* generation over single-atom-bismuth-doped spinel Co_3O_4. *ACS Catalysis* 12: 12432–12443.

16 Mao, Q., Mu, X., Wang, W. et al. (2023). Atomically dispersed Cu coordinated Rh metallene arrays for simultaneously electrochemical aniline synthesis and biomass upgrading. *Nature Communications* 14: 5679.

17 Zhou, P., Lv, X., Tao, S. et al. (2022). Heterogeneous-interface-enhanced adsorption of organic and hydroxyl for biomass electrooxidation. *Advanced Materials* 34 (2204089).

18 Ge, R., Wang, Y., Li, Z. et al. (2022). Selective electrooxidation of biomass-derived alcohols to aldehydes in a neutral medium: promoted water dissociation over a Ni-oxide-supported ruthenium single-atom catalyst. *Angewandte Chemie International Edition* 61 (e202200211).

19 Xu, C., Paone, E., Rodríguez-Padrón, D. et al. (2020). Recent catalytic routes for the preparation and the upgrading of biomass derived furfural and 5-hydroxymethylfurfural. *Chemical Society Reviews* 49: 4273–4306.

20 Li, W., Xu, Y., Wang, G. et al. (2025). Sustainable carbon-based catalyst materials derived from lignocellulosic biomass for energy storage and conversion: atomic modulation and properties improvement. *Carbon Energy* 7: e708.

21 Borghei, M., Lehtonen, J., Liu, L. et al. (2017). Advanced biomass-derived electrocatalysts for the oxygen reduction reaction. *Advanced Materials* 30: 1703691.

22 You, B., Liu, X., Jiang, N., and Sun, Y. (2016). A general strategy for decoupled hydrogen production from water splitting by integrating oxidative biomass valorization. *Journal of the American Chemical Society* 138: 13639–13646.

23 You, B., Jiang, N., Liu, X., and Sun, Y. (2016). Simultaneous H_2 generation and biomass upgrading in water by an efficient noble-metal-free bifunctional electrocatalyst. *Angewandte Chemie International Edition* 55: 9913–9917.

24 Borghei, M., Laocharoen, N., Kibena-Põldsepp, E. et al. (2017). Porous N,P-doped carbon from coconut shells with high electrocatalytic activity for oxygen reduction: alternative to Pt-C for alkaline fuel cells. *Applied Catalysis B: Environmental* 204: 394–402.

25 Liu, X., Zhou, Y., Zhou, W. et al. (2015). Biomass-derived nitrogen self-doped porous carbon as effective metal-free catalysts for oxygen reduction reaction. *Nanoscale* 7: 6136–6142.

26 Wang, P., Zheng, J., Xu, X. et al. (2024). Unlocking efficient hydrogen production: nucleophilic oxidation reactions coupled with water splitting. *Advanced Materials* 36: 2404806.

27 Oh, Y., Theerthagiri, J., Aruna Kumari, M.L. et al. (2024). Electrokinetic-mechanism of water and furfural oxidation on pulsed laser-interlaced Cu_2O and CoO on Ni foam. *Journal of Energy Chemistry* 91: 145–154.

28 Xi, W., Yang, P., Jiang, M. et al. (2024). Electrochemical CO_2 reduction coupled with alternative oxidation reactions: electrocatalysts, electrolytes, and electrolyzers. *Applied Catalysis B: Environmental* 341: 123291.

29 Lin, Z., Wang, L., Jia, T. et al. (2024). ZIF-67-derived FeCoNi-LDH with a 3D nanoflower hierarchical structure for highly efficient oxidation of 5-Hydroxymethylfurfural and coupling seawater splitting hydrogen production. *Chemical Engineering Journal* 481: 148429.

30 Chen, B., Yang, B., Su, Y. et al. (2025). NiFeCo wrinkled nanosheet electrode for selective oxidation of 5-hydroxymethylfurfural to FDCA. *Green Chemistry* 27: 2117–2129.

31 Li, R., Xiang, K., Peng, Z. et al. (2021). Recent advances on electrolysis for simultaneous generation of valuable chemicals at both anode and cathode. *Advanced Energy Materials* 11: 2102292.

32 Lu, X., Wu, K.H., Zhang, B. et al. (2021). Highly efficient electro-reforming of 5-hydroxymethylfurfural on vertically oriented Ni nanosheet/carbon hybrid catalysts: structure-function relationships. *Angewandte Chemie International Edition* 60: 14528–14535.

33 Yang, Z., Wang, S., Wei, C. et al. (2024). Proton transfer mediator for boosting the current density of biomass electrooxidation to the ampere level. *Energy & Environmental Science* 17: 1603–1611.

34 Qian, Q., He, X., Li, Z. et al. (2023). Electrochemical biomass upgrading coupled with hydrogen production under industrial-level current density. *Advanced Materials* 35: 2300935.

35 Zeng, L., Chen, Y., Sun, M. et al. (2023). Cooperative Rh-O_5/Ni(Fe) site for efficient biomass upgrading coupled with H_2 production. *Journal of the American Chemical Society* 145: 17577–17587.

36 Tang, C., Zheng, Y., Jaroniec, M., and Qiao, S.Z. (2021). Electrocatalytic refinery for sustainable production of fuels and chemicals. *Angewandte Chemie International Edition* 60: 19572–19590.

37 Xiang, K., Wu, D., Deng, X. et al. (2020). Boosting H_2 generation coupled with selective oxidation of methanol into value-added chemical over Co hydroxide@hydroxysulfide nanosheets electrocatalysts. *Advanced Functional Materials* 30: 1909610.

38 Zhao, B., Liu, J., Xu, C. et al. (2020). Hollow NiSe nanocrystals heterogenized with carbon nanotubes for efficient electrocatalytic methanol upgrading to boost hydrogen co-production. *Advanced Functional Materials* 31: 2008812.

39 Du, R., Wu, B., Tan, W. et al. (2025). Metallic Ni as electron acceptor modulates the redox of catalytic centers at activated $Ni^0/Ni(OH)_2$ heterojunctions for efficient ethanol electrooxidation. *Angewandte Chemie International Edition* 64: e202510285.

40 Zhao, Q., Wang, L., Fang, L. et al. (2026). Sulfur form regulation and dual-interface effect: catalytic mechanism of Ni-based catalyst in ethanol electrooxidation. *Journal of Colloid and Interface Science* 701: 138689.

41 Ning, S., Jian, Y., Yu, G.Q. et al. (2025). Dynamic electron spring effect in hollow $Fe_2O_3/CoSe_2$ heterostructure enhance ethanol electro-oxidation activity and stability. *Advanced Functional Materials,* 35,: 2509007.

42 Huang, H., Yu, C., Han, X. et al. (2020). Ni, Co hydroxide triggers electrocatalytic production of high-purity benzoic acid over 400 mA cm^{-2}. *Energy & Environmental Science* 13: 4990–4999.

43 Zheng, J., Chen, X., Zhong, X. et al. (2017). Hierarchical porous NC@CuCo nitride nanosheet networks: highly efficient bifunctional electrocatalyst for overall water splitting and selective electrooxidation of benzyl alcohol. *Advanced Functional Materials* 27: 1704169.

44 Lin, Y., Chen, Y., Ren, H. et al. (2024). Inspiration of bimetallic peroxide for controllable electrooxidizing ethylene glycol through modulating surficial intermediates. *Advanced Functional Materials* 34: 2404594.

45 Long, B., Yang, M., and Li, Y. (2025). Coordinating the competitive adsorption of organic and OH^- to accelerate electrooxidation kinetics of biomass-derived nucleophiles. *Angewandte Chemie International Edition* 64: e202508095.

46 Luo, W., Tian, H., Li, Q. et al. (2025). Controlled electrocatalytic glycerol upgrading to glyceraldehyde in near neutral media by Co oxide lattice activation. *Angewandte Chemie International Edition* 64: e202505059.

47 Du, P., Zhang, J., Liu, Y. et al. (2017). Hydrogen generation from catalytic glucose oxidation by Fe-based electrocatalysts. *Electrochemistry Communications* 83: 11–15.

48 Lin, C., Zhang, P., Wang, S. et al. (2020). Engineered porous Co-Ni alloy on carbon cloth as an efficient bifunctional electrocatalyst for glucose electrolysis in alkaline environment. *Journal of Alloys and Compounds* 823: 153784.

49 Chen, G., Li, X., and Feng, X. (2022). Upgrading organic compounds through the coupling of electrooxidation with hydrogen evolution. *Angewandte Chemie International Edition* 61 (e202209014).

50 Han, M., Luo, Y., Xu, L. et al. (2024). Oxygen vacancy boosts nitrogen-centered radical coupling initiated by primary amine electrooxidation. *Journal of the American Chemical Society* 146: 33893–33902.

51 Pan, X., Sun, L., Zhou, Z. et al. (2024). Positive electronic Ni active site enhances N-H/C-H bonds breaking for electrooxidation of amines to nitriles coupling with hydrogen production. *Advanced Energy Materials* 14: 2400374.

52 Zheng, Z., Li, K., Lin, L. et al. (2025). Recent advances on the electrocatalytic oxidation of biomass-derived aldehydes. *Green Energy & Environment* 10: 898–916.

53 Zhang, W., Ge, W., Qi, Y. et al. (2024). Surfactant directionally assembled at the electrode-electrolyte interface for facilitating electrocatalytic aldehyde hydrogenation. *Angewandte Chemie International Edition* 63: e 202407121.

54 Wang, Y., Wang, B., and Lin, S. (2025). Deoxygenative cyanofunctionalization of aldehydes and ketones enabled by electrochemical reduction. *Journal of the American Chemical Society* 147: 36992–36998.

55 Jiang, X., Ma, X., Yang, Y. et al. (2024). Enhancing the electrocatalytic oxidation of 5-hydroxymethylfurfural through cascade structure tuning for highly stable biomass upgrading. *Nano-Micro Letters* 16: 275.

56 Liu, X., Zhang, J., Chen, Y. et al. (2025). Hydrogen bridge-mediated efficient electrooxidation of 5-hydroxymethylfurfural on $Ni(OH)_2$-PO_4-/$Ni_3(PO_4)_2$ heterojunctions. *Angewandte Chemie International Edition* 64 (e202509274).

57 Chen, L., Yang, Z., Hu, Q. et al. (2025). Engineering gradient d-d orbital occupancy to boost substrate adsorption for efficient electrocatalytic biomass valorization. *Angewandte Chemie International Edition* 64 (e202511868).

58 Chen, D., Li, W., Liu, J., and Sun, L. (2025). Bio-inspired proton relay for promoting continuous 5-hydroxymethylfurfural electrooxidation in a flowing system. *Energy & Environmental Science* 18: 3120–3128.

59 Poerwoprajitno, A.R., Gloag, L., Watt, J. et al. (2020). Faceted branched Ni nanoparticles with tunable branch length for high-activity electrocatalytic oxidation of biomass. *Angewandte Chemie International Edition* 59: 15487–15491.

60 Deng, X., Li, M., Fan, Y. et al. (2020). Constructing multifunctional 'Nanoplatelet-on-Nanoarray' electrocatalyst with unprecedented activity towards novel selective organic oxidation reactions to boost hydrogen production. *Applied Catalysis B: Environmental* 278: 119339.

61 Gao, L., Liu, Z., Ma, J. et al. (2020). NiSe@NiO_x core-shell nanowires as a non-precious electrocatalyst for upgrading 5-hydroxymethylfurfural into FDCA. *Applied Catalysis B: Environmental* 261: 118235.

62 Deng, R. and Zhang, Q. (2024). Cu embedded in Co-P nanosheets with super wetting structure for accelerated overall water splitting under simulated industrial conditions. *Advanced Energy Materials* 14: 2401444.

63 Zhou, Z.Q., Chen, C.L., Gao, M.R. et al. (2019). In situ anchoring of a Co_3O_4 nanowire on nickel foam: an outstanding bifunctional catalyst for energy-saving simultaneous reactions. *Green Chemistry* 21: 6699–6706.

64 Lu, Y., Dong, C.L., Huang, Y.C. et al. (2020). Identifying the geometric site dependence of spinel oxides for the electrooxidation of 5-hydroxymethylfurfural. *Angewandte Chemie International Edition* 59: 19215–19221.

65 Chakraborty, B., Beltrán-Suito, R., Hausmann, J.N. et al. (2020). Enabling iron-based highly effective electrochemical water-splitting and selective oxygenation of organic substrates through in situ surface modification of intermetallic iron stannide precatalyst. *Advanced Energy Materials* 10: 2001377.

66 Yang, Y., Kong, L., Xia, Z. et al. (2025). Dual-vacancy-induced selective oxidative cleavage of C_α-C_β bonds for the electrocatalytic depolymerization of lignin. *Journal of the American Chemical Society* 147: 35924–35934.

67 Zhou, H., Li, Z., Xu, S.M. et al. (2021). Selectively upgrading lignin derivatives to carboxylates through electrochemical oxidative C(OH)-C bond cleavage by a Mn-doped Co oxyhydroxide catalyst. *Angewandte Chemie International Edition* 60: 8976–8982.

68 Qi, Y., Liu, B., Qiu, X. et al. (2023). Simultaneous oxidative cleavage of lignin and reduction of furfural via efficient electrocatalysis by P-doped $CoMoO_4$. *Advanced Materials* 35: 2208284.

69 Dong, F., Xu, B., Ma, X. et al. (2025). Advanced triboelectric materials for contact electrocatalytic degradation of pollutants. *Small* 21: 2500369.

70 Seifi, A., Keyikoğlu, R., Karataş, O. et al. (2025). Nanoarchitecturing of CuFeLa layered double hydroxide on graphite felt for photo-electrocatalytic degradation of emerging pollutants. *Journal of Industrial and Engineering Chemistry* 147: 598–607.

71 Sebastian, A., Deepthi, V., Vidhya, B. et al. (2025). Exploring the potential of Cu_3XS_4 (X = Fe, Sn) functionalized reduced graphene oxide for Electrocatalytic hydrogen evolution and photocatalytic degradation of pollutants. *Inorganic Chemistry Communications* 180: 114983.

72 Jian, Z., Heide, M., Yang, N. et al. (2021). Diamond fibers for efficient electrocatalytic degradation of *environmental* pollutants. *Carbon* 175: 36–42.

73 Xu, M., Mao, Y., Song, W. et al. (2018). Preparation and characterization of Fe-Ce co-doped Ti/TiO_2 NTs/PbO_2 nanocomposite electrodes for efficient electrocatalytic degradation of organic pollutants. *Journal of Electroanalytical Chemistry* 823: 193–202.

74 Ma, B., Lu, B., Tang, H. et al. (2024). Square-wave pulsed potential driven electrocatalytic degradation of 4-chlorophenol using Fe-Ni/rGO/PPy@NF three dimensional electrode. *Journal of Hazardous Materials* 480: 136054.

75 Sun, M., Zhao, Y., and Ge, H. (2022). Electrocatalytic degradation of methyl orange using nano-Fe_3O_4 supported on conductive carbon black as particle electrode. *International Journal of Electrochemical Science* 17: 220822.

76 Ru, M., Lang, K., Xiao, X. et al. (2025). Efficient electrocatalytic degradation of tetracycline using $NiCo_2O_4/MnO_2$ composite electrode: synergistic enhancement mechanisms and environmental toxicity assessment. *Journal of Environmental Chemical Engineering* 13: 115947.

77 Xie, L., Wang, P., Li, Y. et al. (2022). Pauling-type adsorption of O_2 induced electrocatalytic singlet oxygen production on N-CuO for organic pollutants degradation. *Nature Communications* 13: 5560.

78 Li, S., Wang, S., Wang, Y. et al. (2023). Doped Mn enhanced NiS electrooxidation performance of HMF into FDCA at industrial-level current density. *Advanced Functional Materials* 33: 2214488.

79 Guo, L., Zhang, X., Gan, L. et al. (2022). Advances in selective electrochemical oxidation of 5-hydroxymethylfurfural to produce high-value chemicals. *Advanced Science* 10 (2205540).

80 Zhou, Z., Xie, Y.-N., Sun, L. et al. (2022). Strain-induced in situ formation of NiOOH species on Co-Co bond for selective electrooxidation of 5-hydroxymethylfurfural and efficient hydrogen production. *Applied Catalysis B: Environmental* 305: 121072.

81 Han, X., Sheng, H., Yu, C. et al. (2020). Electrocatalytic oxidation of glycerol to formic acid by $CuCo_2O_4$ spinel oxide nanostructure catalysts. *ACS Catalysis* 10: 6741–6752.

82 Matsagar, B.M., Yang, R.-X., Dutta, S. et al. (2021). Recent progress in the development of biomass-derived nitrogen-doped porous carbon. *Journal of Materials Chemistry A* 9: 3703–3728.

83 Kwon, Y., Schouten, K.J.P., Van Der Waal, J.C. et al. (2016). Electrocatalytic conversion of furanic compounds. *ACS Catalysis* 6: 6704–6717.

84 Butler, E. and Hayes, K. (2000). Kinetics of the transformation of halogenated aliphatic compounds by iron sulfide. *Environmental Science & Technology* 34: 422–429.

85 Butler, E. and Hayes, K. (1998). Effects of solution composition and pH on the reductive dechlorination of hexachloroethane by iron sulfide. *Environmental Science & Technology* 32: 1276–1284.

86 Li, K., Ma, S., Zou, C. et al. (2023). Unrecognized role of organic acid in natural attenuation of pollutants by Mackinawite (FeS): the significance of carbon-center free radicals. *Environmental Science & Technology* 57: 20871–20880.

87 Fan, J., Liu, J., Cai, Y. et al. (2023). Efficient degradation of tetracycline in FeS-based SR-AOPs process at basic pHs: The overlooked role of metal complexation and redox reaction in persulfate activation. *Chemical Engineering Journal* 466: 143168.

88 Zhu, H., Ma, J., Du, W. et al. (2026). Preparation of N/O Co-doped recyclable catalysts based on activated pine needle-derived carbon-supported FeS nanoparticles via a "turning waste into treasure" strategy and their degradation study on emerging contaminants: DFT calculations and toxicity assessment. *Journal of the Taiwan Institute of Chemical Engineers* 180: 106467.

89 Wang, Q., Wu, Y., Shan, Y., and Dong, W. (2025). Synergizing surface-bound radicals with singlet oxygen to expedite emerging organic contaminant degradation in soil-based heterogeneous surface oxidation. *Environmental Science & Technology* 59: 22901–22913.

90 Huo, Y.-C., Li, W.-W., Chen, C.-B. et al. (2016). Biogenic FeS accelerates reductive dechlorination of carbon tetrachloride by Shewanella putrefaciens CN32. *Enzyme and Microbial Technology* 95: 236–241.

91 Mullet, M., Boursiquot, S., and Ehrhardt, J.-J. (2004). Removal of hexavalent chromium from solutions by mackinawite, tetragonal FeS. *Colloids and Surfaces A: Physicochemical and Engineering Aspects* 244: 77–85.

92 Lyu, H., Zhao, H., Tang, J. et al. (2018). Immobilization of hexavalent chromium in contaminated soils using biochar supported nanoscale Fe sulfide composite. *Chemosphere* 194: 360–369.

93 Dai, J., Zhang, H., Chen, N. et al. (2025). Nitrogen configurations driving efficient elimination of Cr(VI) by nitrogen-doped biochar/zero-valent Fe composites. *Chemical Engineering Journal* 515: 163655.

94 Zhu, Z., Jin, X., Zhou, Y. et al. (2025). Advancing magnetic chitosan-based materials for sustainable heavy metal removal from water: synthesis, regeneration, and future perspectives. *Separation and Purification Technology* 377: 134166.

95 Deng, J., Wang, P., Xu, Z. et al. (2024). Contaminated soil remediation with nano-FeS loaded lignin hydrogel: A novel strategy to produce safe rice grains while reducing cadmium in paddy field. *Journal of Hazardous Materials* 469: 133965.

96 Liu, Y., Huang, Y., Zhang, C. et al. (2020). Nano-FeS incorporated into stable lignin hydrogel: A novel strategy for cadmium removal from soil. *Environmental Pollution* 264: 114739.

97 Kim, E., Kim, J., Chang, Y. et al. (2014). Effects of metal ions on the reactivity and corrosion electrochemistry of Fe/FeS nanoparticles. *Environmental Science & Technology* 48: 4002–4011.

98 He, Y., Wilson, J., and Wilkin, R. (2010). Impact of iron sulfide transformation on trichloroethylene degradation. *Geochimica et Cosmochimica Acta* 74: 2025–2039.

99 Zhou, Y., Qin, W., Sun, X. et al. (2022). Synergistic effects on d-band center via coordination sites of M-N_3P_1 (M = Co and Ni) in dual single atoms that enhances photocatalytic dechlorination from tetrachlorobispheonl A. *Journal of Hazardous Materials* 430: 128419.

100 Nawaz, A., Khalid, A., Qayyum, W. et al. (2025). FeS-based nanocomposites: A promising approach for sustainable environmental remediation – Focus on adsorption and photocatalysis – A review. *Journal of Environmental Management* 373: 123530.

101 Choi, C., Wang, X., Kwon, S. et al. (2023). Efficient electrocatalytic valorization of chlorinated organic water pollutant to ethylene. *Nature Nanotechnology* 18: 160–167.

102 Kim, S., Park, T., and Lee, W. (2015). Enhanced reductive dechlorination of tetrachloroethene by nano-sized mackinawite with cyanocobalamin in a highly alkaline condition. *Journal of Environmental Management* 151: 378–385.

103 Li, C., Liu, Y., Zhuo, Z. et al. (2018). Local charge distribution engineered by Schottky heterojunctions toward urea electrolysis. *Advanced Energy Materials* 8: 1801775.

104 Wang, L., Zhu, Y., Wen, Y. et al. (2021). Regulating the local charge distribution of Ni active sites for the urea oxidation reaction. *Angewandte Chemie International Edition* 60: 10577–10582.

105 Rodrigues, E., Fernandes, C., Alves, O. et al. (2024). Urea Electro-Oxidation Reaction Enhancement via Magneto-Electrocatalysis on Nickel–Iron Oxide Nanoring Catalysts. *ACS Applied Energy Materials* 7: 8526–8535.

106 Lei, C., Chen, Z., Jiang, T. et al. (2024). Ultra-dense supported ruthenium oxide clusters via directed ion exchange for efficient valorization of 5-hydroxymethylfurfural. *Angewandte Chemie International Edition* 63: e202319642.

107 Li, Y., Huang, H., Jiang, M. et al. (2024). Advancements in transition bimetal catalysts for electrochemical 5-hydroxymethylfurfural (HMF) oxidation. *Journal of Energy Chemistry* 98: 24–46.

108 Zhao, Z., Zeng, H., Zhou, Y. et al. (2025). Synergistic catalytic mechanism of cyanide-bridged Ni-Fe bimetallic electrocatalyst for highly efficient urea oxidation. *Applied Catalysis B: Environment and Energy* 375: 125414.

109 Li, D., Zhang, P., Jin, Y. et al. (2025). Heterogeneous Fenton-like catalysts of tetranitro Fe phthalocyanine@Fe_3O_4@fly ash composites for fast decomposition of *environmental* contaminants. *Waste Disposal & Sustainable Energy* 7: 41–56.

110 Chen, J., Wan, J., Li, C. et al. (2022). Synthesis of novel Fe^0-Fe_3O_4/CeO_2/C composite cathode for efficient heterogeneous electro-Fenton degradation of ceftriaxone sodium. *Journal of Hazardous Materials* 437: 129393.

111 Hu, Z., Zhang, W., Zhong, Y. et al. (2024). MXene-derived nano N-TiO_2@MXene loaded with Fe_3O_4 for photo-Fenton synergistic degradation of phenol. *ACS Applied Nano Materials* 7: 20653–20664.

112 Chen, J., Chu, K., Sun, S. et al. (2023). Synthesis of magnetic core-shell Fe_3O_4-Mn_3O_4 composite for degradation of sulfadiazine via peroxymonosulfate activation: characterization, mechanism and toxicity analysis. *Journal of Environmental Chemical Engineering* 11 (109230).

113 Xie, Y., Yi, C., Zhou, Q. et al. (2025). Magnetic organic frame material Fe_3O_4@Au/COF for sensitive SERS detection and photo-Fenton degradation of tetracycline. *Journal of Environmental Chemical Engineering* 13: 401–410.

10

Applications of Iron-Group (Fe, Co, Ni) Metal Compound Nanomaterials in Other Potential Fields

In the previous chapters, we have elaborated on the applications of iron-group (Fe, Co, Ni) metal compound nanomaterials in catalysis and other fields. Owing to their characteristics such as large specific surface area, excellent thermal conductivity, and outstanding chemical reactivity, these nanomaterials have also been widely explored in other areas. This chapter will focus on providing a detailed introduction to their applications in three aspects: sensors, microwave absorption, and drug delivery.

10.1 Sensors

Micro–nano sensors are the core of key innovations in fields ranging from medicine to environmental science. In recent years, with the advancement of micro- and nano-fabrication technologies and the adoption of novel materials, the level of sensor design has been significantly improved, which has also promoted the development of high-performance gas sensors. Herein, gas sensors based on metal oxide semiconductors (MOS) constitute an important part in the field of gas sensing, especially p type MOS gas sensors based on Co_3O_4, NiO, etc., which exhibit excellent selectivity toward volatile organic compounds (VOCs). This section elaborates on MOS gas sensors in detail from four aspects: structure, mechanism, fabrication methods, and applications.

10.1.1 Gas Sensors and MOS Sensors

As core components of intelligent detection systems, gas sensors have garnered significant interest owing to their extensive applicability across diverse fields, including medical care, indoor/outdoor air quality monitoring, environmental science, the automotive sector, and military applications [1, 2]. The performance of gas sensors is assessed using multiple key parameters: sensitivity, selectivity, precision, limit of detection (LOD), resolution, accuracy, reversibility, recovery time, and response time. Furthermore, device miniaturization, power consumption, cost-effectiveness, long-term durability, and compatibility with wireless networks

Iron-Group Metal Compound Nanomaterials: Preparation, Characterization, Structure, and Applications, First Edition. Jinlong Zheng.

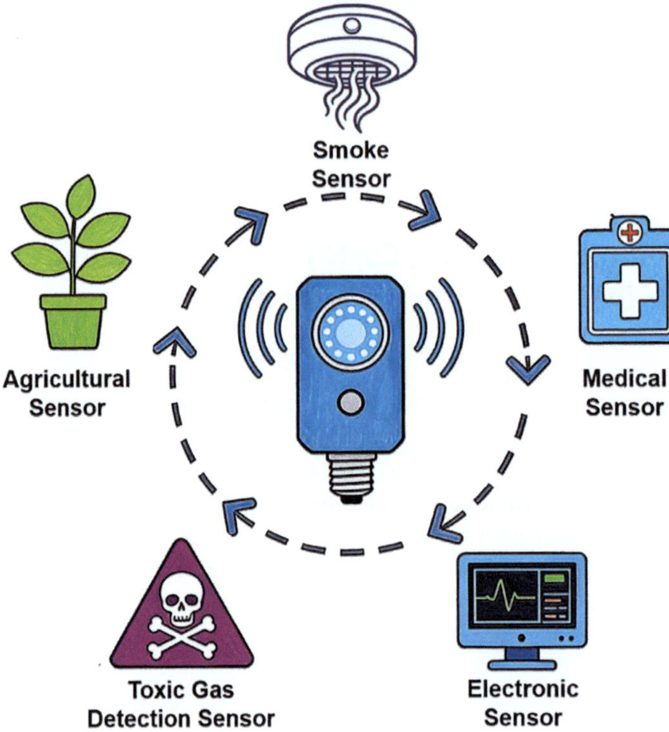

Figure 10.1 Block diagram of the e-nose prototype.

are recognized as critical factors for the successful deployment of sensors in various practical scenarios. Over the past several decades, numerous studies have been carried out on gas-sensing technologies, leading to the development of several distinct detection method branches. As illustrated in Figure 10.1, these approaches have gained increasing attention due to their unique operating mechanisms, fabrication techniques, and promising application prospects.

Among these detection technologies, when metal oxide semiconductor (MOS) sensors were first developed several decades ago, researchers observed that the resistance of Cu_2O is influenced by water vapor adsorption. Nevertheless, the rapid advancement of semiconductor sensors originated from research findings on ZnO and SnO_2 thin films, breakthroughs that led to the creation of the first commercial combustible gas detection sensor based on SnO_2, thereby enabling the practical application of semiconductor sensors. Currently, due to their unique structural characteristics and advantages like low cost, fast response times, and extended service life, metal oxides occupy a crucial position in multiple domains of chemistry, physics, and materials science.

MOS gas sensors stand out as preferred options for applications demanding the detection of low-concentration VOCs (at the ppm level). Additionally, they offer advantages such as straightforward measurement procedures, high durability, and cost-effectiveness, benefits derived from their easy manufacturing process and

Figure 10.2 Cross-sectional view of the MOS sensor, which includes a set of electrodes, a microheater, and a sensing layer fabricated on a thin suspended film using MEMS manufacturing technology.

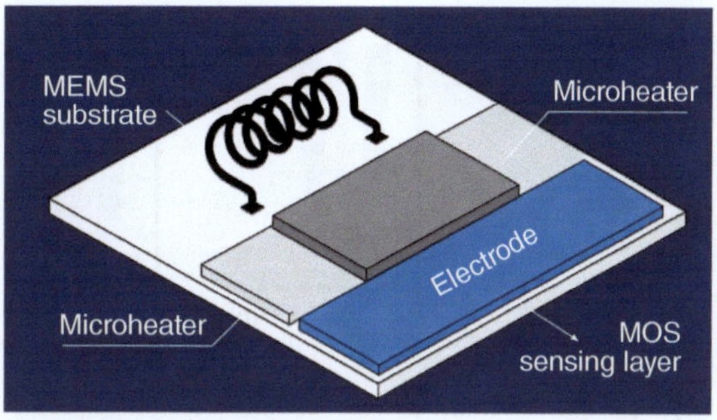

compact size. MOS sensors with a length of several hundred micrometers have been produced in conjunction with microheaters of comparable size, which are designed to supply the optimal operating temperature. It is well-documented that most MOS-based sensors exhibit high stability and anti-poisoning capabilities; consequently, they have emerged as a key technology in numerous household, commercial, and industrial gas-sensing systems.

10.1.2 Structure of MOS Sensors

MOS gas sensors consist of a specific sensing material deposited on a set of electrodes and a microheater required to electrically isolate the insulating layer from the sensing element. Figure 10.2 shows a schematic diagram of the MOS sensor. Advanced microelectromechanical systems (MEMS) manufacturing technology can be used to develop MOS sensors, where the sensing area of the device is constructed on a thin suspended film to achieve an optimal thermal isolation structure.

Since the sensitivity and performance of the device are highly dependent on the interaction between the analyte and the sensing layer, MOS sensors are generally classified into two types: the first type is n-type MOS sensors based on TiO_2, ZnO, SnO_2, and WO_3; the second type is p-type MOS sensors based on NiO, Co_3O_4, Mn_3O_4, and Cr_2O_3 [3, 4]. The thickness of the sensing material ranges from several nanometers to tens of nanometers. The thickness of the receptor layer, temperature, and the catalytic metal particles embedded in it are all factors that affect the sensitivity of MOS sensors. In addition, the change in electrical conductivity varies with the type of MOS used. In n-type MOS, the majority charge carriers are electrons; when it interacts with reducing gases, the electrical conductivity increases, while interaction with oxidizing gases leads to carrier depletion and a subsequent decrease in electrical conductivity. In p-type MOS, the majority charge carriers are holes; the electrical conductivity increases under oxidizing gases and decreases under reducing gases when the number of positive charge carriers (or holes) of the analyte increases.

10.1.3 Mechanism of MOS Sensors

The operation of gas sensors based on MOS mainly relies on chemiresistivity.

The gas-sensing process primarily relies on two key factors: the properties of majority carriers in the film and the redox characteristics of the target gas present

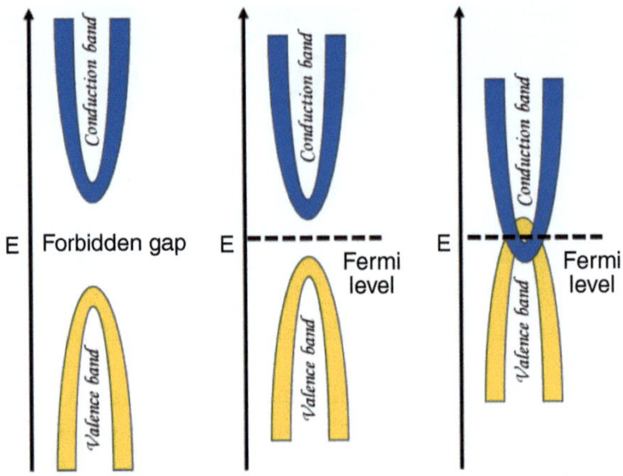

Figure 10.3 Band theory of insulators, semiconductors, and conductors.

in the surrounding environment. Specifically, when the film is exposed to oxidizing gases or acceptor substances, its resistance increases; in contrast, exposure to donor substances or reducing gases leads to a decrease in film resistance. The fundamental theory underlying gas sensing is associated with band theory [5], and Figure 10.3 provides a detailed illustration of this theory.

The sensing mechanism of p-type MOS is determined by variations in surface resistance, which results from changes in the concentration of majority charge carriers triggered by redox reactions. For p-type metal oxides, acceptor energy levels are positioned close to the valence band at room temperature and remain filled—this filling process creates holes within the valence band. When the surface of the metal oxide is exposed to air under room temperature conditions, oxygen molecules are physically adsorbed onto the surface.

Specifically, O^{2-} undergoes chemical adsorption at low temperatures (<150 °C) but disappears quickly, whereas O^- is chemically adsorbed at higher temperatures. O^- and O^{2-} are the most common types of chemically adsorbed ionic species, with O^- being the predominant form of adsorbed oxygen. The chemical reactions that take place after these oxygen ions acquire electrons are outlined below:

$$O_{2(ads)} \leftrightarrow O_{2(ads)} \tag{10.1}$$

$$O_{2(ads)} \leftrightarrow O^-_{2(ads)} T \leq 150\,^\circ C \tag{10.2}$$

$$O_2^-(ads) + e^- \leftrightarrow 2O^-(ads) 150\,^\circ C \leq T \leq 300\,^\circ C \tag{10.3}$$

$$2O^-(ads) + e^- \leftrightarrow O^{2-}(ads) T \geq 300\,^\circ C \tag{10.4}$$

This electron-capturing process from the sample surface gives rise to band bending and an elevation in hole concentration near the interface. As a consequence, a hole accumulation layer forms on the metal oxide surface. When the metal oxide comes into contact with any reducing gas, the gas molecules undergo physical adsorption on the active layer surface and react with the pre-adsorbed oxygen, thereby releasing electrons. During the gas desorption process from the metal oxide surface, the holes in the accumulation layer are further restored, and this restoration causes a decrease in the sensor's resistance. Notably, metal oxides based on n-type semiconductors exhibit an effect that is the reverse of that observed in p-type semiconductors.

10.1.4 Fabrication of MOS Sensors

Silicon micromachining technology is a widely applied technique for manufacturing MOS sensors. Its typical fabrication procedures include preparing a silicon substrate equipped with an insulating layer, a microheater, and a sensing layer, along with depositing electrodes. SiO_2 is a frequently used insulating layer, which is grown to a thickness of several micrometers through thermal oxidation.

Furthermore, as the reactions within the sensing layer require high-temperature conditions, a microheater must be installed at the bottom of the MOS sensor to raise the temperature of the sensing region. The microheater can be produced by forming a Ti/Pt metal layer (10/60 nm thick) on the photomask of the microheater pattern via DC magnetron sputtering, combined with a lift-off process. To avoid electrical conduction between the microheater and the microelectrodes/metal oxide sensing layer, an electron beam evaporation technique is employed to coat a 300-nm thick SiO_2 insulating layer on the microheater.

The deposition of MOS electrodes utilizes the same fabrication method as the microheater, namely DC magnetron sputtering and a lift-off process. These electrodes are typically composed of a Cr/Au layer with a thickness of 10/60 nm. The method for depositing the MOS sensing layer on the fabricated electrodes depends on the sensing material used; different sensing materials correspond to distinct deposition methods. Taking the SnO_2-based MOS sensor as an example, DC magnetron sputtering is needed to deposit a 50-nm thick sensing material. In the final step of the process, reactive ion etching (RIE) is used to etch the substrate and the backside of silicon dioxide, aiming to achieve better temperature isolation and reduce power consumption.

10.1.5 Applications of MOS Sensors

As shown in Figure 10.4, this figure presents the highly harmful VOCs involved in sensing applications and their corresponding chemical formulas. The following sections below provide a detailed summary of the main VOC sensors (targeting toluene, ethanol, and formaldehyde) reported over the past 5 years.

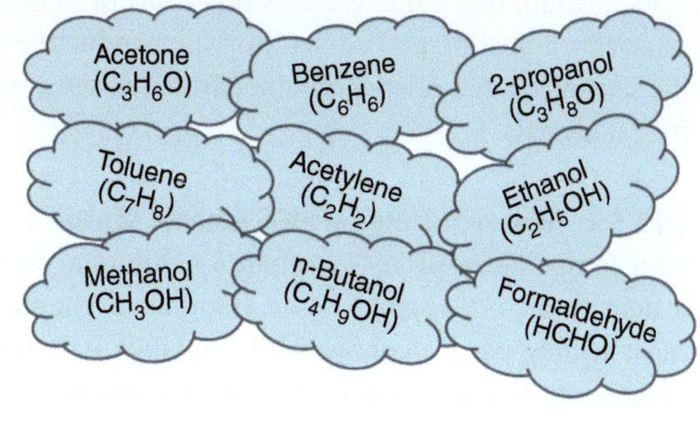

Figure 10.4 Main volatile organic compounds (VOCs) in the atmosphere.

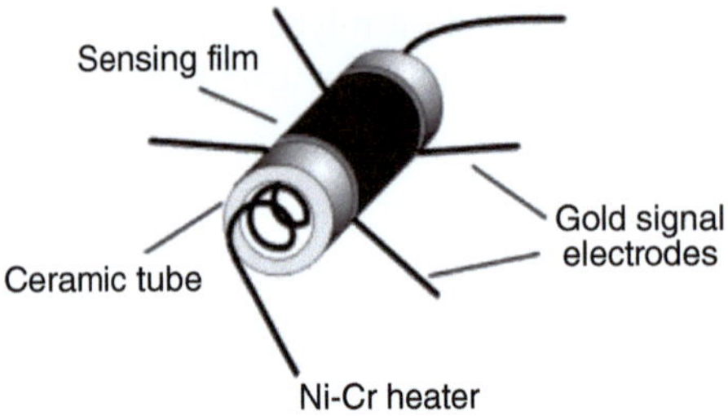

Figure 10.5 The schematic structure of the gas sensor. *Source*: Reproduced from Ref. [8] / with permission of John Wiley & Sons.

10.1.5.1 Research Status of MOS Toluene Sensors

Toluene (C_7H_8) is a colorless liquid with a unique aromatic odor. At 20 °C, it has a density of $0.866\,g\,cm^{-3}$ and a boiling point of 110.7 °C [6]. In terms of toxicity, toluene is similar to benzene. The human organs affected by it mainly include the skin, liver, and kidneys. Additionally, toluene is one of the important biomarkers for lung cancer research [7]. According to WHO guidelines, the indoor air concentration of toluene at 25 °C is 0.07 ppm. This highlights the necessity of a sensitive, reasonable, accurate, simple, and reproducible toluene gas-sensing method. Currently, a variety of gas-sensing technologies based on metal oxides, electrochemistry, etc., can be used for toluene gas sensing in Figure 10.5 [8].

The reaction pathways describing the interaction between toluene gas and n-type as well as p-type MOS are outlined as follows:

$$C_7H_8 + 2O^- \leftrightarrow H_2O + C_7H_6 - O + 2e^-(\text{n-type}) \tag{10.5}$$

$$C_7H_8 + 18O^- \leftrightarrow 4H_2O + 7CO_2 + 18e^-(\text{p-type}) \tag{10.6}$$

Gao et al. developed a NiO-based toluene gas sensor using a hydrothermal approach and combined it with SnO_2 [9]. To meet the performance criteria of the NiO-modified sensor, its gas-sensing properties were optimized. Operating temperature is a critical factor influencing the sensor's response, and this parameter was adjusted to the range of 220–250 °C. At 250 °C, the sensor showed a notably higher response to toluene in comparison to other target gases. When compared with the pure NiO sensor, the modified NiO sensor exhibited a 50-fold increase in response to toluene gas. Due to the incorporated material functioning as a catalyst and its dynamic crystal surface enabling efficient detection of the target gas, this work presents an ultra-sensitive toluene sensor.

As a result, this study extends this concept to develop an ultra-sensitive toluene sensor by adjusting and controlling the conductive channel. Figure 10.6 illustrates the sensing characteristics obtained from the sensors based on the pristine and modified materials.

10.1.5.2 Research Status of MOS Ethanol Sensors

At a temperature of 20 °C, ethanol exhibits a vapor pressure of 43 Torr (T) and a density of $0.789\,g\,cm^{-3}$. As a flammable chemical, ethanol is also one of the most widely used alcohols. More notably, it is utilized in fields including food processing, transportation, biomedicine, and industrial production. Furthermore,

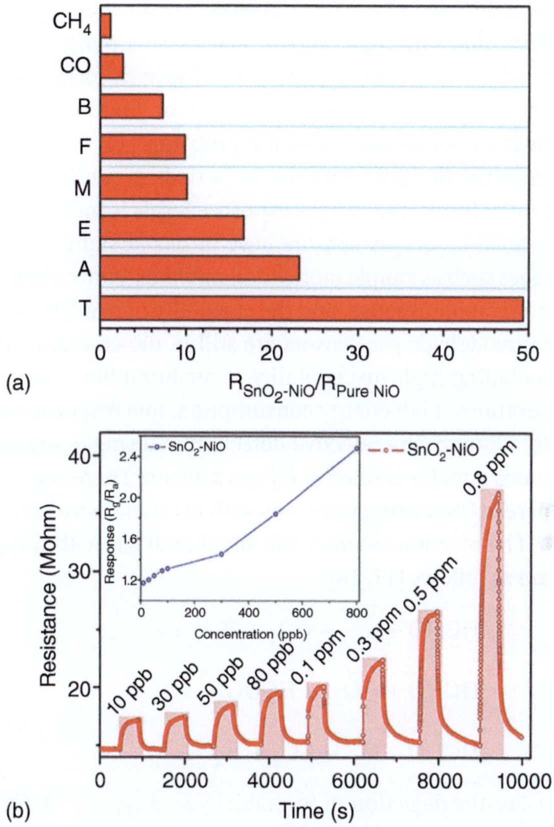

Figure 10.6 (a) Gas response curves of pure NiO and SnO_2-modified NiO at their optimal temperatures; (b) real-time resistance curves and corresponding responses of the SnO_2-modified NiO sensor to low toluene concentrations ranging from 10 ppb to 0.8 ppm at 250 °C. *Source*: Reproduced from Ref. [9] / with permission of John Wiley & Sons.

the consumption of ethanol as a beverage may lead to various problems, including accidents and assaults [10].

The reaction mechanisms of ethanol gas with n-type semiconductors are as follows [11]:

$$C_2H_5OH(gas) + 3O_2(ads) \rightarrow 2CO_2 + 4H_2O + 3e^- \quad (10.7)$$

$$C_2H_5OH(gas) + 6O^-(ads) \rightarrow 2CO_2 + 3H_2O + 6e^- \quad (10.8)$$

$$C_2H_5OH(gas) + 3O_2^-(ads) \rightarrow 2CO_2 + 3H_2O + 6e^- \quad (10.9)$$

Jiss Paul et al. [12] fabricated an interdigital capacitive sensor for ethanol sensing using $CoFe_2O_4$ nanocomposites. This work successfully achieved considerable sensor performance for low-concentration ethanol gas at room temperature. The enhanced sensing capability was realized due to the oxygen adsorbed in the material, which enabled capacitive conduction of the sensing layer and the IDC-based sensor. Zhao et al. invented an ethanol sensor using ZnO nanomaterials synthesized via hydrothermal method [13]. This work achieved outstanding sensor performance, attributed to the material's large aspect ratio (length-to-diameter ratio), oxygen vacancies as surface defects, and one-dimensional (1D) structure.

10.1.5.3 Research Status of MOS Formaldehyde Sensors

Formaldehyde is a colorless VOC with a pungent odor, and it is found in various products used in daily life [14]. However, formaldehyde is widely used as a basic chemical; it is a harmful air pollutant. Similarly, prolonged exposure to formaldehyde can cause serious health problems [15, 16]. Subsequently, the WHO has set an exposure limit of 0.08 ppm for periods exceeding 30 minutes. Therefore, the detection of formaldehyde gas exposure levels is indispensable for a healthy society.

MOS-based gas sensors have attracted significant attention due to their advantages such as simple fabrication methods, convenient operation, high sensitivity, low cost, rapid response, and the capability of real-time detection. However, MOS-based formaldehyde gas sensors are still in the emerging stage and face many problems, including high susceptibility to air humidity, dependence on high operating temperatures, high energy consumption, low response values, and long recovery times. In addition, the selective detection of a single target gas in complex environments poses a major challenge for gas sensors. Therefore, it is of great significance to prepare MOS sensing materials with excellent performance.

The reaction mechanisms of ethanol gas with n-type and p-type semiconductors are as follows [17, 18]:

$$HCHO + O^- \rightarrow CO_2 + H_2O + e^- \tag{10.10}$$

$$HCHO + \frac{1}{2}O_2 \rightarrow HCOOH \tag{10.11}$$

$$HCHO + O_2 \rightarrow CO_2 + H_2O \tag{10.12}$$

For the detection of formaldehyde, a type of VOC gas, there are three main optimization strategies for MOS gas sensors, namely heteroatom doping of MOS gas sensors, heterostructure construction, and preparation of ternary MOS gas sensors, which will be introduced in detail below.

Heteroatom Doping of MOS Gas Sensors Introducing an appropriate amount of heteroatoms into the lattice of MOS materials is an effective method to improve sensing performance. When the size of ions in the lattice is similar to that of heteroions, the heteroatoms will undergo substitutional doping, meaning the impurity atoms directly replace the positions of lattice atoms. The doping of heteroatoms introduces defects in the crystal structure, generates more surface oxygen vacancies, and increases the number of oxygen molecules adsorbed on the material surface. These oxygen molecules are used to capture more free electrons, thereby promoting surface reactions and enhancing the gas-sensing performance of the material.

Wang et al. synthesized a Co-doped In_2O_3 nanorod sensor via a hydrothermal method [19]. As shown in Figure 10.7, which is a schematic diagram of a typical gas sensor in the test circuit, the sensor was found to exhibit a good sensing response to formaldehyde. At 130 °C, the 1% Co-doped In_2O_3 nanorods showed a sensitivity response value of 23.2–10 mg L^{-1} formaldehyde, which was 4.5 times that of the pure In_2O_3 sensor. They attributed the improvement in the gas-sensing performance of this sensor mainly to two aspects: (i) The large specific surface area of the rod-shaped

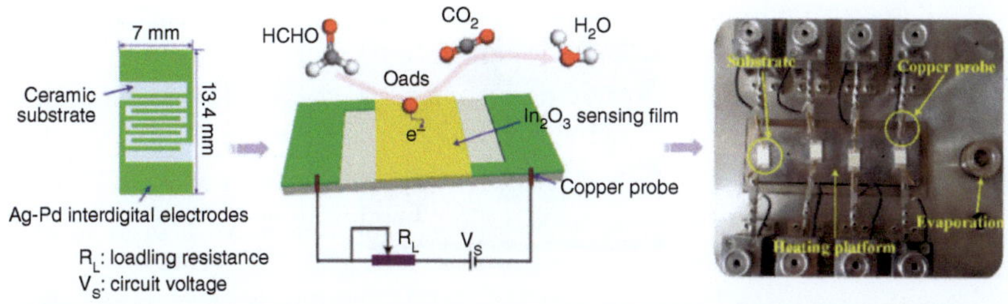

Figure 10.7 Schematic diagram of a typical gas sensor in the test circuit. *Source*: Reproduced from Ref. [19] / with permission of American Chemical Society.

In_2O_3 material provides abundant reaction sites for formaldehyde and accelerates the surface reaction of the material, thereby increasing the response speed. (ii) The incorporation of Co to replace In^{3+} as a donor can effectively increase the number of oxygen vacancies in the crystal structure; accordingly, the number of free electrons that can be captured also increases, which effectively affects the adsorption behavior and thus enhances the performance of the sensing material.

Heterostructure Construction of MOS Gas Sensors Constructing heterojunctions is an important strategy to improve the gas-sensing performance of sensing materials. The ability of heterostructures to enhance the sensing performance of materials is mainly attributed to electronic effects and chemical interactions. The interface formed by the contact of two materials is conducive to increasing the mobility of charge carriers. Heterojunctions in metal oxide sensing materials can be classified into n-n type heterojunctions, p-p type heterojunctions, and n-p type heterojunctions. Among them, n-p heterojunctions have been extensively studied in gas-sensing applications due to their superior performance.

Cao et al. prepared an In_2O_3/Co_3O_4 core–shell structure using electrospinning and hydrothermal methods, and Figure 10.8 shows the corresponding characterization diagram of the material [20]. The sensor exhibited a response of $15.7–10^{-6}$ formaldehyde, while the response of the pure Co_3O_4 sensor was only 2.8. They suggested that compared with the Co_3O_4 sensor, the improved response of the In_2O_3/Co_3O_4 sensor could be partially explained by the reduced hole concentration caused by the formation of p-n junctions.

Analysis indicated that there are three main reasons for the enhanced gas-sensing performance of the sensing material. First, a large number of homogeneous and heterogeneous interfaces are formed between Co_3O_4 and In_2O_3. Electrons can flow from In_2O_3 to Co_3O_4, and the electron concentration of In_2O_3 is reduced through electron–hole recombination. Additionally, an n-p heterointerface is formed at the boundary between In_2O_3 and Co_3O_4. Therefore, compared with pure Co_3O_4, the resistance of the hybrid functional band increases significantly. Further increasing the resistance by constructing the In_2O_3-Co_3O_4 heterointerface helps to improve the sensitivity. Second, based on the mutual diffusion of cations, Co^{2+} and Co^{3+} can replace In^{3+} sites, leading to lattice defects and mutations, thereby increasing the types of adsorbed oxygen and vacancy oxygen. Finally, Co_3O_4 is a good catalyst for

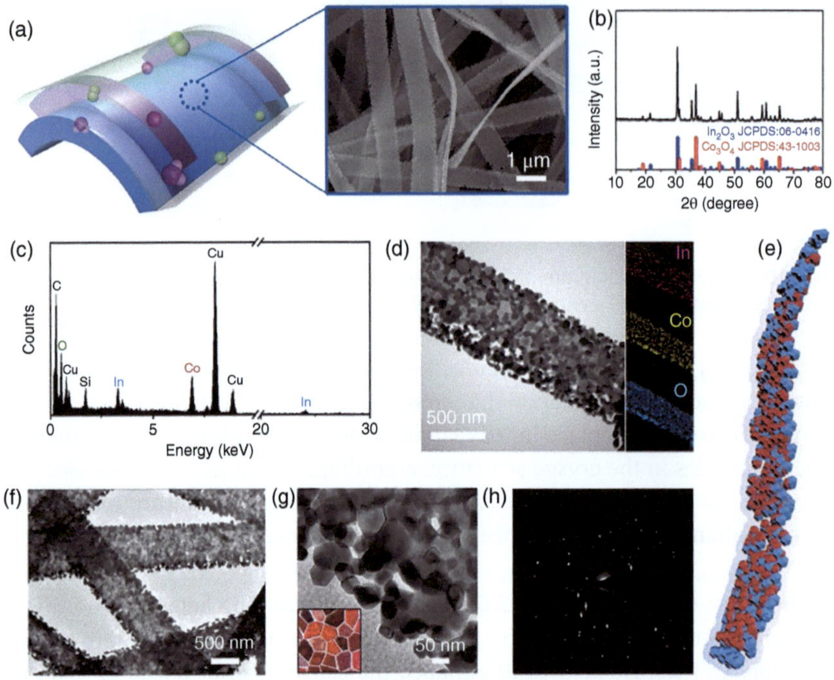

Figure 10.8 (a) SEM image and (b) XRD pattern of the In_2O_3-Co_3O_4 strip; (c) dot mapping and (d) elemental mapping of In, Co, and O in In_2O_3-Co_3O_4; (e) ideal geometric model of the corresponding single strip; (f) TEM image of the In_2O_3-Co_3O_4 strip; (g) magnified TEM image and (h) SAED image. *Source*: Reproduced Ref. [20] / with permission of John Wiley & Sons.

formaldehyde oxidation due to its multivalent characteristics and abundant oxygen adsorption sites.

Preparation of Ternary MOS Gas Sensors In recent years, perovskite-type metal oxides have attracted scholars' research due to their unique photoelectric properties, stable crystal structure, excellent isomerization and electrocatalytic properties, and have been widely used in fields such as catalysis, fuel cells, electrodes, and gas sensors [21–24]. Among them, LaFeO has been proven to be a gas-sensing material with excellent sensing performance [25]. Yang K et al. prepared a layered porous $LaFeO_3$ and mainly studied the effects of F108 structure-directing agent and calcination temperature on optimizing gas-sensing performance [26]. The results showed that $LaFeO_3$ containing 0.5 g of F108 had the best gas-sensing performance: at 125 °C, its gas-sensing response to 50×10^{-6} formaldehyde was 116, and the detection limit was 50×10^{-9}.

To investigate the effects of different structures and morphologies on the sensing performance of MOS materials, Zhang et al. prepared fibrous, spherical, and cage-like Ag-$LaFeO_3$ nanomaterials, as shown in Figure 10.9 [27]. Studies indicated that these materials all exhibited excellent response and selectivity to formaldehyde gas. In addition to perovskite-type ternary metal oxides, ternary oxides such as Zn_2SnO_4, $ZnFe_2O_4$, and $ZnCo_2O_4$ have also been studied for formaldehyde detection. Li et al. synthesized ZnO/Zn_2SnO_4 via a simple hydrothermal method, and the

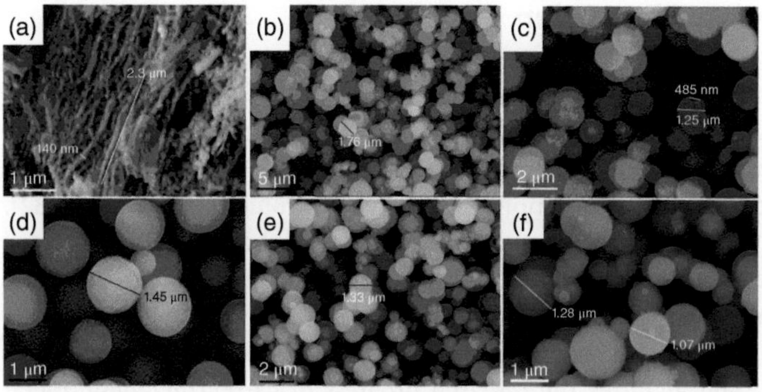

Figure 10.9 SEM images of the as-prepared (a) ALFO fibers, ALFO spheres (b) before calcination and (c) after sintering, (d) carbon spheres (the templates of the cages), and ALFO cages (e) before sintering and (f) after calcination. *Source*: Reproduced from Ref. [27] / with permission of The Royal Society of Chemistry.

prepared sample had a hexagonal layered structure with a diameter of 600–900 nm [28]. Its formaldehyde-sensing performance was studied. At 160 °C, the sensor exhibited a response of 22.5 to formaldehyde (the concentration of formaldehyde is missing in the original text), with a response time and recovery time of 4 seconds and 89 seconds, respectively, and a minimum detection limit of 0.5×10^{-6}. Compared with pristine Zn_2SnO_4 and ZnO, the ZnO/Zn_2SnO_4 composite showed higher sensitivity (increased by 80% and 120%, respectively), better selectivity, and a lower detection limit.

However, MOS sensors still face significant challenges in the practical application of formaldehyde detection. First, although many studies have shown that a variety of materials can selectively detect formaldehyde gas, the specific detection mechanism has not been elaborated in detail. This makes it difficult to design sensors that meet practical needs. Second, the optimal operating temperature of many MOS materials is relatively high, which not only causes substantial energy consumption but also greatly shortens the service life of the sensors. Nevertheless, some studies have found that certain synthesis strategies can effectively reduce the optimal operating temperature; the problem is that these strategies simultaneously prolong the response and recovery times of the sensors. Third, there is an urgent need to further improve the detection limit of MOS-based sensors. Although many reports indicate that MOS-based sensors have a low detection limit for formaldehyde, most of them still fail to meet international standards, which greatly limits the practical application of these materials. Fourth, MOS nanomaterial-based sensors currently still suffer from issues such as poor repeatability and difficulties in industrial mass production. Finally, the problem of humidity interference is also one of the challenges faced by current sensors. Differences in sensitivity under different humidity conditions lead to large errors in the sensors. Therefore, researching and developing advanced materials that can not only meet the requirements of rapid, convenient, and accurate detection but also save costs and achieve a low detection limit remains a frontier and hot topic in current gas sensor research.

10.2 Microwave Absorption

In recent decades, with the rapid development of wireless communication technology and the wide application of electronic devices, electromagnetic radiation has become a severe environmental pollution problem, which seriously threatens human health and hinders the operation of precision electronic equipment [29–32]. In the exploration over the past few decades, we have found that a variety of carbon/magnetic metal composites containing metal elements such as Fe, Co, and Ni exhibit excellent microwave absorption capabilities [33–36]. This section introduces the microwave absorption mechanism and influencing factors of carbon/magnetic metal composites, and analyzes the research progress of carbon/magnetic metal composites in recent years. It discusses in detail the preparation methods and microwave absorption properties of carbon/magnetic metal composites with different compositions, morphologies, and microstructures.

10.2.1 Microwave Absorption Mechanism

When electromagnetic waves strike an absorbing material, the incident energy can be split into three components: reflected energy, absorbed energy, and transmitted energy [37]. Reflected energy refers to a small fraction of electromagnetic waves that bounce off the surface of the absorbing material, a phenomenon caused by the impedance mismatch between the material and free space [38]. Additionally, electromagnetic waves can penetrate into the interior of the absorbing material and interact with it during propagation; the absorbing material is capable of converting electromagnetic energy into thermal energy or other energy forms, thus resulting in energy dissipation [39]. Finally, a portion of the electromagnetic waves will pass through the material.

Therefore, two methods currently exist to enhance microwave absorption: one is to enable incident electromagnetic waves to enter the material's interior instead of being reflected off its surface, and the other is to improve electromagnetic absorption capacity by adjusting electromagnetic parameters. Since incident electromagnetic waves consist of coupled oscillating electric and magnetic fields, materials can weaken electromagnetic wave energy (i.e., achieve absorption attenuation) by interacting with either one or both of these fields [40–42]. Thus, the absorption mechanism of electromagnetic waves can ultimately be described as being dominated by electronic interaction, dominated by magnetic interaction, or involving a combination of both. It can be calculated using the following approach, which relies on the relative complex permittivity (ε_r) and relative complex permeability (μ_r) [43]:

$$Z_{in} = Z_0\sqrt{\frac{\mu_r}{\varepsilon_r}}\tanh\left[j\left(\frac{2\pi fd}{c}\right)\sqrt{\mu_r\varepsilon_r}\right] \tag{10.13}$$

$$RL(dB) = 20\log\left|\frac{(Z_{in} - Z_0)}{(Z_{in} + Z_0)}\right| \tag{10.14}$$

Among them, Z_{in} is the input impedance of the absorber, Z_0 is the impedance of free space, c is the speed of light, d represents the thickness of the absorber, and f is the microwave frequency.

10.2.2 Influencing Factors of Microwave Absorption

10.2.2.1 Size Effect

Material dielectric and magnetic characteristics are directly altered by particle size. As dimensions shrink to the nanometer range, electromagnetic interactions undergo a gradual intensification [44]. On one side, nanomaterials acquire superior electromagnetic performance courtesy of their distinctive size effects, surface configurations, and quantum tunneling phenomena. In the context of microwave-absorbing materials, nanoscale counterparts exhibit lower densities compared to their bulk equivalents. Furthermore, the substantial specific surface area of nano-sized microwave absorbers generates abundant active sites and interfacial regions, which in turn promote enhanced polarization relaxation losses [45].

On the other hand, magnetic metal materials possess excessively high electrical conductivity, leading to eddy current losses induced by electromagnetic waves; this phenomenon compromises the materials' complex permeability at elevated frequencies. However, when particle sizes are reduced below the skin depth threshold, eddy current losses are effectively constrained, resulting in improved absorption capabilities [46]. Additionally, a transition from multi-domain structures (typical of bulk materials) to single-domain configurations occurs in magnetic metals when their diameters fall beneath a critical size [47]. Single-domain materials lack domain wall motion, a feature that confers higher coercivity and consequently amplifies hysteresis losses. For these reasons, researchers are dedicated to the design and development of nano-scale microwave-absorbing materials.

10.2.2.2 Morphology and Microstructure

Conventional particulate absorbing materials feature straightforward microstructures, including spherical, flaky, and rod-like geometries, among others. Intricate microstructures can prolong the propagation path of electromagnetic waves while expanding the contact area between the material and the waves [48]. Ultimately, the design of specific morphologies or microstructures allows for the achievement of tunable dielectric and magnetic properties, as well as improved impedance matching [49].

10.2.3 Regulation of Microwave Absorption Capacity

As is commonly known, electromagnetic waves are composed of rapidly oscillating electric and magnetic fields [50]. Among these, materials primarily relying on magnetic loss include ferrites (such as $CoFe_2O_4$, $MnFe_2O_4$, $ZnFe_2O_4$), magnetic metal oxides (such as Fe_3O_4, Fe_2O_3, CoO), and magnetic metals (Fe, Co, Ni, along with their alloys) [51–61].

While these single-component absorbing materials demonstrate a degree of microwave absorption ability, they fail to meet the standards of ideal microwave absorption. This is attributed to their limited attenuation mechanisms and the challenges in regulating electromagnetic parameters. Currently, two primary strategies are widely employed to enhance the performance of electromagnetic wave-absorbing materials. The first effective approach involves synthesizing composite materials that integrate dielectric and magnetic components by incorporating dielectric materials; this method can significantly boost the microwave absorption performance of single magnetic materials. For instance, researchers including Qiao, Deng, and Bao have respectively combined ZnO (as a dielectric material) with various magnetic metals to prepare Ni/ZnO composites, FeCo/ZnO composites, and ZnO/Co hybrid nanotubes [62–64].

10.2.4 Preparation Methods of Carbon/Magnetic Metal Composites

In the following sections, we present several representative examples to elaborate on each synthesis route, along with the corresponding synthesis mechanisms and the resulting microwave absorption performance of the composites.

10.2.4.1 Arc Plasma Method

By regulating arc discharge to vaporize the electrodes, the target products are deposited either on the cathode or the inner wall of the vacuum chamber. This technology boasts unique advantages: the high-energy plasma field can supply the energy and reaction conditions required for atomic excitation, collision bonding, and material formation, thereby enabling the rapid synthesis of nanostructures.

For instance, through arc plasma evaporation, Co_xNi_{100-x}@C nanoparticles (NPs) fully encapsulated by curved graphite layers were synthesized in situ [65]. As illustrated in Figure 10.10, at elevated temperatures, carbon atoms derived from dissociated ethanol (C_2H_5OH) collide with vaporized metal atoms to form aggregates. As the temperature drops, these aggregates condense upon collision with supersaturated carbon droplets, ultimately forming nanocapsules. Research findings indicate that $Co_{40}Ni_{60}$@C nanocapsules achieve the optimal electromagnetic wave absorption performance at 13.2 GHz, with a minimum reflection loss (RL) of −53 dB.

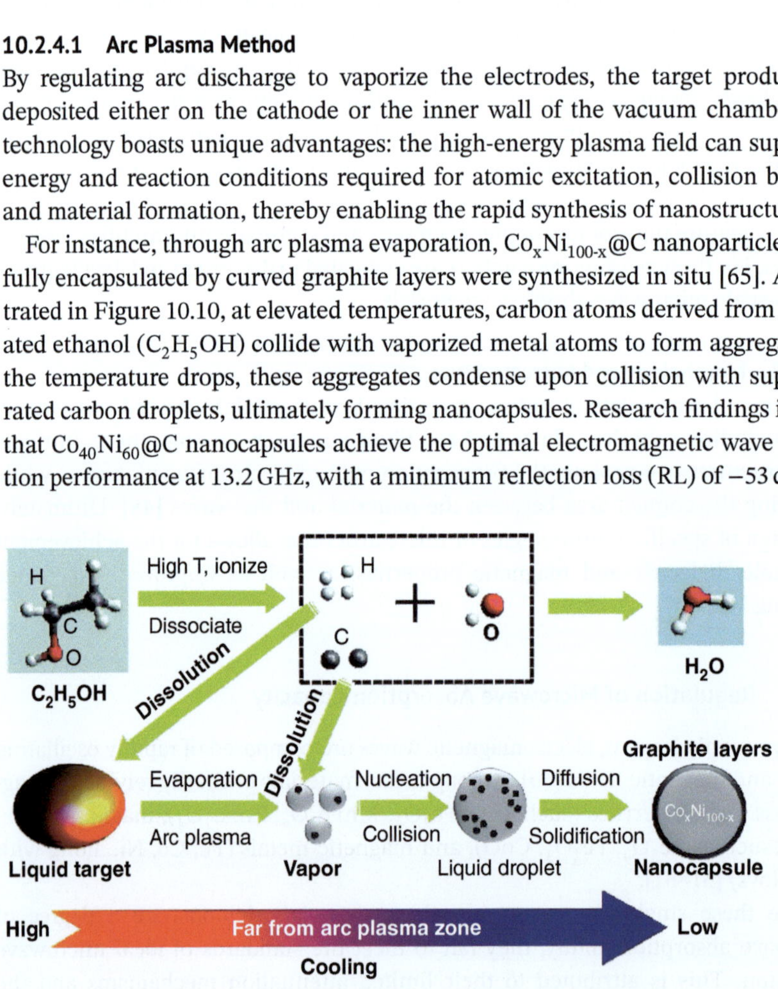

Figure 10.10 Preparation of Co_xNi_{100-x}@C by arc plasma method. *Source*: Reproduced from Ref. [65] / with permission of The Royal Society of Chemistry.

Zhang's research team successfully prepared Co@C NPs with excellent microwave absorption performance using the arc plasma method. In their experiment, a cobalt metal ingot served as the anode, a carbon rod as the cathode, and methane (CH_4) as the gaseous carbon source [66]. Liu et al. integrated the arc plasma method with other techniques to synthesize microporous Co/reduced graphene oxide (rGO) nanocomposites with varying rGO mass fractions [67]. Notably, CoAl NPs were first synthesized via arc melting of ingots in a 0.1 MPa mixed gas environment (50% Ar and 50% H_2). The researchers suggested that the microporous structure can effectively regulate complex permittivity, while rGO enhances impedance matching, both factors collectively contributing to a significant improvement in microwave absorption performance. The sample containing 15 wt% (weight percent) rGO achieved the widest absorption bandwidth of 10.0 GHz, whereas the Co/rGO sample with 10 wt% rGO displayed the highest RL value of −108.0 dB.

While the arc plasma technique offers practical value for fabricating carbon/magnetic metal composites, it suffers from notable drawbacks. For one thing, the yield of NPs generated via this approach is relatively low, confining its application to scientific concept validation rather than large-scale industrial manufacturing. For another, the overall reaction process lacks sufficient controllability, posing difficulties for regulating the quantity and size distribution of the synthesized NPs.

10.2.4.2 Solvothermal/Hydrothermal Method

For example, metal–organic framework (MOF)-derived CoNi@C microspheres can be fabricated using a simple hydrothermal route combined with a post-annealing process, as depicted in Figure 10.11 [68]. Owing to their tunable composition and unique structural features, CoNi@C microspheres demonstrate excellent microwave absorption performance. Zhang et al. synthesized Ni-btc microspheres via an in situ solvothermal method [69]. These hierarchical microspheres exhibit an outstanding reflection loss (RL) intensity of −73.2 dB and an effective absorption bandwidth of 4.8 GHz.

10.2.4.3 Chemical Vapor Deposition Method

Chemical vapor deposition (CVD) stands as one of the most widely utilized technologies for producing one-dimensional nanomaterials and serves as a universal

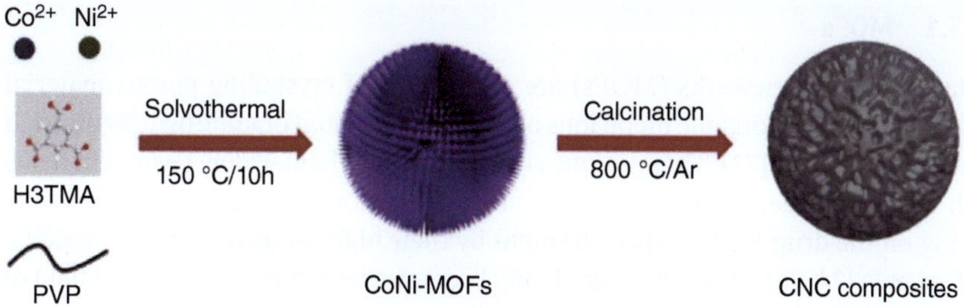

Figure 10.11 Preparation of CoNi@C microspheres by the solvothermal method. *Source:* Reproduced from Ref. [68] / with permission of Elsevier.

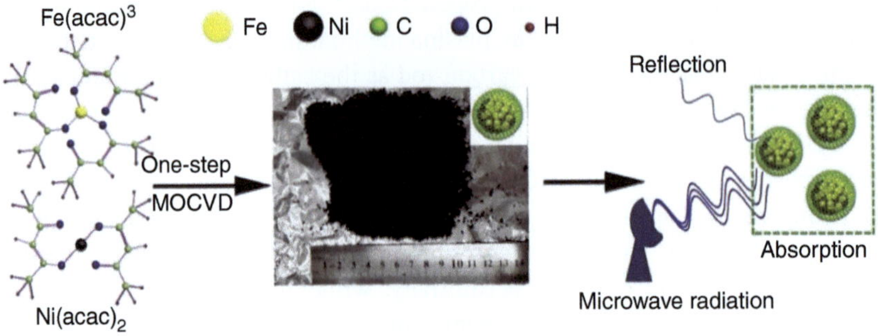

Figure 10.12 Preparation of FeNi–C core–shell nanoparticles by the CVD method. *Source*: Reproduced from Ref. [72] / with permission of Elsevier.

approach for synthesizing carbon nanotube-based materials [70]. A typical CVD system comprises three core components: precursors, substrates, and gases, and it operates under specific conditions of desired growth temperature and deposition pressure. The key process of CVD involves the high-temperature decomposition of organic gases (e.g., methane, ethylene, acetylene), followed by interactions of the decomposed species on the substrate surface, and ultimately the deposition of these species onto the substrate to form composite materials [71].

Kuang et al. obtained carbon nanotubes coated with extremely small FeCo—C NPs (with a diameter of approximately 5 nm) (FeCo—C-CNTs) in an argon atmosphere at 1000 °C using a simple one-step MOCVD method [72]. Subsequent analysis utilizing an equivalent circuit model demonstrated that the incorporation of carbon nanotubes (CNTs) notably boosted the dielectric loss capability of FeCo—C-CNTs composites, thereby endowing them with outstanding microwave absorption properties: a minimum reflection loss (RL) of −79.2 dB and an absorption bandwidth of 6.3 GHz. Following this, the research team further refined the fabrication process through experiments conducted within the temperature range of 600–1000 °C, successfully synthesizing ultra-small FeNi@C core–shell NPs, as illustrated in Figure 10.12.

10.3 Drug Delivery

10.3.1 MOFs

Metal–organic frameworks (MOFs) are a new type of crystalline porous material self-assembled by inorganic metal ions or metal clusters and organic ligands through coordination bonds [73]. Their unique structure makes them widely favored in fields such as gas storage, catalysis, sensing, and lithium-ion batteries. In recent years, relying on the drug-loading space brought by their high porosity and large specific surface area [74, 75], MOFs have gradually become a research hotspot in the field of drug delivery and are regarded as a drug delivery carrier with great potential.

MOFs have a wide variety of metal nodes, with common ones including Cr, Cu, Zr, Al, Zn, Fe, etc. (Figure 10.13) [76–78]. Among them, MOFs constructed

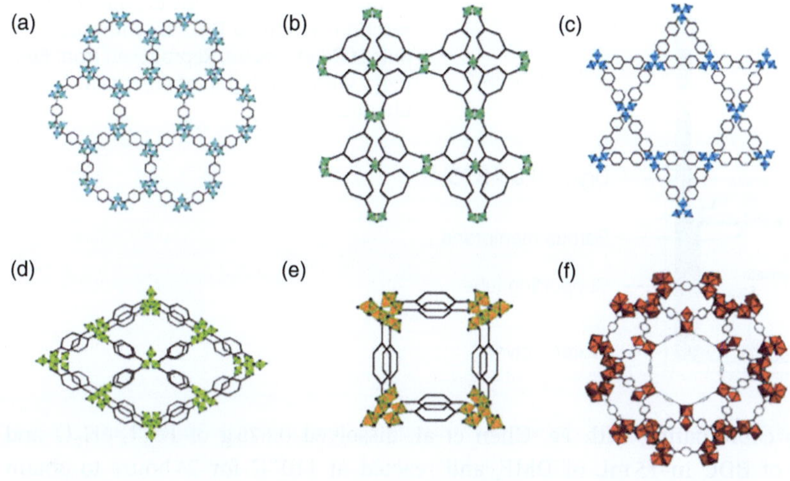

Figure 10.13 The structures of metal–organic frameworks (MOFs). (a) MIL88-Cr; (b) MOF199-Cu; (c) NU1000-Zr; (d) MIL53-Al; (e) MOF5-Zn; (f) MIL101-Fe. *Source:* Reproduced from Ref. [76] / with permission of American Chemical Society.

with metal Fe as nodes (Fe-based MOFs) exhibit better biocompatibility and safety compared to MOFs composed of other metals. Meanwhile, they can also directly exert antitumor effects through approaches such as assisting in inducing ferroptosis. In addition, due to the inherent magnetic responsiveness of Fe, Fe-based MOFs possess natural potential for magnetic targeting and magnetic resonance imaging (MRI). By adjusting their structure and other aspects, they can fulfill the functions of active targeting as well as integration of diagnosis and treatment. Therefore, as a new type of drug carrier with high safety and high added value, Fe-based MOFs have unique advantages in antitumor drug delivery [79].

Zhu et al. obtained 12 Co(II) complexes by adjusting the substituents on the α-N-heterocycle. Among them, cobalt(II) pyridine-2-carboxaldehyde condensed 4-phenyl-3-thiosemicarbazide (C3) exhibited an IC_{50} value of 1.2 μM against A549 lung cancer cells, and its selectivity index (SI) was 8 times higher than that of cisplatin [80]. X-ray diffraction analysis revealed that the square planar coordination structure of C3 is the key factor contributing to its high activity.

10.3.1.1 Preparation Methods of MOFs

The solvothermal method is one of the most common, simple, and effective synthetic methods. Typically, Fe salts (such as $FeCl{\cdot}H_2O$) and organic ligands (such as 1,4-dicarboxybenzene [BDC] and trimesic acid) are dissolved in a suitable solvent, and the synthesis is carried out under high-temperature and high-pressure conditions. Common solvents mainly include N,N-dimethylformamide (DMF), water, methanol, ethanol, dimethyl sulfoxide, etc., among which DMF and water are the most frequently used.

DMF has extremely strong solubility, and the dimethylamine produced by its thermal decomposition during the reaction can assist the deprotonation of ligands,

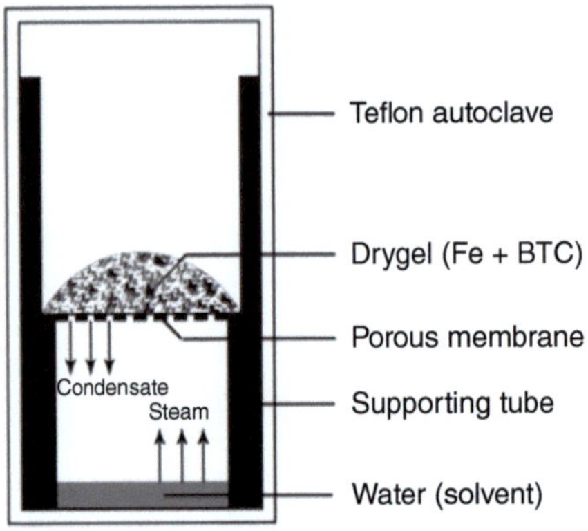

Figure 10.14 Schematic drawing of the reaction system for DGC to synthesize MIL-100(Fe). *Source*: Reproduced from Ref. [84] / with permission of American Chemical Society.

thereby coordinating with Fe. Chen et al. dissolved 0.675 g of $FeCl_3{\cdot}6H_2O$ and 0.225 g of BDC in 15 mL of DMF, and reacted at 110 °C for 24 hours to obtain MIL-101 type Fe-based MOF [81]. However, DMF has certain toxicity issues; therefore, the residual DMF in the pores of MOFs should be thoroughly removed before drug loading. The conventional removal method involves using a suitable solvent for dissolution and washing. To remove residual DMF from the MIL-101 type Fe-based MOF, Chen et al. washed the material with a large amount of ethanol and deionized water. Subsequently, they soaked it in 95% ethanol for 24 hours and dried it under vacuum at 200 °C overnight before proceeding with the subsequent drug-loading process.

Another common solvent is water, and the process of using water as a synthesis solvent is also called the "hydrothermal synthesis method." This project intends to prepare Fe-based MOFs with configurations such as MIL-100 via the Fe^{3+}/trimesic acid-hydrothermal synthesis method, and on this basis, add hydrofluoric acid to regulate their crystal forms. Using water as a solvent offers higher safety performance, but its solubility is limited—the reactants need to have good water solubility, which to a certain extent restricts its application.

The microwave synthesis method is a technique for synthesizing Fe-based MOFs using microwave energy as the energy source, and it also features rapidity and high efficiency. Dong et al. heated the reactant solution to 150 °C using microwaves and successfully synthesized MIL-53 type Fe-based MOF within 10 minutes [82]. Ma et al. conducted a study on the reaction time of preparing MIL-88B-type Fe-based MOF via the microwave synthesis method [83]. They found that even with only 1 minute of microwave irradiation, MIL-88B microcrystals with good crystallinity could be obtained.

The xerogel conversion method is characterized by the fact that the Fe used does not need to be in an ionic state; elemental Fe can also be directly used for synthesis. Additionally, it offers good reaction sustainability and generates little waste. Imteaz et al. employed the xerogel conversion method: 0.341 g of elemental Fe and 0.884 g of trimesic acid were placed in the upper layer of a reaction vessel, 3 mL of water was placed in the lower part, and MIL-100-type Fe-based MOF was directly synthesized at 165 °C in Figure 10.14 [84].

The ultrasonic synthesis method is an efficient and rapid synthesis method. The difference from the solvothermal method lies in that the energy required for the coordination reaction is provided by ultrasonic waves rather than heating. Gordon et al. dissolved 1.35 g of $FeCl_3{\cdot}6H_2O$ and 0.83 g of BDC in 25 mL of DMF, and used probe ultrasound without temperature control [85]. Uniform MIL-53 type Fe-based MOF was synthesized in only 10 minutes, while the solvothermal method usually takes several hours to dozens of hours.

10.3.1.2 Characteristics of MOFs as Antitumor Drug Carriers

Current organic nanomedicines generally suffer from the problem of excessively low drug-loading capacity. For example, the drug-loading capacity of liposomes is usually difficult to exceed 1 wt%, while that of polymeric NPs is only about 6 wt%. This means that patients will ingest a large amount of excipients, leading to an increased physical burden.

MOFs are characterized by high porosity and large specific surface area, and are generally recognized to have higher drug-loading potential [86]. However, the pores of MOFs are usually at the micropore level (pore size less than 2 nm), which limits their drug-loading capacity. The use of organic ligands with longer molecular sizes in spatial structure can result in larger pore sizes; nevertheless, MOFs prepared via this strategy generally have poor stability, and their pores tend to collapse. In contrast, Fe-based MOFs with configurations such as MIL-100 and MIL-101 possess a cage-like structure. Apart from conventional micropores, the inner space of their cages exhibits pore sizes at the mesopore level (2–50 nm) (Figure 10.15), thereby significantly enhancing the drug-loading capacity while ensuring structural stability [88]. Taking busulfan (Bu), an anti-leukemia drug, as an example, the loading capacity of liposomes for Bu is only 0.4 wt%, whereas the drug-loading capacity of MIL-100-type Fe-based MOF for Bu can reach 25 wt%, which is 62.5 times that of liposomes [89]. Gao et al. achieved a 35 wt% drug-loading capacity for the antitumor drug fluorouracil (5-Fu) using MOF-5 type Fe-based MOF [90]. In contrast, the maximum drug-loading capacity of 5-Fu by the chitosan-gold nanoparticle composite system is 11.85 wt%, less than 1/3 of that of Fe-based MOFs. Even micelles, which are known for high drug-loading capacity, can only achieve a drug-loading capacity of 19 wt%, still lower than that of Fe-based MOFs.

Passive targeting refers to the use of biological activities that the material itself does not possess to achieve passive accumulation at tumor sites. Among them, leveraging the size advantage of self-assembled NPs and utilizing the EPR effect to achieve highly selective accumulation in tumor tissues is an important means to realize the integration of nano-diagnosis and treatment. In terms of the EPR effect, NPs with larger particle sizes have better retention ability and lower permeability, while those with smaller particle sizes exhibit better permeability and retention. Therefore, the key lies in achieving a balance between permeability and retention, and this optimal size generally varies depending on the type of tumor [91]. Fe-based MOF materials have excellent size adjustability; by regulating parameters such as synthesis time and raw material concentration, their particle sizes can be effectively controlled, thereby achieving active targeting of tumor sites.

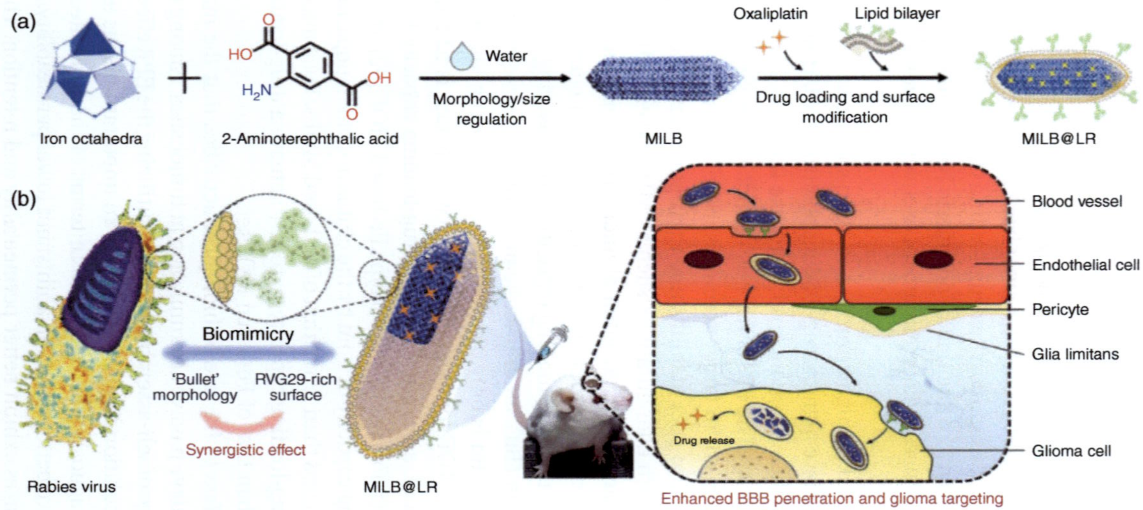

Figure 10.15 (a) Fabrication of MILB@LR. (b) Comprehensive RABV-mimic strategy for enhanced BBB penetration and chemotherapy of Glioma. *Source*: Reproduced from Ref. [87] / with permission of John Wiley & Sons.

The physiological barriers of the organism are important lines of defense for protecting specific tissues and organs, among which the most representative one is the blood–brain barrier (BBB). The BBB protects brain homeostasis by regulating the material exchange between the blood and the brain; however, for conventional nanomedicines, it is an insurmountable barrier that directly hinders their targeting of intracerebral tumors. A major reason why conventional nanomedicines are difficult to cross the BBB is that their spherical morphology leads to weak penetrability. Nevertheless, Qiao et al. found that the bullet-like morphology of MIL-88B-type Fe-based MOF is similar to that of rabies virus, which is adept at invading the brain, thus endowing it with excellent BBB penetration ability [87]. On this basis, the diameter and aspect ratio of the MOF were optimized by adjusting the amount of solvent, and modified with rabies virus proteins to obtain a bionic rabies virus-mimetic Fe-based MOF. This modified MOF has been proven to possess good targeting ability for brain tumors.

MOFs are composed of metal nodes and organic ligands through coordination bonds; therefore, the factors determining their safety mainly include the following three aspects: the difficulty of MOFs degradation in vivo, the safety of metal ions produced after dissociation, and the safety of organic ligands.

Good biodegradability ensures that the carrier decomposes into small-molecule substances within a relatively short period, which are then excreted from the body, reducing the metabolic burden on the organism. The relatively weak binding force of coordination bonds guarantees the good degradability of Fe-based MOFs. Particularly in the acidic and reductive tumor microenvironment, the coordination bonds of Fe-based MOFs are more prone to dissociation. Baati et al. prepared various Fe-based MOFs such as MIL-88A, MIL-88B, and MIL-100, and administered them to rats via intravenous injection. It was found that Fe-based MOFs in the body were rapidly decomposed into Fe ions and organic ligands by the liver and spleen, and excreted through urine and feces, with basically no metabolic toxicity.

Common metal elements composing MOFs include Cr, Cu, Zr, Al, Zn, Fe, etc. Among them, Cr is generally considered to be the most toxic, while Zn and Fe are usually regarded as the two least toxic ones [92]. Cr-based MOFs were the earliest MOFs developed for drug delivery; however, due to the problem of excessively high toxicity, researchers turned to developing Fe-based MOFs to replace them for drug delivery. Zn is generally considered safe, but a study by Tamames-Tabar et al. showed that Zn-based MOFs still have higher toxicity than Fe-based MOFs because Zn^{2+} competitively binds to ion channels, interfering with cell metabolism. As an essential trace element for the human body and a key component of hemoglobin, Fe has higher safety. Ploetz et al. studied the cytotoxicity of MIL-100-type Fe-based MOF, and the results showed that it had almost no effect on the activity of HeLa cells at a concentration of 100 $\mu g\ mL^{-1}$, with its IC_{50} value as high as 3 $g\ mL^{-1}$ [93]. In practical applications, the Fe-based MOF with MIL-88A configuration has been approved by the US Food and Drug Administration (FDA) for use as an iron supplement, which also proves its high safety.

However, the toxicity of drugs is often dose-dependent to a certain extent, and excessive iron can also cause oxidative damage to organs such as the liver and spleen.

Therefore, the intake of iron is a problem that needs to be solved urgently. The introduction of organic ligands into Fe-based MOF materials can reduce the proportion of iron. For example, in MIL-100 series MOF materials, the ratio of Fe to ligands is 3:2. Calculated based on their relative weights, the Fe content is approximately 28.5%, which significantly reduces the amount of Fe used.

In addition, the toxicity of the ligands themselves can also have a certain impact on the safety of MOFs. Fe-based MOF compounds with ligands represented by BDC and trimesic acid are the most widely discovered type of carboxylic acid ligands so far. Their toxicity to rats after administration is $6.4\,g\,kg^{-1}$ and $8.4\,g\,kg^{-1}$, respectively, and they are easy to excrete, making them safer ligands. However, these two compounds have certain irritant effects on the oral cavity, respiratory tract, eyes, and skin. Therefore, how to further optimize them remains a very meaningful research topic.

10.3.2 Nanoparticles

In 1956, Gilchrist applied magnetic NPs to biomedical research for the first time, using the inductive heating of NPs to treat lymph nodes near tumor sites [94]. In 1963, Meyer et al. reported the use of magnetic targeting technology to apply iron oxide NPs to targeted drug delivery [95]. The design and assembly of magnetic drug delivery systems can involve either the surface modification of individual particles or the modification of the hollow and hybrid structures of magnetic NPs. Magnetic NPs can be transformed into core–shell structured magnetic NPs after being coated with silica, gold, or polymers, which is more conducive to their surface modification and drug loading.

Lee et al. designed a drug delivery system based on magnetic core–shell nanoparticles (MCNPs), which achieves the combined effect of chemotherapy and hyperthermia and improves the apoptosis rate of malignant brain tumor cells and metastatic breast cancer cells [96]. Among them, the surface of the magnetic core is coated with a gold nanoshell, which can promote the binding of amphiphilic tail-anchoring peptide (ATAP) and Arg-Gly-Asp (RGD) peptide. Moreover, the magnetic core is crucial for the targeted delivery of the MCNP-ATAP drug delivery system and local hyperthermia. In addition, MCNP-ATAP can enter mitochondria, cause mitochondrial dysfunction, and induce the release of cytochrome C from mitochondria. Cytochrome C binds to Apaf-1 and activates Caspase-3 under the mediation of ATP/dATP, ultimately leading to tumor cell apoptosis. The authors confirmed the excellent tumor cell-killing effect of the MCNP-ATAP drug delivery system both in vitro and in vivo. Iron oxide magnetic NPs can also be used to modify other drug delivery carriers for the purpose of magnetic targeting. For example, carbon nanotubes modified with magnetic NPs have been proven to be effective carriers for delivering antitumor drugs. Chen et al. modified Fe_3O_4 NPs onto carbon nanotubes and confirmed that applying alternating magnetic field (AMF) treatment at the treatment site could double the drug efficacy [97]. This indicates that incorporating magnetic NPs into drug delivery systems can improve therapeutic effects and reduce the dosage of chemotherapeutic drugs. In addition to high drug-loading efficiency and magnetic targeted therapy, hybrid magnetic nanostructures can be used to load multiple drugs for tumor-targeted therapy. Drugs such as doxorubicin and paclitaxel are commonly used in research on magnetic targeted drug delivery. Antibiotics like tetracycline, penicillin, and ciprofloxacin can also be loaded into magnetic drug delivery carriers to achieve local treatment at infection sites. Furthermore, radioisotopes can be used for antitumor therapy by means of magnetic targeting carriers. Studies have proven that magnetic microspheres with β-emitter ^{90}Y can localize to tumor regions and reduce the tumor volume by half. Beyond utilizing their targeting function, iron oxide magnetic NPs can also be used to control drug release. Langer et al. confirmed that under AMF excitation, the drug release from polymer nanocomposites containing iron oxide beads could be increased [98]. Researches have demonstrated that the movement of iron oxide beads under AMF excitation generates "microcracks" in the polymer matrix, thereby enabling the release of drug molecules encapsulated in the nanocomposites.

10.4 Functions and Types

10.4.1 Imaging Function

By constructing a drug delivery system using materials with imaging functionality, it is possible to assist in the diagnosis of tumor conditions while delivering drugs, thereby achieving the integrated function of diagnosis and treatment.

10.4.1.1 Magnetic Resonance Imaging Type

MOFs with paramagnetic metal cations (such as Gd^{3+}, Mn^{2+}, and Fe^{3+}) as metal nodes can be used as contrast agents to enhance the imaging capability of MRI, among which Fe^{3+} with low toxicity is the optimal choice. The Fe ions in the structure of Fe-based MOFs can be regarded as Lewis acids, which can bind to a part of water molecules through strong coordination. Free water molecules can diffuse through the pores of MOFs and exchange with the bound water molecules, thereby affecting the proton relaxation of water and endowing Fe-based MOFs with MRI functionality. Wang et al. used MIL-101 type Fe-based MOFs modified with polylactic acid and polyethylene glycol to load dihydroartemisinin (DHA) for the treatment of cervical cancer in Figure 10.16 [99]. It was found that this MOF not only improved the efficacy of DHA but also exhibited excellent T2-weighted MRI capability. Due to the retention of the MOF at the tumor site by the EPR effect, the imaging duration could last up to 24 hours after injection.

Iron oxide nanoparticles (IONPs) are widely used as contrast agents for noninvasive diagnostic imaging combined with MRI. Currently, a variety of IONPs have been approved for clinical use as diagnostic and imaging agents. Despite the great success of IONPs in clinical research, most approved IONPs have been discontinued. At present, the most clinically applied IONP is Ferumoxytol, whose trade names in the United States and Europe are Feraheme® and Rienso®, respectively. Developed by AMAG Pharmaceuticals®, Ferumoxytol was initially researched as an

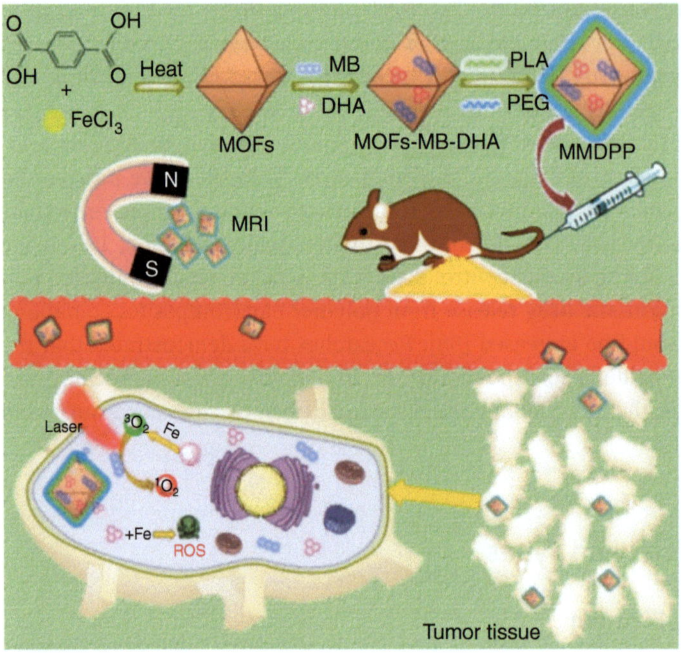

Figure 10.16 Synthetic process and therapeutic mechanism of MOFs-MB-DHA@PLA@PEG nanocomposites. *Source*: Reproduced from Ref. [99] / with permission of American Chemical Society.

MRI contrast agent. However, it was approved by the FDA in 2009 for the treatment of iron deficiency caused by chronic kidney disease and has undergone the most comprehensive clinical studies in both imaging and anemia. At the preclinical level, Ferumoxytol also has a variety of other applications, including radiolabeling of Ferumoxytol and using it as a carrier for chemotherapeutic drugs. Compared with the use of free drugs alone, the in vivo therapeutic effect of the Ferumoxytol drug-loading system is better [100].

10.4.1.2 Fluorescent Imaging Type

After being excited, electrons jump from the ground state to the excited state. However, electrons in the excited state are unstable and return to the ground state by releasing energy. The energy released in the form of fluorescence can be used for fluorescent imaging, which can reflect the state of drugs or the organism in real time. Fe-based MOFs can acquire fluorescent imaging capabilities through multiple strategies. First, Fe-based MOFs with certain configurations (such as MIL-53 type) inherently possess fluorescent imaging capabilities. Second, their excellent modifiability allows many Fe-based MOFs that originally have no fluorescence or weak fluorescence to be surface-modified with fluorescent reagents (e.g., 5-carboxyfluorescein) to gain fluorescent imaging capabilities. Finally, their strong drug-loading capacity enables them to directly load fluorescent agents, such as indocyanine green (ICG), for fluorescent imaging [101]. It is worth noting that the conversion between Fe^{3+} and Fe^{2+} can cause changes in the fluorescence intensity of Fe-based MOFs.

Xi et al. found that ascorbic acid can affect the fluorescence ability of Fe ions and BDC-NH_2 by interfering with the charge transfer between them [102]. When ascorbic acid reduces Fe^{3+} to Fe^{2+}, the charge transfer between Fe and the ligand is inhibited, thereby enhancing the fluorescence of BDC-NH_2. In the presence of AA (ascorbic acid), the fluorescence emission of Fe-MOFs is significantly enhanced, while the oxidase-like activity of Fe-MOFs is obviously inhibited. We further developed a dual-mode analysis platform by combining AA-induced dual-signal generation with a sandwich immunoassay linked to alkaline phosphatase (ALP), where AA is produced by the hydrolysis of ascorbic acid 2-phosphate (AAP) catalyzed by ALP. Therefore, using the cancer biomarker prostate-specific antigen (PSA) as a model analyte to evaluate the performance of the developed dual-mode colorimetric and fluorescent immunoassay. The proposed dual-mode immunoassay platform has many advantages: (i) both qualitative screening by the naked eye and quantitative determination via absorbance/fluorescence measurement can be achieved in a single test; (ii) compared with single-mode analysis, dual-signal reading offers higher detection accuracy and precision; (iii) there is no need for bioconjugation of Fe-MOFs at all, eliminating the cumbersome and time-consuming modification process; (iv) the AA-induced dual-signal generation system is well compatible with the mature ALP-linked immunoassay platform. This suggests that such fluorescence changes may be used to reflect the redox status of the tumor microenvironment, helping to adjust tumor treatment strategies in a timely manner.

10.4.1.3 Multimodal Imaging Type

In practical diagnosis and treatment, a single imaging system often finds it difficult to provide sufficiently comprehensive information. However, based on the nuclear magnetic resonance or fluorescence imaging of Fe-based MOFs, their multimodal imaging function can be achieved through methods such as functional modification or preparation of core–shell structured MOFs. Fan et al. used aminated MIL-101 type Fe-based MOFs to load the immunoadjuvant CpG-ODN (cytosine phosphate-guanosine oligonucleotide), and prepared ICG-CpG@MOF by combining its amino groups with ICG. This nanocarrier (ICG-CpG@MOF) can passively target tumors through the enhanced permeability and retention (EPR) effect, realizing multimodal imaging of tumors (fluorescence, photoacoustic, photothermal, and MR imaging), while achieving light-driven tumor therapy, including photodynamic and photothermal methods. After the NPs enter tumor cells through amplified macropinocytosis, when irradiated with an 808-nm laser, they can kill tumor cells by increasing reactive oxygen species (ROS) accumulation. In addition, due to the photothermal effect of ICG, ICG can directly destroy tumor cells and induce the release of tumor antigens. The released tumor-associated antigens, together with CpG, can induce the conversion of tumor cells from "cold" to "hot" by activating the immune system, thereby significantly enhancing tumor cytotoxicity, achieving a high cure rate with minimal side effects, and greatly improving the ability to diagnose the internal conditions of tumors.

10.4.2 Magnetic Hyperthermia Function

Magnetic Fe_3O_4-MOF combines the excellent responsiveness of Fe_3O_4 to alternating magnetic fields and the safety of MOFs. It is a new type of magnetic hyperthermia agent, suitable for the combined therapy of magnetic hyperthermia and other therapies such as chemotherapy. Xiang et al. used polyvinylpyrrolidone-modified Fe_3O_4-MOF to load DOX (doxorubicin) for the treatment of tongue squamous cell carcinoma in nude mice [103]. They found that under the action of an external magnetic field, Fe_3O_4-MOF could rapidly heat the tumor site to $42.4 \pm 0.3\,°C$ and showed a good magnetic hyperthermia-chemotherapy synergistic antitumor effect with DOX. The proposed intelligent platform demonstrates a synthesis strategy involving the following components: (i) novel porous Fe_3O_4@C-PVP@DOX nanocomposite exhibits efficient MRI contrast enhancement and excellent magnetic heating efficiency; (ii) it has good biocompatibility, favorable DOX loading capacity, and magnetically triggered on-demand drug release capability; (iii) it enables magnetically triggered on-demand combined hyperthermia and chemotherapy for precise cancer treatment. By administering it to CAL27 tumor-bearing mice, it was found that this platform effectively inhibited tumor growth through the synergistic effect of magnetic hyperthermia and chemotherapy. The mechanism indicates that the high proportion of cell apoptosis, tumor necrosis, and low microvessel density in tumors are the reasons for inhibiting tumor growth. Owing to its simple fabrication method, on-demand drug release, and magnetically triggered combined magnetic hyperthermia and chemotherapy, this intelligent nanoplatform holds great potential in the clinical precise treatment of cancer.

Artificially engineered nanozymes show great promise in improving the efficacy of cancer treatment. Herein, we have elaborately designed an advanced sandwich structure of cobalt ferrite-layered double hydroxides (Co—Fe LDHs), which can serve as a superior nanozyme with Co/Fe dual catalytic cycles and also act as a nanocarrier for loading gold NPs and maleimide (MA), so as to realize an "integrated" reactive oxygen species (ROS)-bursting tumor therapy strategy.

The Co—Fe LDHs@Au@MA nanozyme constructed by Li et al. can not only enhance the oxidative stress of tumors through cascade enzyme reactions, but also disrupt the supply of glutathione, further exacerbating the imbalance of the established redox homeostasis [104]. By adjusting the Co/Fe doping amount, the synergistic effect of multi-metallic enzyme-like activities is obtained, thereby optimizing the Co/Fe active sites. In addition, the mild photothermal effect of Au NPs helps to improve the catalytic efficiency of Co—Fe LDHs, and also initiates tumor cell apoptosis through the hyperthermia effect.

Hyperthermia selectively heats tumor tissues to 39–45 °C to inhibit and kill tumors, and is characterized by high safety. In recent years, studies have also found that hyperthermia can produce good synergistic effects with chemotherapy, gene therapy, immunotherapy, and other therapies. However, the insufficient efficiency and precision of traditional heating methods have led to the slow development of hyperthermia and limited therapeutic effects. In contrast, magnetic hyperthermia utilizes the targeting ability of magnetic nanoagents to tumors and the high penetrability of alternating magnetic fields to the human body to generate inductive heat directly at the tumor site. It has the ability to effectively kill deep-seated tumors and is regarded as a highly promising tumor treatment method [105].

10.4.3 Phototherapy Function

Tumor optical therapy includes photothermal therapy (PTT) and photodynamic therapy (PDT). It is a treatment that kills tumors through light-excited photothermal effects or reactive oxygen species (ROS) and has the advantages of high efficiency and low toxicity. Currently, it is a research hotspot in antitumor therapy. The core of the phototherapeutic nano-drug delivery system lies in selecting agents and photosensitizers. Porphyrin macromolecules not only possess photothermal properties but also have natural coordination ability, which enables them to form coordination bonds with iron. Wang et al. utilized this characteristic to prepare Fe-porphyrin MOFs [106]. Under near-infrared light at $1\,\text{W}\,\text{cm}^{-2}$, $50\,\mu\text{g}\,\text{mL}^{-1}$ of Fe-porphyrin MOFs could generate a high temperature of 48.6 °C, exhibiting a tumor photothermal therapy effect. Meanwhile, the Fe atoms in them could promote the production of reactive oxygen species (ROS) through changes in spin states, thereby exerting a photodynamic therapy effect. The combination of the two effects could significantly regress cervical cancer in nude mice without obvious organ damage or inflammatory reactions.

Phototherapy has been explored for a century and has made significant progress over the past few decades with the development of Photoresponsive Drug Delivery Systems (PDDS). Meanwhile, it has shown great potential for drug delivery through precise temporal control, site-specific targeting, and controlled dosage. However, it should be acknowledged that the clinical application of PDDS is still in its infancy. Three major challenges in the field of light-responsive drug delivery have been highlighted, namely accumulation, precision, and penetration, and the strategies to address these challenges have been described in detail. The accumulation of PDDS can be enhanced through a series of strategies, which aim to provide greater distribution in target tissues and more precise targeting of cells or organelles. This facilitates safer and more efficient delivery of active pharmaceutical ingredients (APIs) and real-time control of their biological activities. Currently, the main challenges in this field lie in the effectiveness of targeting strategies and light-responsive strategies in clinical applications. Like many other Drug Delivery Systems (DDS), well-designed PDDS often fail to translate their promising in vitro performance into the desired clinical outcomes. The complexity of the in vivo environment and extensive interactions with various physiological factors may be crucial but are often overlooked. Future research efforts should prioritize incorporating a broader range of in vivo factors into the design considerations of PDDS [107].

Imaging-guided PDDS improves the precision of light irradiation and maximizes the advantages of focused light stimulation through optical or non-optical imaging guidance. This is of great significance for the precise treatment and integrated diagnosis and treatment of many diseases. The main goal is to achieve better imaging accuracy and resolution. The integration of multiple imaging modalities has the potential to synergistically leverage the advantages of different technologies, thereby compensating for the inherent limitations of any single method. In addition, the adoption of emerging imaging theories at the emerging stage can also yield significant benefits.

To overcome the limitations of near-infrared light application in vivo, new photosensitizers (PS) and light-responsive strategies have been developed. As the main driving force for the advancement of PDDS, developing more effective photon energy transfer mechanisms is the key in this field. Future research work targeting medical applications must pay great attention to biocompatibility to reduce potential risks. In addition, by leveraging the inherent deep tissue penetration capabilities of ultrasound and X-rays, their strategic combination with phototherapy and PDDS is a promising approach to overcome the challenge of limited penetration depth.

10.4.4 Intelligent Responsive Release Function

Intelligent responsive nanodrugs can respond to the tumor microenvironment (such as acidic and reductive microenvironments) or external physical stimuli (such as magnetic fields, light irradiation, and ultrasound), thereby controlling drug release. This not only enhances the therapeutic effect but also reduces adverse drug reactions.

10.4.4.1 Tumor Microenvironment-Responsive Type

Tumor microenvironment-responsive nanodrugs mainly utilize the characteristics of the tumor microenvironment, such as acidity and reducibility, to induce changes in the state or structural disintegration of nanocarriers, thereby achieving the goal of targeted drug release in tumor tissues. For Fe-based MOFs constructed with Fe^{3+}, glutathione (GSH) in the tumor microenvironment can change the valence state of Fe ions in them, leading to the disintegration of the MOF structure and enabling responsive release to the tumor microenvironment. The acidic microenvironment of tumors can also accelerate the disintegration of Fe-based MOFs, and modifying their surface with a pH-responsive coating is an excellent strategy. Yang et al. modified Fe-based MOFs with hydroxyapatite (HAp) to obtain pH-responsive Fe-based MOFs, which release the loaded doxorubicin (DOX) in the acidic tumor microenvironment for inhibiting HeLa cells, and can avoid adverse reactions caused by drug release in normal tissues in Figure 10.17 [108].

10.4.4.2 Physical Stimulus-Responsive Type

In addition to responding to the tumor microenvironment, external stimuli such as ultrasound, magnetic fields, and excitation light can also regulate changes in nanocarriers, thereby controlling the drug release site and release timing.

Ultrasound can induce cavitation and collapse of Fe-based MOFs, thereby releasing the drugs encapsulated within them. When Fe-NDC type Fe-based MOFs were loaded with calcein and doxorubicin (DOX), respectively, it was found that without ultrasonic stimulation, the drug release was slow (17.8% released within 10 minutes), while under low-frequency ultrasound intervention, the release rate increased significantly (95.2% released within 10 minutes). Magnetically triggered drug release is another form of physical stimulus-responsive release. The

Figure 10.17 (a–c) SEM and (d–g) TEM images of as-prepared (a and d) Fe_3O_4, (b and e) Fe_3O_4@Fe-MOF, and (c, f, and g) Fe_3O_4@Fe-MOF@HAp and the corresponding elemental maps for Fe (h), O (i), P (j), and Ca (k) signals. The inset in (f) presents the high-resolution TEM image of HAp. *Source*: Reproduced from Ref. [108] / with permission of The Royal Society of Chemistry.

DOX-loaded Fe_3O_4-MOF prepared by Xiang et al. possesses magnetic field responsiveness [103]. Tests showed that the drug release amount of this MOF was only 4% within 6 hours without magnetic field intervention; however, after magnetic field intervention, 7.3% of the drug could be released within 15 minutes, and a cumulative release of 24% could be achieved after 3 rounds of magnetic field intervention.

10.4.4.3 Multiple Stimulus-Responsive Type

Swathi Sudhakar synthesized cobalt-nickel metal–organic frameworks (CNMs) and loaded them with cisplatin a combination of cisplatin (Cis), doxorubicin (Dox), and 5-fluorouracil (5-Fu) (designated as MD-CNM) to develop a novel multidrug delivery system [109]. MD-CNM exhibits high biocompatibility, with an effective drug-loading capacity of $95.44 \pm 4.05\%$ and a sustained release duration of over 96 hours. Furthermore, MD-CNM shows potent cytotoxicity against MDA-MB-231 breast cancer cells, with a half-maximal inhibitory concentration (IC_{50}) of 461 nM, which is 2 to 5 times lower than the IC_{50} values of CNMs loaded with individual drugs (Cis, Dox, and 5-Fu), indicating enhanced therapeutic efficacy. In addition, flow cytometry analysis reveals that MD-CNM induces necrosis (77.59%) and late apoptosis (12.37%), and causes cell cycle arrest at the G_0/G_1 phase, which further confirms its anticancer potential.

Although metal–organic frameworks (MOFs), as a new type of anticancer drug carrier, possess the aforementioned advantages, their research in anticancer drug delivery has just begun, and there are still some shortcomings.

The first issue is that the existing preparation technologies for iron-based MOFs are not yet suitable for industrial production. The solvothermal process requires high temperature and pressure, resulting in high energy consumption and low efficiency. Moreover, it involves the use of highly toxic reagents such as hydrofluoric acid, leading to poor environmental friendliness. Although methods like microwave-assisted synthesis have advantages such as high efficiency and fast speed, they are still limited to laboratory use due to the strict requirements for equipment precision and operating conditions. Therefore, currently, commercialized iron-based MOF products are scarce and expensive, which restricts their practical application. Hence, developing a green, efficient, and eco-friendly iron-based MOF material is a crucial approach to realize its commercialization.

Second, the pores of iron-based MOF materials are vital for their drug-loading capacity. However, the current preparation of iron-based MOFs mainly relies on organic solvents, which brings about problems such as the easy retention of organic solvents in the pore channels and the generation of biological toxicity. On the other hand, when water is used as the medium, issues like low solubility, low ligand selectivity, and low preparation efficiency arise. Additionally, water has high surface tension, which creates pressure inside the pores before drug loading, causing pore collapse and thus a decrease in drug-loading capacity. This project intends to prepare organic–inorganic hybrid materials with non-toxicity, easy volatility, and low surface tension by selecting appropriate solvents and activation modes. Furthermore, a method combining freeze-drying with supercritical CO_2 fluid will be adopted. Leveraging the sublimation effect and the characteristic of supercritical CO_2 fluid (which exerts low pressure inside the pores), efficient loading of porous materials will be achieved.

Third, most of the existing iron-based MOFs are used in fields such as adsorption and catalysis. When selecting these materials, their chemical properties are the primary consideration, which inevitably causes certain irritation to the human body. Therefore, from a pharmaceutical perspective, it is necessary to select more appropriate ligands—especially using compounds with specific functions or antitumor efficacy as organic ligands. This can better exert the synergistic effect between drugs and play an auxiliary role of drugs. Iron-based MOFs have a variety of structural types. Due to differences in structure and other aspects among different types of iron-based MOFs, research on their in vitro and in vivo metabolic processes, biosafety, and biosecurity is still insufficient. Systematic research in this field is urgently needed.

References

1 Qu, J., Chai, Y., and Yang, S.X. (2009). A real-time de-noising algorithm for e-noses in a wireless sensor network. *Sensors (Basel)* 9: 895–908.

2 Refaat, T.F., Ismail, S., Koch, G.J. et al. (2011). Backscatter Lidar validation for atmospheric differential absorption Lidar applications. *IEEE Transactions on Geoscience and Remote Sensing* 49: 572–580.

3 Fine, G.F., Cavanagh, L.M., Afonja, A., and Binions, R. (2010). Metal oxide semi-conductor gas sensors in environmental monitoring. *Sensors (Basel)* 10: 5469–5502.

4 Dey, A. (2018). Semiconductor metal oxide gas sensors: a review. *Materials Science and Engineering B* 229: 206–217.

5 Mirzaei, A., Leonardi, S.G., and Neri, G. (2016). Detection of hazardous volatile organic compounds (VOCs) by metal oxide nanostructures-based gas sensors: a review. *Ceramics International* 42: 15119–15141.

6 Peng, G., Tisch, U., Adams, O. et al. (2009). Diagnosing lung cancer in exhaled breath using gold nanoparticles. *Nature Nanotechnology* 4: 669–673.

7 Kawamura, K., Vestergaard, M.d., Ishiyama, M. et al. (2006). Development of a novel hand-held toluene gas sensor: possible use in the prevention and control of sick building syndrome. *Measurement* 39: 490–496.

8 Zeng, Y., Zhang, T., Wang, L. et al. (2009). Enhanced toluene sensing characteristics of TiO_2-doped flowerlike ZnO nanostructures. *Sensors and Actuators B: Chemical* 140: 73–78.

9 Gao, H., Zhao, L., Wang, L. et al. (2018). Ultrasensitive and low detection limit of toluene gas sensor based on SnO_2-decorated NiO nanostructure. *Sensors and Actuators B: Chemical* 255: 3505–3515.

10 Wang, W., Li, Z., Zheng, W. et al. (2010). Cr_2O_3-sensitized ZnO electrospun nanofibers based ethanol detectors. *Sensors and Actuators B: Chemical* 143: 754–758.

11 Nagarjuna, Y. and Hsiao, Y.-J. (2020). Au doping ZnO nanosheets sensing properties of ethanol gas prepared on MEMS device. *Coatings* 10: 945.

12 Paul, J. and Philip, J. (2020). Inter-digital capacitive ethanol sensor coated with cobalt ferrite nano composite as gas sensing material. *Materials Today: Proceedings* 25: 148–150.

13 Zhao, S., Shen, Y., Yan, X. et al. (2019). Complex-surfactant-assisted hydrothermal synthesis of one-dimensional ZnO nanorods for high-performance ethanol gas sensor. *Sensors and Actuators B: Chemical* 286: 501–511.

14 Usman Ali, S.M., Nur, O., Willander, M., and Danielsson, B. (2010). A fast and sensitive potentiometric glucose microsensor based on glucose oxidase coated ZnO nanowires grown on a thin silver wire. *Sensors and Actuators B: Chemical* 145: 869–874.

15 Bunkoed, O., Davis, F., Kanatharana, P. et al. (2010). Sol–gel based sensor for selective formaldehyde determination. *Analytica Chimica Acta* 659: 251–257.

16 Wang, X., Ding, B., Sun, M. et al. (2010). Nanofibrous polyethyleneimine membranes as sensitive coatings for quartz crystal microbalance-based formaldehyde sensors. *Sensors and Actuators B: Chemical* 144: 11–17.

17 Xu, Y., Tian, X., Fan, Y., and Sun, Y. (2020). A formaldehyde gas sensor with improved gas response and sub-ppm level detection limit based on $NiO/NiFe_2O_4$ composite nanotetrahedrons. *Sensors and Actuators B: Chemical* 309: 127719.

18 Zhang, R., Ma, S.Y., Zhang, J.L. et al. (2020). Enhanced formaldehyde gas sensing performance based on Bi doped Zn_2SnO_4/SnO_2 porous nanospheres. *Journal of Alloys and Compounds* 828: 154408.

19 Wang, Z., Hou, C., De, Q. et al. (2018). One-step synthesis of Co-doped In_2O_3 nanorods for high response of formaldehyde sensor at low temperature. *ACS Sensors* 3: 468–475.

20 Zhang, R., Cao, S., Zhou, T.T. et al. (2020). Rational design and tunable synthesis of Co_3O_4 nanoparticle-incorporating into In_2O_3 one-dimensional ribbon as effective sensing material for gas detection. *Sensors and Actuators B: Chemical* 310: 127695.

21 Kucharczyk, B., Okal, J., Tylus, W. et al. (2019). The effect of the calcination temperature of $LaFeO_3$ precursors on the properties and catalytic activity of perovskite in methane oxidation. *Ceramics International* 45: 2779–2788.

22 Jing, Z., Li, H., and Jiang, Z. (2019). The chemical interaction of support and active phase in sintering resistant $La_{0.8}Ca_{0.2}FeO_3$ perovskite catalysts. *Fuel* 243: 322–331.

23 Wang, G., Sun, J., Zhang, W. et al. (2009). Simultaneous determination of dopamine, uric acid and ascorbic acid with $LaFeO_3$ nanoparticles modified electrode. *Microchimica Acta* 164: 357–362.

24 Chu, X., Zhou, S., Zhang, W., and Shui, H. (2009). Trimethylamine sensing properties of nano-$LaFeO_3$ prepared using solid-state reaction in the presence of PEG400. *Materials Science and Engineering B* 164: 65–69.

25 Xiang, J., Chen, X., Zhang, X. et al. (2018). Preparation and characterization of Ba-doped $LaFeO_3$ nanofibers by electrospinning and their ethanol sensing properties. *Materials Chemistry and Physics* 213: 122–129.

26 Yang, K., Ma, J., Qiao, X. et al. (2020). Hierarchical porous $LaFeO_3$ nanostructure for efficient trace detection of formaldehyde. *Sensors and Actuators B: Chemical* 313: 128022.

27 Zhang, Y., Zhang, J., Zhao, J. et al. (2017). Ag-$LaFeO_3$ fibers, spheres, and cages for ultrasensitive detection of formaldehyde at low operating temperatures. *Physical Chemistry Chemical Physics* 19: 6973–6980.

28 Li, X., Zhang, N., Liu, C. et al. (2021). Enhanced gas sensing properties for formaldehyde based on ZnO/Zn_2SnO_4 composites from one-step hydrothermal synthesis. *Journal of Alloys and Compounds* 850: 156606.

29 Wu, Z., Pei, K., Xing, L. et al. (2019). Enhanced microwave absorption performance from magnetic coupling of magnetic nanoparticles suspended within hierarchically tubular composite. *Advanced Functional Materials* 29: 1901448.

30 Yang, W., Jiang, B., Liu, Z. et al. (2021). Magnetic coupling engineered porous dielectric carbon within ultralow filler loading toward tunable and high-performance microwave absorption. *Journal of Materials Science & Technology* 70: 214–223.

31 Ning, M., Man, Q., Tan, G. et al. (2020). Ultrathin MoS_2 nanosheets encapsulated in hollow carbon spheres: a case of a dielectric absorber with optimized impedance for efficient microwave absorption. *ACS Applied Materials & Interfaces* 12: 20785–20796.

32 Wang, G., Ong, S.J.H., Zhao, Y. et al. (2020). Integrated multifunctional macrostructures for electromagnetic wave absorption and shielding. *Journal of Materials Chemistry A* 8: 24368–24387.

33 Li, X., Yu, L., Zhao, W. et al. (2020). Prism-shaped hollow carbon decorated with polyaniline for microwave absorption. *Chemical Engineering Journal* 379: 122393.

34 Zhou, W., Long, L., and Li, Y. (2019). Mechanical and electromagnetic wave absorption properties of Cf-Si_3N_4 ceramics with PyC/SiC interphases. *Journal Of Materials Science & Technology* 35: 2809–2813.

35 Zhou, M., Gu, W., Wang, G. et al. (2020). Sustainable wood-based composites for microwave absorption and electromagnetic interference shielding. *Journal of Materials Chemistry A* 8: 24267–24283.

36 Liao, H., Li, D., Wang, B. et al. (2021). Microporous $Co_{1-x}Fe_x$@C nanoparticles: strong wideband microwave absorbers for reflection loss less than -20 dB. *Journal of Alloys and Compounds* 856: 158175.

37 Thomassin, J.-M., Lou, X., Pagnoulle, C. et al. (2007). Multiwalled carbon nanotube/poly(ε-caprolactone) nanocomposites with exceptional electromagnetic interference shielding properties. *The Journal of Physical Chemistry C* 111: 11186–11192.

38 Liu, X., Ma, Y., Zhang, Q. et al. (2018). Facile synthesis of Fe_3O_4/C composites for broadband microwave absorption properties. *Applied Surface Science* 445: 82–88.

39 Zhao, B., Shao, G., Fan, B. et al. (2015). Investigation of the electromagnetic absorption properties of Ni@TiO_2 and Ni@SiO_2 composite microspheres with core-shell structure. *Physical Chemistry Chemical Physics* 17: 2531–2539.

40 Green, M. and Chen, X. (2019). Recent progress of nanomaterials for microwave absorption. *Journal of Materiomics* 5: 503–541.

41 Liu, D., Du, Y., Wang, F. et al. (2020). MOFs-derived multi-chamber carbon microspheres with enhanced microwave absorption. *Carbon* 157: 478–485.

42 Zhang, M., Lin, H., Ding, S. et al. (2019). Net-like SiC@C coaxial nanocable towards superior lightweight and broadband microwave absorber. *Composites Part B: Engineering* 179: 107525.

43 Michielssen, E., Sajer, J.M., Ranjithan, S., and Mittra, R. (1993). Design of lightweight, broad-band microwave absorbers using genetic algorithms. *IEEE Transactions on Microwave Theory and Techniques* 41: 1024–1031.

44 Wang, P., Wang, X., Qiao, L. et al. (2018). High-frequency magnetic properties and microwave absorption performance of oxidized Pr_2Co_{17} flakes/epoxy composite in x-band. *Journal of Magnetism and Magnetic Materials* 468: 193–199.

45 Huo, J., Wang, L., and Yu, H. (2009). Polymeric nanocomposites for electromagnetic wave absorption. *Journal of Materials Science* 44: 3917–3927.

46 Guo, Z., Park, S., Hahn, H.T. et al. (2007). Magnetic and electromagnetic evaluation of the magnetic nanoparticle filled polyurethane nanocomposites. *Journal of Applied Physics* 101: 09M511.

47 Wang, L., Xiong, J., Cheng, B. et al. (2022). Cascadable in-memory computing based on symmetric writing and readout. *Science Advances* 8: eabq6833.

48 Wang, Y., Wang, H., Ye, J. et al. (2020). Magnetic CoFe alloy@C nanocomposites derived from ZnCo-MOF for electromagnetic wave absorption. *Chemical Engineering Journal* 383: 123096.

49 Huang, M., Wang, L., Pei, K. et al. (2020). Multidimension-controllable synthesis of MOF-derived Co@N-doped carbon composite with magnetic-dielectric synergy toward strong microwave absorption. *Small* 16: e2000158.

50 Zhu, T., Chang, S., Song, Y.-F. et al. (2019). PVP-encapsulated $CoFe_2O_4$/rGO composites with controllable electromagnetic wave absorption performance. *Chemical Engineering Journal* 373: 755–766.

51 Shang, T., Lu, Q., Chao, L. et al. (2018). Effects of ordered mesoporous structure and La-doping on the microwave absorbing properties of $CoFe_2O_4$. *Applied Surface Science* 434: 234–242.

52 Wang, H., Wang, C., Xiong, Y. et al. (2017). Solvothermal fabrication and growth behavior study of spherical $MnFe_2O_4$ through a bottom-up method on wood substrate with effective microwave absorption. *RSC Advances* 7: 24764–24770.

53 Wu, G., Zhang, H., Luo, X. et al. (2019). Investigation and optimization of $Fe/ZnFe_2O_4$ as a Wide-band electromagnetic absorber. *Journal of Colloid and Interface Science* 536: 548–555.

54 Sun, G., Dong, B., Cao, M. et al. (2011). Hierarchical dendrite-like magnetic materials of Fe_3O_4, γ-Fe_2O_3, and Fe with high performance of microwave absorption. *Chemistry of Materials* 23: 1587–1593.

55 Xie, X., Ni, C., Lin, Z. et al. (2020). Phase and morphology evolution of high dielectric CoO/Co_3O_4 particles with Co_3O_4 nanoneedles on surface for excellent microwave absorption application. *Chemical Engineering Journal* 396: 125205.

56 Zhang, L., Dai, P., Yu, X. et al. (2015). The preparation of Fe_3O_4 cube-like nanoparticles via the ethanol reduction of α-Fe_2O_3 and the study of its electromagnetic wave absorption. *Applied Surface Science* 359: 723–728.

57 Liu, T., Zhou, P.H., Xie, J.L., and Deng, L.J. (2011). The hierarchical architecture effect on the microwave absorption properties of cobalt composites. *Journal of Applied Physics* 110: 033918.

58 Liu, T., Pang, Y., Xie, X. et al. (2016). Synthesis of microporous Ni/NiO nanoparticles with enhanced microwave absorption properties. *Journal of Alloys and Compounds* 667: 287–296.

59 Zhang, N., Huang, Y., Wang, M. et al. (2019). Design and microwave absorption properties of thistle-like CoNi enveloped in dielectric Ag decorated graphene composites. *Journal of Colloid and Interface Science* 534: 110–121.

60 Zhao, B., Zhang, X., Deng, J. et al. (2018). A novel sponge-like 2D Ni/derivative heterostructure to strengthen microwave absorption performance. *Physical Chemistry Chemical Physics* 20: 28623–28633.

61 Liu, T., Pang, Y., Kikuchi, H. et al. (2015). Superparamagnetic property and high microwave absorption performance of FeAl@$(Al, Fe)_2O_3$ nanoparticles induced by surface oxidation. *Journal of Materials Chemistry C* 3: 6232–6239.

62 Lv, H., Ji, G., Wang, M. et al. (2014). FeCo/ZnO composites with enhancing microwave absorbing properties: effect of hydrothermal temperature and time. *RSC Advances* 4: 57529–57533.

63 Qiao, J., Xu, D., Lv, L. et al. (2018). Self-assembled ZnO/Co hybrid nanotubes prepared by electrospinning for lightweight and high-performance electromagnetic wave absorption. *ACS Applied Nano Materials* 1: 5297–5306.

64 Deng, J., Wang, Q., Zhou, Y. et al. (2017). Facile design of a ZnO nanorod–Ni core–shell composite with dual peaks to tune its microwave absorption properties. *RSC Advances* 7: 9294–9302.

65 Wang, H., Dai, Y.Y., Geng, D.Y. et al. (2015). Co_xNi_{100-x} nanoparticles encapsulated by curved graphite layers: controlled in situ metal-catalytic preparation and broadband microwave absorption. *Nanoscale* 7: 17312–17319.

66 Li, Y., Wang, J., Liu, R. et al. (2017). Dependence of gigahertz microwave absorption on the mass fraction of Co@C nanocapsules in composite. *Journal of Alloys and Compounds* 724: 1023–1029.

67 Liao, H., Li, D., Zhou, C., and Liu, T. (2019). Microporous Co/rGO nanocomposites: strong and broadband microwave absorber with well-matched dielectric and magnetic loss. *Journal of Alloys and Compounds* 782: 556–565.

68 Liu, Y., Chen, Z., Xie, W. et al. (2019). Enhanced microwave absorption performance of porous and hollow CoNi@C microspheres with controlled component and morphology. *Journal of Alloys and Compounds* 809: 151837.

69 Liu, D., Du, Y., Xu, P. et al. (2019). Waxberry-like hierarchical Ni@C microspheres with high-performance microwave absorption. *Journal of Materials Chemistry C* 7: 5037–5046.

70 Ruan, C., Li, Z., Zhang, D. et al. (2020). A scalable chemical vapor deposition synthesis of high purity hollow carbon onions. *Carbon* 161: 622–628.

71 Levchenko, I., Ostrikov, K., Keidar, M., and Xu, S. (2005). Microscopic ion fluxes in plasma-aided nanofabrication of ordered carbon nanotip structures. *Journal of Applied Physics* 98: 064304.

72 Kuang, D., Hou, L., Wang, S. et al. (2019). Large-scale synthesis and outstanding microwave absorption properties of carbon nanotubes coated by extremely small FeCo-C core-shell nanoparticles. *Carbon* 153: 52–61.

73 Xia, Q., Wang, H., Huang, B. et al. (2019). State-of-the-art advances and challenges of iron-based metal organic frameworks from attractive features, synthesis to multifunctional applications. *Small* 15: 1803088.

74 Ma, L., Jiang, F., Fan, X. et al. (2020). Metal–organic-framework-engineered enzyme-mimetic catalysts. *Advanced Materials* 32: e2003065.

75 Zhou, Y., Niu, B., Wu, B. et al. (2020). A homogenous nanoporous pulmonary drug delivery system based on metal-organic frameworks with fine aerosolization performance and good compatibility. *Acta Pharmaceutica Sinica B* 10: 2404–2416.

76 Guo, H., Li, X., Qu, D., and Chen, Y. (2022). Research progress on Fe-based metal-organic frameworks in antitumor drug delivery. *Acta Pharmaceutica Sinica* 57: 1252–1262.

77 Lu, Z., Liu, J., Zhang, X. et al. (2020). Node-accessible zirconium MOFs. *Journal of the American Chemical Society* 142: 21110–21121.

78 Wang, Y., Wu, W., Liu, J. et al. (2019). Cancer-cell-activated photodynamic therapy assisted by Cu(II)-based metal–organic framework. *ACS Nano* 13: 6879–6890.

79 Liu, X., Liang, T., Zhang, R. et al. (2021). Iron-based metal–organic frameworks in drug delivery and biomedicine. *ACS Applied Materials & Interfaces* 13: 9643–9655.

80 Zhu, M., Xu, S., Li, G. et al. (2025). Development of a high-efficacy and low-toxicity cobalt(II) agent for targeting inhibition of tumor growth through mitochondrial damage-mediated chemotherapy and immunotherapy. *Journal of Medicinal Chemistry* 68: 13113–13126.

81 Chen, Q., Xu, M., Zheng, W. et al. (2017). Se/Ru-decorated porous metal–organic framework nanoparticles for the delivery of pooled siRNAs to reversing multidrug resistance in Taxol-resistant breast cancer cells. *ACS Applied Materials & Interfaces* 9: 6712–6724.

82 Dong, W., Liu, X., Shi, W., and Huang, Y. (2015). Metal–organic framework MIL-53(Fe): facile microwave-assisted synthesis and use as a highly active peroxidase mimetic for glucose biosensing. *RSC Advances* 5: 17451–17457.

83 Ma, M., Bétard, A., Weber, I. et al. (2013). Iron-based metal–organic frameworks MIL-88B and NH_2-MIL-88B: high quality microwave synthesis and solvent-induced lattice "breathing". *Crystal Growth & Design* 13: 2286–2291.

84 Ahmed, I., Jeon, J., Khan, N.A., and Jhung, S.H. (2012). Synthesis of a metal–organic framework, iron-benezenetricarboxylate, from dry gels in the absence of acid and salt. *Crystal Growth & Design* 12: 5878–5881.

85 Gordon, J., Kazemian, H., and Rohani, S. (2015). MIL-53(Fe), MIL-101, and SBA-15 porous materials: potential platforms for drug delivery. *Materials Science and Engineering: C* 47: 172–179.

86 Wu, M.-X. and Yang, Y.-W. (2017). Metal–organic framework (MOF)-based drug/cargo delivery and cancer therapy. *Advanced Materials* 29: 1606134.

87 Qiao, C., Zhang, R., Wang, Y. et al. (2020). Rabies virus-inspired metal–organic frameworks (MOFs) for targeted imaging and chemotherapy of glioma. *Angewandte Chemie International Edition* 59: 16982–16988.

88 Pham, M.-H., Vuong, G.-T., Vu, A.-T., and Do, T.-O. (2011). Novel route to size-controlled Fe–MIL-88B–NH_2 metal–organic framework nanocrystals. *Langmuir* 27: 15261–15267.

89 Horcajada, P., Chalati, T., Serre, C. et al. (2010). Porous metal–organic-framework nanoscale carriers as a potential platform for drug delivery and imaging. *Nature Materials* 9: 172–178.

90 Gao, X., Cui, R., Song, L., and Liu, Z. (2019). Hollow structural metal-organic frameworks exhibit high drug loading capacity, targeted delivery and magnetic resonance/optical multimodal imaging. *Dalton Transactions* 48: 17291–17297.

91 Liu, R., Xiao, W., Hu, C. et al. (2018). Theranostic size-reducible and no donor conjugated gold nanocluster fabricated hyaluronic acid nanoparticle with optimal size for combinational treatment of breast cancer and lung metastasis. *Journal of Controlled Release* 278: 127–139.

92 Kumar, P., Anand, B., Tsang, Y.F. et al. (2019). Regeneration, degradation, and toxicity effect of MOFs: opportunities and challenges. *Environmental Research* 176: 108488.

93 Ploetz, E., Zimpel, A., Cauda, V. et al. (2020). Metal–organic framework nanoparticles induce pyroptosis in cells controlled by the extracellular pH. *Advanced Materials* 32: 1907267.

94 Gilchrist, R.K., Medal, R., Shorey, W.D. et al. (1957). Selective inductive heating of lymph nodes. *Annals of Surgery* 146: 596–606.

95 Meyers, P.H., Cronic, F., and Nice, C.M.J.A.J.R. Jr., (1963). Experimental approach in the use and magnetic control of metallic IRON particles in the lymphatic and vascular system of dogs as a contrast and isotropic agent. *The American Journal of Roentgenology, Radium Therapy, and Nuclear Medicine* 90: 1068–1077.

96 Shah, B.P., Pasquale, N., De, G. et al. (2014). Core–shell nanoparticle-based peptide therapeutics and combined hyperthermia for enhanced cancer cell apoptosis. *ACS Nano* 8: 9379–9387.

97 Chen, M.-L., He, Y.-J., Chen, X.-W., and Wang, J.-H. (2012). Quantum dots conjugated with Fe_3O_4-filled carbon nanotubes for cancer-targeted imaging and magnetically guided drug delivery. *Langmuir* 28: 16469–16476.

98 Chen, H. and Langer, R. (1997). Magnetically-responsive polymerized liposomes as potential oral delivery vehicles. *Pharmaceutical Research* 14: 537–540.

99 Wang, C., Jia, X., Zhen, W. et al. (2019). Small-sized MOF-constructed multifunctional diagnosis and therapy platform for tumor. *ACS Biomaterials Science & Engineering* 5: 4435–4441.

100 Anselmo, A.C. and Mitragotri, S. (2015). A review of clinical translation of inorganic nanoparticles. *The AAPS Journal* 17: 1041–1054.

101 Cai, W., Gao, H., Chu, C. et al. (2017). Engineering phototheranostic nanoscale metal–organic frameworks for multimodal imaging-guided cancer therapy. *ACS Applied Materials & Interfaces* 9: 2040–2051.

102 Xie, W., Tian, M., Luo, X. et al. (2020). A dual-mode fluorescent and colorimetric immunoassay based on in situ ascorbic acid-induced signal generation from metal-organic frameworks. *Sensors and Actuators B: Chemical* 302: 127180.

103 Xiang, Z., Qi, Y., Lu, Y. et al. (2020). MOF-derived novel porous Fe_3O_4@C nanocomposites as smart nanomedical platforms for combined cancer therapy: magnetic-triggered synergistic hyperthermia and chemotherapy. *Journal of Materials Chemistry B* 8: 8671–8683.

104 Li, S., Zhao, X., Ding, H. et al. (2023). Advanced cobalt-ferrite layered double hydroxides sandwich-structured nanozymes for ROS-bloomed tumor therapy. *Chemical Engineering Journal* 473: 145414.

105 Kumar, C.S.S.R. and Mohammad, F. (2011). Magnetic nanomaterials for hyperthermia-based therapy and controlled drug delivery. *Advanced Drug Delivery Reviews* 63: 789–808.

106 Wang, L., Qu, X., Zhao, Y. et al. (2019). Exploiting single atom iron centers in a porphyrin-like MOF for efficient cancer phototherapy. *ACS Applied Materials & Interfaces* 11: 35228–35237.

107 Yang, Y., Long, K., Chu, Y. et al. (2024). Photoresponsive drug delivery systems: challenges and progress. *Advanced Functional Materials* 34: 2402975.

108 Yang, Y., Xia, F., Yang, Y. et al. (2017). Litchi-like Fe_3O_4@Fe-MOF capped with HAp gatekeepers for pH-triggered drug release and anticancer effect. *Journal of Materials Chemistry B* 5: 8600–8606.

109 Basha, S.S., Sasmal, A., Ariraman, S.B.,.M. et al. (2025). Cobalt–nickel metal–organic frameworks (CNMs) as drug delivery agents for triple-negative breast cancer. *Nanoscale Advances* 7: 5361–5376.

11

Challenges and Prospects

Iron-based metal compound nanomaterials (including Fe, Co, and Ni-based materials and their composites) possess significant application value in batteries, catalysis, environmental remediation, biomedicine, sensing, and other fields due to their unique electronic structures, magnetic properties, and electrochemical performances. They are currently a research focus in materials science. However, several key challenges persist in the development of these materials. In the preparation process, common issues include poor process controllability, difficulty in constructing complex structures, challenges in balancing scale-up with refinement, and insufficient green credentials and economic viability. At the microstructural level, challenges involve the difficulty in analyzing complex crystal structures, quantitative regulation of defects and interfaces, tracking dynamic processes, and obstacles in integrating data from multiple characterization techniques, which restrict the understanding of structure–performance relationships. In the battery field, materials are limited by intrinsic properties, system-specific issues, and interface compatibility. Supercapacitors face material performance bottlenecks, interface optimization challenges, and cost pressures during scale-up. Electrocatalysis and photocatalysis encounter problems such as the difficulty in synergizing activity with stability, poor selectivity, low carrier separation efficiency, and narrow spectral response, respectively. The field of organic oxidation and degradation is challenged by the selectivity of directional conversion, the thoroughness of degradation, and the recyclability of catalysts. Additionally, the sensor, microwave absorption, and drug delivery fields face unique pain points, including performance interference, uneven absorption performance, and insufficient biosecurity, respectively.

This chapter will systematically analyze the core challenges from preparation to application mentioned above, anticipate breakthrough directions in combination with cutting-edge technological trends such as intelligent preparation, precise characterization, and green synthesis, and provide references for the optimization, upgrading, and industrial application of iron-based metal compound nanomaterials.

Iron-Group Metal Compound Nanomaterials: Preparation, Characterization, Structure, and Applications, First Edition. Jinlong Zheng.

11.1 Challenges and Prospects Related to Preparation Methods

11.1.1 Current Core Challenges

11.1.1.1 Poor Controllability of Preparation Process and Low Product Uniformity

The synthesis of iron-based nanomaterials is highly sensitive to key parameters such as temperature, pH value, and precursor concentration. Minor fluctuations can lead to significant deviations in product structure and performance [1, 2]. For example, in the hydrothermal synthesis of nickel–iron layered double hydroxides, slight changes in reaction temperature directly affect the thickness and specific surface area of nanosheets. In the sol–gel preparation of iron oxide, deviations in pH significantly alter gelation kinetics, resulting in a broadened particle size distribution of nanoparticles. Moreover, controlling the purity of the crystal phase is another major challenge. For instance, the synthesis of cobalt tetraoxide is often accompanied by the formation of cobalt monoxide impurities, which reduces the intrinsic catalytic activity of the material. These factors collectively lead to significant performance fluctuations between different product batches, making it difficult to meet the high standards for material uniformity required in industrial applications.

11.1.1.2 Difficulty in Precise Construction of Complex Structures

The controllable preparation of advanced nanostructures, such as core–shell structures, hollow structures, and heterojunctions, faces technical bottlenecks [3]. Taking core–shell structures as an example, it is difficult to precisely control the shell layer thickness, which is prone to incomplete coating or shell rupture, affecting material functionality. The preparation of hollow structures often relies on template methods. Template removal may introduce impurities, and precise regulation of cavity size and wall thickness is quite challenging. For heterostructures, the interfacial bonding strength between different components is often insufficient. During subsequent heat treatment, elemental interdiffusion is likely to occur, destroying the well-defined heterointerfaces and thereby affecting interfacial charge-transfer efficiency.

11.1.1.3 Difficulty in Balancing Scale-Up and Refinement

In laboratories, high-quality materials with regular structures can be obtained through precisely controlled techniques such as dropwise addition and electrodeposition. However, during scale-up production, problems such as uneven mass and heat transfer in reactors become prominent, leading to product agglomeration, loss of structural order, and degradation of key performance indicators. For example, industrial spray processes struggle to achieve the precise control of precursor loading possible with laboratory dropwise addition, resulting in uneven distribution of active sites. Although high-precision technologies like atomic layer deposition can achieve atomic-level control, their high equipment costs and low production capacity limit their application in large-scale production.

11.1.1.4 Insufficient Greenness and Economic Viability

Existing preparation routes often involve toxic and harmful reagents and high-energy-consuming processes. The environmental risks associated with organic solvents in solvothermal methods and the use of strongly corrosive reagents in template removal processes bring environmental pressures and high treatment costs [4, 5]. Meanwhile, some organometallic precursors are expensive, increasing production costs. Furthermore, the R&D process is highly dependent on the "trial-and-error method," lacking real-time monitoring and feedback mechanisms for preparation parameters, as well as effective predictive models linking material structures and properties, leading to long R&D cycles and low efficiency.

11.1.2 Future Development Trends and Prospects

11.1.2.1 Toward Precise and Controllable Intelligent Preparation

Integrating in situ monitoring technology with artificial intelligence algorithms to construct intelligent reaction systems is expected to realize dynamic optimization and closed-loop control of the preparation process. For example, adjusting voltage through real-time current feedback during electrodeposition can stabilize alloy composition. Atomic layer deposition technology can achieve atomic-level thickness control of thin-film materials. Combining the design of crystal phase regulators guided by theoretical calculations can effectively inhibit the formation of impurity phases. Emerging methods, such as supercritical fluid synthesis, leverage a uniform reaction environment to provide a new approach for producing nanoparticles with a narrow particle size distribution.

11.1.2.2 Integrated Innovation to Break Through Structural Regulation Bottlenecks

The collaboration and integration of multiple preparation technologies have opened new paths for the precise construction of complex nanostructures. For example, combining the rapid heating characteristics of Joule flash sintering technology with the spatial confinement effect of template methods can prepare hollow materials with regular structures and inhibit particle agglomeration. Integrating sol–gel processes with 3D printing technology enables the direct fabrication of functional devices with three-dimensional porous structures. Developing template-free self-assembly strategies to spontaneously form hollow or heterostructures by regulating thermodynamic and kinetic processes can simplify the process flow and improve product purity.

11.1.2.3 Promoting Green and Economical Large-Scale Production

Developing continuous flow preparation processes and improving mass and heat transfer efficiency through optimized reactor design are key to achieving stable and consistent mass production of materials. Meanwhile, the exploration of green synthesis routes has attracted increasing attention, including the use of biomass-derived precursors, the development of solvent-free mechanochemical synthesis methods, and the establishment of template recycling systems, aiming to reduce environmental impact and production costs from the source.

11.1.2.4 Building a Data-Driven New R&D Paradigm

By integrating in situ characterization technology with preparation equipment to construct a closed-loop "preparation-monitoring-feedback" system, the material growth process can be regulated in real time, significantly shortening the process optimization cycle. Establishing a correlation database covering preparation parameters, material structures, and application properties, and using machine learning to build predictive models, is expected to fundamentally transform the traditional "trial-and-error" R&D model and realize efficient and directional material design. For different application scenarios, constructing cross-scale technical systems will enable seamless connection from basic research to industrial applications.

11.2 Challenges and Prospects Related to Material Microstructures and Characterization

11.2.1 Current Core Challenges

11.2.1.1 Difficulty in Precise Analysis of Complex Microstructures

Iron-based materials exhibit diverse crystal structure characteristics, such as multiple lattice site occupations in spinel structures, disordered distribution of interlayer ions in layered double hydroxides, and elemental segregation in solid solutions, making the analysis of their fine structures extremely challenging [6–8]. Traditional characterization techniques each have limitations: X-ray diffraction can accurately identify phase composition but struggles to quantitatively distinguish the specific occupancy of Ni^{2+} in tetrahedral and octahedral sites in $NiFe_2O_4$; X-ray photoelectron spectroscopy can provide information on the chemical state of surface elements but has a detection depth limited to a few nanometers and thus cannot effectively characterize bulk structure information; for amorphous or low-crystallinity materials, X-ray absorption fine structure spectroscopy can provide local coordination environment information but cannot describe the overall structural symmetry. These fragmented characterization results lead to a lack of systematic understanding of the structural origins determining the electronic and magnetic properties of materials, making it difficult to construct a complete structural picture, akin to the parable of "blind men touching an elephant."

11.2.1.2 Challenges in Quantitative Characterization and Regulation of Defects and Interfaces

Defects (such as oxygen vacancies, metal vacancies, and dislocations) and heterointerfaces are key structural factors regulating the catalytic, electrochemical, and magnetic properties of materials, but their precise quantification and controllable construction are extremely challenging [3, 7]. For example, the oxygen vacancy concentration in CoO cannot be directly measured by conventional characterization techniques and needs to be indirectly estimated using methods such as electron paramagnetic resonance, with errors usually exceeding 10%; when constructing core–shell heterostructures, it is difficult to clearly distinguish the type of interfacial chemical bonding (covalent bonds, ionic bonds, or physical adsorption), and the interface is highly unstable under thermodynamic or electrochemical driving forces, prone to elemental interdiffusion, which destroys the expected efficient charge transfer channels. Current research usually only stays at the qualitative description level of "defects or interfaces improve performance," but cannot accurately answer key quantitative questions such as "which type of defect, under what interfacial atomic structure, and to what extent each contributes to performance improvement," seriously restricting the precise design of defect engineering.

11.2.1.3 Bottlenecks in Dynamic Process Tracking and Structure–Performance Model Construction

Materials undergo dynamic changes in real working environments, and their structures continuously evolve with reaction conditions. For example, $Ni(OH)_2$ undergoes gradual dehydration and oxidation under oxygen evolution reaction conditions, eventually transforming into NiOOH. However, traditional ex-situ characterization can only capture the initial and final states of the reaction, failing to obtain structural information of intermediate states, thus making it difficult to reveal the true phase transition path and rate-determining step. Precisely because dynamic structural evolution processes are poorly understood and the synergistic effects of multidimensional structural parameters (such as specific surface area, defect concentration, interface area, and lattice strain) are difficult to quantify, it remains impossible to establish an accurate quantitative prediction model for "structure-performance." This cognitive gap results in material design still largely relying on empirical trial-and-error, leading to low R&D efficiency.

11.2.1.4 Limitations of Characterization Techniques Themselves and Difficulties in Data Integration

The challenges stem not only from the structural complexity of materials themselves but also from the inherent limitations of characterization techniques. Sample preparation processes prior to nanomaterial characterization (such as ultrasonic dispersion, conductive coating, and ultrathin sectioning) are prone to introducing structural artifacts, altering the original state of materials, and affecting the authenticity of test results [7]. Although combining multiple techniques is an inevitable choice for in-depth understanding of material structures, different techniques such as X-ray diffraction (XRD), transmission electron microscope (TEM), and X-ray photoelectron spectroscopy (XPS) have varying test conditions (vacuum level, temperature, and sample morphology requirements), resulting in naturally mismatched data in terms of dimensions and scales. Additionally, the absence of unified data analysis standards and systematic error quantification methods makes the effective integration and cross-validation of multisource data extremely difficult. This data silo phenomenon seriously restricts the holistic understanding of material structures.

11.2.2 Future Development Trends and Prospects

11.2.2.1 Breakthroughs in High-Resolution and In Situ Technologies: From "Static Snapshots" to "Dynamic Movies"

With the rapid development of advanced characterization technologies, we will witness revolutionary breakthroughs in the understanding of material structures. High spatial and energy resolution technologies centered on aberration-corrected transmission electron microscopy and synchrotron radiation light sources will enable us to directly "see" fine structural features such as lattice site occupations, defect atomic configurations, and interfacial atomic arrangements at the atomic scale. More importantly, various in situ/operando characterization technologies are increasingly maturing. For example, integrating XRD, TEM, and Raman spectroscopy with electrochemical cells, reaction gas environments, or external magnetic fields allows real-time observation of structural evolution processes of materials under actual working conditions. This technological breakthrough will fundamentally change the way we understand material behavior, shifting from speculation about reaction endpoints to direct tracking of the entire process, thereby enabling the clear identification of true active sites, capture of transient intermediates, and analysis of phase transition paths, providing direct experimental evidence for the rational design of high-performance materials.

11.2.2.2 Integration of Precise Structural Regulation and Intelligent Data Analysis: From "Qualitative Description" to "Quantitative Prediction"

Guided by precise characterization, atomic-level manufacturing technologies (such as atomic layer deposition, plasma etching, and electrochemical atomic processing) will achieve fine regulation of defect types and concentrations, as well as interfacial atomic structures, reaching an unprecedented level of control precision. Meanwhile, artificial intelligence and machine learning technologies will empower the in-depth mining and correlation analysis of massive multidimensional characterization and performance data. By constructing quantitative correlation models of "structural parameters-performance output," we can predict material performance under specific structural designs, thereby inversely guiding the optimization of synthesis paths and realizing rational material design. This "data-driven" research paradigm will significantly reduce R&D costs and cycles, promoting the transformation of material development from traditional "empirical trial-and-error" to modern "theoretical prediction."

11.2.2.3 Improvement of Cross-Scale Characterization Systems: Building a Complete Cognitive Chain

The future development trend is to construct an integrated cross-scale characterization system ranging from macro-performance (such as catalytic activity, battery capacity, and magnetization) to mesoscopic structures (such as pore structure, particle morphology, and interface characteristics), and further to microscopic information (such as atomic arrangement, electronic structure, and chemical bonding). By systematically integrating test data at different spatial and temporal scales, we will be able to fully reveal the inherent logical chain connecting "atomic structure defects → micro-morphology and interfaces → macro-device performance." This comprehensive cognitive system can not only explain the structural origin of material performance but also provide a solid scientific basis for the directional design of materials for specific application scenarios, ultimately promoting the practical application of iron-based metal compound nanomaterials in energy, environment, information, and other fields.

11.3 Challenges and Prospects Related to Material Dimensions

11.3.1 Current Core Challenges

11.3.1.1 Obvious Shortcomings in the Synthesis and Precise Regulation of Low-Dimensional Materials

Research on zero-dimensional iron-based nanoclusters is seriously lagging [2, 9]. Currently, there are very few studies and reports on iron-based single-metal nanoclusters and multi-metal alloy nanoclusters with excellent electrocatalytic performance, and they are still in the initial exploration stage. Meanwhile, the synthesis of zero-dimensional clusters relies on ligands to maintain structural stability, but the influence of ligands on the electrocatalytic activity of iron-based clusters is not yet clear, making it impossible to optimize cluster activity in a targeted manner. Compared with quantum dots, one-dimensional nanomaterials face greater difficulties in the precise control of size, morphology, purity, and chemical composition, leading to relatively slow research progress and difficulty in mass-producing one-dimensional iron-based nanotubes, nanowires, and other materials with uniform structures and stable performance. The synthesis methods for two-dimensional layered iron-based materials are far fewer than those for non-layered materials. Currently, they mainly rely on molecular beam epitaxy and exfoliation methods, lacking diverse and efficient preparation paths, making it difficult to meet the differentiated structural and performance requirements of two-dimensional materials for different application scenarios. These problems collectively restrict the further development of low-dimensional iron-based materials.

11.3.1.2 Multiple Bottlenecks in Performance Stability and Adaptability of Materials for Multi-scenario Applications

In magnetic applications, the magnetic properties of iron-based nanowires and nanoparticles are easily affected by size, morphology, and environment, leading to problems such as coercivity attenuation [1, 9]. The synergistic regulation of magnetic anisotropy is difficult, making it hard to optimize these materials for high-frequency devices and magnetic storage scenarios. In the biomedical field, the biocompatibility and toxicity control of iron-based nanomaterials need further improvement. They are prone to agglomeration and degradation, and targeted modification technology is immature, limiting precise diagnosis and treatment. In energy storage scenarios, iron-based nanomaterials used as electrodes have high volume expansion rates and are prone to structural collapse, resulting in rapid battery capacity decay. In the catalytic field, iron-based catalytic materials are easily affected by reaction conditions, with insufficient stability of active sites and unclear ligand regulation mechanisms, making it difficult to adapt to the requirements of various catalytic reactions. In environmental remediation, these materials are easily affected by working conditions such as water pH value, exhibiting insufficient stability and difficulty in recovery, leading to low reuse efficiency. Additionally, materials with a single structure struggle to adapt to the differentiated needs of multiple scenarios, and the synergistic optimization of multiple properties is challenging [10, 11].

11.3.1.3 Obvious Deficiencies in Large-Scale Preparation and Adaptability to Practical Applications

Existing synthesis methods such as the hydrothermal method, template method, and vapor deposition are mostly suitable for small-scale laboratory preparation, with complex operation processes and high costs. Moreover, it is difficult to ensure the uniformity of material structure and performance during mass production. There is a lack of efficient and low-cost large-scale technologies suitable for industrial production, which seriously restricts the industrialization process of materials. Meanwhile, some materials exhibit excellent performance under ideal laboratory conditions, but in industrial application scenarios facing complex working conditions (such as high temperature, high humidity, and strong corrosion environments), their performance stability and service life significantly decrease. The compatibility with working conditions needs to be further improved to meet actual application requirements.

11.3.2 Future Development Trends and Prospects

11.3.2.1 Innovations in Synthesis Methods and Precise Structural Regulation

Future efforts will focus on breaking through the preparation bottlenecks of iron-based metal compound nanomaterials through innovation in synthesis methods and precise structural regulation. For zero-dimensional iron-based nanoclusters, efficient synthesis paths for iron-based single-metal/multi-metal alloy clusters will be developed, ligand design will be optimized, and ligand-free or weak-ligand stabilization strategies will be explored. Combined with high-resolution characterization technology and theoretical calculations, the interaction mechanism between ligands and active sites will be clarified to achieve precise regulation of cluster activity. For one-dimensional nanomaterials, precise control technologies will be upgraded. By optimizing process parameters of methods like template and hydrothermal synthesis, precise adjustment of size and morphology will be realized, improving material uniformity and laying a foundation for subsequent applications. For two-dimensional layered materials, synthesis methods will be expanded, and new technologies such as space-confined synthesis and ligand-assisted energy control will be developed to enrich the preparation means of two-dimensional iron-based materials and meet the needs of different scenarios.

11.3.2.2 Expansion of Multi-field Applications and Performance Upgrading

The expansion of multi-field applications and performance upgrading will become a core direction for the development of iron-based metal compound nanomaterials. In the field of magnetic materials, the composition and diameter of one-dimensional iron-based nanowires will be precisely regulated, magnetic anisotropy will be optimized, and high-performance materials for applications like high-frequency magnetic devices will be developed to replace rare-earth permanent magnetic materials. In the biomedical field, zero-dimensional iron-based quantum dots will undergo surface functionalization to improve biocompatibility and targeting, expanding in-vivo diagnosis and treatment applications. In the energy storage field, three-dimensional iron-based heterostructures will be used as electrodes, constructing core–shell/hollow structures to alleviate volume expansion and improve cycle stability and energy density. In the catalytic field, the active sites and stability of iron-based clusters and nanoparticles will be optimized to adapt to reactions such as hydrogen evolution reaction (HER)/oxygen evolution reaction (OER) and oxygen reduction reaction (ORR), replacing precious metal catalysts. In environmental remediation, two-dimensional layered double hydroxide (LDH) materials and iron-based metal–organic framework (MOF) derivatives will be used to treat pollutants, and efficient recovery and reuse will be achieved through magnetic separation. In the field of electronic devices, leveraging the excellent electron transport performance of iron-based metalenes and nanowires, flexible electronic and sensor devices will be developed, expanding applications such as wearable devices [12].

11.3.2.3 Strengthening Large-Scale Preparation and Theoretical Guidance

Strengthening large-scale preparation capabilities and theoretical guidance will accelerate the industrialization process of iron-based metal compound nanomaterials. In terms of large-scale preparation, technologies suitable for industrial production will be developed, existing synthesis processes will be optimized, and the continuous and automated transformation of technologies such as hydrothermal and template methods will be promoted to reduce production costs. Meanwhile, new low-cost synthesis paths such as one-step methods and green synthesis will be explored to realize mass material preparation while ensuring uniformity of structure and performance. To address adaptability to practical applications, combined with the needs of different scenarios, the corrosion resistance and stability of materials will be improved through structural design (such as constructing core–shell structures and hollow structures), and the composition and microstructures will be optimized to adapt to complex working conditions and reduce performance attenuation. In terms of theoretical guidance, combined with density functional theory (DFT) calculations and high-resolution characterization technologies such as HAADF-STEM and XPS, the quantitative correlation between structural characteristics (size, morphology, and defects) of iron-based nanomaterials and properties (magnetic properties, biocompatibility, and electron transport performance) will be clarified. Through theoretical calculations, the optimal structure and composition combinations will be predicted, and high-performance iron-based nanomaterials will be designed directionally, reducing experimental blindness and promoting the transition of iron-based metal compound nanomaterials from laboratory research to practical applications in various fields.

11.4 Challenges and Prospects in Battery Applications

11.4.1 Current Core Challenges

11.4.1.1 Limitations of Intrinsic Material Properties

Iron–cobalt–nickel group metal compounds have obvious shortcomings [13–15]. For example, iron-based polyanion oxides and iron–cobalt–nickel-based transition metal sulfides (such as CoS_2 and NiS_2) have poor intrinsic electronic/ionic conductivity, which directly restricts battery rate performance. Iron–cobalt–nickel-based conversion-type materials (such as Co_3O_4) undergo significant volume changes during charge and discharge, which easily lead to electrode pulverization, active material shedding, and rapid capacity decay. Iron–cobalt–nickel-based layered oxides (such as O3-type Na_xTMO_2, where TM is Fe, Co, and Ni) also face an inherent contradiction between air stability and ionic diffusion kinetics. Strengthening the lattice's binding to sodium to improve air stability often hinders sodium ion diffusion, resulting in reduced kinetic performance and making it difficult to balance the two.

11.4.1.2 System-Specific Issues of Batteries

Such materials face differentiated problems in different battery types. In lithium-ion batteries, Co-based cathodes (such as $LiCoO_2$) have high costs and rely on scarce cobalt resources. Although Mn-based spinels (such as $LiMn_2O_4$) have lower costs, they suffer from Mn dissolution, which poisons the anode and shortens cycle life. Iron–cobalt–nickel-based composite anodes also face insufficient synergy with carbon carriers, making it difficult to balance capacity and stability simultaneously. In sodium/potassium-ion batteries, due to the large ionic radius of Na^+/K^+, iron–cobalt–nickel-based layered oxides (such as P2-type $Na_{0.67}Ni_{0.33}Mn_{0.67}O_2$) are prone to phase transitions (such as P2–O2 phase transition) during charge and discharge, causing structural stress and leading to poor cycle stability. Moreover, the diffusion barrier of Na^+/K^+ in iron–cobalt–nickel-based materials is high, affecting rate performance [13, 15]. In lithium–sulfur batteries, although iron–cobalt–nickel group compounds (such as $CoSe_2$) can alleviate the shuttle effect by adsorbing polysulfides, the composite uniformity between some iron–cobalt–nickel-based carriers and sulfur is insufficient, leading to issues such as incomplete polysulfide adsorption and low sulfur utilization [14, 16]. In metal–air batteries, iron–cobalt–nickel-based catalysts (such as Co_3O_4/Fe_2O_3 and Fe–NiCoP) have bifunctional catalytic capabilities for ORR/OER, but their catalytic activity still needs improvement, and active site loss is prone to occur during long-term cycling. Meanwhile, in zinc–air batteries, the interfacial stability between iron–cobalt–nickel-based cathodes and alkaline electrolytes is poor, which easily triggers side reactions, and the problem of zinc dendrite growth on the zinc anode has not been completely solved, affecting battery cycle life.

11.4.1.3 Interface and System Compatibility Issues

Interface and system compatibility issues are also prominent. When iron–cobalt–nickel-based layered cathodes such as $LiNiO_2$ cycle in traditional carbonate electrolytes, unstable solid electrolyte interphase (SEI) films are easily formed, leading to intergranular cracks and aggravating capacity decay. High-voltage iron–cobalt–nickel-based cathodes (such as $LiMn_{1.5}Ni_{0.5}O_4$) lack compatible high-voltage-resistant electrolytes, limiting the improvement of battery voltage. Moreover, the synthesis process of most high-performance iron–cobalt–nickel-based materials is complex (such as in situ phosphidation and the template-assisted hydrothermal method), making large-scale production difficult, and there is an imbalance between "performance and cost," which restricts the industrialization process.

11.4.2 Future Development Trends and Prospects

11.4.2.1 Optimization of Material Structure and Composition

Optimizing material structure and composition is an important development direction. By developing iron–cobalt–nickel-based heterogeneous composite structures (such as Co_3O_4/Fe_2O_3 heterojunctions), the synergistic effect between different components can be utilized to simultaneously improve material conductivity, structural stability, and ionic transport efficiency. Promoting the R&D and cost reduction of high-nickel low-cobalt (such as high-Ni content NMC) and iron-based polyanion materials (such as $LiFePO_4$) can reduce dependence on scarce resources like cobalt and lower material costs. Designing nanoscale functional structures (such as CoP nanosheet arrays, NiS_2@CoS_2 concave nanocubes, and hollow porous FeNi alloy/carbon composite structures) can provide buffer space for volume changes during charge and discharge, shorten ionic transport paths, and improve reaction kinetics.

11.4.2.2 Directions for Key Technology Breakthroughs

Interface engineering can improve the compatibility between electrodes and electrolytes, inhibit side reactions, and enhance cycle stability by optimizing electrolyte composition (such as introducing sodium vacancies to regulate interface characteristics) and applying surface coatings on iron–cobalt–nickel-based materials (such as carbon layers and polydopamine coatings). Strengthening the catalytic performance of iron–cobalt–nickel-based catalysts, increasing the number of active sites, regulating the electronic structure, and reducing the overpotential of ORR/OER reactions through doping (such as Fe-doped NiCoP) and constructing porous structures can improve the energy efficiency of metal–air batteries. Improving the wide-temperature adaptability of iron–cobalt–nickel-based materials with the help of doping and skeleton support strategies can enable batteries to work stably in a wide-temperature range of $-25\sim55°C$ while extending cycle life. For example, developed iron–cobalt–nickel-based composite cathodes have achieved over 100,000 cycles, laying a foundation for long-life battery applications.

11.4.2.3 Expansion of Application Scenarios and Industrialization

In terms of application scenario expansion and industrialization, sodium-ion batteries, combined with iron–cobalt–nickel-based cathodes (such as P2/O3 biphasic composite layered oxides) and anode materials, are expected to replace some lithium-ion batteries in large-scale energy storage scenarios such as grid energy storage and low-speed electric vehicles due to the advantages of abundant sodium resources and low costs. After solving problems such as the shuttle effect and low catalytic efficiency with iron–cobalt–nickel-based materials (such as $CoSe_2$/C sulfur carriers and FeNi alloy/porous carbon catalysts), lithium–sulfur batteries and metal–air batteries can target high energy density demand scenarios such as new energy vehicles and portable electronic devices. Meanwhile, promoting the recycling of iron–cobalt–nickel-based battery materials (such as closed-loop regeneration of $LiCoO_2$) together with green synthesis processes (such as the low-cost hydrothermal method and electrospinning technology) can reduce the environmental impact of the material's entire lifecycle and support the sustainable development of the battery industry.

11.5 Challenges and Prospects in Supercapacitor Applications

11.5.1 Current Core Challenges

11.5.1.1 Bottlenecks of Intrinsic Material Properties

The intrinsic conductivity and cycle stability of iron–cobalt–nickel-based compounds still have significant room for improvement. Some typical materials have low intrinsic conductivity [17]. Even after composite modification, obvious polarization phenomena occur under high-rate working conditions, affecting energy output efficiency. More importantly, during charge–discharge cycles, materials are prone to significant volume changes, which directly lead to the shedding of active materials from the electrode surface, thereby causing continuous capacity decay and severely restricting the long-term service performance of devices.

11.5.1.2 Optimization of Electrode/Electrolyte Interface

The interface compatibility between electrodes and electrolytes is a key factor restricting the overall performance of devices. In aqueous electrolyte systems, iron–cobalt–nickel-based compounds are prone to dissolution and corrosion reactions, leading to a continuous increase in interface impedance and reduction in ionic transport efficiency. In organic electrolyte systems, the wettability of electrode materials is generally poor, resulting in low ionic diffusion rates, which also affects device response speed. In addition, interface side reactions such as electrolyte decomposition and unstable SEI film formation will further aggravate capacity decay and shorten device service life.

11.5.1.3 Process and Cost of Large-Scale Preparation

The preparation processes commonly used in laboratories, such as hydrothermal and solvothermal methods, are mostly small-scale, refined operations, with problems such as long preparation cycles and limited output, making them difficult to directly adapt to industrial mass production needs. For materials with complex structures such as carbon composite systems, the preparation process often requires multiple steps like carbon source coating and high-temperature calcination, and the process complexity directly increases production costs. Meanwhile, during large-scale production, it is difficult to achieve precise control over the composition consistency and performance stability of materials, and the core performance of products from different batches is prone to significant deviations, posing obstacles to practical applications.

11.5.2 Future Development Trends and Prospects

11.5.2.1 Design and Development of New Material Systems

Developing new iron–cobalt–nickel-based compounds with both high conductivity and high stability is a core direction of future research. Among them, ternary metal phosphides show outstanding potential in cycle life and capacity performance due to their excellent conductivity and abundant redox active sites. Iron–cobalt–nickel-based compounds derived from MOFs, leveraging their unique porous structure advantages, can provide sufficient active sites and ionic transport channels, significantly improving electrochemical performance. At the same time, in line with the concept of sustainable development, the R&D of cobalt-free and low-cobalt materials (such as nickel–iron-based and iron–manganese-based hydroxides) has attracted wide attention. Such materials can effectively reduce dependence on scarce resources and greatly control production costs.

11.5.2.2 Advanced Electrolyte and Additive Engineering

The R&D of advanced electrolyte systems and innovation in additive engineering have opened effective paths for solving interface compatibility issues. Ionic liquid electrolytes can significantly broaden the working voltage window of devices and greatly improve energy density. Solid electrolytes can fundamentally avoid the risk of electrolyte leakage and enhance device safety, especially suitable for special application scenarios such as flexible electronics and high-temperature environments. In addition, through additive engineering strategies, introducing specific ions (such as lithium ions and sodium ions) into the electrolyte can precisely regulate the interface reaction process, inhibit material dissolution and side reactions, and significantly improve the cycle stability of devices.

11.5.2.3 Artificial Intelligence-Assisted Material Discovery and Optimization

Artificial intelligence technology is reshaping the R&D paradigm for iron–cobalt–nickel-based compounds, greatly improving R&D efficiency. Using machine learning algorithms to build correlation models of material composition–structure–performance can quickly screen potential optimal material systems and shorten the R&D cycle. Precisely optimizing the nanostructural parameters (such as particle size and pore size distribution) of materials through deep learning technology can achieve directional improvement of electrochemical performance. At the same time, intelligent algorithms can also be applied to the optimization of preparation processes. For example, adjusting reaction parameters through reinforcement learning can effectively shorten the hydrothermal preparation cycle and improve the efficiency and stability of large-scale production.

11.5.2.4 Multifunctional Integration and Intelligent Devices

Multifunctional integration and intelligence are important trends in the future development of devices, which can significantly expand the application scenarios of iron–cobalt–nickel-based compounds. Combining energy storage functions with sensing, catalysis, and other characteristics, composite devices with both supercapacitor and oxygen reduction catalytic functions have been developed, which can be applied in self-powered fuel cell systems and other scenarios. Some materials, combined with magnetic response and other characteristics, have also achieved magnetic field-regulated capacitance performance. On this basis, by integrating impedance sensors and control systems into devices, intelligent functions such as capacity decay monitoring and fault diagnosis can be realized, greatly improving the reliability and convenience of device use.

11.5.2.5 Environmental Friendliness and Full Lifecycle Design

The concept of green and sustainable development has been deeply integrated into the entire R&D and application process of iron–cobalt–nickel-based compounds. Environmentally friendly preparation processes have become a research hotspot. Using biomass carbon sources (such as sugarcane and straw) as composite carriers and green chemical methods such as ultrasound-assisted synthesis can effectively reduce energy consumption and carbon emissions during production. At the same time, the concept of full lifecycle design is promoting the development of material recycling technology. Through efficient recycling processes such as acid dissolution–precipitation, high-proportion recovery of precious metals such as nickel and cobalt can be achieved, building a resource cycle system of "preparation-use-recycling-reuse," which aligns with the requirements of green energy development.

11.6 Challenges and Prospects in Electrocatalysis Applications

11.6.1 Current Core Challenges

11.6.1.1 Difficulty in Synergizing Catalytic Activity and Long-Term Stability, and Limited Practical Application Life

Iron-based metal compound nanocatalysts can exhibit excellent catalytic activity in short-term tests, but during long-term participation in reactions, their performance is prone to decay due to multiple factors, making it difficult to meet the life requirements of practical applications. For example, during the OER process of NiFe LDH, the interlayer structure is easily eroded by the electrolyte and collapses, and the Ni^{3+}

active sites can dissolve and be lost as the reaction progresses, resulting in a significant decay in current density after long-term cycling. The FeP HER catalyst is prone to surface oxidation, forming an FeO_x passivation layer under acidic conditions, which blocks electron transport channels, leading to an increase in the overpotential and a gradual decrease in activity. In addition, the binding force between the catalyst and the carrier is often weak. For example, when $CoSe_2$ nanoparticles are loaded on a carbon carrier, they are prone to agglomeration, resulting in a decrease in the exposure rate of active sites and further shortening the actual service life of the catalyst.

11.6.1.2 Great Difficulty in Regulating Reaction Selectivity and Competitive Reactions Reduce Target Product Yield

Electrocatalytic reactions mostly involve multi-electron transfer processes. The adsorption energies of intermediates in different reaction paths are similar, which easily triggers competitive reactions, leading to low selectivity of target products. For example, in CO_2RR, Fe–N–C catalysts may simultaneously initiate the two-electron path to generate CO and the eight-electron path to generate CH_4, and different paths have different requirements for reaction conditions, making it difficult to achieve directional generation through simple regulation. In ORR, Co_3O_4 catalysts may follow the two-electron path to generate H_2O_2 instead of the target four-electron path to generate H_2O. It is necessary to precisely regulate the Co^{3+}/Co^{2+} ratio in the catalyst to effectively inhibit side reactions. In addition, in biomass electrocatalytic upgrading scenarios, Ni-based catalysts are prone to excessive oxidation of glucose, generating CO_2 instead of the target product gluconic acid, with poor selectivity.

11.6.1.3 Contradiction Between Uniformity and Cost in Large-Scale Preparation, and Difficult to Industrialize

On the laboratory scale, highly dispersible and high-activity iron-based catalysts can be obtained through refined preparation processes such as hydrothermal and electrodeposition methods. However, when these processes are scaled up to industrial production, they face many bottlenecks. For example, when preparing NiFe LDH nanosheets using the kettle-type hydrothermal method, the uneven temperature and concentration distribution inside the reaction kettle leads to poor uniformity in the thickness of the generated nanosheets, and significant performance fluctuations between batches. When Fe/N–C catalysts are prepared by the impregnation method, achieving uniform control of the Fe precursor loading on the carrier is challenging, and some areas will affect the overall catalytic efficiency due to insufficient active sites. At the same time, some preparation processes rely on expensive precursors (such as organometallic compounds) or complex special equipment, resulting in high catalyst production costs, which cannot meet the economic requirements of industrial mass production.

11.6.1.4 Insufficient Catalytic Mechanism Analysis and Lack of Basis for Directional Design

Currently, the identification of active sites and the evolution paths of reaction intermediates for most iron-based catalysts rely on indirect speculation, lacking direct experimental evidence, making it difficult to form a clear understanding of catalytic mechanisms. For example, there is still controversy within the academic community regarding the HER active sites of FeP—whether they are Fe sites, P sites, or Fe–P interfaces—and DFT calculation results often fail to fully match experimental data. During the OER process, the generation sequence and conversion energy barrier of the *OOH intermediate cannot be accurately captured by ex-situ characterization technology, making it impossible to clarify the quantitative relationship between "defect concentration-intermediate adsorption-catalytic activity." In addition, in complex electrolyte environments (such as seawater and industrial wastewater), the specific impact of impurities such as Cl^- and heavy metal ions on the catalytic mechanism is not clear, which greatly restricts the directional design of catalysts with anti-interference capabilities.

11.6.1.5 Poor Adaptability to Complex Application Environments and Insufficient Anti-interference Capabilities

In practical application scenarios, fluctuations in conditions such as electrolyte composition and reaction temperature can significantly impact the performance of iron-based catalysts, leading to insufficient adaptability. For example, in seawater electrolysis, Cl^- in the electrolyte is prone to oxidation reaction at the anode to generate Cl_2, which can corrode the surface of the Co_3O_4 catalyst to form inactive $CoCl_2$, leading to catalyst failure. During the electrocatalytic degradation of industrial wastewater, heavy metal ions such as Cu^{2+} and Pb^{2+} are easily adsorbed on the active sites of Fe–N–C catalysts to form metallic deposits, causing catalyst poisoning. In low-temperature environments, the HER activity of FeNi alloy decreases significantly, mainly because the electrolyte freezes, hindering ion transport, and existing iron-based catalysts generally lack adaptive design for low-temperature environments.

11.6.2 Future Development Trends and Prospects

11.6.2.1 Precise Structural Design to Achieve Synergistic Improvement of Activity and Stability

Through advanced strategies such as defect engineering, interface engineering, and single-atom catalysis, the electronic structure and structural stability of iron-based catalysts can be targeted for optimization to achieve synergistic improvement of catalytic activity and service life [18, 19]. For example, introducing Se vacancies into $CoSe_2$ can not only optimize the adsorption energy of the *H intermediate to improve HER activity but also enhance the binding force between atoms on the catalyst surface and inhibit oxidation failure. Constructing a Fe_3O_4@CoO heterostructure can improve OER activity by utilizing the electron transfer effect at the interface, and the CoO shell can effectively protect the Fe_3O_4 core from electrolyte corrosion. Developing single-atom Fe–N–C catalysts allows Fe atoms to be stabilized through coordination with N atoms to avoid their agglomeration, ensuring high activity after long-term cycling. In addition, strategies such as carbon coating (e.g., CoP@CNTs) and oxide coating (e.g., NiFe LDH@Al_2O_3) will further enhance the anti-dissolution and anti-poisoning capabilities of the catalyst and extend its application life.

11.6.2.2 Multifunctional Catalysts Simplify Reaction Systems and Expand Application Scenarios

Developing bifunctional (e.g., HER/OER) and trifunctional (e.g., HER/OER/ORR) iron-based catalysts can greatly reduce the complexity and cost of electrocatalytic equipment and promote multi-scenario applications. For example, NiCoP nanosheets can simultaneously catalyze water splitting and zinc–air battery reactions without frequent catalyst replacement, simplifying system design. Designing "pollutant degradation-energy conversion" synergistic catalysts (e.g., Fe–N–C@MnO_2) can realize the degradation of pollutants such as phenol at the anode and the simultaneous generation of H_2 at the cathode, achieving a win–win situation of "treating waste with waste" and energy production. In addition, customized catalysts for specific complex scenarios will become a development trend. For example, Ti-doped CoOOH, which is resistant to Cl^- corrosion (suitable for seawater electrolysis), and hydrophilic Fe_2O_3@rGO, which is resistant to heavy metal poisoning (suitable for industrial wastewater treatment), can significantly improve the adaptability of catalysts in complex environments.

11.6.2.3 In-Depth Integration of Mechanism Analysis Technologies to Guide Rational Catalyst Design

The in-depth combination of in situ characterization technology and DFT calculation can break through the limitations of traditional mechanism analysis, establish a clear correlation between "active sites-intermediates-catalytic activity," and provide a basis for rational catalyst design. For example, using in situ XAS technology to track the changes in the Ni–K edge near-edge structure of NiFe LDH during the OER process can accurately confirm Ni^{3+}–O–Fe^{3+} as the true active site. Capturing the characteristic peaks of the *OOH intermediate through in situ Raman and combining DFT to calculate its conversion energy barrier can further optimize the catalyst's electronic structure to reduce the reaction energy barrier. In addition, machine learning technology will help improve the efficiency of mechanism analysis. By building a "structural parameters-catalytic activity" database, catalyst performance can be predicted, replacing the traditional "trial-and-error method," shortening the catalyst R&D cycle, and promoting the transformation of iron-based catalyst design from experience-driven to data-driven.

11.6.2.4 Green and Low-Cost Large-Scale Preparation Technologies to Promote Industrialization

Developing continuous flow preparation processes (such as the continuous flow hydrothermal method and spray pyrolysis) and precisely controlling parameters such as reaction flow rate and temperature gradient can realize the uniform mass production of iron-based catalysts and solve the scale-up problem from laboratory to industry. For example, preparing NiFe LDH nanosheets using a continuous flow reactor can effectively improve reaction uniformity and reduce performance fluctuations between batches. Selecting bio-based precursors (such as lignin-derived carbon) and solvent-free preparation processes (such as the mechanochemical method) can reduce the use of toxic solvents and energy consumption, significantly lowering production costs. At the same time, catalyst recycling technology will be gradually improved. For example, recovering Fe_3O_4@Co_3O_4 catalysts through magnetic separation and regenerating Ni-based catalysts through electrolysis can realize catalyst recycling and further reduce the cost of industrial applications.

11.6.2.5 Construction of Cross-Scenario Application Systems to Expand the Boundaries of Electrocatalytic Technology

Coupling iron-based catalysts with renewable energy (such as solar and wind) can build "green electricity-green hydrogen" and "green electricity-high-value chemicals" conversion systems, contributing to the achievement of the carbon neutrality goal. For example, developing a solar-driven Fe–N–C catalytic CO_2RR system can directionally convert CO_2 into high-value carbon-based fuels (such as CH_4), realizing the recycling of carbon resources [20]. Designing distributed seawater electrolysis devices, powered by renewable energy such as wave energy, and producing H_2 with corrosion-resistant iron-based catalysts (such as Ti-CoOOH) can meet the energy needs of special areas such as islands. In addition, the integration of electrocatalytic systems with energy storage devices (such as zinc–air battery–water electrolysis integrated devices) will realize the closed loop of energy "storage-conversion," greatly improving energy efficiency and practicality of the system, and promoting the extension of electrocatalytic technology to broader scenarios.

11.7 Challenges and Prospects in Photocatalysis Applications

11.7.1 Current Core Challenges

11.7.1.1 Dilemma of Carrier Separation and Transport

Iron-based materials generally suffer from inefficient separation and rapid transport of photogenerated electron–hole pairs [21, 22]. The electron and hole migration rates in such materials are significantly different, and inherent defects and impurity energy levels are prone to exist in the lattice. These sites become "traps" for carrier recombination, leading to the recombination of a large number of photogenerated carriers in the bulk phase or on the surface before reaching the surface-active sites, resulting in a very low proportion of carriers effectively participating in the reaction. At the same time, nanoscale iron-based materials are prone to agglomeration, forming high contact resistance between particles and lacking continuous and unobstructed charge transport channels, which further hinders charge migration and makes it difficult to improve the photocatalytic reaction rate. In addition, some materials themselves have low carrier mobility. Even without obvious agglomeration, the transport of charges inside the material is restricted, exacerbating recombination loss.

11.7.1.2 Limitation of Spectral Response Range

Most iron-based materials have insufficient band gap adaptability, leading to a narrow spectral response range. Such materials mostly only respond to ultraviolet light, which accounts for a very small proportion of sunlight, and lack effective absorption capacity for visible and near-infrared light, which account for a higher proportion, resulting in the waste of a large amount of solar energy resources. Although traditional doping modification strategies attempt to narrow the band gap by introducing heterogeneous elements to expand the spectral response range, there is often an irreconcilable contradiction—excessive doping is prone to form new carrier recombination centers inside the material, which instead weakens the carrier separation efficiency, making it difficult to balance the expansion of spectral response and the improvement of catalytic activity. At the same time, the band structures of different iron-based materials have inherent differences, and the conduction and valence band positions of some materials are unreasonably designed. Even if visible light is absorbed, it is difficult to meet the redox potential requirements of specific reactions.

11.7.1.3 Slow Surface Reaction Kinetics

The surface reaction kinetic characteristics of iron-based materials have become a key factor restricting the overall efficiency. On the one hand, the density of surface-active sites of such materials is generally low, especially the limited number of coordination-unsaturated active centers, resulting in weak adsorption capacity for reactant molecules and inability to efficiently capture substrates and activate reactions. On the other hand, the conversion of key intermediates in photocatalytic reactions faces high energy barriers. The bond breaking and recombination of some intermediates require energy exceeding the redox capacity of photogenerated carriers, leading to the reaction being easily stuck in the intermediate stage and unable to efficiently generate target products. In addition, the surface-active sites of iron-based materials have poor selectivity and are prone to adsorbing multiple reaction intermediates simultaneously, triggering side reactions, leading to product diversification, further reducing the generation efficiency of target products, and increasing the difficulty of subsequent separation and purification.

11.7.1.4 Insufficient Photocorrosion Resistance and Stability

Structural stability under long-term illumination is a basic requirement for photocatalytic materials to realize practical applications, while iron-based materials generally suffer from weak photocorrosion resistance. Under continuous illumination conditions, photogenerated holes are prone to oxidize metal atoms in the material lattice, destroying the integrity of the crystal structure and leading to the loss of catalytically active sites. Some iron-based materials may also dissolve in aqueous reaction systems, releasing metal ions to pollute the reaction medium while simultaneously destroying their own structure. In addition, by-products generated during the reaction (such as carbon species) are prone to deposit on the material surface, blocking active sites and reducing the contact probability between reactants and active centers. Long-term illumination may also trigger phase transition of the material, generating inactive phases, further exacerbating the decay of catalytic activity, and making it difficult to meet the requirements for long-term stable operation of materials in practical applications.

11.7.1.5 Adaptability Challenges in Practical Application Scenarios

The excellent performance under laboratory conditions is difficult to directly transfer to practical scenarios, and iron-based photocatalytic materials face many limitations in application conditions. In scenarios such as water purification, suspended solids and colloidal particles in actual water bodies hinder light penetration, resulting in limited light utilization depth, and catalysts inside the reactor cannot fully absorb light energy. At the same time, factors such as coexisting ions and pH fluctuations in water bodies change the charge state and active site properties on the catalyst surface, leading to a significant decrease in catalytic activity. In addition, the design of existing photocatalytic reactors has many deficiencies. Catalysts in suspended reactors are prone to agglomeration, leading to uneven illumination and poor mass transfer. The catalyst layer in fixed-bed reactors is often too thick, increasing mass transfer resistance and making it difficult to balance light energy utilization efficiency and reaction kinetic rate. Some iron-based catalysts are sensitive to conditions such as reaction temperature and medium composition, which further limits their adaptability in complex practical scenarios.

11.7.2 Future Development Trends and Prospects

11.7.2.1 Precise Regulation of Material Structure to Enhance Charge Separation and Spectral Response

Through diversified structural design strategies, the problems of low carrier separation efficiency and narrow spectral response range can be solved simultaneously[23, 24]. In terms of charge separation optimization, constructing heterojunction structures (such as S-scheme and Z-scheme heterojunctions) can utilize the built-in electric field at the interface to directionally regulate the carrier migration path and inhibit the direct recombination of electrons and holes. Loading suitable cocatalysts can form charge trapping sites on the material surface, extend the carrier lifetime, and improve the probability of their participation in reactions. Constructing three-dimensional conductive network composite materials can effectively reduce the contact resistance between nanoparticles, build continuous charge transport channels, and accelerate charge migration. In terms of spectral response expansion, strategies such as nonmetal doping, quantum dot sensitization, and plasma enhancement have become important directions. By modifying the material's band structure and introducing intermediate energy levels, the spectral response range can be broadened, realizing efficient utilization of visible and near-infrared light. In addition, the introduction of single-atom catalysis technology can not only improve the density of active sites through atomic-level dispersion but also regulate the electronic structure of materials, synchronously optimizing charge separation and spectral absorption performance.

11.7.2.2 Synergistic Improvement of Surface Reaction and Stability

Through the synergistic efforts of active site engineering and anti-corrosion design, the surface reaction kinetics and long-term stability of iron-based materials can be significantly improved. In terms of surface reaction optimization, guided by DFT calculations, doping modification or active site modification can precisely regulate the electronic structure and adsorption energy of active sites, reduce the energy barrier of intermediate conversion, and improve the reaction rate and target product selectivity. The application of a single-atom catalysis strategy can maximize the utilization rate of active sites and, at the same time, enhance the adsorption and activation capacity for specific reactants through the synergistic effect between the carrier and metal atoms. In terms of stability improvement, surface coating technology can form a dense physical barrier on the material surface to prevent the oxidation of the lattice by photogenerated holes and the erosion of the reaction medium. Element doping can regulate the crystal phase structure of the material, inhibit light-induced phase transition, and maintain crystal structure stability. Designing self-healing mechanisms to repair lattice defects generated during illumination using photogenerated electrons or reducing species in the reaction system can effectively reduce the photocorrosion rate and extend the material's service life.

11.7.2.3 Optimization of Practical Application Systems and Technology Implementation

Focusing on the core needs of practical application scenarios, through reactor innovation, composite system construction, and adaptability design, the practicality of iron-based photocatalytic technology can be greatly improved. In terms of reactor design, the combination of microfluidic technology and photocatalytic reactors can utilize the advantages of microscale channels to enhance mass transfer efficiency and light penetration capacity, realizing precise regulation of reaction conditions. New reactor structures (such as multilayer stacked and light-guided) can improve the depth and uniformity of light utilization, adapting to high-turbidity and high-concentration actual wastewater treatment scenarios. In terms of composite system construction, composite technologies such as "photocatalysis-adsorption" and "photocatalysis-biodegradation" can exert synergistic effects, enriching pollutants through adsorbent materials or synergistic metabolism of microorganisms, improving the overall treatment efficiency, and simplifying the catalyst recovery process. In terms of adaptability design, developing iron-based catalysts with pH adaptability and strong anti-interference capabilities can break through the strict restrictions of traditional materials on reaction conditions and adapt to different medium environments. Coupling solar concentrator technology, waste heat recovery systems, etc., can further improve energy utilization efficiency and provide strong support for the large-scale application of photocatalytic technology.

11.8 Challenges and Prospects in Organic Oxidation and Degradation Applications

11.8.1 Current Core Challenges

11.8.1.1 Poor Selectivity of Biomass Directional Conversion

Biomass has complex and diverse components, including various functional groups such as benzene rings in lignin, hydroxyl groups in cellulose, and aldehyde and polyhydroxyl groups in glucose [25]. The physical and chemical properties of the three major components (cellulose, hemicellulose, and lignin) are significantly different. A single catalyst is difficult to accurately identify and break specific chemical bonds, which not only leads to a variety of products but also easily triggers degradation and condensation side reactions between components. Especially, the condensation reaction of lignin seriously affects the overall conversion efficiency. The functional group structures of different types of biomass are significantly different, requiring targeted matching of multiple catalysts to meet conversion needs, which increases the complexity of process design and raises equipment and operating costs, greatly hindering the realization of large-scale unified treatment.

11.8.1.2 Insufficient Thoroughness of Pollutant Degradation

Refractory organic pollutants generally have complex and stable chemical structures, such as the conjugated π-electron system of aromatic compounds, the unique conjugated structure of heterocyclic compounds, and the macromolecular skeleton of long-chain aliphatic compounds. These characteristics make them difficult to completely decompose by conventional technologies. During the degradation process, such pollutants are prone to incomplete mineralization, generating intermediate products with higher toxicity. Emerging pollutants such as antibiotics and endocrine disruptors have particularly limited removal effects in traditional processes. The reduction products of some inorganic pollutants may redissolve under specific environmental conditions, requiring additional treatment steps, which not only prolong the process chain but also may trigger secondary pollution risks.

11.8.1.3 Insufficient Catalyst Stability and Recyclability

Products such as organic acids and aldehydes generated from biomass degradation, as well as impurities such as heavy metal ions and halogen ions contained in pollutants, are prone to occupying the active sites of the catalyst through strong adsorption, or destroying the catalyst's crystal structure and corroding the surface, leading to catalyst poisoning and deactivation. At the same time, catalyst recovery faces multiple challenges. Nanocatalysts are prone to agglomeration. Adsorbent materials such as biochar have high energy consumption during regeneration, and repeated regeneration leads to the loss of surface functional groups and a significant decrease in adsorption capacity. Traditional recovery processes such as centrifugation and filtration are not only complex but also have high energy consumption and low separation efficiency, which further limits the reuse of catalysts and increases the overall treatment cost.

11.8.1.4 Poor Compatibility of Synergistic Treatment Systems

There are inherent conflicts between the reaction requirements of biomass upgrading and pollutant degradation. Biomass oxidation usually requires mild conditions, such as neutral pH and room temperature, while pollutant degradation often relies on more extreme environments such as acidic conditions. Biomass adsorption requires catalysts to have a high specific surface area, while pollutant reduction requires high requirements for catalyst conductivity. It is difficult to balance the structural requirements of catalysts for both. In addition, the kinetic rates of the two types of reactions are mismatched, and there is competitive adsorption in the system. Electrolyte components may also interact to generate inactive species or precipitates. These contradictions seriously reduce the stability of the reaction system and affect the overall treatment effect.

11.8.1.5 Limited Practicality of Industrial Applications

Existing biomass conversion and pollutant degradation processes generally have the problem of high energy consumption. Some processes require additional oxidants, light, or other auxiliary conditions, which greatly increase reagent costs. Most processes are still batch operations with limited processing capacity, which cannot meet large-scale application needs. The technology conversion cycle is long, and it is difficult to transition laboratory results to engineering applications. The reactor design is not perfect, and the mass transfer efficiency is low, leading to insufficient catalyst utilization and further increasing energy consumption pressure. At the same time, small and medium-sized enterprises have limited R&D investment and insufficient cross-field technical collaboration, and the rigid requirements of policies on treatment efficiency and resource utilization rate form a structural contradiction between technical supply and market demand, restricting the industrialization process.

11.8.2 Future Development Trends and Prospects

11.8.2.1 High-Efficiency Directional Catalysts Empower Biomass Valorization

By regulating the electronic structure and spatial configuration of catalyst active sites, combining "protective chemistry" strategies and polymerization inhibition technologies, and inhibiting degradation and condensation side reactions during biomass conversion, precise breaking and directional conversion of chemical bonds can be realized [26]. Develop new catalysts, such as core–shell structures and doped modifications, to enhance the selectivity for specific reactions through pore size sieving effects and complexation, thereby improving the utilization efficiency of biomass resources. Design multifunctional catalysts to adapt to the conversion needs of different types of biomass by adjusting the reaction potential, and simultaneously couple technologies such as pyrolysis and anaerobic fermentation to build an integrated "separation-oxidation" process, reducing the process complexity of multi-substrate treatment.

11.8.2.2 Innovation of Pollutant Resource-Oriented Degradation Technology

Break through traditional degradation ideas, develop integrated "degradation-conversion" technologies, couple pollutants with reactions such as biomass conversion and CO_2 reduction, and generate high-value products while completely degrading pollutants, realizing "turning waste into treasure." Use the synergistic effect of microorganisms and bioelectrochemical systems to enhance the decomposition efficiency of refractory organic matter, track the reaction process in real time with in situ monitoring technology, optimize reaction conditions, and avoid the generation of toxic intermediates. By constructing multi-active site interfaces and heterojunction structures, promote free radical generation and charge transfer, improve the thoroughness and safety of pollutant degradation, and form a synergistic utilization system of pollutants and biomass.

11.8.2.3 Optimization and Upgrading of Anti-toxicity and Recycling Processes

Introduce substances such as polymer coatings and polyethylene glycol on the catalyst surface through surface modification technology, reducing interaction between impurities and active sites and improving anti-toxicity performance. Develop magnetic separation and self-regenerating catalysts to achieve rapid separation and recovery through magnetism, or realize in situ regeneration of active sites through electrochemical reduction, pyrolysis, etc., to extend service life. Optimize the regeneration process of materials such as biochar to reduce regeneration energy consumption and functional group loss, and improve separation technology to simplify the recovery process, enhance catalyst reuse rate, and reduce operating costs.

11.8.2.4 Improvement of Compatibility of Synergistic Treatment Systems

Develop catalysts suitable for a wide range of conditions, such as amphoteric oxides, to maintain stable activity under different pH and temperature environments and resolve conflicts in reaction conditions. Design partitioned reactors such as dual-chamber and microchannel reactors to isolate different reaction regions through proton exchange membrane modification, avoid competitive adsorption and electrolyte interference, and improve system stability. Optimize the structural characteristics of catalysts through interface regulation technologies such as heterojunction construction and functionalization design, and adjust reaction parameters to achieve kinetic matching, so that different requirements such as high specific surface area and high conductivity can be compatible, and synergistic treatment efficiency can be improved.

11.8.2.5 Efficient and Economical Development of Industrial Processes

Promote continuous flow reactors to replace traditional batch processes, and combine microchannel design to enhance mass transfer efficiency to meet large-scale treatment needs. Couple renewable energy such as solar photovoltaic for power supply to reduce the energy consumption pressure of electrocatalytic processes, and use industrial by-products to prepare catalysts to reduce raw material costs. Optimize the reactor structure, improve mass transfer efficiency by adding stirrers, aeration devices, etc., and strengthen catalyst utilization. Strengthen cross-field technical collaboration and technical support for small and medium-sized enterprises, shorten the technology conversion cycle, and promote the large-scale implementation of low-cost and high-efficiency processes under policy guidance to achieve a win–win situation of environmental and economic benefits.

11.9 Challenges and Prospects in Other Application Fields (Sensors, Microwave Absorption, and Drug Delivery)

11.9.1 Current Core Challenges

11.9.1.1 Sensor Field: Prominent Performance Limitations and Practical Obstacles

Metal oxide semiconductor (MOS) sensors are easily affected by interfering components in complex gas environments and struggle to accurately identify a single target gas [27]. Especially in scenarios where multiple volatile organic compounds

(VOCs) coexist, the response signal of the target gas is easily masked. Specific gas sensors (such as formaldehyde sensors) generally have high energy consumption and need to operate at relatively high temperatures to achieve stable detection, which not only increases energy consumption but also shortens the service life of the device. Changes in environmental humidity significantly impact the conductivity of the sensing layer, easily leading to large errors in detection results and making it difficult to adapt to humid or dynamically changing humidity environments. The detection limit of most sensors cannot meet international safety standards, making it difficult to accurately capture low-concentration target gases. Moreover, sensor preparation relies on precision manufacturing processes, resulting in poor product repeatability, low yield, and high industrial mass production costs, which restrict large-scale application in practical scenarios.

11.9.1.2 Microwave Absorption Field: Coexisting Performance Balance and Preparation Challenges

Single-component iron-based nanomaterials have poor impedance matching performance and struggle to simultaneously meet the comprehensive requirements of strong absorption intensity, broad absorption bandwidth, and minimal thickness. They either have high absorption intensity but narrow bandwidth, or wide bandwidth but weak absorption capacity, failing to meet the microwave absorption requirements in complex scenarios. The mainstream preparation methods for carbon/magnetic metal composite materials each have limitations. Some methods have extremely low yields and are only suitable for laboratory research. Some methods require the use of toxic solvents, which are prone to environmental residues and safety hazards. Others rely on high temperatures and precision equipment, resulting in high preparation costs, which are not conducive to large-scale production. The material stability is insufficient, and nanoparticles are prone to agglomeration, leading to a decrease in the specific surface area of the material, weakened polarization relaxation loss, and significant attenuation of microwave absorption performance over time, which cannot meet long-term use requirements.

11.9.1.3 Drug Delivery Field: Safety and Function Need to Be Breached

The synthesis of Fe-based MOFs mostly relies on high-temperature and high-pressure conditions or toxic reagents, making industrial scale-up difficult. Moreover, organic solvents are prone to remain in the pores of the material, which may cause biological toxicity and require complex cleaning and drying processes, further increasing preparation costs and process complexity [28, 29]. The biocompatibility and targeting are poor. Some Fe-based MOFs are toxic to normal cells at certain concentrations, and their biosafety needs improvement. Nanoparticles are easily cleared by the human reticuloendothelial system, with weak active targeting ability, making it difficult to effectively accumulate at tumor sites and unable to achieve precise drug delivery. The precision of drug release regulation is low. Existing responsive drug carriers have problems such as delayed response and uneven release. Under the action of trigger signals, precise control over the amount of drug release is difficult, and nonspecific leakage is prone to occur in normal tissues, affecting treatment effects and increasing the risk of adverse reactions.

11.9.2 Future Development Trends and Prospects

11.9.2.1 Sensor Field: Moving Toward Precision and Low Power Consumption

Continuous performance optimization: For example, through technical means such as heterojunction construction and heteroatom doping, special electronic structures can be constructed, or the number of active sites on the material surface can be increased, significantly improving the selectivity and sensitivity of the sensor, enabling it to accurately identify low-concentration target gases and meet strict detection standards. Upgraded adaptability to working conditions: For example, developing low-dimensional sensing layers combined with advanced microfabrication technology can effectively reduce the operating temperature and energy consumption of the sensor, extending the service life of the device. Through surface functionalization modification, the anti-interference ability of the sensor to environmental factors such as humidity can be enhanced, expanding its applicable scenario range. Expanding application scenarios: For example, in the medical field, biosensors based on iron-based nanomaterials can be developed for the detection of disease markers, assisting in early disease diagnosis. In environmental monitoring, portable sensors can realize real-time monitoring of harmful gases in the atmosphere and industrial waste gases, providing data support for environmental protection decisions. In the food industry, they can also be used for the detection of food freshness and harmful substances to ensure food safety.

11.9.2.2 Microwave Absorption Field: Driven by Structural Innovation and Green Processes

Improved performance balance: For example, designing new microstructures (such as yolk-shell structures and hierarchical porous structures) can extend the transmission path of electromagnetic waves inside the material, enhance the synergistic effect of dielectric loss and magnetic loss, and achieve comprehensive optimization of absorption performance, meeting the diverse needs of electronic equipment, electromagnetic protection, stealth technology, and other fields. Green innovation in preparation processes: Promote biomass carbon composite processes by using natural biomass resources to prepare microwave absorption materials, thereby reducing raw material costs and environmental burdens. Develop new efficient and low-consumption preparation technologies to shorten reaction time, reduce energy consumption, and promote the industrial mass production process of microwave absorption materials. Expansion of emerging application fields: With the development of technologies such as 5G communication and artificial intelligence, the demand for microwave absorption materials in emerging fields such as electromagnetic compatibility of electronic equipment and electromagnetic protection of smart wearable devices will continue to grow. In addition, in the national defense and military fields, lightweight and high-performance microwave absorption materials will also play an important role in the research and development of stealth equipment.

11.9.2.3 Drug Delivery Field: Moving Toward Safety and Precision

Green preparation and improved safety: For example, developing green and efficient synthesis methods for Fe-based MOFs, avoiding the use of toxic reagents and residual organic solvents, and reducing preparation costs and safety risks [30]. Further improve the biocompatibility of Fe-based MOFs and reduce toxicity to normal cells through ligand optimization and surface modification. Enhanced precise delivery capabilities: For example, designing multi-stimuli responsive drug carriers can accurately trigger drug release according to the tumor microenvironment or external physical signals, improve the release efficiency of drugs at tumor sites, and reduce nonspecific leakage. Combined with targeting molecular modification, the active targeting ability of nanocarriers to tumor cells can be enhanced, improving the effect of tumor treatment. Accelerated advancement of integrated diagnosis and treatment: For example, combining Fe-based MOF drug carriers with imaging functions can realize the integration of drug distribution monitoring, tumor diagnosis, and treatment, providing strong support for clinical precise treatment.

References

1 Shah, S.A., Ahmad, W., Anwar, M. et al. (2025). Synthesis, properties, and applications of Fe_3O_4 and Fe_3O_4-based nanocomposites: a review. *Applied Catalysis O: Open* 203: 207049.

2 Zhou, L., Zheng, X.Y., Wang, L.L. et al. (2025). Noble metal/iron-group metal compound composite catalysts: characterization, synthesis, and electrocatalytic application. *Small* 21: e08616.

3 Hao, J., Wu, K.L., Lyu, C.J. et al. (2023). Recent advances in interface engineering of Fe/Co/Ni-based heterostructure electrocatalysts for water splitting. *Materials Horizons* 10: 2312–2342.

4 Mondal, P., Anweshan, A., and Purkait, M.K. (2020). Green synthesis and environmental application of iron-based nanomaterials and nanocomposite: a review. *Chemosphere* 259: 127509.

5 Karunakaran, S., Ramanujam, S., and Gurunathan, B. (2018). Green synthesised iron and iron-based nanoparticle in environmental and biomedical application: a review. *IET Nanobiotechnology* 12: 1003–1008.

6 Zhang, A., Liang, Y.X., Zhang, H. et al. (2021). Doping regulation in transition metal compounds for electrocatalysis. *Chemical Society Reviews* 50: 9817–9844.

7 Wu, J.W., Liu, N.Y., Li, F.S. et al. (2025). Iron group elements (Fe, Co, Ni) in Electrocatalytic applications: evaluation, characterization and prospects. *Coordination Chemistry Reviews* 525: 216343.

8 Wang, X.W. (2021). Atomic layer deposition of iron, cobalt, and nickel chalcogenides: progress and outlook. *Chemistry of Materials* 33: 6251–6268.

9 Van de Walle, A., Figuerola, A., Espinosa, A. et al. (2023). Emergence of magnetic nanoparticles in photothermal and ferroptotic therapies. *Materials Horizons* 10: 4757–4775.

10 Ji, D.W., Yang, H., Zhang, Q. et al. (2024). Iron series metal-organic frameworks and their composite nanomaterials: controllable synthesis and clean energy applications. *Nano Energy* 125: 109559.

11 Kifle, G.A., Huang, Y., Xiang, M.H. et al. (2022). Heterogeneous activation of peroxygens by iron-based bimetallic nanostructures for the efficient remediation of contaminated water. A review. *Chemical Engineering Journal* 442: 136187.

12 Li, Z., Deng, S., Yu, H. et al. (2022). Fe–Co–Ni trimetallic organic framework chrysanthemum-like nanoflowers: efficient and durable oxygen evolution electrocatalysts. *Journal of Materials Chemistry A* 10: 4230–4241.

13 Guo, Y., Huang, Z.W., Hu, Y.Q. et al. (2025). Recent advances in iron-based heterostructure anode materials for sodium ion batteries. *Acta Physico-Chimica Sinica* 41: 100022.

14 Sun, X.X., Liu, S.K., Sun, W.W., and Zheng, C.M. (2023). Emerging multifunctional iron-based nanomaterials as polysulfides adsorbent and sulfur species catalyst for lithium-sulfur batteries—a mini-review. *Chinese Chemical Letters* 34: 107501.

15 Chen, X., Han, L., Zhao, Y. et al. (2024). Insights in iron-based polyanion electrode materials for advanced sodium-ion batteries. *Chemical Engineering Journal* 501: 157600.

16 Yang, Y., Sun, B.W., Sun, Z.Q. et al. (2024). Recent advances and strategies of metal phosphides for accelerating polysulfide redox and regulating Li plating. *Coordination Chemistry Reviews* 510: 215836.

17 Feng, T., Liu, G.H., Li, G. et al. (2023). Engineering iron-rich nanomaterials for supercapacitors. *Chemical Engineering Journal* 473: 145045.

18 Yang, Y.Q., Huang, M.F., Gao, B.B. et al. (2025). Preparation of Fe, Co, Ni-based single atom catalysts and the progress of their application in electrocatalysis. *Microstructures* 5: 2025001.

19 Yuan, Y.R., Wang, C.J., Yang, Y.Q. et al. (2025). Advancements in iron-based single atom catalysts and their applications in electrocatalysis. *Journal of Power Sources* 644: 237149.

20 Karnaukhov, T., Likozar, B., and Kostyniuk, A. (2025). Advancements in thermo and photothermal CO_2 hydrogenation to light olefins using fe-based catalysts: current progress and future directions. *Carbon Energy* 7: e70036.

21 Yang, M., Ma, G.Z., Yang, H.L. et al. (2023). Advanced strategies for promoting the photocatalytic performance of $FeVO_4$ based photocatalysts: a review of recent progress. *Journal of Alloys and Compounds* 941: 168995.

22 Jha, B.K., Chaule, S., and Jang, J.H. (2024). Enhancing photocatalytic efficiency with hematite photoanodes: principles, properties, and strategies for surface, bulk, and interface charge transfer improvement. *Materials Chemistry Frontiers* 8: 2197–2226.

23 Arif, N., Zafar, M.N., Batool, M. et al. (2024). Recent advances and perspectives on iron-based photocatalysts. *Journal of Materials Chemistry C* 12: 12653–12691.

24 Sharma, S., Jakhar, P., and Sharma, H. (2023). $CuFe_2O_4$ nanomaterials: current discoveries in synthesis, catalytic efficiency in coupling reactions, and their environmental applications. *Journal of the Chinese Chemical Society* 70: 107–127.

25 Zhao, Z.T., Ding, J., Luo, G. et al. (2025). Current challenges in nano-engineered biomass valorization: a comprehensive review from the whole procedure of biomass fermentation perspective. *Advances in Applied Energy* 18: 100219.

26 Wu, H.J., Zheng, X.Y., Liu, J.J. et al. (2025). Research progress of transition metal catalysts for electrocatalytic EG oxidation. *Nanoscale* 17: 13145–13173.

27 Adampourezare, M., Hasanzadeh, M., Hoseinpourefeizi, M.A., and Seidi, F. (2023). Iron/iron oxide-based magneto-electrochemical sensors/biosensors for ensuring food safety: recent progress and challenges in environmental protection. *RSC Advances* 13: 12760–12780.

28 Gong, Y., Hu, X.Y., Chen, M., and Wang, J. (2024). Recent progress of iron-based nanomaterials in gene delivery and tumor gene therapy. *Journal of Nanobiotechnology* 22: 309.

29 Bhatia, I. and Khanchandani, S. (2025). Stimuli-responsive iron-based metal-organic frameworks (Fe-MOFs) for drug delivery and theranostics in breast cancer: a critical review. *Journal of Drug Delivery Science and Technology* 114: 107387.

30 Mosupi, K., Masukume, M., Weng, G.M. et al. (2025). Recent advances in Fe-based metal-organic frameworks: structural features, synthetic strategies and applications. *Coordination Chemistry Reviews* 529: 216467.

Index

Iron-Group Metal Compound Nanomaterials: Preparation, Characterization, Structure, and Applications, First Edition. Jinlong Zheng.

e

f

n

q

r